T0176294

Mathematical Methods for Physical and Analytical Chemistry

Mathematical Methods for Physical and Analytical Chemistry

David Z. Goodson
Department of Chemistry & Biochemistry
University of Massachusetts Dartmouth

A JOHN WILEY & SONS, INC., PUBLICATION

The text was typeset by the author using LaTex (copyright 1999, 2002–2008, LaTex3 Project) and the figures were created by the author using gnuplot (copyright 1986–1993, 1998, 2004, Thomas Williams and Colin Kelley).

Published by John Wiley & Sons, Inc., Hoboken, New Jersey.
Published simultaneously in Canada.

Library of Congress Cataloging-in-Publication Data is available.

ISBN 978-0-470-47354-2

Printed in the United States of America.

10 9 8 7 6 5 4 3 2 1

To Betsy

Contents

*This section treats an advanced topic. It can be skipped without loss of continuity.

Part IV. Linear Algebra

Preface

This is an intermediate level post-calculus text on mathematical and statistical methods, directed toward the needs of chemists. It has developed out of a course that I teach at the University of Massachusetts Dartmouth for third-year undergraduate chemistry majors and, with additional assignments, for chemistry graduate students. However, I have designed the book to also serve as a supplementary text to accompany undergraduate physical and analytical chemistry courses and as a resource for individual study by students and professionals in all subfields of chemistry and in related fields such as environmental science, geochemistry, chemical engineering, and chemical physics. I expect the reader to have had one year of physics, at least one year of chemistry, and at least one year of calculus at the university level. While many of the examples are taken from topics treated in upper-level physical and analytical chemistry courses, the presentation is sufficiently self contained that almost all the material can be understood without training in chemistry beyond a first-year general chemistry course.

Mathematics courses beyond calculus are no longer a standard part of the chemistry curriculum in the United States. This is despite the fact that advanced mathematical and statistical methods are steadily becoming more and more pervasive in the chemistry literature. Methods of physical chemistry, such as quantum chemistry and spectroscopy, have become routine tools in all subfields of chemistry, and developments in statistical theory have raised the level of mathematical sophistication expected for analytical chemists. This book is intended to bridge the gap from the point at which calculus courses end to the level of mathematics needed to understand the physical and analytical chemistry professional literature.

Even in the old days, when a chemistry degree required more formal mathematics training than today, there was a mismatch between the intermediate-level mathematics taught by mathematicians (in the one or two additional math courses that could be fit into the crowded undergraduate chemistry curriculum) and the kinds of mathematical methods relevant to chemists. Indeed, to cover all the topics included in this book, a student would likely have needed to take separate courses in linear algebra, differential equations, numerical methods, statistics, classical mechanics, and quantum mechanics.

Condensing six semesters of courses into just one limits the depth of coverage, but it has the advantage of focusing attention on those ideas and techniques most likely to be encountered by chemists. In a work of such breadth yet of such relatively short length it is impossible to provide rigorous proofs of all results, but I have tried to provide enough explanation of the logic and underlying strategies of the methods to make them at least intuitively reasonable. An annotated bibliography is provided to assist the reader interested in additional detail. Throughout the book there are sections and examples marked with an asterisk (*) to indicate an advanced or specialized topic. These starred sections can be skipped without loss of continuity.

Part I provides a review of calculus. The first four chapters provide a brief overview of elementary calculus while the next three chapters treat, in relatively more detail, topics that tend to be shortchanged in a typical introductory calculus course: numerical methods, complex numbers, and Taylor series. Parts II (Statistics), III (Differential Equations), and IV (Linear Algebra) can for the most part be read in any order. The only exceptions are some of the starred sections, and most of Chapter 20 (Schrödinger's Equation), which draws significantly on Part III as well as Part IV. The treatment of statistics is somewhat novel for a presentation at this level in that significant use is made of Monte Carlo simulation of random error. Also, an emphasis is placed on robust methods of estimation. Most chemists are unaware of this relatively new development in statistical theory that allows for a more satisfactory treatment of outliers than does the more familiar Q-test.

Exercises are included with each chapter, and answers to many of them are provided in an appendix. Many of the exercises require the use of a computer algebra system. The convenience and power of modern computer algebra software systems is such that they have become an invaluable tool for physical scientists. However, considering that there are various different software systems in use, each with its own distinctive syntax and its own enthusiastic corps of users, I have been reluctant to make the main body of the text too dependent on computer algebra examples. Occasionally, when discussing topics such as statistical estimation, Monte Carlo simulation, or Fourier transform that particularly require the use of a computer, I have presented examples in *Mathematica*. I apologize to users of other systems, but I trust you will be able to translate to your system of choice without too much trouble.

I thank my students at UMass Dartmouth who have been subjected to earlier versions of these chapters over the past several years. Their comments (and complaints) have significantly shaped the final result. I thank various friends and colleagues who have suggested topics to include and/or have read and commented on parts of the manuscript—in particular, Dr. Steven Adler-Golden, Professor Bernice Auslander, Professor Gerald Manning, and Professor Michele Mandrioli. Also, I gratefully acknowledge the efforts of the anonymous reviewers of the original proposal to Wiley. Their insightful and thorough critiques were extremely helpful. I have followed almost all of their suggestions. Finally, I thank my wife Betsy Martin for her patience and wisdom.

DAVID Z. GOODSON

Newton, Massachusetts
May, 2010

List of Examples

Greek Alphabet

Letters	Name	Trans- literation	Letters	Name	Trans- literation
A, α	alpha	a	N, ν	nu	n
B, β	beta	b	Ξ, ξ	xi	x
Γ, γ	gamma	g	O, o	omicron	o
Δ, δ	delta	d	Π, π	pi	p
E, ϵ	epsilon	e	P, ρ	rho	r
Z, ζ	zeta	z	Σ, σ, ς	sigma	s
H, η	eta	e	T, τ	tau	t
Θ, θ	theta	th	Υ, υ	upsilon	u
I, ι	iota	i	Φ, ϕ	phi	ph
K, κ	kappa	k	X, χ	chi	kh
Λ, λ	lambda	l	Ψ, ψ	psi	ps
M, μ	mu	m	Ω, ω	omega	o

Part I

Calculus

Chapter 1

Functions: General Properties

This chapter provides a brief review of some basic ideas and terminology from calculus.

1.1 Mappings

A ***function*** is a ***mapping*** of some given number into another number. The function $f(x) = x^2$, for example, maps the number 3 into the number 9,

$$3 \xrightarrow{x^2} 9.$$

The function is a rule that indicates the destination of the mapping. An ***operator*** is a mapping of a function into another function.

Example 1.1. *Contrasting the concepts of function and operator.* The operator $\frac{d}{dx}$ maps $f(x) = x^2$ into $f'(x) = 2x$,

$$x^2 \xrightarrow{\frac{d}{dx}} 2x.$$

The first-derivative *function* $f'(x) = 2x$ applied, for example, to the number 3 gives $3 \xrightarrow{f'} 6$. In contrast, the operator $\frac{d}{dx}$ applied to the number 3 gives

$$3 \xrightarrow{\frac{d}{dx}} 0,$$

as it treats "3" as a function $f(x) = 3$ and "0" as a function $f(x) = 0$.

In principle, a mapping can have an ***inverse***, which undoes its effect. Suppose g is the inverse of f. Then

$$g(f(x)) = x. \tag{1.1}$$

For the example $f(x) = x^2$ we have the mappings $3 \xrightarrow{f} 9 \xrightarrow{g} 3$. The effect of performing a mapping and then performing its inverse mapping is to map the value of x back to itself.

For the function x^2 the inverse is the square root function, $g(y) = \sqrt{y}$. To prove this, we simply note that if we let y be the result of the mapping f (that is, $y = x^2$), then

$$g(f(x)) = \sqrt{x^2} = x.$$

Graphs of x^2 and $\sqrt{y}$ are compared in Fig. 1.1. Note that the graph of $\sqrt{y}$ can be obtained by reflecting[1] the graph of x^2 through the diagonal line $y = x$.

[1] The *reflection* of a point through a line is a mapping to the point on the opposite side such that the new point is the same distance from the line as was the original point.

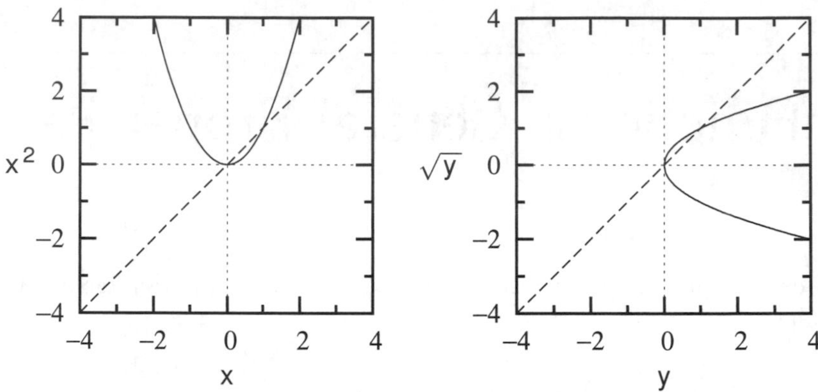

Figure 1.1: Graph of $y = x^2$ and its inverse, $\sqrt{y}$. Reflection about the dashed line ($y = x$) interchanges the function and its inverse.

Fig. 1.1 illustrates an interesting fact: An inverse mapping can in some cases be multiple valued. x^2 maps 2 to 4, but it also maps -2 to 4. The mapping f in this case is *unique*, in the sense that we can say with certainty what value of $f(x)$ corresponds to any value x. The inverse mapping g in this case is not unique; given $y = 4$, g could map this to $+2$ or to -2. In Fig. 1.1, values of the variable y for $y > 0$ each correspond to two different values of $\sqrt{y}$. This function has two ***branches***. On one branch, $g(y) = |\sqrt{y}|$. On the other, $g(y) = -|\sqrt{y}|$.

The inverse of $f(u)$ is designated by the symbol $f^{-1}(u)$. This can be confusing. Often, the indication of the variable, "(u)," is omitted to make the notation less cumbersome. Then, f^{-1} can be the inverse, $f^{-1}(u)$, or the reciprocal, $f(u)^{-1} = 1/f(u)$. Usually these are not equivalent. If $f(u) = u^2$, the inverse is $f^{-1} = f^{-1}(u) = \sqrt{u}$ while the reciprocal is $f^{-1} = f(u)^{-1} = u^{-2}$. Which meaning is intended must be determined from the context.

1.2 Differentials and Derivatives

A function $f(x)$ is said to be ***continuous*** at a specified point x_0 if the limit $x \to x_0$ of $f(x)$ is finite and has the same value whether it is approached from one direction or the other. Calculus is the study of continuous change. It was developed by Newton[2] to describe the motions of objects in response to change in time. However, as we will see in this book, its applications are much broader.

The basic tool of calculus is the ***differential***, an infinitesimal change in a variable or function, indicated by prefixing a "*d*" to the symbol for the

[2]English alchemist, physicist, and mathematician Isaac Newton (1642-1727). Calculus was also developed, independently and almost simultaneously, by the German philosopher, mathematician, poet, lawyer, and alchemist Gottfried Wilhelm von Leibniz (1646-1716).

quantity that is changing. If x is changed to $x+dx$, where dx is "infinitesimally small," then $f(x)$ changes to $f+df$ in response. The formal definition of the differential of f is

$$df = \lim_{\Delta x\to 0}[f(x+\Delta x) - f(x)]. \tag{1.2}$$

This is usually written

$$df = f(x+dx) - f(x), \tag{1.3}$$

where $f(x+dx)-f(x)$ is an abbreviation for the left-hand side of Eq. (1.2).

The basic idea of differential calculus is that the response to an infinitesimal change is *linear*. In other words, df is proportional to dx; that is,

$$df = f'dx, \tag{1.4}$$

where the proportionality factor, f', is called the ***derivative*** of f. Solving Eq. (1.4) for f', we obtain $f' = df/dx$. We now have three different notations for the derivative,

$$f', \qquad \frac{df}{dx}, \qquad \frac{d}{dx}f,$$

all of which mean the same thing. The choice of notation is a matter of convenience. f' is very concise and allows for convenient indication of the function's variable, for example, $f'(3)$. The fractional notation df/dx is particularly convenient for calculations in which this derivative is expressed in terms of other derivatives. However, to indicate the function's variable requires the awkward notation $\left.\frac{df}{dx}\right|_{x=3}$. The operator notation $\frac{d}{dx}f$ is commonly used in advanced mathematics as it can simplify theoretical analyses. The operation of calculating a derivative is called ***differentiation***.[3]

It is important not to confuse the concepts of derivative and differential. The derivative is a number, describing the rate of change of the function. In contrast, the differential has no numerical value. It is a theoretical construct that describes the smallest imaginable amount of change, smaller in magnitude than any number yet not quite zero. The usefulness of the differential is in mathematical derivations. The key idea is that while the numerical values of dx and df are undefined, their ratio df/dx can have a defined value.[4]

Example 1.2. *Numerical approximation of a derivative.* Consider the derivative of $f(x) = x^2$ at the point $x = 3$. Let us approximate dx with the numerical value 0.01. Then[5]

$$df \approx (x+0.01)^2 - x^2 = (3.01)^2 - 3^2 = 0.0601,$$

and $f'(3) = df/dx \approx 0.0601/0.01 = 6.01$. This is quite close to the exact value $f'(3) = 6$ that we obtain from the analytical formula $f'(x) = 2x$.

Example 1.2 suggests that the derivative can be evaluated as a limit in which a finite change in the variable becomes infinitesimal. Let Δx be some finite

---

[3] Perhaps this is why students new to the subject so often confuse the words "differential" and "derivative"!

[4] There is no *guarantee* the ratio has a defined value. This is discussed in Section 1.5.

[5] The symbol "$\approx$" means "approximately equal to."

but small change in x. Then

$$\frac{df}{dx} = \lim_{\Delta x \to 0} \frac{f(x + \Delta x) - f(x)}{\Delta x}, \tag{1.5}$$

which can be taken as the definition of a derivative. This equation can be used to derive rules for calculating derivatives of analytic expressions.

Example 1.3. *The derivative of x^2.*

$$\lim_{\Delta x \to 0} \frac{(x + \Delta x)^2 - x^2}{\Delta x} = \lim_{\Delta x \to 0} \frac{x^2 + 2x\Delta x + (\Delta x)^2 - x^2}{\Delta x} = \lim_{\Delta x \to 0} (2x + \Delta x) = 2x.$$

There are useful theorems concerning differentials and derivatives of combinations of two functions. Let $f(x)$ and $g(x)$ be two arbitrary functions.

Theorem 1.2.1. *For the sum of two functions:*

$$d(f + g) = df + dg, \qquad \frac{d}{dx}(f + g) = \frac{df}{dx} + \frac{dg}{dx}. \tag{1.6}$$

Theorem 1.2.2. *For the product of two functions:*

$$d(fg) = fdg + gdf\ , \qquad \frac{d}{dx}(fg) = f\frac{dg}{dx} + g\frac{df}{dx}. \tag{1.7}$$

Theorem 1.2.3. *For a function of a function, $f(g(x))$:*

$$df(g) = \frac{df}{dg}dg, \qquad \frac{df}{dx} = \frac{df}{dg}\frac{dg}{dx}. \tag{1.8}$$

Theorem 1.2.3 is called the ***chain rule***.

Example 1.4. *The chain rule.* Consider $f = g^{-5}$. The derivative $\frac{df}{dg}$ is $(-5)g^{-6}$. Suppose that $g = 1 + x^2$. Then $\frac{dg}{dx} = 2x$ and, according to the chain rule,

$$\frac{df}{dx} = \frac{d}{dx}(1 + x^2)^{-5} = \frac{dg}{dx}\frac{df}{dg} = (2x)(-5)g^{-6} = -\frac{10x}{(1 + x^2)^6}.$$

Given that $df = f'dx$, it follows that $dx = df/f'$. This is true as long as f' is not equal to zero. Dividing each side by df, we obtain the derivative of x as a function of f:

Theorem 1.2.4. *For all x such that $f'(x) \neq 0$,* $\dfrac{dx}{df} = \dfrac{1}{\frac{df}{dx}}$.

It is usually the case that a function responds more strongly to a change in its variables in some regions than in others. We expect in general that f' is also a function, and it can be of interest to consider the rate of change of the derivative of a derivative,

$$f''(x) = \frac{d}{dx}f'(x), \qquad f'''(x) = \frac{d}{dx}f''(x),$$

which defines the *second derivative*, the *third derivative*, etc. The nth derivative is defined as

$$f^{(n)}(x) = \frac{d}{dx}f^{(n-1)}(x) \,. \tag{1.9}$$

The superscript indicates the number of primes (e.g., $f^{(3)} = f'''$).[6] We can also use the notations

$$f^{(n)}(x) = \frac{d^n f}{dx^n} = \frac{d^n}{dx^n} f \,. \tag{1.10}$$

1.3 Partial Derivatives

The extension of the concepts of differentials and derivatives to multivariable functions is straightforward, but we must take into account that the various variables can be varied independently of each other. Consider a function $f(x, y)$, of two variables. The response to the change $(x, y) \to (x + dx, y + dy)$ is $f \to f + df$, where

$$df = \frac{\partial f}{\partial x}\, dx + \frac{\partial f}{\partial y}\, dy \,. \tag{1.11}$$

The proportionality factors $\frac{\partial f}{\partial x}$ and $\frac{\partial f}{\partial y}$ are called ***partial derivatives*** with respect to x and y. $\frac{\partial f}{\partial x}$ of $f(x, y)$ is calculated in the same way as $\frac{df}{dx}$ of $f(x)$ except that y is treated as a constant. Eq. (1.5) is modified as follows:

$$\frac{\partial f}{\partial x} = \lim_{\Delta x \to 0} \frac{f(x + \Delta x, y) - f(x, y)}{\Delta x} . \tag{1.12}$$

It is a common practice to add subscripts to derivatives and differentials to indicate any variables being held constant. For example, the partial derivative given by Eq. (1.12) can be designated as $\left(\frac{\partial f}{\partial x}\right)_y$. Eq. (1.11) can be written

$$df = \left(\frac{\partial f}{\partial x}\right)_y dx + \left(\frac{\partial f}{\partial y}\right)_x dy \,. \tag{1.13}$$

Example 1.5. *Differential of Gibbs free energy of reaction.* Consider a chemical reaction A $\to$ B. Whether the reaction can occur spontaneously is determined by the sign of the differential dG of the Gibbs free energy of the mixture of A and B, which is a function $G(T, p, n_A, n_B)$. (The reaction is spontaneous if $dG < 0$.) The variables are temperature, pressure, and numbers of moles of A and B. dG can be written

$$dG = \left(\frac{\partial G}{\partial T}\right)_{p,n_A,n_B} dT + \left(\frac{\partial G}{\partial p}\right)_{T,n_A,n_B} dp + \left(\frac{\partial G}{\partial n_A}\right)_{T,p,n_B} dn_A + \left(\frac{\partial G}{\partial n_B}\right)_{T,p,n_A} dn_B \,.$$

[6] The parentheses are included in the superscript to distinguish from f raised to a power. If $f = x^2$, then $f^{(2)} = \frac{d}{dx}f' = 2$ while $f^2 = (x^2)(x^2) = x^4$.

For a process in which y is held constant, we get from Eq. (1.13)

$$df_y = \left(\frac{\partial f}{\partial x}\right)_y dx_y + \left(\frac{\partial f}{\partial y}\right)_x dy_y = \left(\frac{\partial f}{\partial x}\right)_y dx_y\,,$$

because dy_y is, by definition, zero. It follows that

$$\frac{df_y}{dx_y} = \left(\frac{\partial f}{\partial x}\right)_y . \tag{1.14}$$

df_y/dx_y is an alternative notation for the partial derivative.

With more than one variable, there is more than one kind of second derivative. The change in $\frac{\partial f}{\partial x}$ in response to a change in x or y, respectively, is described by

$$\frac{\partial^2 f}{\partial y \partial x} = \frac{\partial}{\partial y}\frac{\partial f}{\partial x}, \qquad \frac{\partial^2 f}{\partial x^2} = \frac{\partial}{\partial x}\frac{\partial f}{\partial x}. \tag{1.15}$$

The order in which partial derivatives are evaluated has no effect:

Theorem 1.3.1. *For a function f of two variables x and y, $\frac{\partial}{\partial y}\frac{\partial f}{\partial x} = \frac{\partial}{\partial x}\frac{\partial f}{\partial y}$.*

Consider a process in which x and y are changed in such a way that the value of f remains constant. Then $df = 0$, which implies that

$$\left(\frac{\partial f}{\partial x}\right)_y dx_f + \left(\frac{\partial f}{\partial y}\right)_x dy_f = 0\,.$$

Solving for dy_f we obtain $dy_f = -\left[\left(\frac{\partial f}{\partial x}\right)_y \Big/ \left(\frac{\partial f}{\partial y}\right)_x\right] dx_f$. Therefore,

$$\left(\frac{\partial y}{\partial x}\right)_f = \frac{dy_f}{dx_f} = -\left(\frac{\partial f}{\partial x}\right)_y \Big/ \left(\frac{\partial f}{\partial y}\right)_x = -\left(\frac{\partial f}{\partial x}\right)_y \left(\frac{\partial y}{\partial f}\right)_x . \tag{1.16}$$

This is usually written in the following more easily remembered form:

Theorem 1.3.2. *For a function f of two variables x and y,*

$$\left(\frac{\partial x}{\partial y}\right)_f \left(\frac{\partial y}{\partial f}\right)_x \left(\frac{\partial f}{\partial x}\right)_y = -1\,. \tag{1.17}$$

This is called the ***triple product rule.*** Note the minus sign!

Example 1.6. *Demonstration of the triple product rule.* The physical state of a substance can be described in terms of *state variables* pressure, molar volume, and temperature. For an ideal gas, these variables are related to each other by the ideal-gas equation of state,

$$pV_\mathrm{m}/RT = 1, \tag{1.18}$$

where R is a constant. Let us use this equation to demonstrate Theorem 1.3.2:

$$\frac{\partial V_\mathrm{m}}{\partial T} = \frac{R}{p}, \qquad \frac{\partial T}{\partial p} = \frac{V_\mathrm{m}}{R}, \qquad \frac{\partial p}{\partial V_\mathrm{m}} = -\frac{RT}{V_\mathrm{m}^2},$$

$$-1 = \left(\frac{\partial V_\mathrm{m}}{\partial T}\right)_p \left(\frac{\partial T}{\partial p}\right)_{V_\mathrm{m}} \left(\frac{\partial p}{\partial V_\mathrm{m}}\right)_T = \frac{R}{p}\frac{V_\mathrm{m}}{R}\left(-\frac{RT}{V_\mathrm{m}^2}\right) = -\frac{RT}{pV_\mathrm{m}},$$

which agrees with Eq. (1.18).

The subscript on the partial derivatives should be omitted only if it is obvious from context which variables are held constant. With multivariable functions there may be alternative ways to choose the variables. For example, the molar Gibbs free energy G_m of a pure substance depends on p, V_m, and T, but these three variables are related to each other by an equation of state, so that the values of any two of the variables determine the third. Thus, G_m is really a function of just two variables. We can choose whichever pair of variables (p, V_m), (p, T), or (V_m, T) is most useful for a given application. It is important in thermodynamics to indicate the variable held constant. $\left(\frac{\partial G_\mathrm{m}}{\partial V_\mathrm{m}}\right)_p$ is not the same as $\left(\frac{\partial G_\mathrm{m}}{\partial V_\mathrm{m}}\right)_T$.

Given three variables (x, y, u) and an equation such that one variable can be expressed in terms of the other two, we can derive a multivariable analog of the chain rule. Consider a process in which x and u are changed with y held constant. Let $dy = 0$ in Eq. (1.11) and then divide each side of the equation by du_y. Thus we obtain the following:

Theorem 1.3.3.

$$\left(\frac{\partial f}{\partial u}\right)_y = \left(\frac{\partial f}{\partial x}\right)_y \left(\frac{\partial x}{\partial u}\right)_y . \tag{1.19}$$

This is valid only if the same variable is held constant in all three derivatives.

1.4 Integrals

The ***integral***, indicated by the symbol $\int$, is the operator that is the inverse of the derivative:

$$f' \xrightarrow{\int} f + c \xrightarrow{\frac{d}{dx}} f'.$$

The integral mapping is not unique. Because the derivative of a constant is zero, c can be any constant. c is called a ***constant of integration***.

To be precise, the operator $\int$ operates on differentials,

$$\int df = f + c. \tag{1.20}$$

However, for a function $f(x)$ we can substitute $f'dx$ for df, according to Eq. (1.4), which gives

$$\int f'dx = f(x) + c. \tag{1.21}$$

It is in this sense that the operator $\int$ maps f' to $f + c$.

Eqs. (1.20) and (1.21) are examples of ***indefinite integrals***. In contrast,

$$\int_{x_1}^{x_2} f'dx = f(x_2) - f(x_1), \tag{1.22}$$

where x_1 and x_2 are specified values, is called a ***definite integral***. The

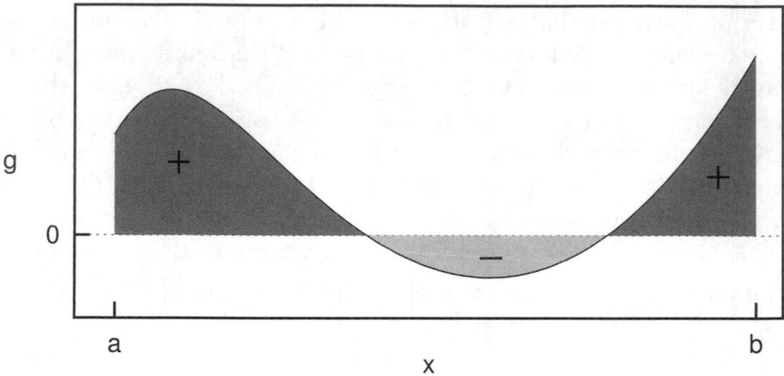

Figure 1.2: The value of the definite integral $\int_a^b g(x)dx$ is the sum of the areas under the curve but above the x-axis, minus the area below the x-axis but above the curve.

definite integral maps a function f' into a *constant* while the indefinite integral maps f' into another *function.* To solve an indefinite integral one needs additional information, in order to assign a value to c.

Example 1.7. *Integrals of x.* Consider the indefinite integral $\int xdx$. We seek a function f such that $f'(x) = x$. We know that the derivative of x^2 is $2x$. Therefore, the derivative of $\frac{1}{2}x^2$ is equal to x. Thus, $\frac{1}{2}x^2$ is one solution for $\int xdx$. However, the derivative of $\frac{1}{2}x^2 + 1$ is also equal to x, as is the derivative of $\frac{1}{2}x^2 + 1.5$. The function $\frac{1}{2}x^2 + c$ for any constant c is an acceptable solution for the *in*definite integral $\int xdx$.

Now consider the definite integral $\int_4^6 xdx$. This is evaluated according to Eq. (1.22) using any acceptable solution for the indefinite integral. For example,

$$\int_4^6 xdx = \tfrac{1}{2}6^2 - \tfrac{1}{2}4^2 = 18 - 8 = 10\,,$$

or

$$\int_4^6 xdx = (\tfrac{1}{2}6^2 + 1) - (\tfrac{1}{2}4^2 + 1) = 19 - 9 = 10\,.$$

The solution for the definite integral is unique. The integration constant cancels out.

The definite integral is the kind of integral most commonly seen in science applications. A remarkable theorem, called the ***fundamental theorem of calculus***,[7] provides an alternative interpretation of what it represents:

Theorem 1.4.1. *The definite integral $\int_a^b g(x)dx$ of a continuous function g is equal to the area under the graph of g from a to b.*

This is illustrated in Fig. 1.2. If g is negative anywhere in the interval, then the area *above* the graph but below zero is counted as "negative area" and subtracted from the total. In Chapter 5 we will use this theorem to develop an important practical technique for evaluating definite integrals.

[7]This theorem is attributed to the Scottish mathematician James Gregory, who proved a special case of it in 1668. Soon after that, Newton proved the general statement.

The following theorems give useful rules for manipulating integrals:

Theorem 1.4.2.

$$\int_a^b g(x)dx = -\int_b^a g(x)dx. \tag{1.23}$$

Theorem 1.4.3. *A definite integral is equal to the sum of integrals over different ranges,*

$$\int_{x_1}^{x_3} g(x)dx = \int_{x_1}^{x_2} g(x)dx + \int_{x_2}^{x_3} g(x)dx. \tag{1.24}$$

Theorem 1.4.4. *The integral of a sum of functions is the sum of the separate integrals,*

$$\int [g(x) + h(x)]dx = \int g(x)dx + \int h(x)dx. \tag{1.25}$$

For the integral of a *product* of functions, $\int g(x)w(x)dx$, there is in general no simple formula. However, if one knows the integral of w alone, that is, if one knows a function $W(x) = \int w(x)dx$, such that $W' = w$, then it may be possible to evaluate the integral of the product gw using ***integration by parts***:

Theorem 1.4.5.

$$\int g(x)w(x)dx = g(x)W(x) - \int g'(x)W(x)dx, \tag{1.26}$$

where $W(x) = \int w(x)dx$.

Proof. This follows from the formula for the differential of a product, $d(gW) = g\,dW + Wdg$. Note that $dW = W'dx = wdx$ and $dg = g'dx$. Writing $g\,dW = d(gW) - Wdg$ and integrating gives the result. □

Integration by parts is useful if the integral of w and the integral of $g'W$ can both be evaluated.

Example 1.8. *Integration by parts: A simple example.* Consider

$$I = \int_a^b xe^x dx.$$

If it were not for the factor of x multiplying the exponential, we could evaluate the integral quite easily. The function e^x has the special property of being equal to its derivative:

$$\tfrac{d}{dx}e^x = e^x, \qquad d(e^x) = \left(\tfrac{d}{dx}e^x\right) dx = e^x dx.$$

Therefore,

$$\int_a^b e^x dx = \int_a^b d(e^x) = e^x|_a^b = e^b - e^a.$$

Let us work through the proof of Theorem 1.4.5 using xe^x as an example:

$$d(xe^x) = xd(e^x) + e^x dx = x\left(\tfrac{d}{dx}e^x\right) dx + e^x dx = xe^x dx + e^x dx,$$

$$xe^x dx = d(xe^x) - e^x dx.$$

$$\int_a^b xe^x dx = \int_a^b d(xe^x) - \int_a^b e^x dx = xe^x|_a^b - \int_a^b e^x dx.$$

We obtain the answer $I = be^b - ae^a - e^b + e^a$.

Derivatives with respect to parameters in the integration ranges of definite integrals can be calculated as follows:

Theorem 1.4.6.
$$\frac{d}{db}\int_a^b f(x)dx = f(b) \qquad \text{(for constant } a\text{)}, \tag{1.27a}$$
$$\frac{d}{da}\int_a^b f(x)dx = -f(a) \qquad \text{(for constant } b\text{)}. \tag{1.27b}$$

The following theorem shows how to integrate a function $f(x,y)$ with respect to one variable while taking a derivative with respect to the other.

Theorem 1.4.7.
$$\frac{d}{dy}\int_{a(y)}^{b(y)} f(x,y)dx = \int_{a(y)}^{b(y)} \left(\frac{\partial f}{\partial y}\right)_x dx + f\big(b(y),y\big)\frac{db}{dy} - f\big(a(y),y\big)\frac{da}{dy}\,. \tag{1.28}$$

If the integration ranges do not depend on y, then we have simply
$$\frac{d}{dy}\int_a^b f(x,y)dx = \int_a^b \left(\frac{\partial f}{\partial y}\right)_x dx. \tag{1.29}$$

The analytic calculation of a derivative is straightforward (although perhaps tedious) but the analytic calculation of an integral is usually more difficult and often impossible. Values for definite integrals over certain special ranges (typically $-\infty$ to ∞ or 0 to 1) are sometimes available even when no result is available for the indefinite integral. The standard reference is Gradsteyn and Ryzhik,[8] an extensive and well-organized collection of indefinite and definite integrals. However, it is common in science applications to encounter integrals that are not available in tables because they are impossible to evaluate in terms of elementary mathematical functions. It is often necessary in such cases to resort to numerical approximation methods. This topic is addressed in Section 5.3. Analytic techniques are discussed in Chapter 4.

A straightforward way to extend the concept of integration to multivariable functions is the ***multidimensional integral***,[9] which in two dimensions is[10]
$$\int\!\!\int f(x,y)dxdy = \int \left[\int f(x,y)dx\right] dy\,. \tag{1.30}$$

This is carried out in two steps. For the integral in the brackets, y is treated as a constant while x is the variable. The function resulting from that integration is then integrated with x treated as constant and y as the variable. The double

[8] For a translation from the original Russian, with corrections and some additional material, see I. S. Gradshteyn, I.M. Ryzhik, A. Jeffrey, and D. Zwillinger, *Table of Integrals, Series, and Products*, 6th ed. (Academic Press, San Diego, 2000). The Gradshteyn-Ryzhik tables have been incorporated into computer algebra software packages.

[9] "Dimension" in this context means the number of variables.

[10] The notation $\int dy \int dx\, f(x,y)$ can also be used.

integral thus defined is the inverse of the operator $\frac{\partial^2}{\partial x \partial y}$. Instead of a *constant* of integration, c, there are two *functions* of integration, $c_1(x)$ and $c_2(y)$,

$$\frac{\partial^2 f}{\partial x \partial y} \xrightarrow{\int\int} f + c_1(x) + c_2(y) \xrightarrow{\frac{\partial^2}{\partial x \partial y}} \frac{\partial^2 f}{\partial x \partial y}. \tag{1.31}$$

The second derivative operator eliminates c_1 and c_2 because each is a function of just one variable. The definite integral in two dimensions is

$$\int_{y_1}^{y_2} \int_{x_1}^{x_2} f(x,y)\,dx\,dy = \int_{y_1}^{y_2} \left[\int_{x_1}^{x_2} f(x,y)\,dx \right] dy. \tag{1.32}$$

The integration in brackets eliminates x as a variable, giving a function only of y. If f is continuous, the order of integration does not matter:

Theorem 1.4.8. *(Fubini's theorem.) If f is continuous everywhere within the area over which it is being integrated, then*

$$\int_{y_1}^{y_2} \left[\int_{x_1}^{x_2} f(x,y)\,dx \right] dy = \int_{x_1}^{x_2} \left[\int_{y_1}^{y_2} f(x,y)\,dy \right] dx. \tag{1.33}$$

Because the integral is defined as an operator that operates on a differential, the integrand should always contain a differential. Without a differential it may be unclear which coordinate is being integrated over. For example, in the expression

$$\int\int f(x,y,z)\,dy\,dz \tag{1.34}$$

the integration is over y and z, but x is a parameter that remains unevaluated in the result. The coordinates in the differentials are ***dummy variables***—they are not present in the final result. x in Eq. (1.34) is called a ***free variable***. The expression

$$\int\int f(x,a,b)\,da\,db$$

is equivalent to (1.34). Symbols used for dummy variables are arbitrary.

Example 1.9. *Dummy variables.* Consider

$$I_{\rm a} = \int_2^3 \int_2^3 x^2 y^3 dx dy = \int_2^3 x^2 \tfrac{1}{4}(3^4 - 2^4) dx = \tfrac{1}{3}(3^3 - 2^3)\tfrac{1}{4}(3^4 - 2^4) = 1235/12.$$

The result is a number. The value of I does not depend on the dummy variables x and y, which are simply placeholders within the integral. Compare this with

$$I_{\rm b} = \int_2^3 x^2 y^3 dx = y^3 \int_2^3 x^2 dx = \tfrac{1}{3}(3^3 - 2^3)y^3 = \tfrac{19}{3} y^3,$$

a function of y, and

$$I_{\rm c} = \int_2^3 x^2 y^3 dy = x^2 \int_2^3 y^3 dy = \tfrac{1}{4}(3^4 - 2^4)x^2 = \tfrac{65}{4} x^2,$$

a function of x.

1.5 Critical Points

An important application of derivatives is to find extrema. An ***extremum*** is a point at which a function has a minimum or a maximum compared with the points in the immediate neighborhood. Consider $f(x) = (x-2)^2$, shown in the left-hand panel of Fig. 1.3. f decreases and then increases as x goes from 0 to 4. If df is negative when dx is positive (i.e., the graph has downward slope), then $f' = df/dx$ must be negative. If df and dx are both positive (upward slope), then f' is positive. For $(x-2)^2$, the derivative initially is negative and then becomes positive. A minimum of $f(x)$ occurs at the value x_0 where the function stops decreasing. There the rate of change is zero, and hence $f'(x_0) = 0$. In this case, $f'(x) = 2(x-2)$ implies that $x_0 = 2$.

Keep in mind that there can exist situations in which an extremum exists but its location is *not* given by the equation $f'(x) = 0$. Furthermore, there are situations in which $f'(x_0) = 0$ but x_0 is not an extremum. Such cases are illustrated, respectively, by the other two panels of Fig. 1.3. The function $g(x) = |x-2|$, in the center panel, has $g'(x) = -1$ for $x < 2$ and $g'(x) = +1$ for $x > 2$, but $g'(x)$ is undefined at the minimum, $x = 2$; the limit $x \to 2$ of $g'(x)$ is different depending on the direction of approach. The function $h(x) = (x-2)^3$, in the right-hand panel, has $h'(x) = 3(x-2)^2$, with $h'(2)$ well defined and equal to zero, yet it is clear from its graph that $x = 2$ is not an extremum. The reason is that the second derivative, $h''(x) = 6(x-2)$ is also zero at $x = 2$, which causes the curvature to switch from concave downward to concave upward. We will define a ***critical point*** as any point at which the first derivative is zero or is discontinuous. (A derivative with value ∞ is also considered "discontinuous.") A point at which the concavity switches is called an ***inflection point***. The second derivative (if it exists) is zero is at an inflection point.

Any extremum is a critical point, but not all critical points are extrema. Methods for locating extrema will be discussed in Sections 5.6 and 12.4. If $f'(x_0) = 0$ and $f''(x_0) \neq 0$, then the sign of the second derivative at x_0 allows us to distinguish between a maximum and a minimum. If $f''(x_0) > 0$, then the derivative goes from negative to positive as x passes through x_0, which means that x_0 is a minimum. If $f''(x_0) < 0$, then x_0 is a maximum.

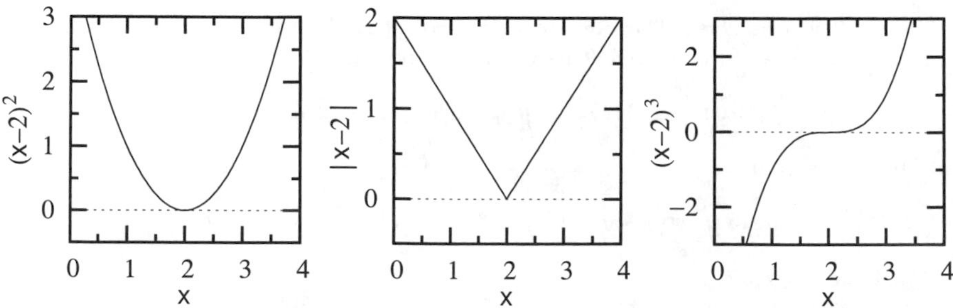

Figure 1.3: Three kinds of critical points.

Example 1.10. *The critical temperature.* Critical inflection points have an important role in chemical thermodynamics. This figure shows pressure p of a gas as function of V_{m} with temperature held constant at various values. At high T the ideal gas law, $p = (RT)V_{\mathrm{m}}^{-1}$, gives a reasonable description. In that case, $p' = -(RT)V_{\mathrm{m}}^{-2}$ is negative everywhere, $p'' = (2RT)V_{\mathrm{m}}^{-3}$ is positive everywhere, and $p(V_{\mathrm{m}})$ is everywhere concave upward. At intermediate T, intermolecular attraction makes the curves less steep at medium V_{m}, but at low V_{m} intermolecular repulsion always makes the curves very steep. The result is a region of V_{m} where the curve is concave downward. The transition occurs at an inflection point. At low T, as we decrease V_{m}, the curve abruptly ends—the gas condenses into a liquid. The lowest T at which condensation cannot occur is called the ***critical temperature***, T_{c}. The graph of p vs. V_{m} at T_{c} is characterized by a critical inflection point $(V_{\mathrm{c}}, p_{\mathrm{c}})$, at which $p'(V_{\mathrm{c}}) = 0$ and $p''(V_{\mathrm{c}}) = 0$.

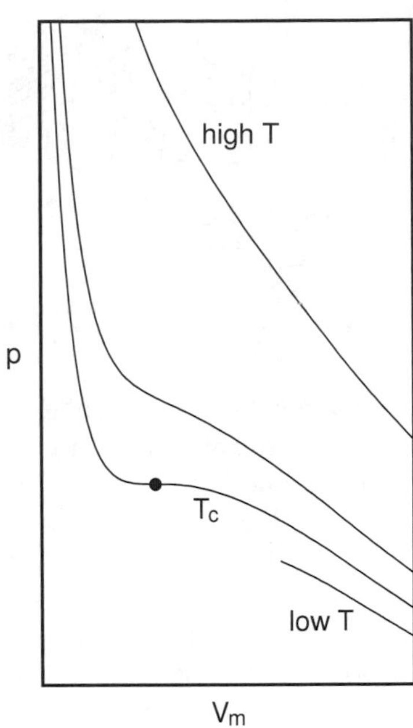

Now consider a function of two variables, $f(x, y)$. A point (x, y) at which both first partial derivatives are zero is called a ***stationary point***. Being a stationary point is not a sufficient condition for a point to be an extremum. The following conditions imply a minimum or a maximum:

$$\frac{\partial f}{\partial x} = 0, \qquad \frac{\partial f}{\partial y} = 0, \qquad \frac{\partial^2 f}{\partial x^2}\frac{\partial^2 f}{\partial y^2} - \left(\frac{\partial^2 f}{\partial x \partial y}\right)^2 > 0\,; \tag{1.35}$$

a minimum if $\partial^2 f/\partial x^2 + \partial^2 f/\partial y^2 > 0$, a maximum if $\partial^2 f/\partial x^2 + \partial^2 f/\partial y^2 < 0$. Now suppose there is a point at which

$$\frac{\partial f}{\partial x} = 0, \qquad \frac{\partial f}{\partial y} = 0, \qquad \frac{\partial^2 f}{\partial x^2}\frac{\partial^2 f}{\partial y^2} - \left(\frac{\partial^2 f}{\partial x \partial y}\right)^2 < 0. \tag{1.36}$$

This is called a ***saddle point***, because the function resembles a saddle in that neighborhood. $\frac{\partial^2 f}{\partial x^2}\frac{\partial^2 f}{\partial y^2} - \left(\frac{\partial^2 f}{\partial x \partial y}\right)^2$ is called the ***Hessian***[11] of f.

Example 1.11. *A saddle point.* Consider $f(x, y) = x^2 + 3xy + y^2$:

$$\frac{\partial f}{\partial x} = 2x + 3y, \qquad \frac{\partial f}{\partial y} = 3x + 2y, \qquad \frac{\partial^2 f}{\partial x^2}\frac{\partial^2 f}{\partial y^2} - \left(\frac{\partial^2 f}{\partial x \partial y}\right)^2 = -5. \tag{1.37}$$

$(x, y) = (0, 0)$ qualifies as a saddle point. The saddle shape is clearly seen in Fig. 1.4.

[11] Named after Prussian mathematician Otto Hesse (1811-1874), who was well known as a teacher and textbook author.

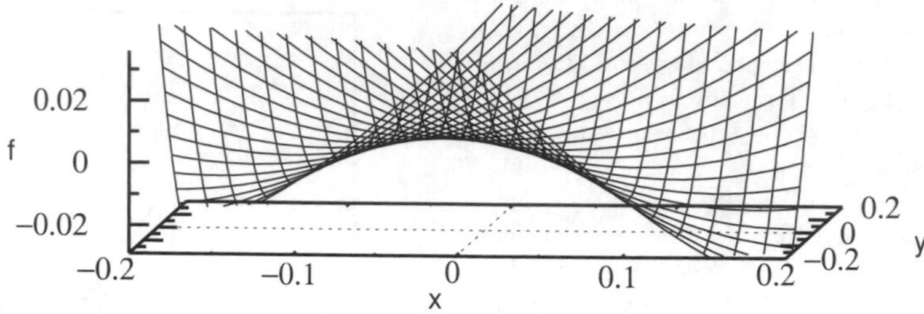

Figure 1.4: Graph of $f = x^2 + 3xy + y^2$, with a saddle point at (0,0).

Saddle points of a molecular potential energy function are important in the theory of chemical reaction rates. The Arrhenius expression for the temperature dependence of a reaction rate constant[12] is

$$k = Ae^{-E_\mathrm{a}/k_\mathrm{B}T}. \tag{1.38}$$

The most important parameter here is the activation energy, E_a. This is the difference in energy between the reactant molecules and the transition state. The transition state is the highest-energy configuration of the reacting molecules along the minimum-energy path between reactants and products. The transition state corresponds to a saddle point.

As an illustration, consider a reaction between an atom and a diatomic

$$\mathrm{A} + \mathrm{BC} \to \mathrm{AB} + \mathrm{C}. \tag{1.39}$$

To simplify the analysis, let us assume that the A — B — C configuration must be linear or no reaction takes place. (For many reactions of interest it turns out that this is a reasonably accurate approximation.) Let q be the distance between atoms A and B, and let r be the distance between B and C. The molecular potential energy function is a two-dimensional surface, $V(q, r)$, with qualitatively the same shape as Fig. 1.4. The transition state coordinates, designated as $(q^\ddagger, r^\ddagger)$, are a saddle point of V. The activation energy equals the difference between the potential energy at the transition state and the potential energy $V_\mathrm{BC}(r)$ of the diatomic BC molecule, that is,

$$E_\mathrm{a} = V(q^\ddagger, r^\ddagger) - V_\mathrm{BC}(r_\mathrm{eq}), \tag{1.40}$$

where r_eq is the equilibrium bond distance of molecule BC.

Usually, chemical reactions depend on more than just two coordinates. The potential energy function is then a ***hypersurface***. Suppose for example that we remove the restriction that the A + BC collision be linear. Then we need a third coordinate, s, which could be defined as the distance between A and C.

[12] k is the rate constant while $k_\mathrm{B} = N_\mathrm{A} R$ is Boltzmann's constant. (N_A is Avogadro's number.) A and E_a are usually assumed to be at least approximately independent of T.

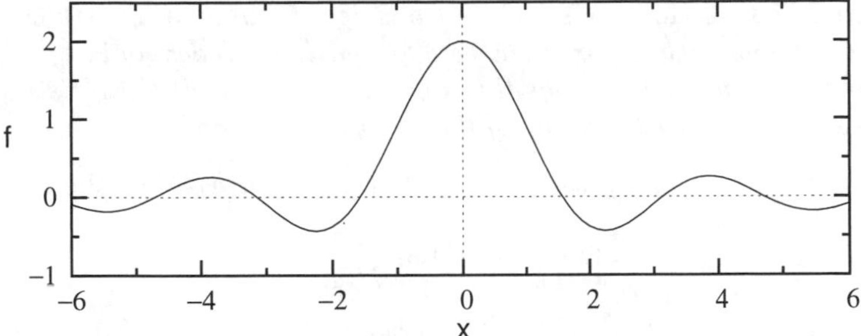

Figure 1.5: $f = \sin(2x)/x$. There is a removable singularity at $x = 0$, but this is not evident from the graph of the function.

Then $V(q, r, s)$ is a hypersurface in a four-dimensional space (the dimensions being q, r, s, and V). Because we humans are accustomed to existing in an apparently three-dimensional space, we find it difficult to visualize surfaces in four dimensions. It is possible to locate saddle points and extrema of hypersurfaces using numerical methods but if the dimensionality is high then this can be a challenging computational problem.[13]

A point at which a function and/or its first derivative do not have a defined value is a ***singular point***, also called a ***singularity***. An example was the critical point in $|x - 2|$ at $x = 2$, where the derivative had no defined value. Another common example is a function such as $(x - 2)^{-1}$ at $x = 2$, in which the function and its derivative become infinite. A singular point at which a function becomes infinite in proportion to $1/(x - c)^n$, where n is a positive integer and c is a constant, is called a ***pole*** of order n.

In some cases the existence of a singular point is not obvious from a graph of the function. Consider

$$f(x) = \frac{\sin(2x)}{x}. \tag{1.41}$$

$\sin(2x)$ is zero at $x = 0$. Thus $f(0) = 0/0$, which has no clear meaning. However, as is clear from the graph of the function, Fig. 1.5, the value of $f(x)$ in the limit $x \to 0$ from either direction is the well-defined value $f = 2$. A ***removable singularity*** is a singular point at which a function is undefined but continuous, such that the function can be made nonsingular by replacing its limit at the singular point with a finite numerical value. The value of a function in the limit of a removable singularity can often be obtained from the following theorem, called ***L'Hospital's rule***:[14]

[13]See F. Jensen, *Introduction to Computational Chemistry* (Wiley, New York, 1999), Section 14.5, and G. Henkelman, G. Jóhannesson, and H. Jónsson, "Methods for Finding Saddle Points and Minimum Energy Paths," in *Theoretical Methods in Condensed Phase Chemistry*, ed. S. D. Schwartz (Springer-Verlag, New York, 2002), pp. 269-300, for overviews of various methods that are available.

[14]The name is pronounced "lowpeetall." It comes from French nobleman Guillaume de L'Hospital, who published a calculus textbook in 1696 in which this theorem first appeared. However, the theorem was derived by L'Hospital's teacher, Johann Bernoulli (1667-1748). Bernoulli was a leading practitioner of calculus (which had only recently been invented), applying it to problems in physics, chemistry, and medicine, as well as pure mathematics.

Theorem 1.5.1 (L'Hospital's Rule). *Consider the limit of $g(x)/h(x)$ when both g and h approach 0, or when both g and h approach infinity. If this limit exists, then $g'(x)/h'(x)$ has the same limit, provided that g' is nonzero throughout some interval containing the point in question.*

Expressed symbolically, if $g(a) = h(a) = 0$ or $g(a) = h(a) = \infty$, then

$$\lim_{x\to a} \frac{g(x)}{h(x)} = \lim_{x\to a} \frac{g'(x)}{h'(x)}. \tag{1.42}$$

For $f(x)$ of Eq. (1.41), $g'(x) = 2\cos(2x)$, $h'(x) = 1$, $g'(0) = 2$, and $\lim_{x\to 0} f(x) = 2$. It is sometimes necessary to apply the rule repeatedly, with higher-order derivatives, in order to arrive at a well-defined limit.

Exercises

1.1 Is $\sqrt[3]{x}$ a multiple-valued function? Illustrate with plots analogous to Fig. 1.1.

1.2 Use Eq. (1.5) to estimate the value of $f'(7)$ for the function $f(x) = x^4/(\sqrt{x} + x^3)$. Choose an arbitrary small (but not quite zero) value for Δx.

1.3 Given that $f = pqr$, express df in terms of dp, dq, and dr.

1.4 Given that $f = p(q(r(x)))$, express df in terms of dx.

1.5 Given that $f = p(y)$ where $y(x) = 1 + x^2$, express df in terms of dx.

1.6 Consider the function $f(x, y) = 1 + x + 3xy + y^2$.

(a) Calculate the partial derivatives $\frac{\partial f}{\partial x}$, $\frac{\partial f}{\partial y}$, $\frac{\partial^2 f}{\partial x \partial y}$, $\frac{\partial^2 f}{\partial x^2}$, and $\frac{\partial^2 f}{\partial y^2}$.

(b) Demonstrate that $\frac{\partial^2 f}{\partial y \partial x}$ gives the same result as $\frac{\partial^2 f}{\partial x \partial y}$.

(c) Suppose that y is some unknown function of x. Express $\frac{df}{dx}$ in terms of y and y'.

1.7 Given that[15]

$$g = \sum_{j=0}^{23} c_j y_j^j \quad \text{and} \quad f = \sum_{k=0}^{23} (y_k - g)^2,$$

calculate $\frac{\partial g}{\partial c_n}$ and $\frac{\partial f}{\partial c_n}$ where n is an arbitrary integer in the range $0 \le n \le 23$. (The y_j's are constants.)

1.8 Evaluate the integral $\int_a^b x^3 e^x dx$. *(Hint: Use integration by parts three times.)*

1.9 Show that Theorems 1.4.2 and 1.4.6 follow from Eq. (1.22).

1.10 Calculate the value of $f(x) = 1 - 2x + x/\sin(3x)$ in the limit $x \to 0$ and check your answer by evaluating f at a very small numerical value of x.

1.11 Calculate the value of $f(x) = \sin^2(5x)/x^2$ in the limit $x \to 0$.

1.12 Use L'Hospital's rule to prove e^x goes to infinity faster than does any power of x.

1.13 Use L'Hospital's rule to evaluate $\lim_{u\to 0} u^{-2} e^{-1/u}$. *(Hint: $u^{-2}e^{-1/u} = u^{-2}/e^{1/u}$.)*

---

[15] The summation operator $\sum_{j=j_{\min}}^{j_{\max}}$ performs a sum with the index incremented in steps of 1. For example, $\sum_{j=1}^{4} j = 1 + 2 + 3 + 4 = 10$, $\sum_{j=1}^{4} (5 - j)x^j = 4x + 3x^2 + 2x^3 + x^4$, $\sum_{j=1}^{4} a^2 = a^2 + a^2 + a^2 + a^2 = 4a^2$, and so on.

Chapter 2

Functions: Examples

We now cxamine several families of functions that commonly occur in physical science.

2.1 Algebraic Functions

One of the simplest mathematical operations is to raise a number to an integer power, multiplying it by itself a given number of times,[1]

$$x^n = \overbrace{x \cdot x \cdot x \cdot \quad \cdots \quad \cdot x}^{n \text{ times}} = \prod_{j=1}^{n} x \,. \tag{2.1}$$

The value of x^0 is defined as 1. We define a negative power, x^{-n}, as n factors of $1/x$ multiplied together. ***Algebraic*** functions are functions that can be constructed from powers and their inverses.

The power of a product is the product of the separate powers,

$$(xy)^n = x^n y^n. \tag{2.2}$$

The power of a sum is given by the ***binomial theorem***:

Theorem 2.1.1.

$$(x+y)^n = x^n + nx^{n-1}y + \frac{n(n-1)}{2}\,x^{n-2}y^2 + \frac{n(n-1)(n-2)}{3\cdot 2}\,x^{n-3}y^3$$
$$+ \frac{n(n-1)(n-2)(n-3)}{4\cdot 3\cdot 2}\,x^{n-4}y^4 + \cdots + y^n. \tag{2.3}$$

It is convenient to write this as

$$(x+y)^n = \sum_{j=0}^{n} \binom{n}{j} x^{n-j} y^j , \tag{2.4}$$

in terms of the ***binomial coefficients***,

$$\binom{n}{j} = \frac{n!}{(n-j)!\,j!} \,, \tag{2.5}$$

and the ***factorial***,

$$n! = 1 \cdot 2 \cdot 3 \cdot \quad \cdots \quad \cdot (n-1) \cdot n = \prod_{j=1}^{n} j \,, \tag{2.6}$$

with $0! = 1$.

[1]The product symbol Π is the multiplicative analog of the summation symbol Σ.

The derivative of x^n is

$$\frac{d}{dx}x^n = nx^{n-1}, \tag{2.7}$$

This can be derived from the binomial theorem, as shown in the following example.

Example 2.1. *Derivation of a derivative formula.* Let us derive this for the case $f(x) = x^3$. In response to a change in the variable, $f(x)$ is mapped to $f(x + dx)$,

$$f(x + dx) = (x + dx)^3 = x^3 + 3x^2dx + 3x(dx)^2 + (dx)^3. \tag{2.8}$$

The differential is

$$df = f(x + dx) - f(x) = [3x^2 + 3xdx + (dx)^2]dx. \tag{2.9}$$

However, if dx is infinitesimal then $3xdx$ is infinitely smaller than $3x^2$, and $3x^2 + 3xdx \approx 3x^2$. $(dx)^2$ is smaller still, by another factor of infinity. It follows that $df = 3x^2dx$, which agrees with Eq. (2.7).

Eq. (2.7) also allows us to calculate integrals. We know that

$$\int \frac{d}{dx}x^n dx = x^n + a.$$

Substituting the formula for the derivative, and letting $k = n - 1$, we obtain

$$\int x^k dx = \frac{1}{k+1}x^{k+1} + b. \tag{2.10}$$

(b is an arbitrary constant.) This works for negative k as well as positive, except for $k = -1$.

The inverse of x^n is the ***nth root***, $\sqrt[n]{y}$. The case $n = 2$ is called the ***square root***, the case $n = 3$, the ***cube root***.

Example 2.2. *The cube root of* -125. Note that $(-5)(-5)(-5) = (-5)(25) = -125$. Therefore, $\sqrt[3]{-125} = -5$.

A function of the form cx^n, where x is a variable, c is a constant, and n is an integer, is called a ***monomial***. Any function that is constructed only from monomials and/or nth roots is called an *algebraic* function. A ***polynomial*** is any sum of monomials. For example,

$$f(x) = c_0 + c_1x + c_2x^2 + c_3x^3 + \cdots + c_nx^n \tag{2.11}$$

is a polynomial of ***degree*** n. A polynomial of degree $n = 1$,

$$f(x) = c_0 + c_1x,$$

is called a ***linear*** function, with ***slope*** c_1 and ***y-intercept*** c_0. Polynomials of degree 2, 3, 4, and 5 are referred to as ***quadratic***, ***cubic***, ***quartic***, ***quintic***.

A *root* of a function[2] $f(x)$ is a value of x at which $f(x) = 0$. (Do not confuse this with the "nth root.") Solving for the root is trivial for a linear

[2]Sometimes called a *zero* of the function.

polynomial but becomes more difficult with increasing polynomial degree. For a quadratic polynomial,

$$f(x) = ax^2 + bx + c,$$

the roots are given by the ***quadratic formula***,[3]

$$x = \frac{1}{2a}\left(-b \pm \sqrt{b^2 - 4ac}\right). \tag{2.12}$$

Analytic formulas also exist for the roots of cubic and quartic polynomials, but they are much more complicated.[4] However, with a computer one can compute roots of polynomials of arbitrary degree using numerical methods. The topic of polynomial roots will be examined in more detail in Chapter 6.

2.2 Transcendental Functions

All functions that are not algebraic are said to be ***transcendental***. Some appear frequently enough in scientific and engineering applications that they have been given names. The most common of these are the logarithm (and its inverse, the exponential) and the circular functions. Many other named functions that have been studied by mathematicians are available in computer algebra packages. They are called "special" functions.

2.2.1 Logarithm and Exponential

The definitions of x^2 as $x \cdot x$, and of x^3 as $x \cdot x \cdot x$ are easy to understand, but what about x^a where a is not an integer? The answer to this question will come from an unlikely source. Let us define the ***logarithm*** as

$$\ln(x) = \int_1^x \frac{1}{u}\, du. \tag{2.13}$$

u is a dummy variable. The free variable, x, is the upper range of the definite integral. All we have really done here is give a name to $\int x^k dx$ in the case $k = -1$ for which it cannot be expressed in terms of algebraic functions. Eq. (2.10) with $k = -1$ fails, due to division by zero. It is common to write $\ln x$ instead of $\ln(x)$ as long as the argument x is not so complicated that the notation would be ambiguous. $\ln ab$, for example, is ambiguous because it could mean either $\ln(ab)$ or $(\ln a)b$.

Fig. 2.1 compares $f(x) = \ln x$ with $f(x) = x$. $\ln x$ passes through zero at $x = 1$. It goes to infinity in the limit $x \to \infty$ but more slowly than does x. It is negative for $0 < x < 1$ and drops to negative infinity in the limit $x \to 0$. From Eq. (2.13) it follows immediately that the derivative of the logarithm is

$$\frac{d}{dx} \ln x = \frac{1}{x}. \tag{2.14}$$

[3]Be careful of roundoff error when evaluating this numerically.

[4]See W. H. Press *et al.*, *Numerical Recipes: The Art of Scientific Computing* (Cambridge University Press, Cambridge, 2007), Section 5.6.

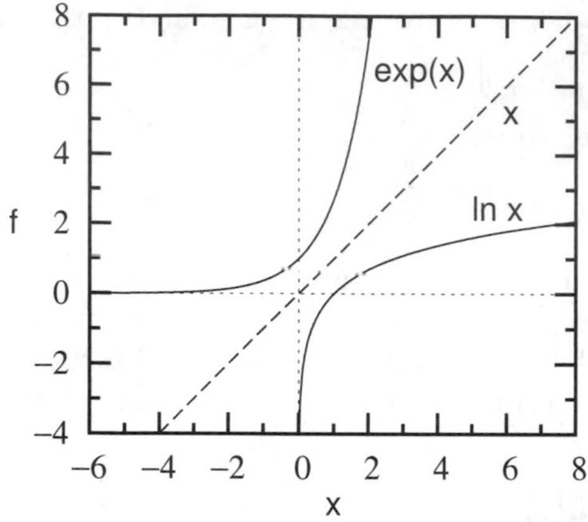

Figure 2.1: $f(x) = \ln x$ and $f(x) = \exp(x)$ (solid curves) compared with $f(x) = x$ (dashed line).

The ***exponential*** function, with symbol $\exp(y)$, is the inverse of $\ln x$:

$$\exp(\ln x) = x, \qquad \ln\,\exp(y) = y, \tag{2.15}$$

$\exp(x)$ is also shown in Fig. 2.1. In the limit $x \to \infty$, $\exp(x)$ goes to infinity much faster than x. In the limit $x \to -\infty$ it goes to zero. The numerical values of $\exp(x)$ are easy to compute, thanks to a formula derived by Euler,[5]

$$\exp(x) = 1 + \frac{1}{1!}x + \frac{1}{2!}x^2 + \frac{1}{3!}x^3 + \cdots = \sum_{j=0}^{\infty} \frac{1}{j!}x^j. \tag{2.16}$$

Clearly, $\exp(0) = 1$. The numerical value of $\exp(1)$, given the symbol e, is called *Euler's number*,

$$e = 1 + \frac{1}{1!} + \frac{1}{2!} + \frac{1}{3!} + \cdots = \sum_{j=0}^{\infty} \frac{1}{j!} \approx 2.71828\,. \tag{2.17}$$

Another formula for calculating $\exp(1)$ had been discovered previously by Jakob Bernoulli:[6]

$$e = \lim_{n\to\infty} (1 + 1/n)^n. \tag{2.18}$$

Characteristic properties of the logarithm are

$$\ln(ab) = \ln a + \ln b \tag{2.19}$$

and

$$\ln(b^a) = a \ln b\,. \tag{2.20}$$

Characteristic properties of the exponential are

$$\exp(a)\exp(b) = \exp(a + b) \tag{2.21}$$

and

$$[\exp(b)]^a = \exp(ab), \tag{2.22}$$

[5] The Swiss mathematician Leonhard Euler (1707-1783), perhaps the most prolific of all mathematicians, made very many important contributions to both pure and applied mathematics and to mathematical physics. His surname is pronounced "*OY-ler*." We will derive this formula for the exponential in Chapter 7.

[6] Swiss mathematician Jakob Bernoulli (1654-1705). His younger brother Johann and his nephew Daniel were also prominent mathematicians.

which can be derived from the properties of the logarithm. From Eq. (2.19), we have

$$\ln\left[\exp(a)\exp(b)\right] = \ln\ \exp(a) + \ln\ \exp(b) = a + b.$$

Taking the exponential of both sides gives

$$\exp\big(\ln\left[\exp(a)\exp(b)\right]\big) = \exp(a)\exp(b) = \exp(a+b).$$

Similarly,

$$\ln\left[\exp(b)^a\right] = a\ln\ \exp(b) = ab,$$

$$\exp\big(\ln\left[\exp(b)^a\right]\big) = \exp(b)^a = \exp(ab).$$

If we set $b = 1$ in Eq. (2.22), we find that the numerical value of $\exp(n)$ for integer n is obtained by raising the number e to the power n. Hence,

$$\exp(n) = e^n \tag{2.23}$$

for integer n. For noninteger exponent b, e^b is *defined* as $\exp(b)$. The notations e^x and $\exp(x)$ can be considered to be equivalent.[7]

We can use the exponential function to give meaning to the concept of a noninteger power. According to Eqs. (2.22) and (2.15),

$$e^{a\ln x} = \left(e^{\ln x}\right)^a = x^a. \tag{2.24}$$

Thus, we define x^a for noninteger a as $\exp(a\ln x)$. a is called the ***exponent***.

Example 2.3. *Noninteger powers.* Let us calculate $2^{0.345}$ using Eq. (2.24):

$$2^{0.345} = [\exp(\ln 2)]^{0.345} = \exp[(0.345)\ln 2]\,.$$

This is probably how your calculator carries this out. It has built-in algorithms for computing values of exp and ln.

Using Eqs. (2.24), (2.21), and (2.22), one can derive the following:

$$x^a x^b = x^{a+b}, \qquad (x^b)^a = x^{ab}, \qquad \sqrt[n]{x} = x^{1/n}. \tag{2.25}$$

The proofs are left as exercises.

According to Eq. (2.24),

$$10^a = e^{a\ln 10}. \tag{2.26}$$

Let us define the ***logarithm to base 10***, with symbol[8] $\log_{10} x$, as the inverse of 10^x. In other words, $\log_{10} x$ is the function such that

$$x = 10^{\log_{10} x}. \tag{2.27}$$

[7]The symbol e^x is more concise and is the one most commonly used. $\exp(x)$ is preferred if the argument x is a bulky expression.

[8]Mathematicians sometimes use the notation $\log x$, with no subscript, to mean $\ln x$, while scientists and engineers sometimes use $\log x$ to mean $\log_{10} x$. This can be confusing.

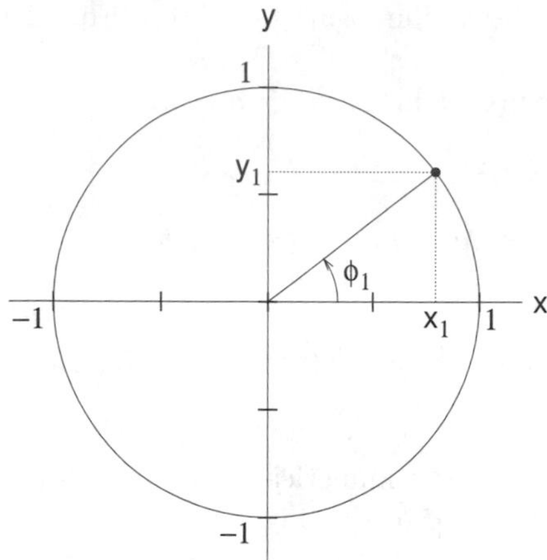

Figure 2.2: Coordinates of the unit circle, $(x_1, y_1) = (\cos\phi_1, \sin\phi_1)$.

To distinguish this new kind of logarithm from the previous one, the term ***natural logarithm*** is often used for ln. Equating Eq. (2.27) to our previous expression, $x = \exp(\ln x)$, we obtain a relation between our two different logarithms,

$$10^{\log_{10} x} = e^{\ln x}. \tag{2.28}$$

Taking the base-10 logarithm of both sides of this equation, one finds that

$$\log_{10} x = (\log_{10} e) \ln x. \tag{2.29}$$

Taking the natural logarithm of both sides of Eq. (2.28), one finds that

$$\ln x = (\ln 10) \log_{10} x. \tag{2.30}$$

2.2.2 Circular Functions

Consider a circle of unit radius, shown in Fig. 2.2. Any point on the circle can be specified by the coordinates x and y. Alternatively, any point could be specified, more economically, with a single coordinate—the angle ϕ between the x-axis and the line drawn from the origin to the point. Let us define the ***cosine*** and ***sine*** functions $\cos\phi$ and $\sin\phi$ as the values of x and y, respectively, corresponding to a given angle ϕ for a unit circle. By convention, angles are measured counterclockwise. Angular coordinates can be specified in terms of radians or in terms of degrees. A full rotation from the x-axis back to the x-axis corresponds to 2π radians or $360°$. The degree symbol ° indicates that the unit is the degree. If it is absent, assume the unit is the radian.

Having defined the sine and cosine, we can now define other circular[9] functions, such as the ***tangent***, ***cotangent***, ***secant***, and ***cosecant***,

$$\tan\phi = \frac{\sin\phi}{\cos\phi}, \quad \cot\phi = \frac{\cos\phi}{\sin\phi}, \quad \sec\phi = \frac{1}{\cos\phi}, \quad \csc\phi = \frac{1}{\sin\phi}. \tag{2.31}$$

[9]Circular functions are often called "trigonometric functions."

The names for the inverses of the circular functions are prefixed with "arc." For example, the inverse of the cosine is the arccosine, such that

$$\arccos(\cos\phi) = \phi, \tag{2.32}$$

and so on. The notation $\cos^{-1}\phi$ is also often used for the inverse, but in most contexts it is best to avoid this as it could be misinterpreted as $1/\cos\phi$.

Example 2.4. *Solve* $\phi = \arctan(-1)$. Take the tangent of each side,

$$\tan\phi = \tan(\arctan(-1)) = -1.$$

We are looking for angles ϕ such that $\sin\phi = -\cos\phi$. $\sin\phi$ and $\cos\phi$ are the y and x coordinates, respectively, of points on the unit circle. From the figure at right, we can see that there are two angles at which $y = -x$, namely, $\phi = \frac{3}{4}\pi$ and $\phi = -\frac{1}{4}\pi$.

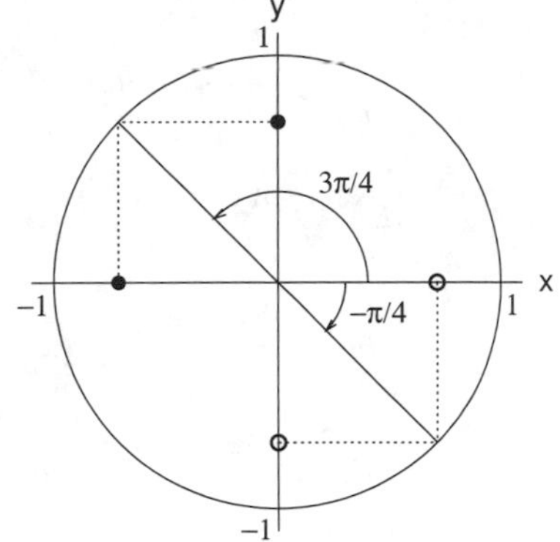

The derivatives of the circular functions are given in Table 2.1.

Closely related to the circular functions are the ***hyperbolic*** functions. The defining relations for the hyperbolic cosine and hyperbolic sine (often pronounced "*cosh*" and "*sinch*," for short) are

$$\cosh x = \tfrac{1}{2}\left(e^{x} + e^{-x}\right), \qquad \sinh x = \tfrac{1}{2}\left(e^{x} - e^{-x}\right). \tag{2.33}$$

By analogy we define

$$\tanh x = \frac{\sinh x}{\cosh x}, \quad \coth x = \frac{\cosh x}{\sinh x}, \quad \operatorname{sech} x = \frac{1}{\cosh x}, \quad \operatorname{csch} x = \frac{1}{\sinh x}, \tag{2.34}$$

(pronounced "*tanch*," "*cotanch*," "*seech*," "*coseech*") and inverses arccosh, arcsinh, etc. There is a deep connection between the hyperbolic and circular functions, which we will see later when we discuss the complex number system.

Derivatives are given in Table 2.2. Note that the circular and hyperbolic functions follow similar but not quite identical patterns. The most common use for the hyperbolic functions is to evaluate certain common integrals. For example, we see from Table 2.2 that $\int_0^{0.5} \frac{1}{1-u^2}du = \operatorname{arctanh}(0.5) - \operatorname{arctanh}(0)$.

For a circle of arbitrary radius r we have

$$x = r\cos\phi, \qquad y = r\sin\phi. \tag{2.35}$$

Substituting these into the Pythagorean theorem,

$$r^2 = x^2 + y^2, \tag{2.36}$$

we obtain an important relation between the sine and cosine,

$$\sin^2\phi + \cos^2\phi = 1. \tag{2.37}$$

Table 2.1: Derivatives of the circular functions.

$$\frac{d}{d\theta}\sin\theta = \cos\theta \qquad \frac{d}{d\theta}\cos\theta = -\sin\theta$$

$$\frac{d}{d\theta}\tan\theta = \sec^2\theta \qquad \frac{d}{d\theta}\cot\theta = -\csc^2\theta$$

$$\frac{d}{d\theta}\sec\theta = \sec\theta\tan\theta \qquad \frac{d}{d\theta}\csc\theta = -\csc\theta\cot\theta$$

$$\frac{d}{dy}\arcsin y = \frac{1}{\sqrt{1-y^2}}, \qquad \left(-\frac{\pi}{2} \le \arcsin y \le \frac{\pi}{2}\right)$$

$$\frac{d}{dy}\arccos y = -\frac{1}{\sqrt{1-y^2}}, \qquad (0 \le \arccos y \le \pi)$$

$$\frac{d}{dy}\arctan y = \frac{1}{1+y^2}, \qquad \left(-\frac{\pi}{2} < \arctan y < \frac{\pi}{2}\right)$$

$$\frac{d}{dy}\text{arccot } y = -\frac{1}{1+y^2}, \qquad (0 \le \text{arccot } y \le \pi)$$

$$\frac{d}{dy}\text{arcsec } y = \frac{1}{y\sqrt{y^2-1}}, \qquad \left(0 \le \text{arcsec } y < \frac{\pi}{2},\ -\pi \le \text{arcsec } y < -\frac{\pi}{2}\right)$$

$$\frac{d}{dy}\text{arccsc } y = -\frac{1}{y\sqrt{y^2-1}}, \qquad \left(0 < \text{arccsc } y \le \frac{\pi}{2},\ -\pi < \text{arccsc } y \le -\frac{\pi}{2}\right)$$

A negative angle means a clockwise rotation. It is obvious, then, from Fig. 2.2 that the sine is an odd function and the cosine is an even function,[10]

$$\sin(-\phi) = -\sin\phi, \qquad \cos(-\phi) = \cos\phi. \tag{2.38}$$

Using geometry one can derive identities for angle addition,

$$\sin(\alpha+\beta) = \sin\alpha\cos\beta + \cos\alpha\sin\beta, \tag{2.39}$$

$$\cos(\alpha+\beta) = \cos\alpha\cos\beta - \sin\alpha\sin\beta, \tag{2.40}$$

with the special cases

$$\sin(2\alpha) = 2\sin\alpha\cos\alpha, \qquad \cos(2\alpha) = \cos^2\alpha - \sin^2\alpha\,. \tag{2.41}$$

2.2.3 Gamma and Beta Functions

The factorial, $n! = n \cdot (n-1) \cdot (n-2) \cdot 1$, is defined for non-negative integer n. We need the restriction to non-negative integers because otherwise the sequence n, $n-1$, etc., would miss the final value of 1 and never end. The

[10] An even function has the property $f(-x) = f(x)$, an odd function, $f(-x) = -f(x)$.

Table 2.2: Derivatives of the hyperbolic functions.

$$\frac{d}{dx}\sinh x = \cosh x \qquad \frac{d}{dx}\cosh x = \sinh x$$

$$\frac{d}{dx}\tanh x = \operatorname{sech}^2 x \qquad \frac{d}{dx}\coth x = -\operatorname{csch}^2 x$$

$$\frac{d}{dx}\operatorname{sech} x = -\operatorname{sech} x \tanh x \qquad \frac{d}{dx}\operatorname{csch} x = -\operatorname{csch} x \coth x$$

$$\frac{d}{dy}\operatorname{arcsinh} y = \frac{1}{\sqrt{y^2+1}}$$

$$\frac{d}{dy}\operatorname{arccosh} y = \frac{1}{\sqrt{y^2-1}}\,, \quad (y > 1 \text{ and } \operatorname{arccosh} y > 0)$$

$$\frac{d}{dy}\operatorname{arctanh} y = \frac{1}{1-y^2}\,, \quad (y^2 < 1)$$

$$\frac{d}{dy}\operatorname{arccoth} y = \frac{1}{1-y^2}\,, \quad (y^2 > 1)$$

$$\frac{d}{dy}\operatorname{arcsech} y = -\frac{1}{y\sqrt{1-y^2}}\,, \quad (0 < y < 1 \text{ and } \operatorname{arcsech} y > 0)$$

$$\frac{d}{dy}\operatorname{arccsch} y = -\frac{1}{|y|\sqrt{1+y^2}}\,,$$

gamma function, $\Gamma(x)$, indicated by the capital of the Greek letter *gamma*, generalizes the factorial to noninteger x.

Note that

$$n! = n(n-1)!\,. \tag{2.42}$$

Thus if we know the value of 6! then we can calculate 7! as $7 \cdot (6!)$. This is an example of a *recursion*, a sequence of mathematical expressions in which each term in the sequence is obtained from previously computed terms. Let us require that $\Gamma(x)$ satisfy the recursion relation

$$\Gamma(x+1) = x\Gamma(x). \tag{2.43}$$

If x is a positive integer, then comparison of Eqs. (2.43) and (2.42) shows that[11]

$$\Gamma(n+1) = n!\,. \tag{2.44}$$

Eq. (2.43) specifies a property we would like the function to have, but it does not tell us how to calculate the function. Euler proposed the following

[11] Note that $n!$ is *not* equal to $\Gamma(n)$. This is easy to forget.

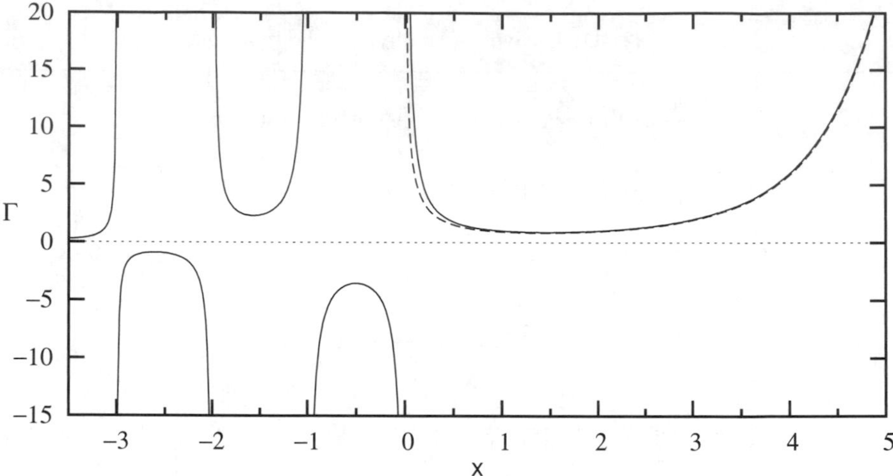

Figure 2.3: The gamma function (solid curve) and Stirling's formula (dashed curve).

definition:

$$\Gamma(x) = \int_0^\infty e^{-t}\, t^{x-1} dt, \tag{2.45}$$

for any $x > 0$. This works for noninteger x as well as for non-negative integers. To demonstrate that this satisfies Eq. (2.43), use integration by parts:

Example 2.5. *Integral representation of the gamma function.* Note that x is a parameter in the integral while t is the dummy variable for the integration. For example,

$$\Gamma(4) = \int_0^\infty e^{-t} t^3 dt.$$

Using integration by parts,

$$\Gamma(4) = -t^3 e^{-t}\big|_0^\infty + \int_0^\infty e^{-t}(3t^2)dt = 0 + 3\int_0^\infty e^{-t} t^{(3-1)} dt = 3\,\Gamma(3),$$

which is consistent with Eq. (2.43).

Although Eq. (2.45) is valid only for positive x, we can define $\Gamma(x)$ for negative x using the recursion relation in the form $\Gamma(x) = \Gamma(x+1)/x$. For example,

$$\Gamma\big(-\tfrac{3}{2}\big) = \frac{\Gamma\big(-\tfrac{1}{2}\big)}{(-3/2)} = \frac{\Gamma\big(\tfrac{1}{2}\big)}{(-1/2)(-3/2)}\,.$$

The gamma function is plotted in Fig. 2.3. Note that it becomes infinite at $0, -1, -2, -3, \ldots$. One can show that $\Gamma(x)$ has a first-order pole at each of these values (Exercise 2.11). For increasing positive x, the function increases very rapidly. In the limit $x \to \infty$ it can be approximated as

$$\Gamma(x) \approx \sqrt{\frac{2\pi}{x}}\left(\frac{x}{e}\right)^x. \tag{2.46}$$

Eq. (2.46) is called ***Stirling's formula***,[12] and, in fact, it remains accurate down to rather small values of x, as shown in the figure.

The gamma function is an example of a *special function*, which is a term used for any function that is not an algebraic, logarithmic, exponential, or circular function but occurs often enough to have been studied and given a name. The integral in Eq. (2.45) occurs quite often in statistical analysis and in quantum mechanics. In general, $\Gamma(x)$ is computed using high-precision approximation methods.[13] It can be evaluated analytically when x is any non negative integer, in terms of the factorial, but also for any half-integer value. Let us make the substitution $t = u^2$, $dt = \frac{dt}{du}du = 2udu$ within the integral. Then

$$\Gamma\left(\tfrac{1}{2}\right) = \int_0^\infty e^{-t}t^{-1/2}dt = 2\int_0^\infty e^{-u^2}du\,. \tag{2.47}$$

The integral $\int_0^\infty e^{-u^2}du$ can be evaluated using a clever trick involving an additional change of variable. Its value turns out to be $\sqrt{\pi}/2$. (This will be derived in Section 4.2.) We thus arrive at the result

$$\Gamma\left(\tfrac{1}{2}\right) = \sqrt{\pi}\,. \tag{2.48}$$

Using the recursion relation, we obtain $\Gamma(3/2) = \Gamma(1/2+1) = \frac{1}{2}\sqrt{\pi}$. Similarly, we can evaluate $\Gamma(5/2)$, $\Gamma(7/2)$, etc. We can similarly evaluate any of the following "Gaussian" integrals:

$$\int_0^\infty e^{-u^2}u^n du = \frac{1}{2}\int_0^\infty e^{-t}t^{(n-1)/2}dt = \tfrac{1}{2}\Gamma\left(\tfrac{n}{2}+\tfrac{1}{2}\right)\,. \tag{2.49}$$

$\Gamma(x)$ can be thought of as a special case of a more general function,

$$\Gamma(a,b) = \int_b^\infty e^{-t}\,t^{a-1}dt, \tag{2.50}$$

which is called the ***incomplete gamma function***, with $\Gamma(x) = \Gamma(x,0)$. The function

$$\gamma(a,b) = \Gamma(a) - \Gamma(a,b) = \int_0^b e^{-t}t^{a-1}dt\,, \tag{2.51}$$

with lowercase gamma, is also (confusingly!) called the "incomplete gamma function."[14] The ***regularized incomplete gamma functions*** are defined as

$$P(a,b) = \gamma(a,b)/\Gamma(a)\,, \qquad Q(a,b) = \Gamma(a,b)/\Gamma(a)\,. \tag{2.52}$$

---

[12]Named for the Scottish mathematician James Stirling (1692-1770).

[13]$\Gamma(x+1)$ is available on scientific calculators, where it is often called $x!$.

[14]Some authors refer to $\gamma(a,b)$ as the "incomplete gamma function" and $\Gamma(a,b)$ as the "complement of the incomplete gamma function" while others use the reverse, with $\gamma(a,b)$ as the "complement." To avoid confusion, still other authors call $\gamma(a,b)$ the "lower incomplete gamma function" and $\Gamma(a,b)$ the "upper incomplete gamma function."

If we make the change of variable $t = u^2$, $dt = 2udu$, then

$$\gamma(a,b) = 2\int_0^{b^{1/2}} e^{-u^2} u^{2a-1} du \; , \tag{2.53}$$

which is useful because integrals of this kind often occur when physical problems are treated with statistical methods, as in the following example.

Example 2.6. *The kinetic molecular theory of gases.* A central result of James Clerk Maxwell's "kinetic molecular" theory of the ideal gas is that the fraction of molecules with speeds less than some given speed s_0 is

$$f_{s_0} = \int_0^{s_0} e^{-\kappa s^2} s^2 ds \Big/ \int_0^{\infty} e^{-\kappa s^2} s^2 ds \; ,$$

where $\kappa = M/(2RT)$, in terms of the molecular weight M (mass per mole), the absolute temperature T (in kelvins), and the gas constant R. Let us make the change of variable $t = \kappa s^2$. Then $dt = 2\kappa s ds$, $s = (t/\kappa)^{1/2}$, and

$$\int_0^{s_0} e^{-\kappa s^2} s^2 ds = \frac{1}{2\kappa}\int_0^{s_0} e^{-\kappa s^2} s\,(2\kappa s ds) = \frac{1}{2\kappa}\int_0^{\kappa s_0^2} e^{-t}(t/\kappa)^{1/2} dt = \frac{1}{2\kappa^{3/2}}\,\gamma\big(\tfrac{3}{2}, \kappa s_0^2\big).$$

The same change of variable in the denominator gives $\Gamma(\frac{3}{2})/2\kappa^{3/2}$. Therefore,

$$f_{s_0} = P\big(\tfrac{3}{2}, \kappa s_0^2\big) = \gamma\big(\tfrac{3}{2}, \kappa s_0^2\big)/\Gamma\big(\tfrac{3}{2}\big) = 2\gamma\big(\tfrac{3}{2}, \kappa s_0^2\big)/\sqrt{\pi}.$$

Consider nitrogen gas, $M = 28.0$ g/mol, at a temperature of 295 K;

$$\kappa = \frac{28.0\text{ g}}{\text{mol}}\,\frac{1}{2}\,\frac{\text{K mol}}{8.314\text{ J}}\,\frac{\text{J}}{\text{kg m}^2\text{ s}^{-2}}\,\frac{\text{kg}}{1000\text{g}}\,\frac{1}{295\text{K}} = 5.71\times10^{-6}\ \text{s}^2/\text{m}^2\,.$$

The fraction of molecules moving slower than, for example, 100 km/h (i.e. 27.8 m/s) is $2\gamma\big(\frac{3}{2}, 0.00440\big)/\sqrt{\pi} = 2.19\times10^{-4}$. The value $\gamma\big(\frac{3}{2}, 0.00440\big) = 1.941\times10^{-4}$ can be obtained from the following *Mathematica*[15] statement:

```
Gamma[3/2] − Gamma[3/2, 0.00440]
```

Only about 2 out of every 10,000 molecules are moving slower than 100 km/h.

The integrals $\int_0^x u^n e^{-cu^2}du \,/\, \int_0^\infty e^{-cu^2}du$ and $\int_x^\infty u^n e^{-cu^2}du \,/\, \int_0^\infty e^{-cu^2}du$ often appear in statistical analysis. The cases with $n = 0$ and $c = 1$ are called the ***error function***, erf x, and the ***complementary error function***, erfc x,

$$\text{erf}\,x = \frac{2}{\sqrt{\pi}}\int_0^x e^{-u^2}du\,, \qquad \text{erfc}\,x = \frac{2}{\sqrt{\pi}}\int_x^\infty e^{-u^2}du\,. \tag{2.54}$$

Another special function that appears occasionally in science applications is the ***beta function***

$$B(p,q) = \frac{\Gamma(p)\Gamma(q)}{\Gamma(p+q)}\,. \tag{2.55}$$

(The capital of the Greek letter beta is identical to the Roman letter B.) One

[15] *Mathematica* is a computational mathematics software system, copyright 1988-2008 by Wolfram Research, Inc. "*Mathematica*" is a registered trademark of Wolfram Research.

can show[16] that the beta function has the following integral representations:

$$B(p,q) = \int_0^1 t^{p-1}(1-t)^{q-1}dt = \int_0^\infty s^{p-1}(1+s)^{-p-q}ds \tag{2.56}$$

for $p > 0$ and $q > 0$. We can generalize this to define the ***incomplete beta function***,

$$B_b(p,q) = \int_0^b t^{p-1}(1-t)^{q-1}dt = \int_0^{b/(1-b)} s^{p-1}(1+s)^{-p-q}ds. \tag{2.57}$$

The ***regularized incomplete beta function***[17] is

$$I_b(p,q) = B_b(p,q)/B(p,q)\,. \tag{2.58}$$

The beta functions play an important role in statistical analysis.

2.3 Functionals

A ***functional*** is a function of functions. In other words, it is a mapping F,

$$f \xrightarrow{F} g, \tag{2.59}$$

that takes as input a function f instead of a number, and gives as output a function or number g. We have already seen an example of a functional, namely, the integral, which we could write as

$$F[f] = \int_a^b f(x)dx. \tag{2.60}$$

It maps a function f into a number. (Note that square brackets, rather than parentheses, enclose the argument of a functional.) In Section 1.4 the integral was treated as an operator. In fact, the concepts of operator, with notation Ff, and functional, $F[f]$, are often equivalent.[18]

Functional notation is most commonly used when one is trying to determine an optimal choice of function from among all possible trial functions of a given form. For example, suppose one wanted to determine the best linear function $f(x) = c_0 + c_1 x$ with which to fit a set of experimental data $\{(x_i, f_i)\}$. The functional F could be defined as some numerical measure of errors $|f(x_i) - f_i|$ between the predicted values $f(x_i)$ and the actual values f_i. The best choice of f would be the one that minimizes $F[f]$. This will be discussed in detail in Chapter 10. Note that in this case it would be completely equivalent to consider F to be a function $F(c_0, c_1)$ of the parameters rather than a functional $F[f]$. The functional perspective would be more convenient if we

---

[16] Using methods described in Chapter 4. See Exercise 4.12.

[17] Some authors call $I_b(p,q)$ the "beta function" rather than $B_b(p,q)$, while some others call $I_b(p,q)$ the "regularized beta function," omitting "incomplete."

[18] Operator is a broader concept. An operator can operate on mathematical entities other than functions.

also wanted to consider nonlinear functional forms, such as $f(x) = c_0 e^{c_1 x}$ or $f(x) = (c_0 + c_1 x)^{-1}$, as well as the linear form.

Another example occurs in quantum chemistry. The electronic energy of a molecule depends on the distribution of electron probability density around the atomic nuclei. This density, ρ, is a function of the coordinates of each of the N electrons,

$$\rho(x_1, y_1, z_1, x_2, y_2, z_2, \ldots, x_N, y_N, z_N) \,.$$

The electronic energy of the molecule, $E[\rho]$, depends on the form of the function ρ but not on the specific locations of the electrons at any given moment. E is a functional of ρ, but *not* a function of the (x_j, y_j, z_j). The exact solution for the function ρ is generally not known. Various forms for ρ can be proposed and then the value of the energy functional can be computed to see what energy would be implied by the choice of ρ.

Exercises

2.1 Evaluate the product $\prod_{k=2}^{6}(1 - 1/k)$.

2.2 Use the binomial formula to (a) derive the formula $dx^n/dx = nx^{n-1}$ for any arbitrary positive integer n and (b) derive the formula $dx^n/dx = nx^{n-1}$ for any *negative* integer n.

2.3 Use the quadratic formula to find all the roots of the equation $13q^2 - q^4 = 36$. *(Hint: This equation has four different roots.)*

2.4 Express $\log_{16} x$ in terms of $\ln x$.

2.5 Derive Eqs. (2.25).

2.6 Use integration by parts to calculate $\gamma(2, 10)$.

2.7 Calculate the first derivative of each of the following.
(a) $4/x^3$ (b) $4x^{1/3}$ (c) $(4x - 5)^{-3}$ (d) $\ln(4x + 3)$ (e) $\exp(-x^2/3)$

2.8 Calculate the first derivative of each of the following.
(a) $\sin^2(\theta/3)$ (b) $\cos\big(\arctan(x)\big)$ (c) $\operatorname{arccsc}(x^{-1/2})$ (d) $(1 - \sin^2\theta)^{-3/2}$

2.9 Calculate the first derivative of each of the following.
(a) $\log_{10} x$ (b) $\log_{47} x$ (c) $\sqrt{\ln x}$ (d) x^x (e) x^{x-1}

2.10 (a) Determine the value of $\Gamma\big(\frac{7}{2}\big)$ recursively, starting with $\Gamma\big(\frac{1}{2}\big) = \pi^{1/2}$. (b) Similarly, determine the value of $\Gamma\big(-\frac{7}{2}\big)$.

2.11 (a) Prove that $\Gamma(x)$ has a first-order pole at $x = 0$. (b) Use the recursion relation for the gamma function to prove that $\Gamma(x)$ has a first-order pole at every negative integer value of x.

2.12 Calculate the following derivatives.
(a) $\frac{\partial}{\partial a}\Gamma(a, b)$ (Express the answer as an integral.) (b) $\frac{\partial}{\partial b}\Gamma(a, b)$ (c) $\frac{\partial^2}{\partial a \partial b}\Gamma(a, b)$

2.13 Express the binomial coefficient in terms of a beta function.

2.14 Derive the approximation $\ln(x!) = x \ln x - x$.

Chapter 3

Coordinate Systems

This chapter begins with a discussion of different ways to specify points in three-dimensional space and then generalizes the notion of a "point" to include spaces of dimension other than 3, to nonspatial coordinates, and to spaces with dimensionality reduced by a constraint. Then we see how changing the coordinate system affects differential operators.

3.1 Points in Space

The ***Cartesian coordinate system***[1] specifies the location of a point relative to three mutually perpendicular axes, $\hat{x}$, $\hat{y}$, and $\hat{z}$, that intersect at the ***origin*** of the coordinate system. The location of any point is specified by an ordered set of three numbers, (x, y, z), that represent positions on the axes of the perpendicular projections of the point, with the origin at (0,0,0).

The Cartesian system is usually the most convenient of the coordinate systems, but there are many situations in which some other coordinate system is more convenient. The ***cylindrical coordinate system*** describes a point in terms of coordinates (ρ, ϕ, z), where z is the same as in Cartesian coordinates but ρ and ϕ are circular polar coordinates in the xy-plane. This is illustrated in the left-hand panel of Fig. 3.1. The equations for the transformations between cylindrical and Cartesian coordinates are

$$x = \rho \cos\phi\,, \quad y = \rho \sin\phi\,, \quad z = z\,, \tag{3.1}$$

$$\rho = (x^2 + y^2)^{1/2}, \quad \phi = \arctan(y/x)\,. \tag{3.2}$$

The coordinate ϕ ranges from 0 to 2π. These are convenient coordinates for problems with cylindrical symmetry.

The ***spherical polar coordinate system*** is like the cylindrical system except that z is replaced by a second angular coordinate, θ, the angle between the z-axis and the line from the origin to the point in question. It is important to note that θ ranges only from 0 to π (180°). This is because an angle greater than this can be described in terms of an angle within the range 0 to π after a rotation about the z-axis by π. Using the fact that $\rho = r \sin\theta$, we obtain the transformation equations

$$x = r\cos\phi \sin\theta, \quad y = r \sin\phi \sin\theta, \quad z = r\cos\theta, \tag{3.3}$$

$$r = (x^2 + y^2 + z^2)^{1/2}, \quad \theta = \arctan(\rho/z). \tag{3.4}$$

[1]Named after the French philosopher and mathematician René Descartes (1596-1650), who pioneered the field of analytical geometry.

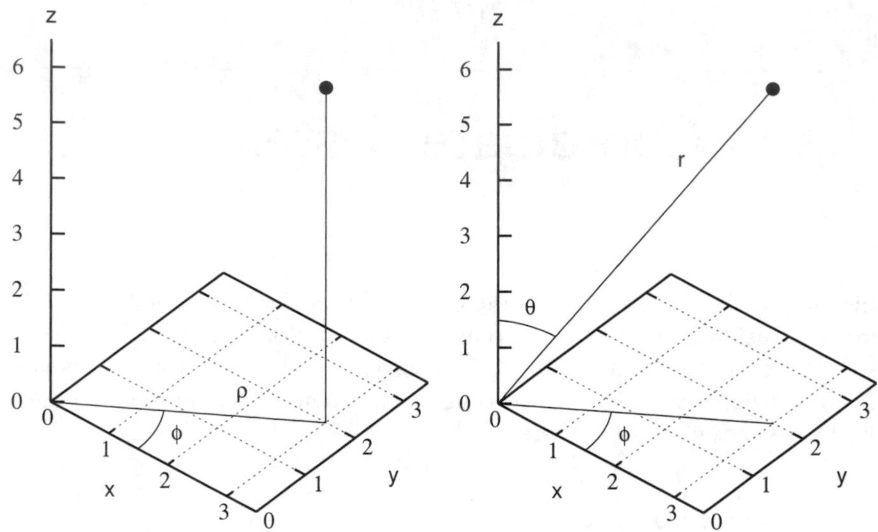

Figure 3.1: Cylindrical and spherical polar coordinate systems. The point $(x, y, z) = (3, 2, 6)$ in Cartesian coordinates corresponds to (ρ, ϕ, z) in cylindrical coordinates, with $\rho = (3^2 + 2^2)^{1/2} = 3.6056$ and $\phi = \arctan(2/3) = 0.5880$. In spherical coordinates it is given by (r, ϕ, θ), with $r = (3^2 + 2^2 + 6^2)^{1/2} = 7$ and $\theta = \arctan(\rho/6) = 0.5411$.

Spherical polar coordinates are shown in the right-hand panel of Fig. 3.1.

Example 3.1. *Kinetic energy in spherical polar coordinates.* The kinetic energy of a particle of mass m moving on a trajectory $\big(x(t), y(t), z(t)\big)$ as a function of time t is, in Cartesian coordinates,

$$T = \frac{1}{2}m\left[\left(\frac{dx}{dt}\right)^2 + \left(\frac{dy}{dt}\right)^2 + \left(\frac{dz}{dt}\right)^2\right]. \tag{3.5}$$

Let us express this in terms of spherical polar coordinates. To simplify the notation we abbreviate the derivative with respect to time as a dot, for example, $dx/dt = \dot{x}$. Then

$$T = \tfrac{1}{2}m(\dot{x}^2 + \dot{y}^2 + \dot{z}^2).$$

The time derivatives of the coordinates are given by

$$\dot{x} = \frac{dx}{dt} = \frac{dr}{dt}\cos\phi\sin\theta + r\frac{d\cos\phi}{dt}\sin\theta + r\cos\phi\frac{d\sin\theta}{dt}$$

$$= \dot{r}\cos\phi\sin\theta - r\dot{\phi}\sin\phi\sin\theta + r\dot{\theta}\cos\phi\cos\theta, \tag{3.6a}$$

$$\dot{y} = \dot{r}\sin\phi\sin\theta + r\dot{\phi}\cos\phi\sin\theta + r\dot{\theta}\sin\phi\cos\theta, \tag{3.6b}$$

$$\dot{z} = \dot{r}\cos\theta - r\dot{\theta}\sin\theta. \tag{3.6c}$$

Substituting these into the expression for T, collecting terms according to the powers of the derivatives that multiply them (for example, $\dot{r}^2$, $\dot{\theta}^2$, $\dot{r}\dot{\phi}$, etc.), and then simplifying the result using the identities $\cos^2\phi + \sin^2\phi = 1$ and $\cos^2\theta + \sin^2\theta = 1$, we obtain

$$T = \tfrac{1}{2}m\dot{r}^2 + \tfrac{1}{2}mr^2\dot{\theta}^2 + \tfrac{1}{2}mr^2(\sin^2\theta)\dot{\phi}^2. \tag{3.7}$$

We will see applications of this equation in subsequent chapters. As a simple example, consider a "stationary" object on the surface of the earth moving along with the rotation of the planet. Let the z-axis be the axis of earth's rotation, along a line from the South Pole to the North Pole. Then the only motion is in the coordinate ϕ; the other two coordinates are constant, which means their derivatives, $\dot{r}$ and $\dot{\theta}$, are zero. The kinetic energy of the object is $T = \frac{1}{2}mr^2(\sin^2\theta)\dot{\phi}^2$. If the object is at the equator $(\theta = \pi/2)$ then $T = \frac{1}{2}mr^2\dot{\phi}^2$ is at its maximum.

3.2 Coordinate Systems for Molecules

Consider the atoms of a diatomic molecule. We could describe their locations in terms of the three coordinates of each atom, which would be a "point" in the six-dimensional space $(x_1, y_1, z_1, x_2, y_2, z_2)$. This is called the ***laboratory coordinate system.*** However, this probably provides more information than we need. Suppose that we are only interested in the vibrational motion of the molecule. That can be described in terms of a single coordinate r_{12} that gives the distance between the two atoms,[2]

$$r_{12} = \sqrt{(x_1 - x_2)^2 + (y_1 - y_2)^2 + (z_1 - z_2)^2}. \tag{3.8}$$

If our interest is only in the molecular vibrations, then we probably do not care about the orientation of the molecular axis or about how the molecule as a whole is moving about the laboratory. An ***internal coordinate systcm*** describes only the relative positions of the atoms within a molecule, without regard to orientation or to collective motion of the entire molecule.

Molecules are usually described in terms of three kinds of motion: *translational, rotational,* and *vibrational.*[3] ***Translational motion*** is motion of the molecule as a whole. To describe it mathematically, we specify Cartesian coordinates (X, Y, Z) of some point that moves along with the molecule. This point then serves as the origin of the internal coordinate system. We could, if we wanted, choose the point to be the location of one of the atoms. However, the usual practice is to choose the mass-weighted average of the locations of all the atoms. This point is called the ***center of mass.***

Example 3.2. *Center of mass of a diatomic molecule.* For a homonuclear diatomic molecule, the center of mass is at the halfway point of the line connecting the two nuclei. For a heteronuclear diatomic, with atomic masses m_A and m_B, the center of mass lies closer to the heavier atom. To see where this point lies in an internal coordinate system, let us specify an internal $\tilde{z}$-axis passing through the nuclei, with atom A at $\tilde{z} = 0$ and atom B at some fixed value r_{12}. Let $m = m_A + m_B$ be the total mass of the molecule. Then the center of mass is at

$$\tilde{z}_{CM} = (m_B/m)r_{12} \,. \tag{3.9}$$

For carbon monoxide, $\tilde{z}_{CM} = (16.0/28.0)r_{12} = 0.571\, r_{12}$, close to the halfway point. With the origin set at the center of mass, carbon is at $\tilde{z} = -0.571\, r_{12}$ while oxygen is at $\tilde{z} = +0.429\, r_{12}$ For hydrogen iodide, $\tilde{z}_{CM} = 0.992\, r_{12}$, very close to the iodine.

---

[2] r_{12} is pronounced "r one two," not "r twelve."

[3] We are treating each atom a single point, ignoring that atoms consist of nuclei and electrons.

Let $(\tilde{x}, \tilde{y}, \tilde{z})$ be internal coordinates of atom j in a Cartesian system with origin at an arbitrary point within the molecule, such as the position of one of the atoms. Note that we are using tildes ($\tilde{}$) to distinguish internal coordinates from laboratory coordinates. The position of the center of mass within this internal coordinate system is $(\tilde{x}_{\rm CM}, \tilde{y}_{\rm CM}, \tilde{z}_{\rm CM})$,

$$\tilde{x}_{\rm CM} = \frac{1}{m}\sum_{j=1}^{N} m_j\tilde{x}_j\,, \quad \tilde{y}_{\rm CM} = \frac{1}{m}\sum_{j=1}^{N} m_j\tilde{y}_j\,, \quad \tilde{z}_{\rm CM} = \frac{1}{m}\sum_{j=1}^{N} m_j\tilde{z}_j\,. \qquad (3.10)$$

m_j is the mass of atom j and $m = \sum_{j=1}^{N} m_j$ is the mass of the molecule.

Example 3.3. *Center of mass of a planar molecule.* Consider the formyl chloride molecule, CHClO. Position the coordinate system such that the molecule lies in the $\tilde{x}\tilde{y}$-plane with carbon at the origin and oxygen on the $\tilde{y}$-axis. Use typical bond lengths $R_{\rm C=O} = 1.2$ Å, $R_{\rm C-H} = 1.1$ Å, $R_{\rm C-Cl} = 1.8$ Å, and assume bond angles of 120°. The C, O, H, and Cl atoms are, respectively, at

$$(0,0,0), \quad (0, R_{\rm C=O}, 0),$$
$$\big(-R_{\rm C-H}\cos(\pi/6),\, -R_{\rm C-H}\sin(\pi/6),\, 0\big),$$
$$\big(R_{\rm C-Cl}\cos(\pi/6),\, -R_{\rm C-Cl}\sin(\pi/6),\, 0\big),$$

For the masses, we use the atomic weights 12, 16.0, 1.0, and 35.0 u. The result is

$$(\tilde{x}_{\rm CM}, \tilde{y}_{\rm CM}, \tilde{z}_{\rm CM}) = (0.84 \text{ Å},\, -0.20 \text{ Å},\, 0),$$

as indicated in the figure. The center of mass is skewed toward the chlorine, which is the heaviest of the four atoms. In the center-of-mass coordinate system the center of mass is the origin, the carbon atom is at $(0,0,0) - (\tilde{x}_{\rm CM}, \tilde{y}_{\rm CM}, \tilde{z}_{\rm CM}) = (-0.84 \text{ Å}, 0.20 \text{ Å}, 0)$, the oxygen atom is at $(-0.84 \text{ Å}, 1.40 \text{ Å}, 0)$, etc.

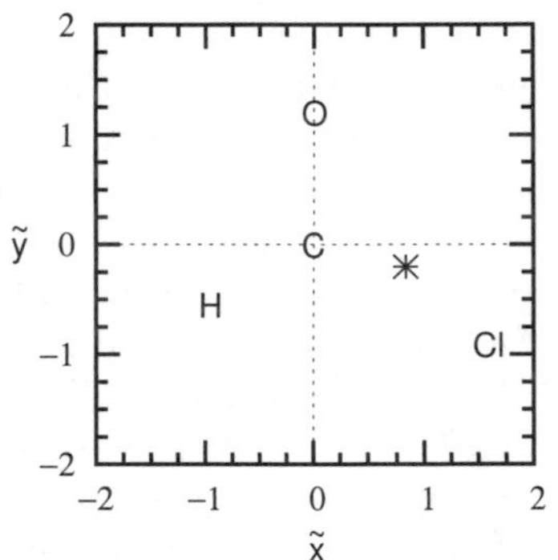

The average of the atomic positions without the mass weightings is called the ***centroid*** of the molecule. It is used for describing properties that do not depend on the atomic masses. For example, the origin of the dipole moment vector is usually placed at the centroid.

Rotation is circular motion in a plane perpendicular to a specified axis of rotation. In general, any change of orientation can be analyzed in terms of circular rotations about three different axes. Therefore, any change in orientation can be referred to as "rotational" motion. The rotational axes themselves must be mutually perpendicular and must all pass through the center of mass, but for nonlinear molecules they are otherwise arbitrary.[4] For a *linear* molecule, we can choose one of the axes to be the axis containing the all the atoms. On a molecular length scale it is reasonable to consider the atoms as point particles with no extension out of the axis. In that case, rotation

[4]There is a particular choice of rotational axes, called the *principal* axes, that leads to a very simple expression for the rotational kinetic energy. This is discussed in Section 19.2.

about that axis does not contribute to the kinetic energy. Two rotational axes are sufficient to specify the rotations of a linear molecule.

How many vibrational coordinates are there? Let us count all the other coordinates. There are three Cartesian coordinates for each of the N atoms, for a total of $3N$. This must be equal to the sum of the number of translational, rotational and vibrational coordinates. Subtract 3, for $\tilde{x}_{\mathrm{CM}}$, $\tilde{y}_{\mathrm{CM}}$, and $\tilde{z}_{\mathrm{CM}}$, and another 3 for the three rotation angles, and we are left with a remainder of $3N - 6$ vibrational coordinates for a nonlinear molecule (or $3N - 5$ for a linear molecule). These vibrations constitute the motion described by the internal coordinate system.

Example 3.4. *Coordinates for a bent triatomic molecule.* The laboratory coordinate system in this case is nine dimensional while the internal coordinate system is three dimensional. The internal coordinates could be chosen as the three internuclear distances, (r_{12}, r_{23}, r_{13}), or as (r_{12}, r_{23}, χ) in terms of the bond distances r_{12} and r_{23} and the bond angle χ.

3.3 Abstract Coordinates

It is often useful to generalize the concept of a "point" beyond simply a location in physical space. Although physical space is three dimensional, we can easily write down on paper coordinate systems for spaces with other dimensionalities. Consider, for example, the motion of hydrogen atoms adsorbed on the surface of a piece of metal. Because the atoms are constrained to remain on the surface, it can be a reasonable approximation to ignore the motion in the z direction. The migrations of an H atom can then be fully described in terms of the coordinates (x, y), in a two-dimensional space.

Coordinate systems with more than three dimensions are also quite common. Consider a collection of N identical atoms. Label the atoms with an index $j = 1, 2, 3, \ldots$. The position of the jth atom in three-dimensional Cartesian coordinates is (x_j, y_j, z_j). We can consider each configuration of the full set of atoms as a single point in a space of dimension $3N$, with coordinate

$$(q_1, q_2, q_3, \ldots, q_{3N}) = (x_1, y_1, z_1, x_2, y_2, z_2, \ldots, x_N, y_N, z_N). \tag{3.11}$$

The mathematical expressions for some properties of the system are simplified by this approach. For example, the internal energy of this system in the absence of forces is the sum of the kinetic energies of each particle,

$$\begin{aligned} U =& \frac{m}{2}\left[\left(\frac{dx_1}{dt}\right)^2 + \left(\frac{dy_1}{dt}\right)^2 + \left(\frac{dz_1}{dt}\right)^2\right] + \frac{m}{2}\left[\left(\frac{dx_2}{dt}\right)^2 + \left(\frac{dy_2}{dt}\right)^2 + \left(\frac{dz_2}{dt}\right)^2\right] \\ &+ \cdots + \frac{m}{2}\left[\left(\frac{dx_N}{dt}\right)^2 + \left(\frac{dy_N}{dt}\right)^2 + \left(\frac{dz_N}{dt}\right)^2\right]. \end{aligned} \tag{3.12}$$

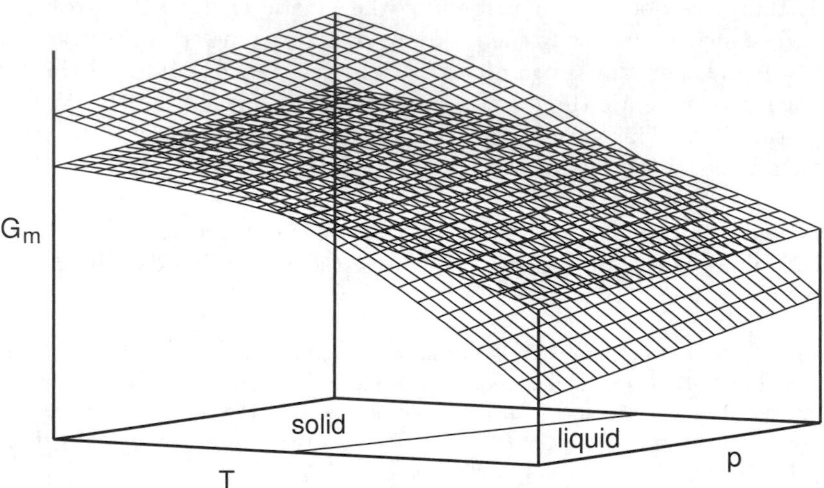

Figure 3.2: Gibbs free energy, and phase diagram (at base of figure), for solid and liquid of a one-component system as function of temperature and pressure. At higher temperature the liquid is the phase with the lower $G_{\rm m}$.

We can express this more simply as the kinetic energy of a single "particle" in $3N$-dimensional space,

$$U = \sum_{i=1}^{3N} \frac{1}{2} m \left(\frac{dq_i}{dt} \right)^2 , \tag{3.13}$$

with the sum over the dimensions. Note that the subscripts here label the different dimensions, not the different particles.

In thermodynamics we use a more abstract concept of coordinate. For example, the molar Gibbs free energy $G_{\rm m}$ is a function of pressure p and temperature T. It is useful to picture (T, p) values as points in a plane and $(T, p, G_{\rm m})$ as points in a three-dimensional space. Each phase of the system (e.g., solid, liquid, gas) corresponds to a different function $G_{\rm m}(T, p)$. The phase with the lowest $G_{\rm m}$ value at a given (T, p) is the one that will exist when the system comes to equilibrium under those conditions. If two phases have equal $G_{\rm m}$ then they can coexist in equilibrium.

A convenient way to represent this is a ***phase diagram***, which is a plot in the (T, p) plane of the curves that correspond to the intersection of $G_{\rm m}$ functions for two different phases. These curves show the values of (T, p) at which two phases can coexist. This is illustrated by Fig. 3.2. Two intersecting surfaces are shown. One is $G_{\rm m}(T, p)$ for the solid and the other is $G_{\rm m}(T, p)$ for the liquid. The phase diagram is the plot shown at the base of the figure, showing the phase coexistence curve at the intersection of the two surfaces.

Example 3.5. *The triple point.* Is it possible for three phases to coexist? A third phase would correspond to a third surface in Fig. 3.2. The intersection of the third surface with the second will give a different coexistence curve $p(T)$. The three phases can coexist if this second curve intersects the first $p(T)$ curve in the two-dimensional projection at the base of the figure. Typically, there is at most a single point of intersection. This is called the "triple" point, because three phases coexist there.

The concept of a point can be further generalized to a space of functions. This is useful in quantum mechanics, in which the fundamental equations involve operators that map one function into another function rather than a number into another number. A "point" in such a space is an ordered set of *functions*, rather than an ordered set of *numbers*. This analogy will be worked out in detail in Chapters 16 and 17.

3.4 Constraints

3.4.1 Degrees of Freedom

The dimensionality of a coordinate system may be smaller than it seems. Consider some $f(x, y)$ subject to a constraint that expresses a relationship between x and y. Suppose, for example, that f describes some property of a particle constrained to the circumfrence of a circle of radius 3. Then

$$x^2 + y^2 = 9. \tag{3.14}$$

x and y are no longer independent coordinates; a change in one implies a change in the other. The true dimensionality of a function is equal to the minimum number of coordinates whose values must be specified in order to completely determine the value of the function. Each of these dimensions is called a ***degree of freedom***. The number of degrees of freedom is equal to the total number of coordinates minus the number of constraint equations.

In the thermodynamics of a chemical system at equilibrium, the number of degrees of freedom typically is much smaller than the number of coordinates.

Example 3.6. *Number of degrees of freedom for a mixture of liquids.* Consider a closed container partially filled with a mixture of water and ethanol. Our variables are the mole fractions $X_{\text{water,liq}}$ and $X_{\text{ethanol,liq}}$ for the liquid, $X_{\text{water,vap}}$ and $X_{\text{ethanol,vap}}$ for the vapor, the vapor's molar volume V_{m}, the total pressure p of the vapor, and the temperature T. Thus, we have seven variables. We can eliminate V_{m} using the equation of state $V_{\text{m}} = RT/p$. Thus we eliminate V_{m} as a degree of freedom. Consider the vapor pressures, which according to Dalton's law are $p_{\text{water}} = X_{\text{water,vap}}\, p$, and $p_{\text{ethanol}} = X_{\text{ethanol,vap}}\, p$. At first the liquid will evaporate, and the vapor pressures will increase. Eventually, the rate of condensation for each component becomes equal the rate of evaporation, and the system will be in equilibrium. Thus, the equilibrium vapor pressures are not degrees of freedom; they are determined by the temperature and by the relative amounts of water and ethanol in the liquid. The total pressure is the sum of the two partial pressures; it is not a degree of freedom. Furthermore, the mole fractions in the liquid must sum to unity. (If the liquid is 14% ethanol, then it must be 86% water.) In conclusion, we have only two degrees of freedom: the temperature and a single mole fraction.

The following theorem is due to Gibbs:[5]

Theorem 3.4.1. *(Gibbs phase rule.) For a system at equilibrium with C components and P phases, the number of thermodynamic degrees of freedom is*

$$F = C - P + 2\,. \tag{3.15}$$

Proof of Gibbs phase rule. Given C components and P phases, the state of the system is described by P times C mole fractions and two additional state variables (typically pressure and temperature, with molar volume having been eliminated by an equation of state); a total of $PC + 2$ variables. For each phase we have P constraints from the requirement that the sum of the mole fractions must equal unity. Furthermore, in order for the system to be in equilibrium, each component must have the same value for G_{m} for each of the phases. For some given value of G_{m} for phase α, we have constraint equations $G_{\mathrm{m}}(\alpha) = G_{\mathrm{m}}(\beta)$, $G_{\mathrm{m}}(\alpha) = G_{\mathrm{m}}(\gamma)$, etc.; $(P-1)$ equations for each component, for a total of $(P-1)C$ constraints. The number of degrees of freedom is the number of variables minus the number of constraints,

$$F = (PC + 2) - [P + (P-1)C] = C - P + 2\,.$$

□

The most direct approach to dealing with constrained coordinate systems is simply to eliminate one of the coordinates using the constraint equation. For motion on the circle of radius 3, we can solve for y in terms of x, substitute into f the equation $y = \sqrt{9 - x^2}$, and then define a new function of a single coordinate,

$$g(x) = f\big(x, \sqrt{9 - x^2}\,\big). \tag{3.16}$$

Another approach, which is not always feasible, but when feasible can greatly simplify the analysis, is to transform to a new coordinate system in which the constraint causes one coordinate to have a constant value. In this example, in polar coordinates (r, ϕ) the constraint can be expressed simply as $r = 3$. A new function, of only a single coordinate, can be defined as

$$h(\phi) = f(3\cos\phi, 3\sin\phi). \tag{3.17}$$

3.4.2 Constrained Extrema*

Occasionally it is necessary to find the extremum of a function subject to a constraint. One approach to such problems is to explicitly reduce the dimensionality, as just described, and then set the derivative with respect to the independent coordinate equal to zero.

[5] J. Willard Gibbs (1839-1903), American mathematician, engineer, physicist, and theoretical chemist, is largely responsible for laying the foundations of the field of chemical thermodynamics.

Example 3.7. *Extrema of a two-coordinate function on a circle: Using the constraint to reduce the number of degrees of freedom.* Consider the function

$$f(x,y) = y^2 - 2xy + 3x^2 + 60. \tag{3.18}$$

Where are the extrema of f on the circle of radius 3? In other words, if we impose the constraint $x^2 + y^2 = 9$ on the (x, y) values, where are the minima and maxima of $f(x, y)$? There are two minima and two maxima, as shown in the figure:

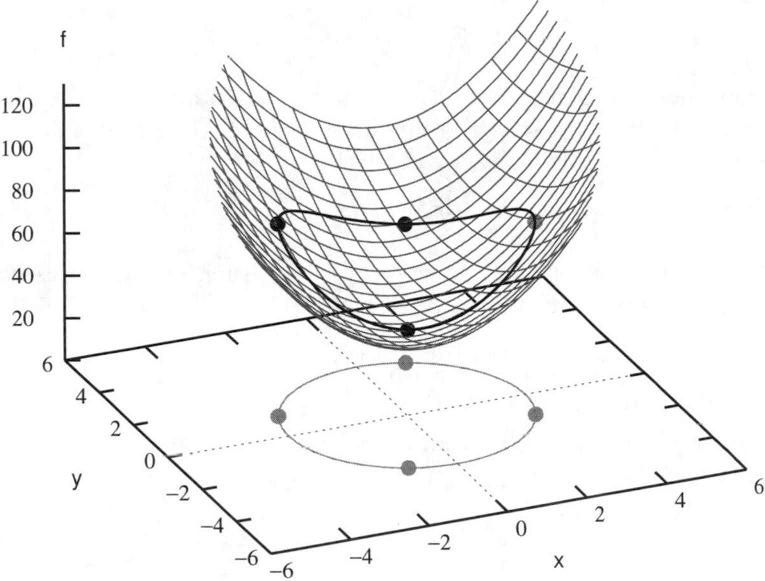

The constraint can be rearranged to express y as a function of x,

$$y(x) = \pm\sqrt{9-x^2}\,, \qquad f(x, y(x)) = 2x^2 \mp 2x\sqrt{9-x^2} + 69.$$

The extrema are the x values that solve the equation

$$0 = \frac{df}{dx} = 4x \mp 2(9-x^2)^{1/2} \mp 2x[(-2x)\tfrac{1}{2}(9-x^2)^{-1/2}] = 4x \mp 2\sqrt{9-x^2} \pm 2x^2/\sqrt{9-x^2}\,.$$

It is possible to solve this for x, but it involves some messy algebra.

Let us instead use polar coordinates, with $f(x, y) = h(\phi)$ according to Eq. (3.17). Then

$$h(\phi) = 3^2(\sin^2\phi - 2\cos\phi\sin\phi + 3\cos^2\phi) + 60 = 3^2(1 + 2\cos^2\phi - \sin 2\phi) + 60,$$

using the identities $\sin^2\phi + \cos^2\phi = 1$ and $2\cos\phi\sin\phi = \sin 2\phi$. The extremum condition is

$$0 = \frac{dh}{d\phi} = 9(-4\sin\phi\cos\phi - 2\cos 2\phi)) = -18(\sin 2\phi + \cos 2\phi).$$

It follows that $\sin 2\phi = -\cos 2\phi$, which implies $2\phi = \arctan(-1)$. The solutions are[6]

$$\phi = 3\pi/8,\ 7\pi/8,\ 11\pi/8,\ \text{and } 15\pi/8.$$

This agrees with the four points shown in the figure, with $(x, y) = (3\cos\phi, 3\sin\phi)$.

The analysis in terms of polar coordinates implicitly takes into account the circular symmetry of the problem and thereby simplifies the analysis.

---

[6] See Example 2.4.

An alternative approach is the ***method of undetermined multipliers***. This is a technique developed by Lagrange. Suppose that we want to find an extremum of a function $f(x, y)$ subject to a constraint

$$q(x, y) = c, \tag{3.19}$$

where q is some specified function and c is some constant. The extremum condition is

$$0 = df = \frac{\partial f}{\partial x}dx + \frac{\partial f}{\partial y}dy. \tag{3.20}$$

(Although the total differential, df, is zero, the individual partial derivatives with respect to x and y are generally not zero.) The fact that q is constant implies that $dq = 0$. Therefore, we can write

$$df - \lambda dq = 0 \tag{3.21}$$

for *any* arbitrary value of a parameter λ. Substituting in the expressions for df and dq in terms of dx and dy, this becomes

$$\left(\frac{\partial f}{\partial x} - \lambda\frac{\partial q}{\partial x}\right) dx + \left(\frac{\partial f}{\partial y} - \lambda\frac{\partial q}{\partial y}\right) dy = 0. \tag{3.22}$$

This equation must be valid for any arbitrary changes dx and dy. This can be true only if the terms in parentheses are each equal to zero,

$$\frac{\partial f}{\partial x} - \lambda\frac{\partial q}{\partial x} = 0, \qquad \frac{\partial f}{\partial y} - \lambda\frac{\partial q}{\partial y} = 0. \tag{3.23}$$

These two equations and the constraint equation $q(x, y) = c$ give three equations for the three unknowns λ, x, and y.

Example 3.8. *Extrema of a two-coordinate function on a circle: Using the method of undetermined multipliers.* Let us revisit the previous example. Let $q(x, y) = x^2 + y^2$. Then,

$$0 = \frac{\partial f}{\partial x} - \lambda\frac{\partial q}{\partial x} = -2y + 6x - 2\lambda x \quad \Rightarrow \quad y = (3 - \lambda)x\,,$$
$$0 = \frac{\partial f}{\partial y} - \lambda\frac{\partial q}{\partial y} = 2y - 2x - 2\lambda y \quad \Rightarrow \quad x = (1 - \lambda)y\,.$$

Substituting the one equation into the other gives

$$y = (3 - \lambda)(1 - \lambda)y \quad \Longrightarrow \quad \lambda^2 - 4\lambda + 2 = 0\,.$$

There are two solutions for λ, according to the quadratic formula: $\lambda = 2 \pm \sqrt{2}$. Substituting this into the constraint equation gives

$$9 = x^2 + y^2 = (1 - \lambda)^2 y^2 + y^2 = 2(2 \pm \sqrt{2}\,)y^2,$$
$$y^2 = \frac{9}{2(2 \pm \sqrt{2}\,)}\,\frac{2 \mp \sqrt{2}}{2 \mp \sqrt{2}} = \frac{9}{2}\,\frac{2 \mp \sqrt{2}}{4 - 2} = \frac{9}{4}(2 \mp \sqrt{2}\,),$$
$$y = \frac{3}{2}\sqrt{2 \mp \sqrt{2}} \quad \text{or} \quad y = -\frac{3}{2}\sqrt{2 \mp \sqrt{2}},$$

with the sign of the "$\pm$" chosen according to the sign in the solution for λ. The corresponding solutions for x follow from $x = (1 - \lambda)y$. The results for the extrema, after some algebraic simplifying, are

$$\left(-\frac{3}{2}\sqrt{2+\sqrt{2}}, \frac{3}{2}\sqrt{2-\sqrt{2}}\right), \quad \left(\frac{3}{2}\sqrt{2+\sqrt{2}}, -\frac{3}{2}\sqrt{2-\sqrt{2}}\right),$$
$$\left(\frac{3}{2}\sqrt{2-\sqrt{2}}, \frac{3}{2}\sqrt{2+\sqrt{2}}\right), \quad \left(-\frac{3}{2}\sqrt{2-\sqrt{2}}, -\frac{3}{2}\sqrt{2+\sqrt{2}}\right).$$

Numerical evaluation confirms that these are equivalent to the solutions found in the previous example.

The real advantage of Lagrange's method is seen with problems in which the constraint equation cannot easily be solved analytically for one of the variables, and for problems with a large number of constraints. With more than one constraint, one simply includes additional multipliers, one for each constraint. With two constraints, for example,

$$\frac{\partial f}{\partial x} - \lambda_1 \frac{\partial q_1}{\partial x} - \lambda_2 \frac{\partial q_2}{\partial x} = 0\,, \quad \frac{\partial f}{\partial y} - \lambda_1 \frac{\partial q_1}{\partial y} - \lambda_2 \frac{\partial q_2}{\partial y} = 0\,. \tag{3.24}$$

This method is often used in such fields as engineering, meteorology, and economics, and occasionally in physical chemistry.

3.5 Differential Operators in Polar Coordinates

$f(r, \phi)$ has the differential

$$df = \left(\frac{\partial f}{\partial r}\right) dr + \left(\frac{\partial f}{\partial \phi}\right) d\phi\,. \tag{3.25}$$

Let us consider the variables r and ϕ as functions $r(x, y)$ and $\phi(x, y)$. Then they have the differentials

$$dr = \left(\frac{\partial r}{\partial x}\right) dx + \left(\frac{\partial r}{\partial y}\right) dy\,, \qquad d\phi = \left(\frac{\partial \phi}{\partial x}\right) dx + \left(\frac{\partial \phi}{\partial y}\right) dy\,. \tag{3.26}$$

Substituting these into Eq. (3.25) and collecting terms, we find that

$$df = \left(\frac{\partial r}{\partial x}\frac{\partial f}{\partial r} + \frac{\partial \phi}{\partial x}\frac{\partial f}{\partial \phi}\right) dx + \left(\frac{\partial r}{\partial y}\frac{\partial f}{\partial r} + \frac{\partial \phi}{\partial y}\frac{\partial f}{\partial \phi}\right) dy. \tag{3.27}$$

The factor in parentheses multiplying dx is, by definition, $\partial f/\partial x$, and the factor multiplying dy is $\partial f/\partial y$. Thus, we have derived an expression for the differential operators $\partial/\partial x$ and $\partial/\partial y$ under the transformation $(x, y) \to (r, \phi)$, from Cartesian to circular polar coordinates:

$$\frac{\partial}{\partial x} = \left(\frac{\partial r}{\partial x}\right)_y \frac{\partial}{\partial r} + \left(\frac{\partial \phi}{\partial x}\right)_y \frac{\partial}{\partial \phi}, \quad \frac{\partial}{\partial y} = \left(\frac{\partial r}{\partial y}\right)_x \frac{\partial}{\partial r} + \left(\frac{\partial \phi}{\partial y}\right)_x \frac{\partial}{\partial \phi}, \tag{3.28}$$

with subscripts x and y included to keep track of which variable is held constant. The expressions for the polar coordinates in terms of Cartesians are

$$r = (x^2 + y^2)^{1/2}, \qquad \phi = \arctan\frac{y}{x}. \tag{3.29}$$

The derivatives we need are[7]

$$\left(\frac{\partial r}{\partial x}\right)_y = \frac{1}{2}\frac{2x}{(x^2+y^2)^{1/2}} = \frac{x}{r} = \cos\phi, \tag{3.30}$$

$$\left(\frac{\partial r}{\partial y}\right)_x = \frac{y}{r} = \sin\phi, \tag{3.31}$$

$$\left(\frac{\partial \phi}{\partial x}\right)_y = \left(-\frac{y}{x^2}\right)\frac{1}{1+(y/x)^2} = -\frac{y}{x^2+y^2} = -\frac{\sin\phi}{r}, \tag{3.32}$$

$$\left(\frac{\partial \phi}{\partial y}\right)_x = \left(\frac{1}{x}\right)\frac{1}{1+(y/x)^2} = \frac{x}{x^2+y^2} = \frac{\cos\phi}{r}. \tag{3.33}$$

It follows that

$$\frac{\partial}{\partial x} = \cos\phi\frac{\partial}{\partial r} - \frac{\sin\phi}{r}\frac{\partial}{\partial\phi}, \qquad \frac{\partial}{\partial y} = \sin\phi\frac{\partial}{\partial r} + \frac{\cos\phi}{r}\frac{\partial}{\partial\phi}. \tag{3.34}$$

Eqs. (3.34) can then be used to obtain transformations for second derivatives:

$$\begin{aligned}\frac{\partial^2 f}{\partial x^2} &= \left(\cos\phi\frac{\partial}{\partial r} - \frac{\sin\phi}{r}\frac{\partial}{\partial\phi}\right)\left(\cos\phi\frac{\partial f}{\partial r} - \frac{\sin\phi}{r}\frac{\partial f}{\partial\phi}\right)\\ &= \cos^2\phi\frac{\partial^2 f}{\partial r^2} - \cos\phi\sin\phi\frac{\partial}{\partial r}\left(\frac{1}{r}\frac{\partial f}{\partial\phi}\right)\\ &\quad - \frac{\sin\phi}{r}\frac{\partial}{\partial\phi}\left(\cos\phi\frac{\partial f}{\partial r}\right) + \frac{\sin^2\phi}{r^2}\frac{\partial^2 f}{\partial\phi^2},\end{aligned} \tag{3.35}$$

and similarly for the other two second partial derivatives.

The same strategy can be used for the transformation $(x, y, z) \to (r, \phi, \theta)$.

$$\frac{\partial}{\partial x} = \left(\frac{\partial r}{\partial x}\right)_{y,z}\frac{\partial}{\partial r} + \left(\frac{\partial \theta}{\partial x}\right)_{x,z}\frac{\partial}{\partial\theta} + \left(\frac{\partial \phi}{\partial x}\right)_{y,z}\frac{\partial}{\partial\phi}, \tag{3.36}$$

etc., lead to

$$\frac{\partial}{\partial x} = \sin\theta\cos\phi\frac{\partial}{\partial r} + \frac{\cos\phi\cos\theta}{r}\frac{\partial}{\partial\theta} - \frac{\sin\phi}{r\sin\theta}\frac{\partial}{\partial\phi}, \tag{3.37a}$$

$$\frac{\partial}{\partial y} = \sin\theta\sin\phi\frac{\partial}{\partial r} + \frac{\sin\phi\cos\theta}{r}\frac{\partial}{\partial\theta} + \frac{\cos\phi}{r\sin\theta}\frac{\partial}{\partial\phi}, \tag{3.37b}$$

$$\frac{\partial}{\partial z} = \cos\theta\frac{\partial}{\partial r} - \frac{\sin\theta}{r}\frac{\partial}{\partial\theta}. \tag{3.37c}$$

[7]It would *not* be correct to use $\frac{\partial x}{\partial r} = \frac{\partial}{\partial r} r\cos\phi = \cos\phi$, $\frac{\partial r}{\partial x} = \frac{1}{\cos\phi}$. This gives $(\partial r/\partial x)_\phi$, with ϕ held constant, not $(\partial r/\partial x)_y$, with y held constant. We need to first express the new coordinates r and ϕ in terms of x and y, calculate the derivatives with x or y held constant, and then express the result in terms of r and ϕ.

The operator

$$\nabla^2 = \frac{\partial^2}{\partial x^2} + \frac{\partial^2}{\partial y^2} + \frac{\partial^2}{\partial z^2}, \tag{3.38}$$

called the ***Laplacian***,[8] is of fundamental importance in quantum mechanics. For problems with spherical symmetry, such as the electronic structure of an atom, it is useful to express the Laplacian in spherical polar coordinates. Using Eqs. (3.37) it is straightforward, albeit tedious, to derive the following:

$$\frac{\partial^2 f}{\partial x^2} + \frac{\partial^2 f}{\partial y^2} + \frac{\partial^2 f}{\partial z^2} = \frac{1}{r^2}\frac{\partial}{\partial r}\left(r^2 \frac{\partial f}{\partial r}\right) + \frac{1}{r^2 \sin\theta}\frac{\partial}{\partial\theta}\left(\sin\theta \frac{\partial f}{\partial\theta}\right) + \frac{1}{r^2\sin^2\theta}\frac{\partial^2 f}{\partial\phi^2}.$$

This means that the operator is

$$\nabla^2 = \frac{1}{r^2}\frac{\partial}{\partial r}\left(r^2 \frac{\partial}{\partial r}\right) + \frac{1}{r^2 \sin\theta}\frac{\partial}{\partial\theta}\left(\sin\theta \frac{\partial}{\partial\theta}\right) + \frac{1}{r^2\sin^2\theta}\frac{\partial^2}{\partial\phi^2}. \tag{3.39}$$

Exercises

3.1 Express the Cartesian coordinates the point $(x, y, z) = (1, 3, 2)$ in terms of (a) cylindrical coordinates and (b) spherical polar coordinates.

3.2 Express the spherical polar coordinates the point $(r, \phi, \theta) = (17, 5\pi/4, \pi/2)$ in terms of (a) Cartesian coordinates and (b) cylindrical coordinates.

3.3 Derive the transformation equations from a four-dimensional Cartesian coordinate system (x, y, z_1, z_2) to a hyperspherical polar coordinate system $(r, \phi, \theta_1, \theta_2)$. *(Hint: θ_1 and θ_2 both range from 0 to π.)*

3.4 Derive Eq. (3.7) from Eqs. (3.6).

3.5 Taking into account that the earth is rotating, what is your kinetic energy when sitting at rest? Express your answer in units of $\text{J} = \text{kg m}^2/\text{s}^2$. (Earth's radius is 6.37×10^6 m.)

3.6 Express kinetic energy in cylindrical coordinates.

3.7 In Example 3.3, what are the coordinates of the hydrogen and of the chlorine in the center-of-mass coordinate system?

3.8 Calculate the distance between the center of mass and the individual atoms of a water molecule. (Use bond lengths of 0.958 Å and a bond angle of 104.5°.)

---

[8]In honor of the French physicist Pierre-Simon Laplace (1749-1827), who developed the mathematical foundations of theoretical astronomy. Mathematical techniques that he presented in his monumental tome *Mécanique Céleste*, completed in 1825, have proved invaluable in astronomy as well as in most other fields of applied mathematics.

3.9 Calculate $\nabla^2 \sin\theta \sin\phi \, e^{-2r}$.

3.10 Derive ∇^2 in cylindrical coordinates.

3.11 Derive the expressions for $\partial/\partial r$, $\partial/\partial\phi$, and $\partial/\partial\theta$ in terms of Cartesian coordinates.

3.12 Prove that $r^{-2}\frac{d}{dr}\left(r^2\frac{d}{dr}\right)f$ is equivalent to $r^{-1}\frac{d^2}{dr^2}(rf)$.

Chapter 4

Integration

4.1 Change of Variables in Integrands

In contrast to differentiation, there is no general systematic procedure for the analytic evaluation of an integral. There are various standard forms that can be memorized or looked up in tables, but it is very common to encounter an integral that is not in a standard form. Sometimes it is possible to transform an integral into a standard form with an algebraic substitution.

4.1.1 Change of Variable: Examples

Example 4.1. *Integrals involving linear polynomials.* Consider

$$I_{\mathrm{a}} = \int_0^{10} \frac{1}{4x+7}\, dx. \tag{4.1}$$

This resembles the integral expression for the natural logarithm,

$$\ln c = \int_1^c \frac{1}{x}\, dx\,,$$

except for the factors 4 and 7 in the denominator and the range of integration starting at 0 instead of 1. Let us define a new coordinate u as a function of x according to

$$u = 4x + 7.$$

The differential is

$$du = \frac{du}{dx}\, dx = 4dx,$$

which implies that $dx = du/4$. The range of integration transforms as follows:

$$u(0) = 7, \qquad u(10) = 47.$$

Substituting into Eq. (4.1) gives

$$I_{\mathrm{a}} = \frac{1}{4}\int_7^{47} \frac{1}{u}\, du.$$

This is still not quite what we want, because the lower integration limit is not 1. We can remedy this by separating it into the sum of two integrals with different ranges. Note that

$$\int_1^{47} \frac{1}{u}\, du = \int_1^{7} \frac{1}{u}\, du + \int_7^{47} \frac{1}{u}\, du.$$

The last term on the right-hand side is equal to $4I_{\mathrm{a}}$. Therefore,

$$I_{\mathrm{a}} = \frac{1}{4}\left[\ln(47) - \ln(7)\right].$$

Consider another example,

$$I_{\mathrm{b}} = \int_0^{10} \frac{1}{(4x+7)^2}\, dx. \tag{4.2}$$

The same change of variable gives

$$I_{\mathrm{b}} = \frac{1}{4}\int_7^{47} u^{-2} du = -\frac{1}{4}\, u^{-1}\big|_7^{47} = \frac{1}{4}\left(\frac{1}{7} - \frac{1}{47}\right).$$

Example 4.2. *Integral of reciprocal of a product of linear polynomials.* Reciprocals of products of linear polynomials can be transformed into sums of integrals that can be evaluated as in the previous example. The trick is to use the ***method of partial fractions***. Consider

$$\int \frac{1}{(a-x)(b-x)^2}\, dx \tag{4.3}$$

for given constants a and b. Let us introduce arbitrary parameters α, β, and γ by noting that

$$\frac{\alpha}{a-x} + \frac{\beta+\gamma x}{(b-x)^2} = \frac{\alpha(b^2-2bx+x^2)+(\beta+\gamma x)(a-x)}{(a-x)(b-x)^2}. \tag{4.4}$$

We can make this look like the integrand by requiring that

$$1 = \alpha(b^2-2bx+x^2)+(\beta+\gamma x)(a-x). \tag{4.5}$$

Expanding this and collecting terms according to powers of x gives three equations,

$$1 = b^2\alpha + a\beta, \qquad 0 = -2b\alpha - \beta + a\gamma, \qquad 0 = \alpha - \gamma,$$

which yield numerical values for α, β, and γ. Thus we obtain

$$\int \frac{1}{(a-x)(b-x)^2}\, dx = \alpha \int \frac{1}{a-x}\, dx + \int \frac{\beta+\gamma x}{(b-x)^2} dx,$$

which can be solved using the substitutions $u = a - x$ and $v = b - x$.

Example 4.3. *An integral involving a product of an exponential and an algebraic function.* Consider

$$I_{\rm d} = \int_0^{10} (x+7)^{1/2} e^{-x/3} dx\,. \tag{4.6}$$

Let us try the change of variable $u = x + 7$, $dx = du$,

$$I_{\rm d} = \int_7^{17} u^{1/2} e^{-(u-7)/3} du = e^{7/3} \int_7^{17} u^{1/2} e^{-u/3} du\,.$$

This is beginning to look like an incomplete gamma function, $\Gamma(a,b) = \int_b^\infty e^{-t}\, t^{a-1} dt$. Let us make a second substitution, $t = u/3$, $u = 3t$, $du = 3dt$,

$$I_{\rm d} = e^{7/3} \int_{7/3}^{17/3} (3t)^{1/2} e^{-t} 3dt = 3^{3/2} e^{7/3} \int_{7/3}^{17/3} t^{1/2} e^{-t} dt = 3^{3/2} e^{7/3} \int_{7/3}^{17/3} t^{3/2-1} e^{-t} dt.$$

Now all we need to do is rearrange the integration range, as follows:

$$\begin{aligned} I_{\rm d} &= 3^{3/2} e^{7/3} \left(\int_{7/3}^{\infty} t^{3/2-1} e^{-t} dt - \int_{17/3}^{\infty} t^{3/2-1} e^{-t} dt \right) \\ &= 3^{3/2} e^{7/3} \left[\Gamma\left(\tfrac{3}{2}, \tfrac{7}{3}\right) - \Gamma\left(\tfrac{3}{2}, \tfrac{17}{3}\right) \right]. \end{aligned} \tag{4.7}$$

Example 4.4. *Integration by parts with change of variable.* Consider

$$I_{\rm e} = \int_a^b \frac{x^3}{(1+3x^2)^6}\, dx\,. \tag{4.8}$$

This can be evaluated using integration by parts. We need to express the integrand as the product of two functions. Let us make the inspired choice

$$g = x^2, \qquad w = \frac{x}{(1+3x^2)^6}\,. \tag{4.9}$$

We can integrate w with the substitution $u = 1 + 3x^2$, $du = 6xdx$,

$$W = \int w(x)dx = \frac{1}{6}\int u^{-6}du = -\frac{1}{30}u^{-5} = -\frac{1}{30(1+3x^2)^5}\,.$$

Integration by parts gives

$$I_e = \int_a^b gwdx = gW|_a^b - \int_a^b g'Wdx = -\frac{x^2}{30(1+3x^2)^5}\bigg|_a^b + \frac{1}{30}\int_a^b \frac{2x}{(1+3x^2)^5}\,dx$$

To evaluate the last integral, again we make the substitution $u = 1 + 3x^2$;

$$\frac{1}{30}\int_a^b \frac{2x}{(1+3x^2)^5}\,dx = \frac{1}{90}\int_a^b \frac{6xdx}{(1+3x^2)^5} = \frac{1}{90}\int_{u(a)}^{u(b)} u^{-5}du = -\frac{1}{90\cdot 4}\,u^{-4}|_{u(a)}^{u(b)}\,,$$

$$I_e = -\frac{1}{30}\left[\frac{b^2}{(1+3b^2)^5} - \frac{a^2}{(1+3a^2)^5}\right] - \frac{1}{360}\left[\frac{1}{(1+3b^2)^4} - \frac{1}{(1+3a^2)^4}\right]\,.$$

These examples give the impression that integration is a nonsystematic collection of inspired tricks. This impression is accurate. There is no systematic procedure for analytically evaluating integrals.[1] Fortunately, modern computer algebra software packages are quite adept at finding substitutions to express integrals in terms of elementary or special functions. Keep in mind, however, that it is very easy to construct integrals that are *impossible* to express in terms of any of the standard elementary or special functions.

4.1.2 Jacobian Determinant

In an integral over two-dimensional Cartesian coordinates, $\int f(x,y)dxdy$, the differential factor $dxdy$ represents the area of an infinitesimal rectangle, as illustrated by the left-hand panel of Fig. 4.1. In three dimensions the differential factor is $dxdydz$, which represents the volume of an infinitesimal box.

When the integration variable is changed, the differential factor must be transformed appropriately. In one dimension the transformation of the differential is straightforward; if the old coordinate x is a function $x(u)$ of the new coordinate u, then $dx = \frac{dx}{du}du$, as in the preceding examples. Consider, however, a two-dimensional integral, with differential $dxdy$. Suppose the old coordinates are functions $x(u)$ and $y(v)$. Then we can transform dx and dy individually, with the result $dxdy = \frac{dx}{du}\frac{dy}{dv}dudv$, but, this works only if x depends *only* on u and y depends *only* on v. Often, this will not be the case. Consider a transformation from Cartesians (x,y) to polar coordinates (r,ϕ),

$$x = r\cos\phi, \qquad y = r\sin\phi\,. \tag{4.10}$$

Now $x(r,\phi)$ and $y(r,\phi)$ are each functions of *both* of the new coordinates.

The differential $dxdy$ is the area of an infinitesimal rectangle that is traversed by increasing x to $x + dx$ and y to $y + dy$, a rectangle of width dx and height dy. For the differential in terms of r and ϕ, we need the infinitesimal area traversed by increasing r to $r + dr$ and ϕ to $\phi + d\phi$. Two examples of this

[1]There is a systematic procedure for *numerical approximation* of integrals, which will be described in Section 5.3.

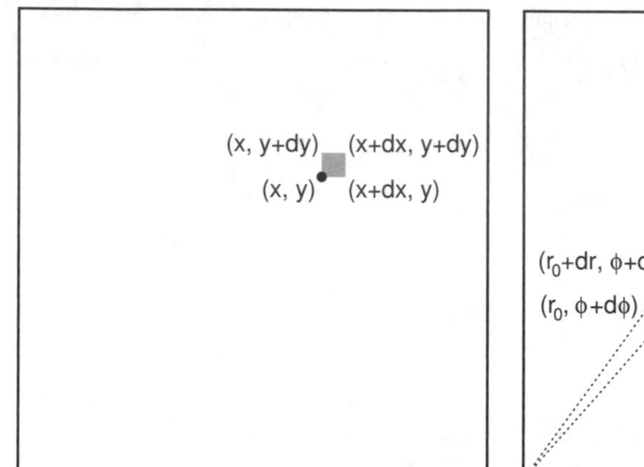

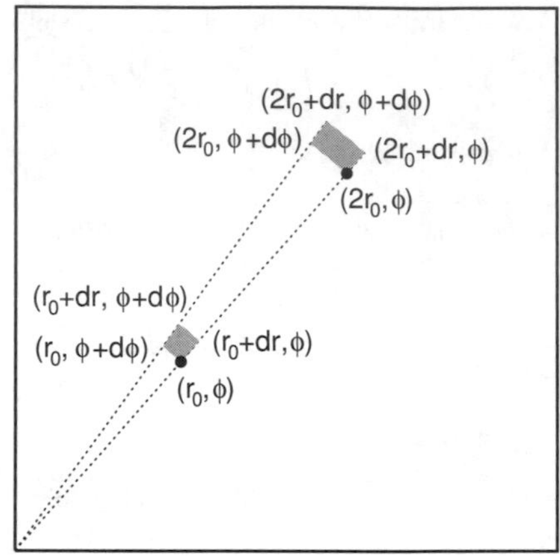

Figure 4.1: Differential areas. The infinitesimals are portrayed as finite quantities so that they can be seen in the figures.

are shown in the right-hand panel of Fig. 4.1, with $r = r_0$ and with $r = 2r_0$ for some arbitrary value of r_0. It is important to note that the area in this case depends on the value of r. In fact, it is proportional to r. The shaded region at $2r_0$ has twice the area of the region at r_0. For the transformation $(x, y) \to (r, \phi)$ the correct transformation of the differential factor is

$$dxdy \to rdrd\phi. \tag{4.11}$$

We can prove this as follows: The area of a circle of radius r is πr^2. This would be the area corresponding to an angular displacement of 2π. The area of a wedge with an angular displacement of $d\phi$ is the fraction $d\phi/2\pi$ of πr^2. The area of the shaded region between r and $r + dr$ is

$$\frac{d\phi}{2\pi}\,\pi(r + dr)^2 - \frac{d\phi}{2\pi}\,\pi r^2 = \frac{d\phi}{2}\,\left[(r^2 + 2rdr + drdr) - r^2\right] = \frac{d\phi}{2}\,(2rdr + drdr).$$

The doubly infinitesimal term $drdr$ is negligible compared with $2rdr$.

This kind of analysis becomes more complicated for more complicated coordinate transformations and for higher dimensionality. A simple general expression for the differential factor for a change of coordinate system was derived by Jacobi.[2] It involves a factor J called the ***Jacobian determinant.*** A two-dimensional *determinant,*

$$\begin{vmatrix} a & b \\ c & d \end{vmatrix},$$

is a commonly used notation for $ad - bc$. A three-dimensional determinant is defined as

$$\begin{vmatrix} a & b & c \\ d & e & f \\ g & h & j \end{vmatrix} = a\begin{vmatrix} e & f \\ h & j \end{vmatrix} - b\begin{vmatrix} d & f \\ g & j \end{vmatrix} + c\begin{vmatrix} d & e \\ g & h \end{vmatrix}. \tag{4.12}$$

[2]Carl Jacobi (1804-1850), a distinguished Prussian mathematician and teacher.

(Determinants are discussed in detail in Section 18.5.) Now suppose that we are transforming the Cartesians (x, y, z) to some coordinate system (u, v, w). Jacobi proved that the differential volume element in the new coordinate system is given by

$$dxdydz = |J|dudvdw, \tag{4.13}$$

where $|J|$ is the absolute value[3] of

$$J = \begin{vmatrix} \frac{\partial x}{\partial u} & \frac{\partial x}{\partial v} & \frac{\partial x}{\partial w} \\ \frac{\partial y}{\partial u} & \frac{\partial y}{\partial v} & \frac{\partial y}{\partial w} \\ \frac{\partial z}{\partial u} & \frac{\partial z}{\partial v} & \frac{\partial z}{\partial w} \end{vmatrix} . \tag{4.14}$$

Consider the case of cylindrical coordinates (r, ϕ, z). Evaluation of the partial derivatives of Eqs. (3.1) gives

$$J = \begin{vmatrix} \cos\phi & -r\sin\phi & 0 \\ \sin\phi & r\cos\phi & 0 \\ 0 & 0 & 1 \end{vmatrix} = r(\cos^2\phi + \sin^2\phi) = r . \tag{4.15}$$

Therefore, in cylindrical coordinates,

$$dxdydz \to rdrd\phi dz . \tag{4.16}$$

For spherical polar coordinates, defined by Eqs. (3.3), it is straightforward to show that

$$dxdydz \to r^2 \sin\theta \, drd\phi d\theta . \tag{4.17}$$

4.2 Gaussian Integrals

Consider the Gaussian integrals,

$$I_n = \int_0^\infty x^n e^{-x^2} dx \tag{4.18}$$

for arbitrary integer n. Integrals of this form occur very frequently in chemistry applications. Let us first treat the case $n = 1$, which can be integrated directly. The derivative of e^{-x^2} is $(-2x)e^{-x^2}$. Therefore,

$$\begin{aligned} I_1 &= \int_0^\infty xe^{-x^2} dx \\ &= -\frac{1}{2}\int_0^\infty (-2x)e^{-x^2} dx = -\frac{1}{2}\int_0^\infty \left(\frac{d}{dx}e^{-x^2}\right) dx = -\tfrac{1}{2}\, e^{-x^2}\Big|_0^\infty \\ &= -\tfrac{1}{2}\left(\lim_{x\to\infty} e^{-x^2} - e^0\right) = -\tfrac{1}{2}(0-1) = \tfrac{1}{2} . \end{aligned} \tag{4.19}$$

Consider next the case $n = 0$, which is more complicated because it cannot be written as an integral of a derivative. Instead of trying to evaluate I_0

[3] Here we have two different, unrelated, uses of vertical bars. The bars mean "absolute value" if they contain a single number, such as $|J|$, but if they enclose a square array of numbers then they designate a determinant.

directly, we will instead evaluate I_0^2 and then take the square root. The first step is to note that

$$\int_{-\infty}^{\infty} e^{-x^2} dx = 2\int_0^{\infty} e^{-x^2} dx\,. \tag{4.20}$$

e^{-x^2} is an even function, because $e^{-(-x)^2} = e^{-x^2}$. Its value at $-x$ is the same as at $+x$. Therefore, the area under the graph of the function from $-\infty$ to 0 is the same as the area from 0 to $+\infty$. Now consider I_0^2. Because x is a dummy variable, we can give it any symbol we want. We can write

$$\begin{aligned} 4I_0^2 &= \left(\int_{-\infty}^{\infty} e^{-x^2} dx\right)\left(\int_{-\infty}^{\infty} e^{-x^2} dx\right) \\ &= \left(\int_{-\infty}^{\infty} e^{-x^2} dx\right)\left(\int_{-\infty}^{\infty} e^{-y^2} dy\right) \\ &= \int_{-\infty}^{\infty}\int_{-\infty}^{\infty} e^{-x^2}e^{-y^2} dxdy = \int_{-\infty}^{\infty}\int_{-\infty}^{\infty} e^{-(x^2+y^2)} dxdy\,, \end{aligned} \tag{4.21}$$

with the symbol x replaced by y in just one of the two integrals. Let us now switch to polar coordinates. This leads to a significant simplification, because

$$x^2 + y^2 = r^2(\cos^2\phi + \sin^2\phi) = r^2. \tag{4.22}$$

Using $dxdy = rdrd\phi$, and changing the ranges $(-\infty,\infty)$ for x and y to $[0,\infty)$ for r and $[0, 2\pi]$ for ϕ, we have

$$\begin{aligned} 4I_0^2 &= \int_0^{2\pi}\int_0^{\infty} e^{-r^2} rdrd\phi = \left(\int_0^{2\pi} d\phi\right)\left(-\frac{1}{2}\right)\int_0^{\infty}(-2r)e^{-r^2} dr \\ &= (2\pi - 0)(-\tfrac{1}{2})(e^{-\infty} - e^0) = \pi\,, \end{aligned} \tag{4.23}$$

and therefore

$$I_0 = \sqrt{\pi}/2\,. \tag{4.24}$$

The general results, with integer exponent $n \geq 0$ and including in the exponential an arbitrary positive constant factor a, are

$$\int_0^{\infty} x^{2n+1}e^{-ax^2} dx = \frac{1}{2}\frac{n!}{a^{n+1}} \tag{4.25a}$$

and

$$\int_0^{\infty} x^{2n}e^{-ax^2} dx = \frac{1}{2}\frac{(2n-1)!!}{(2a)^n}\left(\frac{\pi}{a}\right)^{1/2}, \tag{4.25b}$$

with

$$(-1)!! = 1; \qquad (2n-1)!! = 1\cdot 3\cdot 5\cdots(2n-1), \text{ for } n \geq 1. \tag{4.26}$$

Eqs. (4.25) can be derived recursively starting from I_0 or I_1, either by using integration by parts or, less laboriously, by making a change of variable to express the integral in terms of a gamma function.

Finally, consider

$$\int_{-\infty}^{\infty} e^{-x^2+bx} dx \tag{4.27}$$

for arbitrary b. We can express this as a Gaussian integral in the standard form using a technique called ***completing the square***. Let us write

$$x^2 - bx = x^2 - bx + \alpha - \alpha,$$

where α is some arbitrary parameter, and then let

$$x^2 - bx + \alpha = (x - \beta)^2,$$

where β is another parameter. Expanding out the right-hand side, we find that $\beta = b/2$ and $\alpha = \beta^2 = b^2/4$, and

$$x^2 - bx = (x - b/2)^2 - b^2/4. \tag{4.28}$$

The change of variable $u = x - b/2$ gives

$$\int_{-\infty}^{\infty} e^{-x^2+bx} dx = e^{b^2/4} \int_{-\infty}^{\infty} e^{-u^2} du = 2e^{b^2/4} I_0 = \sqrt{\pi}\, e^{b^2/4}. \tag{4.29}$$

The integral $\int_0^\infty e^{-x^2+bx} dx$ can be treated similarly but the result will include the error function, Eq. (2.54).

4.3 Improper Integrals

An ***improper integral*** has at least one of the following properties:

1. At least one of the ranges of integration is infinite.
2. The integrand is infinite at some point within the range of integration.

Let us first consider an integral with just the first property,

$$I_\text{a} = \int_5^\infty e^{-x} dx. \tag{4.30}$$

We can integrate this exactly with elementary methods:

$$I_\text{a} = -e^{-x}\Big|_5^\infty = -e^{-\infty} + e^{-5} = e^{-5},$$

with the understanding that $e^{-\infty}$ is an abbreviation for the limit

$$e^{-\infty} = \lim_{c\to\infty} e^{-c} = \lim_{c\to\infty} \frac{1}{e^c}, \tag{4.31}$$

which equals zero.

Now consider the second category of improper integrals, in which the integrand goes to $+\infty$ or $-\infty$ inside the integration range. Let us compare the following:

$$I_\text{b} = \int_0^7 \frac{1}{(x-4)^2}\, dx \quad \text{and} \quad I_\text{c} = \int_0^7 \frac{1}{(x-4)}\, dx.$$

The integrands are plotted in Figs. 4.2 and 4.3. The value of the integral is

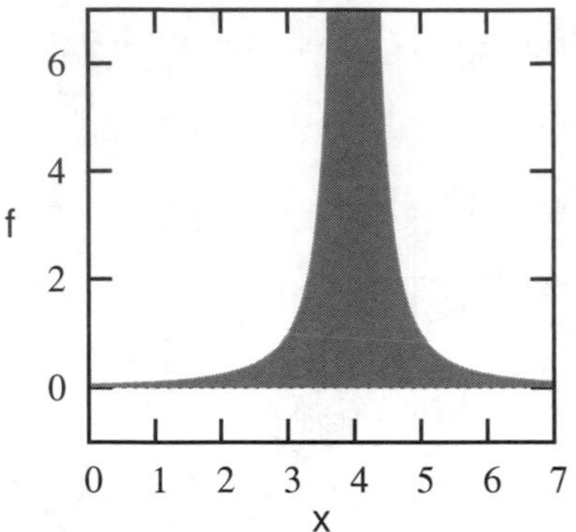

Figure 4.2: $f_{\rm b}(x) = (x-4)^{-2}$, the integrand of divergent integral $I_{\rm b}$.

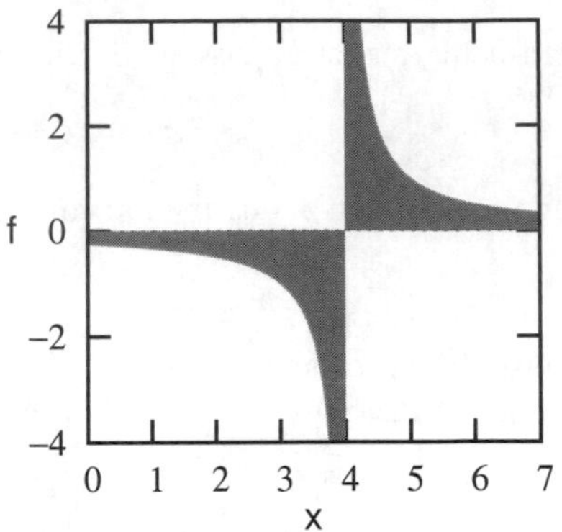

Figure 4.3: $f_{\rm c}(x) = (x-4)^{-1}$, the integrand of the integral $I_{\rm c}$.

equal to the area of the shaded region. The area under the plot of $(x-4)^2$ is infinite, due to the second-order pole at $x = 4$. Therefore, the integral I_b is not just improper—it is *divergent.* It does not have a finite numerical value.

The integral $I_{\rm c}$ is more interesting. It goes to positive infinity on the right of the singular point but it goes to negative infinity on the left of the singular point. Could these two infinities cancel each other out to give a finite result? An integral of this kind can be given meaning as follows:

Definition. Consider a function $f(x)$ with a singular point $x_{\rm s}$ such that $\lim_{\epsilon\to 0} f(x_{\rm s}-\epsilon) = \pm\infty$ and $\lim_{\epsilon\to 0} f(x_{\rm s}+\epsilon) = \mp\infty$. The ***Cauchy principal value***[4] of the integral $\int_{x_{\rm min}}^{x_{\rm max}} f(x)dx$ with $x_{\rm min} < x_{\rm s} < x_{\rm max}$ is

$$\lim_{\epsilon\to 0}\left[\int_{x_{\rm min}}^{x_{\rm s}-\epsilon} f(x)dx + \int_{x_{\rm s}+\epsilon}^{x_{\rm max}} f(x)dx\right],$$

if this limit is finite.

Example 4.5. *Cauchy principal value.* For the integral $I_{\rm c}$ we have

$$I_{\rm c} = \lim_{\epsilon\to 0}\left[\int_0^{4-\epsilon}\frac{1}{x-4}\,dx + \int_{4+\epsilon}^{7}\frac{1}{x-4}\,dx\right],$$

Making the change of variable $p = x-4$,

$$\int_{4+\epsilon}^{7}\frac{1}{x-4}\,dx = \int_\epsilon^3 p^{-1}dp = \int_1^3 p^{-1}dp - \int_1^\epsilon p^{-1}dp = \ln 3 - \ln\epsilon.$$

Making the change of variable $q = 4-x$,

$$\int_0^{4-\epsilon}\frac{1}{x-4}\,dx = \int_4^\epsilon q^{-1}dq = -\int_\epsilon^4 q^{-1}dq = -\int_1^4 q^{-1}dq + \int_1^\epsilon q^{-1}dq = -\ln 4 + \ln\epsilon.$$

Adding these together, we obtain

$$I_c = \ln 3 - \ln 4 = \ln(3/4).$$

We do not need to take the limit $\epsilon\to 0$ because all ϵ dependence has canceled out.

[4]The French mathematician Augustin-Louis Cauchy (1789-1857) made fundamental contributions to a wide variety of mathematical subjects. He was known for his insistence on careful proofs, which set a standard for rigor in mathematical analysis. The name is pronounced *kōshee.*

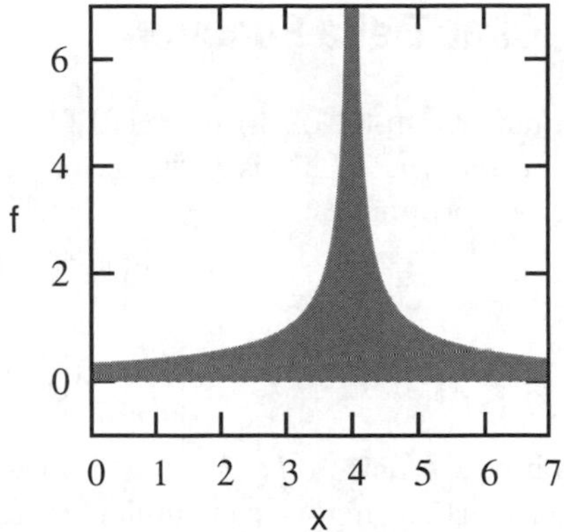

Figure 4.4: The function $f_\mathrm{d}(x) = (x-4)^{-4/5}$, which is the integrand of the integral I_d of Example 4.7. This is an example of an integrable singularity.

Example 4.6. *A divergent integral.* Let us attempt to calculate a Cauchy principal value for I_b:

$$\int_{4+\epsilon}^{7} \frac{1}{(x-4)^2}\,dx = -(x-4)^{-1}\Big|_{4+\epsilon}^{7} = \frac{1}{\epsilon} - \frac{1}{4},$$

$$\int_{0}^{4-\epsilon} \frac{1}{(x-4)^2}\,dx = -(x-4)^{-1}\Big|_{0}^{4-\epsilon} = -\frac{1}{3} + \frac{1}{\epsilon}.$$

Adding these together gives $2/\epsilon - 7/12$, which becomes infinite in the limit $\epsilon \to 0$.

Suppose that we were to ignore the presence of a singularity and naively try to evaluate I_b in the usual fashion:

$$\int_0^7 \frac{1}{(x-4)^2}\,dx \stackrel{?}{=} -(x-4)^{-1}\Big|_0^7 = -\frac{1}{3} - \frac{1}{4} = -\frac{7}{12}. \tag{4.32}$$

This result is manifestly nonsense. It claims that the integral has a finite *negative* value. Clearly this is impossible, as the integrand is everywhere positive. It is important to examine the integrand for singularities before trying to evaluate an integral!

Example 4.7. *Another example of a Cauchy principal value.* Consider

$$I_\mathrm{d} = \int_0^7 \frac{1}{(x-4)^{4/5}}\,dx.$$

Its integrand is plotted in Fig. 4.4. Let us calculate its Cauchy principal value:

$$\int_{4+\epsilon}^{7} (x-4)^{-4/5} dx = 5(x-4)^{1/5}\Big|_{4+\epsilon}^{7} = 5\left(3^{1/5} - \epsilon^{1/5}\right),$$

$$\int_{0}^{4-\epsilon} (x-4)^{-4/5} dx = 5(x-4)^{1/5}\Big|_{0}^{4-\epsilon} = 5\left[(-\epsilon)^{1/5} - (-4)^{1/5}\right) = 5\left(4^{1/5} - \epsilon^{1/5}\right).$$

Adding these up gives

$$I_\mathrm{d} = \lim_{\epsilon\to 0} 5\left(3^{1/5} + 4^{1/5} - 2\epsilon^{1/5}\right) = 5\left(3^{1/5} + 4^{1/5}\right),$$

which is finite. We might have expected this to be divergent, on account of the qualitative resemblance of the graph of the integrand to that of I_b. However, in this case the peak at the singular point is much more narrow. The integrand does become infinite at $x = 4$ but the area under the curve is in fact finite. The peak becomes narrower faster than it grows higher.

4.4 Dirac Delta Function

As a final example, of an especially improper integral, consider the ***Dirac delta function***.[5] This is a "function" $\delta(x)$ that is defined to have the following two properties:

$$\int_{-\infty}^{\infty} \delta(x) f(x) dx = f(0), \tag{4.33}$$

$$\int_{-\infty}^{\infty} \delta(x) dx = 1, \tag{4.34}$$

for arbitrary function f. At first glance, the existence of a function δ with these properties might seem unlikely—for Eq. (4.33) to hold, $\delta(x)$ must be zero at all values of x except in the infinitesimal neighborhood of the single point $x = 0$, but in order for the area under $\delta(x)$ in this infinitesimal region to be finite, to satisfy Eq. (4.34), the value $\delta(0)$ must be infinite. Nevertheless, one can construct $\delta(x)$ from a conventional but sharply peaked function as a limit in which the peak becomes infinitely high and infinitesimally narrow.

Fig. 4.5 shows

$$g(x, a) = e^{-x^2/a^2} \Big/ \int_{-\infty}^{\infty} e^{-u^2/a^2} du \,. \tag{4.35}$$

Consider this as a finite approximation to $\delta(x)$. By construction it satisfies Eq. (4.34), and in the limit of small a it becomes sharply peaked at $x = 0$. The integral in the denominator is a Gaussian integral, which we can evaluate exactly:

$$\int_{-\infty}^{\infty} e^{-u^2/a^2} du = a\sqrt{\pi},$$

according to Eq. (4.25b). We can consider an integral with a delta function in the integrand as a shorthand notation for the operation of taking the limit[6]

$$\int_{-\infty}^{\infty} \delta(x) f(x) dx = \lim_{a \to 0} \frac{1}{a\sqrt{\pi}} \int_{-\infty}^{\infty} e^{-x^2/a^2} f(x) dx. \tag{4.36}$$

This definition of $\delta(x)$ is not unique. Many other functions have the desired limit, such as

$$g(x, a) = \frac{e^{-|x/a|}}{2a} \quad \text{or} \quad g(x, a) = \frac{\sin(x/a)}{\pi x},$$

and will work just as well as Eq. (4.35). The Dirac delta is easily generalized to three dimensions:

$$\delta(x, y, z) = \delta(x)\delta(y)\delta(z). \tag{4.37}$$

[5]Developed by the English physicist P. A. M. Dirac in his excellent and still very useful textbook, *The Principles of Quantum Mechanics* (Oxford University Press, Oxford, 1930).

[6]Strictly speaking, $\delta(x)$ is not a function but a *distribution* (also called a *generalized function*), because it is well defined only when it appears in an integral. The formal theory of distributions was developed by mathematicians after the fact, in the 1930's, in response to the enthusiastic embrace of $\delta(x)$ by quantum physicists.

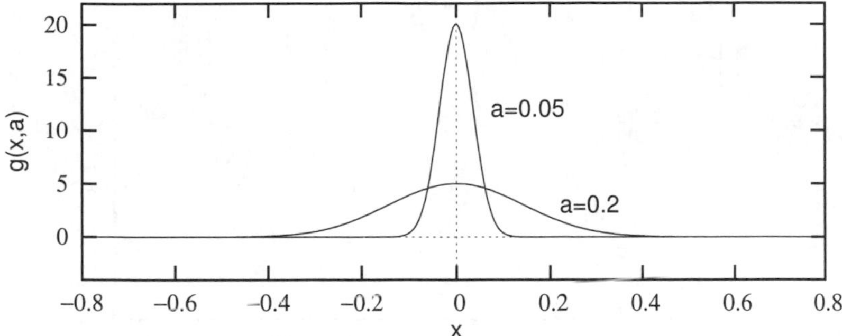

Figure 4.5: $g(x,a) = e^{-(x/a)^2}/a\sqrt{\pi}$, a model for $\delta(x)$ in the limit $a \to 0$.

Example 4.8.* *Quantum mechanical applications of the Dirac delta function.* One use for the Dirac delta is as a model for a short-range potential energy function, such as in the scattering of one particle off another in the gas phase, in which the effect of the collision is felt only in the immediate vicinity of the target particle. Another use is to make the mathematics of quantum mechanics reduce to that of classical mechanics in the classical limit. The mathematical formulation of quantum mechanics looks very different from that of classical mechanics. Consider a particle subject to a potential energy function $V(x, y, z)$. If the particle is at a point (x_0, y_0, z_0) then the potential energy in classical mechanics is simply $V(x_0, y_0, z_0)$. In quantum mechanics we cannot specify the particle's position. Instead, we calculate the expectation value

$$\langle V \rangle = \int_{-\infty}^{\infty} V(x, y, z)\, |\Psi(x, y, z)|^2 \, dxdydz, \tag{4.38}$$

where $|\Psi^2|dxdydz$ is the probability of finding the particle in an infinitesimal neighborhood of (x, y, z). $\langle V \rangle$ is the average of the unpredictable individual values of V that would be obtained by measuring V repeatedly. To obtain the classical result as a quantum expectation value, replace $|\Psi^2|$ with $\delta(x - x_0, y - y_0, z - z_0)$. Eq. (4.33) implies

$$\int_{-\infty}^{\infty} V(x, y, z)\delta(x - x_0, y - y_0, z - z_0)\, dxdydz = V(x_0, y_0, z_0), \tag{4.39}$$

which is indeed the classical result. Thus, classical mechanics can be expressed in terms of the mathematics of quantum mechanics.

4.5 Line Integrals

The multidimensional integral $\int\int f(x, y)dxdy$ is not the only way to extend the concept of integration to multivariable functions. An alternative approach is a ***line integral.*** Consider some function $f(x, y)$ and a path starting at a point (x_0, y_0) and ending at a point (x_1, y_1). Two such paths are illustrated in Fig. 4.6, labeled C_a and C_b. A line integral is an integral evaluated along a path. It can be evaluated by expressing x and y as functions of a single parameter. For example, the path C_a in the figure can be represented by the functions $x(u) = 1 + 4u$ and $y(u) = 1 + 2u$ with the parameter u going from

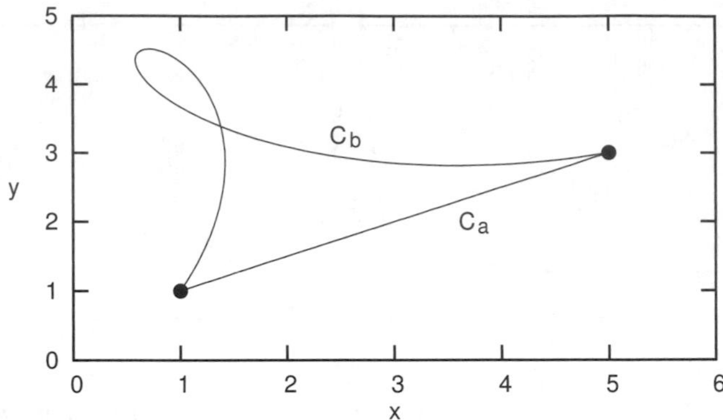

Figure 4.6: Examples of paths, C_a and C_b, between the points $(x_0, y_0) = (1,1)$ and $(x_1, y_1) = (5,3)$.

0 to 1. C_b is $x(u) = 6u + 2\cos(3\pi u/2) - 1$ and $y(u) = 4u + 2\sin(3\pi u/2) + 1$. The path is a constraint that reduces the number of degrees of freedom by one. The line integral is a one-dimensional integral over the variable u,

$$\int_C f dx dy = \int_0^1 f\big(x(u), y(u)\big)\, du\,. \tag{4.40}$$

The "C" subscript on the integral symbol represents a path of integration in the two-dimensional space of x and y. The integral on the right-hand side is a conventional integral with a single variable u ranging from 0 to 1.

The integral is an operator that maps *df* to f. For a one-dimensional function $f(x)$, with specified value $f(x_0)$ at some point x_0, we have[7]

$$\int df = \int_{x_0}^{x} f'(x) dx = f(x) - f(x_0), \tag{4.41}$$

which means that

$$f(x) = f(x_0) + \int df\,. \tag{4.42}$$

Thus, $\int$ constructs a function from its differential. If the derivative, $f'(x)$ in Eq. (4.41), exists for all points between x_0 and x, then $f(x)$ exists and is given by Eq. (4.42). For a line integral in two dimensions, can we similarly define f as $f(x,y) = f(x_0, y_0) + \int df$? In other words, can we use the line integral to construct a unique function $f(x,y)$ from a differential? Sometimes we can but not always.

Let us examine more carefully the concept of a differential in a space of two dimensions. For an arbitrary function $f(x,y)$ of two variables, the differential exists and is given by $df = \frac{\partial f}{\partial x} dx + \frac{\partial f}{\partial y} dy$ as long as the derivatives $\partial f/\partial x$ and $\partial f/\partial y$ exist. But consider the converse. Given a differential

$$df = A(x,y) dx + B(x,y) dy\,, \tag{4.43}$$

is there always a function $f(x,y)$ for any choices of the functions A and B?

[7] *Within* the integral, x is a dummy variable. We can call it x or u or anything else. In the integration *range*, x is a free variable that represents the actual argument x of $f(x)$.

The answer is *no*. The mixed second derivative must be independent of the order in which the derivatives are calculated, and this imposes a constraint on A and B. Eq. (4.43) implies that $A = \partial f/\partial x$ and $B = \partial f/\partial y$, which if true implies

$$\frac{\partial^2 f}{\partial y \partial x} = \left(\frac{\partial A}{\partial y}\right)_x, \qquad \frac{\partial^2 f}{\partial x \partial y} = \left(\frac{\partial B}{\partial x}\right)_y, \tag{4.44}$$

and these must be equal to each other. The resulting constraint is

$$\left(\frac{\partial A}{\partial y}\right)_x = \left(\frac{\partial B}{\partial x}\right)_y. \tag{4.45}$$

This is called ***Euler's test for exactness***. An ***exact differential*** is a differential for which there exists a corresponding $f(x, y)$. To say the function "exists" is to say that for given differential df we obtain a well defined mapping of (x, y) to some number $f(x, y)$. For a path starting at some point (x_0, y_0), this would mean that the value of

$$f(x, y) = f(x_0, y_0) + \int_C df \tag{4.46}$$

would depend only on the values of (x, y) and (x_0, y_0), not on the path C.

The differential of f as function of u is

$$df = A\big(x(u), y(u)\big)\frac{dx}{du}\,du + B\big(x(u), y(u)\big)\frac{dy}{du}\,du\,. \tag{4.47}$$

If Euler's test is satisfied, and dx/du and dy/du exist throughout the path, then $\int_C df$ for a path C from a point (x_0, y_0) to a point (x_1, y_1) will equal $f(x_1, y_1) - f(x_0, y_0)$ regardless of the details of the route through the xy-plane. However, even if df does not correspond to an $f(x, y)$, the differential may still be a useful concept. An inexact differential does correspond to a function, defined by Eq. (4.46), but its value depends on the path C; it is not a function of the initial and final points alone.

In thermodynamics, a function corresponding to an exact differential is called a ***state function*** while a function corresponding to an inexact differential is called a ***path function***. Functions such as energy and entropy are state functions, and this fact is used to great advantage to build up the theory from differentials in the two-dimensional space (T, p). For example, Hess's law, which states that the enthalpy change for a reaction is equal to the sum of enthalpy changes for any sum of reactions that have the same net effect as the reaction of interest, follows directly from the fact that enthalpy is a state function; its value depends only on the initial state and the final state of the process. However, the work resulting from a thermodynamic process is a path function. Its value depends on the particular path in (T, p) between the initial and final conditions.

A special case of a line integral is a ***cyclic integral***, indicated by the symbol $\oint$. This a line integral over a path that ends where it began. Any cyclic integral of an exact differential is equal to zero, because, by definition, $(x_1, y_1) = (x_0, y_0)$ for a cyclic path.

Exercises

4.1 Evaluate the following: (a) $\int_a^b x^2 e^{2x^3} dx$ (b) $\int_0^2 \frac{x}{8-3x} dx$ (c) $\int_a^b \sin\theta \cos^2\theta d\theta$

4.2 Evaluate the following integrals using integration by parts:

$$\text{(a)} \int_a^b xe^{-3x} dx \qquad \text{(b)} \int_a^b x^2 e^{-3x} dx \qquad \text{(c)} \int_a^b x^3 e^{-3x^2} dx$$

4.3 Complete the analysis of Example 4.2 to evaluate $\int \frac{1}{(a-x)(b-x)^2} dx$.

4.4 Evaluate the following integrals:

$$\text{(a)} \int_{-1}^{\infty} e^{-x} dx \quad \text{(b)} \int_{-\infty}^{1} e^{-x} dx \quad \text{(c)} \int_{-2}^{2} (2x+5)^{-3} dx \quad \text{(d)} \int_{-4}^{4} (2x+5)^{-3} dx \quad \text{(e)} \int_{1}^{\infty} x^{-1} dx$$

4.5 Derive Eqs. (4.25). Use a recursive procedure, starting with the results for $n = 0$.

4.6 Evaluate $\int_0^\infty e^{-x^2+bx} dx$. Express your answer in terms of the error function.

4.7 Evaluate the integral $\frac{1}{\sqrt{\pi}} \int_{-\infty}^{\infty} e^{-x^2+14x-2} dx$.

4.8 Transform $dxdy$ to the coordinate system (σ, τ) where $x = \sigma\tau$ and $y = (\tau^2 + \sigma^2)/2$. ($\sigma$ and τ are called *parabolic* coordinates.)

4.9 Derive the differential volume element for spherical polar coordinates, Eq. (4.17).

4.10 Use Euler's test to verify that $df = 2xydx + x^2dy$ is an exact differential and then integrate over each of the two paths in Fig. 4.6 to demonstrate that the result is the same for each path.

4.11 Use Euler's test to verify that $df = y^2dx + x^2dy$ is an inexact differential. Integrate over each of the two paths in Fig. 4.6 to demonstrate that the results are not the same.

4.12 In this exercise you will derive the integral representation of the beta function,

$$B(p,q) = \int_0^1 u^{p-1}(1-u)^{q-1} du\,.$$

(a) Write $\Gamma(p)\Gamma(q)$ as a two-dimensional integral over dummy variables t and s.

(b) Using $t = x^2$ and $s = y^2$, obtain a two-dimensional integral over x and y.

(c) Transform to polar coordinates r and ϕ with ranges of integration $0 \le r < \infty$ and $0 \le \phi \le \pi/2$. (Why this range for ϕ?)

(d) Make the substitution $u = \cos^2\phi$.

(e) Show that the result is equal to $\Gamma(p+q) \int_0^1 u^{p-1}(1-u)^{q-1} du$.

4.13 Evaluate the following: (a) $\int_0^1 x^{3/2}(1 - x/4)^{3/4} dx$ (b) $\int_0^1 x^{3/2}(2 - x)^{3/4} dx$. *(Hint: Incomplete beta functions).*

4.14 Derive the alternative integral representation for the beta function,

$$B(p,q) = \int_0^\infty v^{p-1}(1+v)^{-p-q} dv\,.$$

[Hint: Make the change of variable $v = u/(1-u)$.]

4.15 Evaluate $\int_{-\infty}^{\infty} (1 + \tanh x)^{2/17} \sqrt{3x^5 - 2x^3 + x^2 + 4}\, e^{-x^2/2} \delta(x) dx$, where $\delta(x)$ is the Dirac delta function. *(Hint: This problem is very easy.)*

Chapter 5

Numerical Methods

An analytic solution, with the answer obtained as an expression that explicitly shows the dependence of a function on the variables, is often just a luxury in science applications. Often the best we can do is obtain the answer as a number, from an experiment or from a computer calculation. This chapter gives an introduction to numerical methods of calculus.

5.1 Interpolation

Suppose we have a set of values f_j at N discrete values of a variable x_j. These comprise a data set $\{(x_1, f_1), (x_2, f_2), (x_3, f_3), \dots (x_N, f_N)\}$. Let us construct a function that reproduces them.[1] This is called ***interpolation***. We seek a function $f(x)$ such that $f(x_j) = f_j$ for each j. There are infinitely many functions that can pass through all the points. If we have some physical theory that leads us to expect a particular kind of functional form, then that can be used to choose an appropriate $f(x)$. Otherwise, the usual practice is to use a polynomial, to keep the mathematics simple.

Given N values (x_j, f_j), there is a unique polynomial of degree $N-1$ that interpolates between the points. It can be constructed using a method called ***Lagrange interpolation***. Let us begin with $N = 2$ to demonstrate. The Lagrange interpolant is then

$$f(x) = \frac{x - x_2}{x_1 - x_2} f_1 + \frac{x - x_1}{x_2 - x_1} f_2 \,. \tag{5.1}$$

By construction, at x_1 the second term is zero while the first term is equal to f_1, so that $f(x_1) = f_1$ as required. Similarly, at x_2 the first term is zero while the second term gives f_2. To make it obvious that this is a polynomial of degree $N - 1 = 1$, collect terms according to the power of the variable x:

$$f(x) = \left(\frac{x_2 f_1 - x_1 f_2}{x_2 - x_1} \right) + \left(\frac{f_2 - f_1}{x_2 - x_1} \right) x \,.$$

The terms in parentheses are constants calculated from the data points.

This method is straightforward to generalize to arbitrary N:

$$\begin{aligned} f(x) =& \frac{(x - x_2)(x - x_3) \cdots (x - x_N) f_1}{(x_1 - x_2)(x_1 - x_3) \cdots (x_1 - x_N)} + \frac{(x - x_1)(x - x_3) \cdots (x - x_N) f_2}{(x_2 - x_1)(x_2 - x_3) \cdots (x_2 - x_N)} \\ & + \cdots + \frac{(x - x_1)(x - x_2) \cdots (x - x_{N-1}) f_N}{(x_N - x_1)(x_N - x_2) \cdots (x_N - x_{N-1})} \,. \end{aligned} \tag{5.2}$$

[1] Alternative strategies of function approximation, in which the function is not required to exactly agree with the known points, will often give better results than the methods that are described in this section. See Chapter 10.

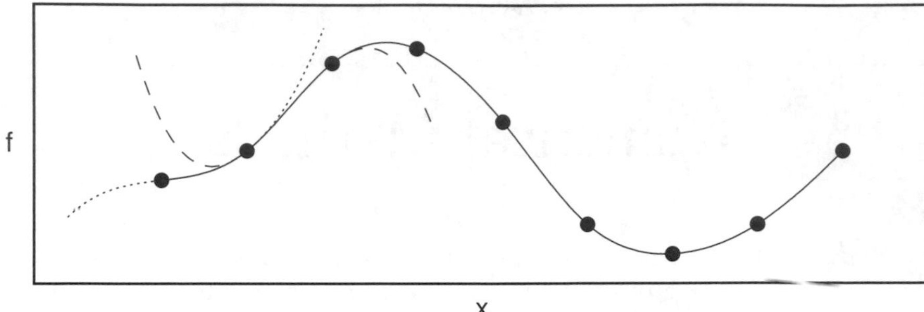

Figure 5.1: Cubic spline interpolation. The first two splines (the dotted curve and the dashed curve) are shown beyond their respective intervals.

The advantage of Lagrange interpolation is that it gives a simple analytical formula. It has a couple of disadvantages. If N is large then the interpolant tends to have unphysical oscillations between the known values. Also, the interpolant is most accurate when neighboring x_i are close to each other, but this leads to poor precision due to roundoff error in the $(x_i - x_j)$ factors in the denominators. For example, we get a precision of 7 significant figures from

$$2.123456 - 1.123456 = 1.000000 = 1.000000 \times 10^0$$

but only one significant figure from

$$1.123457 - 1.123456 = 0.000001 = 1 \times 10^{-6}.$$

The failings of Lagrange interpolation for large N come from the fact that it is asking a lot of a *single* function to expect it work over a wide range of x. An alternative is to use ***splines***. These are a sequence of functions that give a piecewise fit over small intervals with a different function for each interval.

The spline functions are required to satisfy the following conditions:

- Adjacent splines have the same value at their bordering point.
- Adjacent splines have the same values of the first derivative at their bordering point.

In other words, we require that the interpolant and its first derivative be continuous. This ensures a smooth interpolation through each point, as shown in Fig. 5.1.

The functional forms of the splines are arbitrary, but usually polynomials are used. Cubic polynomials seem to work best in practice. This is the method of ***cubic splines***. Let

$$s(x) = a + bx + cx^2 + dx^3, \tag{5.3}$$

with a different set of coefficients $\{a, b, c, d\}$ for each of the $N - 1$ intervals, and let each interval consist just of the region between two adjacent points. $s(x)$ has four parameters; we need four conditions to determine them. The requirement $s(x_i) = f_i$ at each border gives two conditions. The requirement that the derivatives be continuous provides the other two. This is sufficient to specify splines for all intervals except those between x_1 and x_2 and between x_{N-1} and x_N. These border the endpoints and therefore have no adjacent spline with which to require the continuity of the derivatives at x_1 and at x_N. The usual choice for the last two conditions is $s''(x_1) = 0$ and $s''(x_N) = 0$. Solving for the spline coefficients involves tedious algebra that is easily handled by a computer. Spline interpolation is more complicated than Lagrange interpolation but it is well suited for computer calculations. An efficient algorithm is developed in the following example.

Example 5.1. *Cubic splines algorithm.** For cubic splines, the algorithm as described above would seem to require that we solve a set of $4N-4$ simultaneous equations. However, we can reduce this to a much smaller set using a clever trick. Suppose that we were given the values of the second derivatives, f_j'', as well as the values f_j of the function itself. Because the $s_j(x)$ are cubic polynomials, the $s_j''(x)$ are linear. They are given by the Lagrange interpolants

$$s_j''(x) = A_j(x) f_j'' + B_j(x) f_{j+1}'', \quad A_j(x) = \frac{x_{j+1} - x}{x_{j+1} - x_j}, \quad B_j(x) = \frac{x - x_j}{x_{j+1} - x_j}, \tag{5.4}$$

which ensure that $s_j''(x_j) = f_j''$ and $s_j''(x_{j+1}) = f_{j+1}''$. We also require $s_j(x_j) = f_j$ and $s_j(x_{j+1}) = f_{j+1}$. Let

$$s_j(x) = A_j(x) f_j + B_j(x) f_{j+1} + C_j(x) f_j'' + D_j(x) f_{j+1}'', \tag{5.5a}$$

where

$$C_j(x) = \tfrac{1}{6}(x_{j+1} - x_j)^2 (A_j^3 - A_j), \quad D_j(x) = \tfrac{1}{6}(x_{j+1} - x_j)^2 (B_j^3 - B_j). \tag{5.5b}$$

Clearly, C_j and D_j are zero at both x_j and x_{j+1}. It is straightforward to show that $C_j'' = A_j$ and $D_j'' = B_j$, and this ensures that Eq. (5.4) is satisfied, because the first two terms in Eq. (5.5a) are linear and are therefore destroyed by the second derivative. The fact that A_j and B_j are linear implies that s_j will be cubic, as expected.

Thus we have explicit solutions for the $N-1$ splines. The only problem is that usually we do not know the f_j''. We are free to choose $f_1'' = 0$ and $f_N'' = 0$. For the rest of the f_j'', we can solve for them by imposing the requirement that the first derivative be continuous: $s_j'(x_{j+1}) = s_{j+1}'(x_{j+1})$. This gives a set of $N-2$ equations for the $N-2$ unknowns $f_2'', f_3'', \ldots, f_{N-1}''$. When interpolating through a large number of points, this is much more efficient than solving the original set of $4N-4$ equations.

5.2 Numerical Differentiation

Given an explicit expression for $f(x)$, one can calculate $f'(x)$ analytically using standard methods of calculus. However, this is often impracticable, for example, if one only has an empirical data set of discrete points (x_i, f_i). Another situation, common in physical chemistry, is that $f(x)$ can be computed at any given x using a computer algorithm or a numerical simulation. All we get is numerical values of $f(x)$, not an explicit expression.

One approach to numerically calculating the derivative is to use

$$f'(x) \approx \frac{f(x+h) - f(x)}{h}, \tag{5.6}$$

which is based on Eq. (1.5), but with a small but nonzero value of h. This requires two evaluations of f, at x and at $x+h$, for each calculated value of $f'(x)$. The smaller the value of h, the closer we are to the limit in Eq. (1.5) and therefore the higher the accuracy, but as h becomes very small, $f(x+h)$ and $f(x)$ become nearly equal, and the numerator $f(x+h)-f(x)$ loses precision.

It turns out that the optimal balance between precision and accuracy is obtained when

$$h \approx \sqrt{|f(x)/f''(x)|\,\epsilon}\,, \tag{5.7}$$

where ϵ is the roundoff error in computing the numerator. This expression is not directly useful in practice, if we do not know $f''(x)$, but it does tell us that the error in f' is proportional to $\epsilon^{1/2}$. If we instead use a slight variation on Eq. (5.6),

$$f'(x) \approx \frac{f(x+h/2)-f(x-h/2)}{h}\,, \tag{5.8}$$

the optimal h turns out to be proportional to $\epsilon^{2/3}$, which is an improvement—for given ϵ we can use a smaller h. (If ϵ is less than 1, then $\epsilon^{2/3}$ will be less than $\epsilon^{1/2}$.) This has the same $h \to 0$ limit as Eq. (5.6) but with less loss of precision. In most cases Eq. (5.8) is to be preferred over Eq. (5.6).

Suppose that you are willing to compute (or measure) three values of f for each computation of f', say, $f(x_1)$, $f(x_1+h)$, and $f(x_1+2h)$. Then you can construct a three-point Lagrange interpolation and analytically derive the derivative of the interpolant. According to Eq. (5.2) with $N=3$,

$$\begin{aligned} f(x) = {}& \frac{(x-x_2)(x-x_3)}{(x_1-x_2)(x_1-x_3)}f(x_1) + \frac{(x-x_1)(x-x_3)}{(x_2-x_1)(x_2-x_3)}f(x_2) \\ & + \frac{(x-x_1)(x-x_2)}{(x_3-x_1)(x_3-x_2)}f(x_3)\,. \end{aligned} \tag{5.9}$$

Letting $x_2 = x_1 + h$ and $x_3 = x_1 + 2h$, taking the derivative with respect to x, and then evaluating the resulting expression at the point x_1 gives

$$f'(x_1) = \frac{1}{h}\left[2f(x_1+h) - \frac{3}{2}f(x_1) - \frac{1}{2}f(x_1+2h)\right].$$

Any value of x_1 will result in the same equation. Therefore, the three-point expression for arbitrary x is

$$f'(x) = \frac{1}{h}\left[2f(x+h) - \frac{3}{2}f(x) - \frac{1}{2}f(x+2h)\right]. \tag{5.10}$$

Using a four-point interpolant one can derive a four-point expression for the derivative, and so on. However, if the points contain random error (which is to be expected if they come from experimental measurements), then you should probably fit a function to the points using a statistical analysis, as will be described in Chapter 10 and Section 21.4.1, rather than using interpolation. Statistical methods become more reliable as the number of points is increased.

Applications of numerical derivatives in physical and analytical chemistry are quite varied. The following example shows an application to spectroscopy.

Example 5.2. *Derivatives of spectra.* Consider the following simulated absorption spectra. The filled circles represent measured values of absorbed intensity at discrete evenly spaced frequencies:

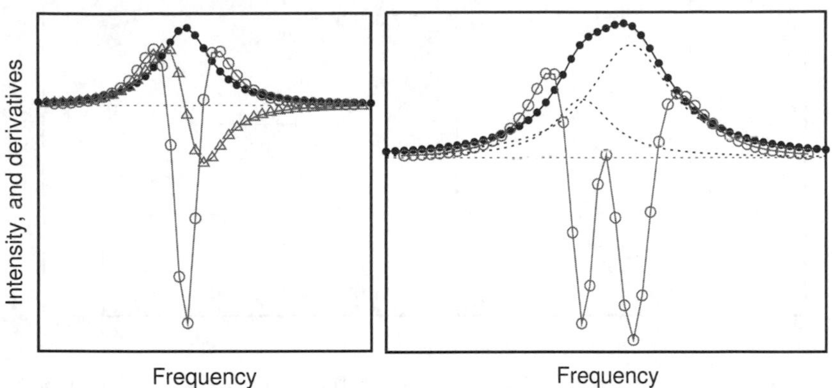

The left-hand panel has a single absorption peak. The triangles and open circles show the first and second derivatives, respectively, computed numerically. For the first derivatives, Eq. (5.10) was used. The second derivatives were computed using the second derivative of a five-point Lagrange interpolant. The first derivative passes through zero at the center of the original peak. This is a reasonable method for precisely determining the peak frequency. The second derivative has a negative peak at the same frequency as the original peak, but is narrower because it must pass through zero at the two inflection points.

The fact that the second derivative peaks are narrower suggests a technique for resolving overlapping peaks. Consider the right-hand panel, which simulates a pair of broad overlapping spectral peaks. The dotted curves are the two signals that were summed to simulate the total spectrum. An experiment would give only the sum, from which it would not be obvious where the underlying peaks are located. The second derivative clearly shows the locations of the two underlying peaks. (In practice, the presence of random noise will often make this technique impractical, because the derivatives amplify the noise. A technique for resolving overlapping peaks in the presence of noise will be described in Section 21.4.2.)

5.3 Numerical Integration

Many of the integrals encountered in physical science cannot be solved in terms of elementary functions. For this reason, numerical methods for evaluating integrals are very important. ***Quadrature*** is the most commonly used numerical integration technique. It is based on Theorems 1.4.1 (the fundamental theorem of calculus) and 1.4.3 (the sum of definite integrals). Let us divide the integration range into N equal intervals where N is an arbitrary positive integer. Label the border points of the intervals as $x_0, x_1, x_2, \ldots, x_N$. The size of each interval is

$$\Delta x = (x_N - x_0)/N, \tag{5.11}$$

with border points $x_j = x_{j-1} + \Delta x$. The full integral from x_0 to x_N can be

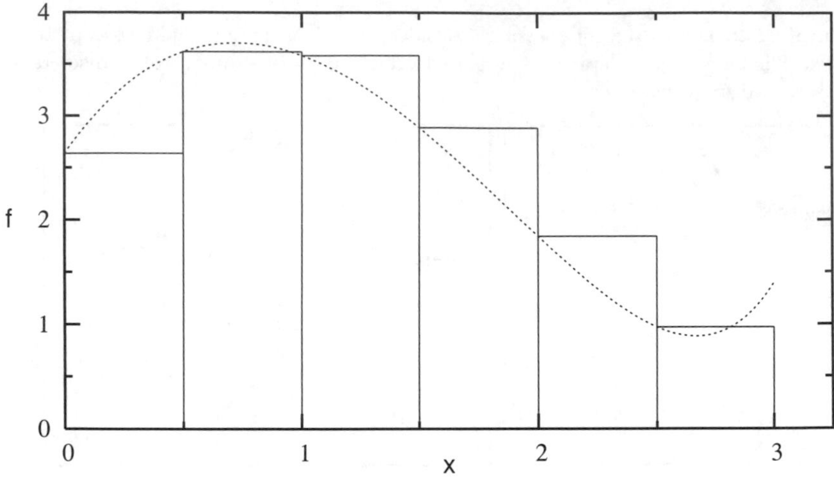

Figure 5.2: $f(x) = (4 + 5x - 5x^2 + x^3)e^{(x^2-5)/12}$, and rectangles such that the sum of their areas approximates $\int_0^3 f(x)dx$.

broken up into a sum of N integrals over the separate intervals,

$$\begin{aligned}\int_{x_0}^{x_N} f(x)dx &= \int_{x_0}^{x_0+\Delta x} f(x)dx + \int_{x_1}^{x_1+\Delta x} f(x)dx + \cdots + \int_{x_{N-1}}^{x_{N-1}+\Delta x} f(x)dx \\ &= \sum_{j=0}^{N-1} \int_{x_j}^{x_j+\Delta x} f(x)dx. \qquad (5.12)\end{aligned}$$

We can simplify this with the substitution $h = x - x_j$ in each integral, so that

$$\int_{x_0}^{x_N} f(x)dx = \sum_{j=0}^{N-1} \int_0^{\Delta x} f(x_j + h)dh. \qquad (5.13)$$

Let us approximate $f(x)$ with its value $f(x_j)$ at the edge of each interval,

$$f(x_j + h) \approx f(x_j)\,. \qquad (5.14)$$

This gives

$$\int_0^{\Delta x} f(x_j + h)dh \approx f(x_j) \int_0^{\Delta x} dh = f(x_j)\Delta x,$$

and

$$\int_{x_0}^{x_N} f(x)dx \approx \sum_{j=0}^{N-1} f(x_j)\Delta x. \qquad (5.15)$$

The meaning of Eq. (5.15) is illustrated by Fig. 5.2. The area under the curve is approximated by the sum of the areas of rectangles. The area of the rectangle beginning at x_j is its height, $f(x_j)$, multiplied by its width, Δx. Eq. (5.15) is the simplest example of a quadrature.[2]

[2]The word "quadrature" comes from the Latin prefix *quadri-*, meaning "four." The integral is being approximated using four-sided geometrical figures.

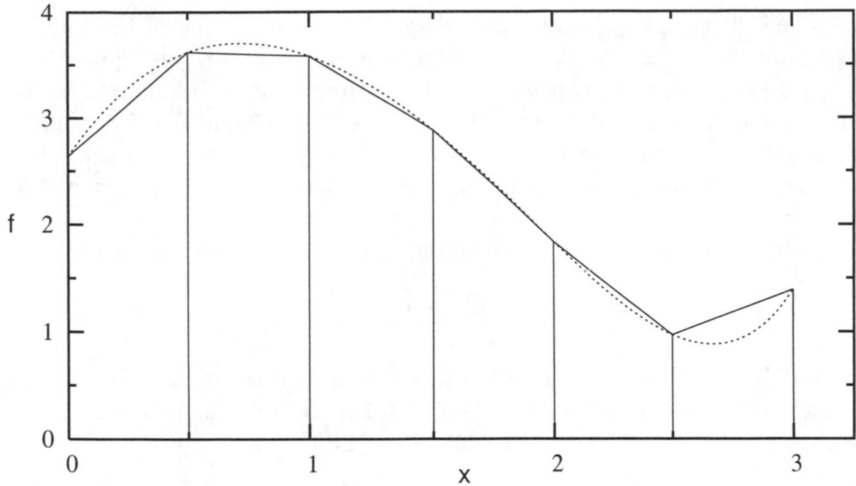

Figure 5.3: $f(x) = (4 + 5x - 5x^2 + x^3)e^{(x^2-5)/12}$, and trapezoids such that the sum of their areas approximates $\int_0^3 f(x)dx$.

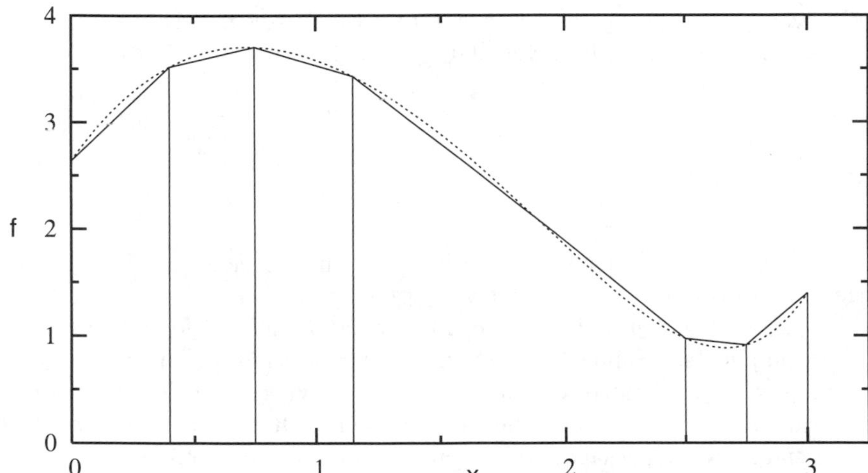

Figure 5.4: $f(x) = (4 + 5x - 5x^2 + x^3)e^{(x^2-5)/12}$, and trapezoids to approximate $\int_0^3 f(x)dx$ with step sizes adapting to the curvature.

The accuracy can be improved by tinkering with Eq. (5.14). Instead of a constant value in each interval, we could use a straight line between each endpoint, as in Fig. 5.3. The regions are now trapezoids instead of rectangles. The algorithm is the same as before except that in place of Eq. (5.14) we use

$$f(x_j + h) \approx [f(x_j) + f(x_{j+1})]/2. \tag{5.16}$$

This usually reduces the error. Comparing Figs. 5.2 and 5.3, one sees that for this function trapezoids are more accurate in all regions except from 2.5 to 3.0, where the rectangle is better, and from 0.5 to 1.0, where both methods are about equal in accuracy. Both methods give the exact value of the integral in the limit $N \to \infty$, $\Delta x \to 0$. Sophisticated variants of the quadrature algorithm have been developed that adjust the value of the step size Δx, choosing a smaller value of Δx in such regions as 2.5 to 3.0 in Fig. 5.3. Fig. 5.4 demonstrates how this can work. The x_j values are chosen such that their density is lower where the function is almost linear and higher where curvature is greater.

Applying quadrature to improper integrals requires extra care. Consider

$$I = \int_5^\infty e^{-x^2} dx.$$

We could begin by specifying the number of quadrature intervals, N, but would immediately encounter a problem with Eq. (5.11), which gives $\Delta x = \infty$, because x_N is infinite. We could replace Eq. (4.30) with

$$I = \lim_{c \to \infty} I(c), \qquad I(c) = \int_5^c e^{-x^2} dx, \tag{5.17}$$

and then compute $I(c)$ for a sequence of increasingly large values of c, which would eventually converge to the exact result. However, this is inefficient. Δx, though finite, will still be very large. A better approach is to make the change of variable

$$x = u^{-1}, \qquad dx = -du/u^2. \tag{5.18}$$

Then we have

$$I = -\lim_{c \to 0} \int_c^{1/5} u^{-2} e^{-1/u^2} du, \tag{5.19}$$

which can be efficiently computed with quadrature, because the integration range remains confined to the narrow region $0 < u \le 1/5$.

The case of an integrand going to infinity *within* the range of integration is more complicated. Simple quadrature routines, with evenly spaced x intervals can give unpredictable results, depending on whether or not the edge of a quadrature interval falls near the singular point. Quadrature algorithms in commercial software packages will usually be able to adapt the step sizes in such a way as to ensure convergence to the Cauchy principal value. However, if the algorithm needs to find the singularity by trial and error, the rate of convergence can be slow. If the locations of integrand infinities are known in advance, this information can be used to mitigate their effects. Sophisticated quadrature routines in computer algebra software packages will allow for known singular points to be specified.

Numerical integration over more than one integration variable is considerably more difficult. A simple two-dimensional version of Eq. (5.15) is

$$\int_{y_0}^{y_N} \int_{x_0}^{x_N} f(x,y) dx dy \approx \int_{y_0}^{y_N} \left[\sum_{j=0}^{N-1} f(x_j, y) \Delta x \right] dy \approx \sum_{k=0}^{N-1} \left[\sum_{j=0}^{N-1} f(x_j, y_k) \Delta x \right] \Delta y. \tag{5.20}$$

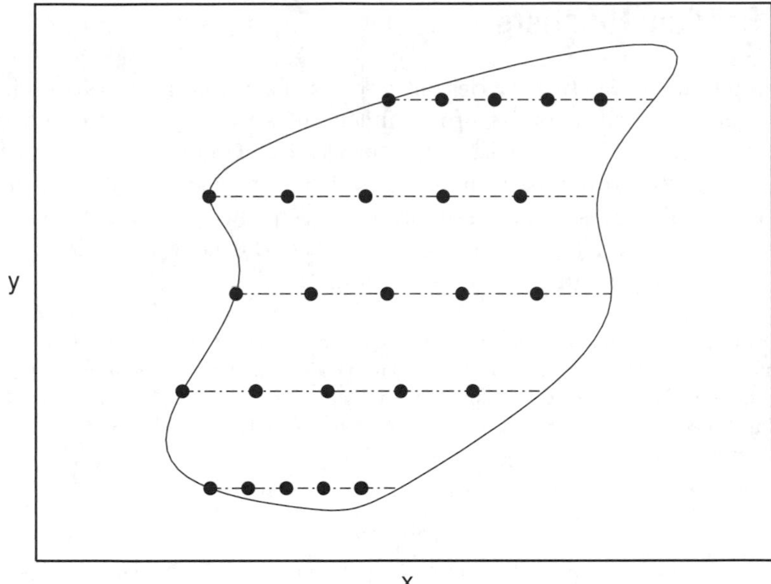

Figure 5.5: Quadrature evaluation points for an irregularly shaped region with evenly spaced values of y and x.

For each of the N values of y_k we must evaluate f at the N different points (x_0, y_k), (x_1, y_k), (x_2, y_k),..., (x_{N-1}, y_k). This means we will need a total of N^2 evaluations of f. For a three-dimensional integral we would need N^3 evaluations. If we require a large value of N, for a highly accurate quadrature, then the computational cost increases very rapidly with increasing dimension. $N = 100$ is not unreasonable for an accurate one-dimensional quadrature. In three dimensions we would need $N = 100^3 = 10^6$ for comparable accuracy.

It is important also to note that Eq. (5.20) applies only to integration over a *rectangular* region of the xy-plane. This is because the same range of x, from x_0 to x_N, is used for every value of y_k. It is quite common in practical applications to encounter integrals over nonrectangular regions. Then it is necessary to express the range of x as a function of y, for example, with

$$\int_{y_0}^{y_N} \int_{x_0(y)}^{x_N(y)} f(x, y) dx dy \approx \sum_{k=0}^{N-1} \left[\sum_{j=0}^{N-1} f(x_{j,k}, y_k) \Delta x_k \right] \Delta y, \tag{5.21}$$

$$\Delta x_k = [x_N(y_k) - x_0(y_k)]/N, \qquad x_{j,k} = x_0(y_k) + j \Delta x_k \, , \tag{5.22}$$

as illustrated by Fig. 5.5. One might also want to make the number of x intervals a function $N_x(y)$, changing with y so that Δx_k will not vary so much.

5.4 Random Numbers

A computer is a device that carries out a series of arithmetic operations in the exact sequence specified by its programmer. It is ironic that scientists very often use computers to model *random* processes. Computers can generate sequences of ***pseudorandom*** numbers, which can then be used to model random natural processes. A pseudorandom sequence is not truly random; however, the relation between one number and the next is made to be so complicated that it might as well be random.[3]

Example 5.3. *A simple random number generator.** We need an operation that maps a number to a value that is not obviously related to it. Consider the ***modulo*** operator, which gives the remainder of one number divided by another. For example, 11 modulo 4 is 3, because $2 \cdot 4 = 8$ is as close as we can get to 11, and $11 - 8 = 3$. This is abbreviated 11 mod 4 or 11 % 4.

Now consider a sequence of numbers q_0, q_1, q_2,..., where

$$q_{j+1} = (aq_j + b) \% M , \tag{5.23}$$

where a, b, and M are some given integers. For example, let us choose $a = 96$, $b = 66$, and $M = 10$. We also need to choose the first member of the sequence, q_0, which is called the ***seed***. Let us seed the sequence with $q_0 = 5$. Then $q_1 = 546 \% 10 = 6$, $q_2 = 642 \% 10 = 2$, and so on. Here are the first 10 values of q_j:

5 6 2 8 4 0 6 2 8 4

This is not a very good random number generator, because the sequence repeats itself once the number 6 reappears at q_6. In general, the *maximum* sequence length before a repeat is M, because the value given by the modulo operator is always less than or equal to the devisor. Therefore, the usual practice is to choose M to be extremely large. Even so, if a and b are not chosen with care, the sequence can display significant nonrandomness. By convention, random number generators output the value q_{j+1}/M, which is always a non-negative real number less than 1.

Before generating a random number sequence, the generator must be *seeded* with a value for q_0, the first number in the sequence. You can specify a seed yourself or let the computer choose it. (Typically, if you do not specify a seed, one will be generated using an algorithm involving the time on the computer's clock.) The advantage of specifying the seed yourself is that the subsequent sequence will be reproduceable. This makes it possible to exactly reproduce your analysis.

Random number generators have a wide variety of applications, some of which will be described in subsequent chapters. One use is to model a large collection of molecules, assigning random values to the initial position and velocity of each molecule. Another is to model the effects of random error in experimental measurements. Computational methods based on random number sequences are called ***Monte Carlo*** methods.[4]

[3]We will follow the common practice of referring to a sequence as "random" with the understanding that if the numbers are computer-generated they are actually pseudorandom.

[4]Named for the Monacan resort town of Monte Carlo, famous for its gambling casino.

Example 5.4. *Monte Carlo integration.** Consider the approximation

$$\int_a^b f dx \approx (b-a)\bar{f}, \tag{5.24}$$

in which the integral is replaced with a single large rectangle. Let the height $\bar{f}$ be the average value of $f(x)$ evaluated at a random collection of points $x_1, x_2, \ldots, x_N$,

$$\bar{f} = \frac{1}{N}\sum_{j=1}^{N} f(x_j). \tag{5.25}$$

This equation offers an alternative to conventional quadrature.

Let us integrate $f(x) = x^2$ over the range 0 to 7. We can generate a sequence of random x values between 0 and 7 by multiplying the output of a random number generator by 7. A sample computation is as follows:

x_j	3.79561	5.16422	1.11814	6.94839	1.98451	4.50620
$f(x_j)$	14.4067	26.6692	1.2502	48.2802	3.9383	20.3059
x_j	2.96532	6.54798	5.29557	0.87060	4.46397	0.67338
$f(x_j)$	8.7931	42.8760	28.0431	0.7579	19.9270	0.4534

The average of the 12 values of f is 17.9751, which gives $\int_0^7 x^2 dx \approx 7 \cdot 17.9751 = 125.826$, which is about 10% off the exact result $\frac{1}{3}x^3\big|_0^7 = 114.333$.

Usually the accuracy of Monte Carlo integration is poorer than that of conventional quadrature for a given computational cost. However, it can be the preferred method if the integrand is highly oscillatory or, in multidimensional integration, if the integration region is so complicated that it is not feasible to express the ranges of the inner variables in terms of the outer variables.

5.5 Root Finding

Consider the equation

$$f(x) = 0$$

for some arbitrary function f. The values of x that satisfy this equation are called the ***roots*** of f. If f is not very complicated, it may be possible to solve this equation analytically. However, it is often the case that an analytical solution is not feasible. A common situation is that the value of f can be obtained at discrete values of x, through measurement or computer calculation, but the function $f(x)$ as an explicit analytical expression is unknown. We will consider two different numerical procedures for searching for roots, one of which is quite dependable but slow, requiring a relatively large number of values of f. The other is much faster but is less dependable, in the sense that it will sometimes fail to find the root.

With either method, the first step is to try to identify a range of x known to contain the root of interest. If one has in advance a general idea of where the root should be, then a method for identifying a search range is to choose a value $x = a$ to the left of the expected point and a value $x = b$ to the right

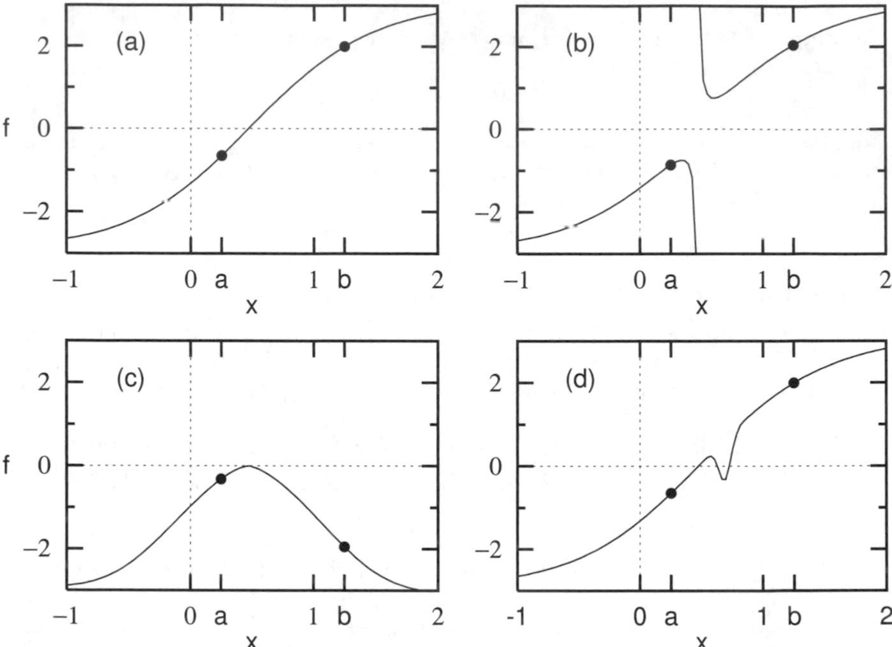

Figure 5.6: Examples of attempts to bracket a root: (a) A nonsingular function with an isolated root. (b) A singular function—bracketing seems to imply the interval contains a root but it in fact does not. (c) A root that is also an extremum, and therefore cannot be bracketed. (d) Nearby roots, difficult to bracket individually.

such that $f(a)$ and $f(b)$ have opposite signs. A root is said to be ***bracketed*** in an interval $a < x < b$ if $f(a)$ and $f(b)$ are opposite in sign. If the function is continuous and nonsingular in this interval, then it must be the case that it passes through zero at least once somewhere between a and b. This is illustrated by Fig. 5.6, which shows various situations often encountered in practice. Panel (a) shows a "well-behaved" function, with a single isolated root. The function in panel (b) has a first-order pole. Clearly, if a function has a singular point in the range under consideration, bracketing will not guarantee the existence of a root. Panel (c) shows a situation in which a root exists but it cannot be bracketed, because the root is an extremum of the function. Finally, panel (d) shows a case of three roots very close together. While it is possible in this case to bracket them individually, in practice the bracketing interval (as in this case) will probably contain three roots.

Suppose that we have bracketed the root between a and b. Then the slow but dependable method for finding the root is ***bisection***. This is guaranteed to converge to a root (or an odd-degree pole) within the bracketing interval. The idea is simple: Evaluate f at the point $c = (a + b)/2$, halfway between a and b. If $f(c)$ has the same sign as $f(a)$ then replace a with c; if it has the same sign as $f(b)$, then replace b with c. Continue until the size of the interval drops below some specified numerical value of error tolerance.

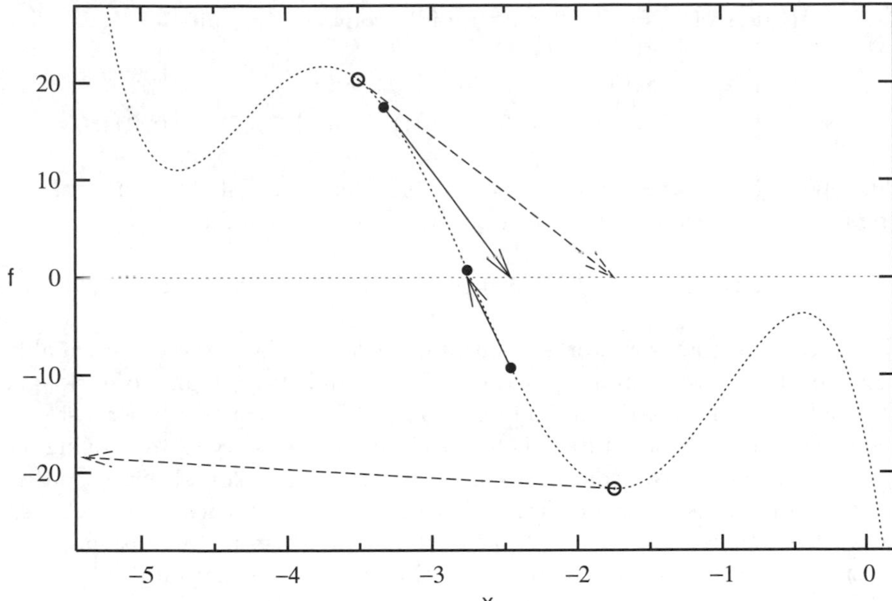

Figure 5.7: An illustration of the Newton-Raphson method for finding a root. One choice of starting point gives rapid convergence (the filled circles) while another choice diverges (the open circles).

The second technique is the ***Newton-Raphson method.***[5] This method is not as general as bisection, because it requires that we be able to compute both the function value and *the value of the first derivative* at arbitrary points, but it is capable of giving extremely fast convergence when used appropriately. It is based on the definition of the derivative, Eq. (1.5), which we write as an approximation

$$f(x + \Delta x) \approx f(x) + f'(x)\Delta x \tag{5.26}$$

for small but finite Δx, without taking the limit $\Delta x \to 0$. Let x_j be our current best estimate of the root and let $x_j + \Delta x$ be our new improved estimate. If this is to yield a root, then

$$0 = f(x_j + \Delta x) \approx f(x_j) + f'(x_j)\Delta x,$$

which implies that we should increment x_j by $\Delta x = -f(x_j)/f'(x_j)$, so that our new estimate is

$$x_{j+1} = x_j - f(x_j)/f'(x_j). \tag{5.27}$$

This represents an extrapolation of the tangent line from the point $(x_j, f(x_j))$ to the point x_{j+1} at which it reaches the x-axis, as illustrated in Fig. 5.7, which shows results for two different starting points. The filled circles correspond to

[5] This method was invented by Newton in the 1670's but not published by him until many years later. It is often simply called "Newton's method." It was first published in 1691 in a book by English mathematician and philosopher of religion Joseph Raphson (1648?-1715?).

an initial guess of $x_0 = -3.32$, with rapidly convergent results:

$$(-3.32,\ 17.4849) \to (-2.4604,\ -9.2588) \to (-2.7569,\ 0.7404) \to (-2.7357,\ -0.0001)$$

The open circles correspond to $x_0 = -3.50$, which is only slightly farther from the root. However, now the points quickly fly off the graph:

$$(-3.50,\ 20.3746) \to (-1.7466,\ -21.6775) \to (-25.81,\ 2 \times 10^{11})$$

Thus, Newton-Raphson works beautifully if the initial guess is reasonably close to the root but can fail miserably if the initial guess is not close enough. It should be considered a method for "polishing" a root, not a general stand-alone method. A good strategy is to first bracket the root you are looking for and then attempt a Newton-Raphson step. If the result is questionable, either because it lies outside the brackets or because it gives a suspiciously small step size, then replace that step with a bisection. Continue in this fashion, first attempting Newton-Raphson, but using bisection if necessary, until $|x_{j+1} - x_j|$ falls below the desired error tolerance.

5.6 Minimization*

Given a reasonably simple function $f(x)$ for which you have an explicit expression, the minima can be determined analytically by finding the values of x satisfying the equation $f'(x) = 0$ and then testing each one to see if it is a minimum or a maximum. In principle, then, one might presume that a *numerical* search for extrema should be carried out by applying numerical root finding methods to the first derivative. However, this approach would be inefficient. Finding an extremum is actually an easier problem for numerical analysis than is root finding. This is because to find a minimum all you really need to do is slide "downhill" until you reach the bottom of a well in the function. In root finding there is no comparable criterion for knowing for sure in which direction to go in order to move toward the root.

It is important to note, however, that a function may have more than one minimum. Strictly speaking, a point at which $f' = 0$ and $f'' > 0$ is called a ***local minimum***. The local minimum with the lowest value of f is called the ***global minimum***. Finding a local minimum is relatively straightforward. Finding the global minimum can be much more challenging. A maximum of f is a minimum of $-f$. Therefore, it is sufficient just to discuss minimization. The same method applied to $-f$ yields a maximum.

Consider the problem of finding a local minimum of a function f of a single coordinate x. We will endeavor to get by with as few evaluations of f as is feasible, under the assumption that f is costly to compute or measure. We begin with arbitrary guesses a_0, u_0, and b_0 to bracket the minimum, with

$a_0 < u_0 < b_0$. The values a_0 and b_0 will be brackets if $f(u_0)$ is less than $f(a_0)$ and $f(b_0)$. Next, we interpolate a parabola between the points $(a_0, f(a_0))$, $(b_0, f(b_0))$, $(u_0, f(a_0))$ and find the minimum of the parabola. This is easily done by expressing the interpolating polynomial in the form

$$p_0(x) = \alpha_0 + \kappa_0(x - v_0)^2,$$

which obviously has its minimum at $x = v_0$. We have three equations,

$$p(a_0) = f(a_0), \quad p(b_0) = f(b_0), \quad p(u_0) = f(u_0), \tag{5.28}$$

from which we can determine the three unknowns α_0, κ_0, and v_0. Subtracting the second equation from the first eliminates α_0, giving

$$\kappa_0 \left[(a_0 - v_0)^2 - (b_0 - v_0)^2\right] = f(a_0) - f(b_0)\,. \tag{5.29}$$

Subtracting the third equation from the first gives

$$\kappa_0 \left[(a_0 - v_0)^2 - (u_0 - v_0)^2\right] = f(a_0) - f(u_0)\,. \tag{5.30}$$

Dividing one by the other eliminates κ_0. Solving for v_0, we find

$$v_0 = \frac{1}{2}\,\frac{(u_0^2 - b_0^2)f(a_0) + (b_0^2 - a_0^2)f(u_0) + (a_0^2 - u_0^2)f(b_0)}{(u_0 - b_0)f(a_0) + (b_0 - a_0)f(u_0) + (a_0 - u_0)f(b_0)}\,. \tag{5.31}$$

It is important to check that our parabola is concave, with a minimum, rather than convex, with a maximum. For this purpose we calculate κ_0 from Eq. (5.30). We want it to be positive. If it turned out negative, we would make a new choice of a_0, u_0, b_0 in the hope of getting a positive κ_0.

We now compute $f(v_0)$ and compare it with the previous three computed values. Of these four points, the lowest three that bracket the minimum are assigned the new names $(a_1, f(a_1))$, $(u_1, f(u_1))$, $(b_1, f(b_1))$ and a new parabola is calculated, from which we obtain a new predicted minimum v_1. This procedure is continued, giving v_2, v_3, and so on until $|v_{n+1} - v_n|$ drops below the desired error tolerance.

This algorithm usually works quite well. However, if the underlying function looks rather different from a parabola, then it is possible that the v_j will not converge adequately. In that case a safer but slower method can be substituted, in which v_j is chosen according to the ***golden mean***,[6]

$$2\Big/\left(1 + \sqrt{5}\right) \approx 0.618\,.$$

Suppose that u_j is closer to b_j than to a_j. Then we can expect that the true

[6]This ratio, also called the *golden ratio* or the *golden section*, was greatly valued by the ancient Greeks for its aesthetic properties. The division of an interval into two parts one of which is equal to the total length times the golden mean was considered to be especially pleasing. A reasonable, though not rigorous, argument can be made for its use in searching for minima. See W. H. Press *et al.*, *Numerical Recipes: The Art of Scientific Computing* (Cambridge University Press, Cambridge, 2007), Chap. 10.

minimum is more likely in the larger region between a_j and u_j than in the smaller region between u_j and a_j. But where within this region is it most likely to be? a_j is just an outer limit for the minimum, not an estimate of its location. This suggests the new estimate should lie closer to u_j than to a_j. How much closer? For this we use the golden mean,

$$v_j = a_j + (0.618)\,(u_j - a_j)\,. \tag{5.32}$$

If u_j is closer to a_j, then we use

$$v_j = b_j - (0.618)\,(b_j - u_j)\,. \tag{5.33}$$

The algorithm described here, developed by Brent,[7] is reliable and efficient for finding a local minimum in one dimension. A golden-mean step is substituted for a parabolic step if either of the following conditions holds:

1. If the amount of movement in a parabolic step is not less half that of the second-to-last step, and that step was also a parabolic step. In other words, if $|v_j - v_{j-1}|/2 > |v_{j-2} - v_{j-3}|$, then the parabolic steps are deemed to be converging too slowly.

2. If v_j lies outside the brackets a_j and b_j.

Example 5.5. *Using Brent's method to determine the bond distance of the nitrogen molecule.* The equilibrium bond distance of a diatomic molecule is the value R_e of the internuclear distance R such that the electronic energy of the molecule, $V(R)$ is at a minimum. V can be computed at specified values of R using methods of quantum chemistry, but an exact analytical solution for $V(R)$ is not available.

Our goal is to obtain R_e to four significant figures. We begin with arbitrary values 1.0, 1.3, and 1.5 Å as a_0, u_0, and b_0. We find that $V(u_0)$ is less than $V(a_0)$ and $V(b_0)$. a_0 and b_0 bracket a mininum. We solve for the parabola through these three points, calculate its minimum v_0 and the factor κ_0, check that κ_0 is positive (it is), and proceed with a new parabolic step using $a_1 = a_0$, $u_1 = u_0$, $b_1 = v_0$. We replace b_0 with v_0 because $f(u_0) < f(v_0) < f(b_0)$; the values a_0, v_0 are a narrower bracket than a_0, b_0. We continue in this way until $v_4 = 1.235$, at which point we find that $|v_4 - v_3| = 0.014$, which is converging too slowly. We use instead $v_4 = a_4 + (0.618)(u_4 - a_4) = 1.154$, and then continue taking parabolic steps. The remaining steps converge satisfactorily according to Brent's criteria. At $v_8 = 1.1498$ we have $|v_8 - v_7| = 0.0050$. This equals our error tolerance. We conclude that $R_e = 1.150$ Å. The sequence of points and parabolas is shown in Fig. 5.8. Note the jump from v_3 to v_4, a golden-mean step.

In this example we could have obtained faster convergence using higher-order polynomials instead of parabolas. For example, a quartic Lagrange interpolant through a_0, u_0, b_0, v_0, and v_1 predicts a minimum at 1.210, which is somewhat better than $v_2 = 1.264$ from the parabolic steps. However, higher-order polynomials can have more than one minimum, and this can lead to disaster. They should be used for minimization only with great care, when you have in advance a good idea of where to expect the minimum.

[7] Australian mathematician Richard Brent. See R. P. Brent, *Algorithms for Minimization without Derivatives* (Prentice-Hall, Englewood Cliffs, NJ, 1973), Chap. 7.

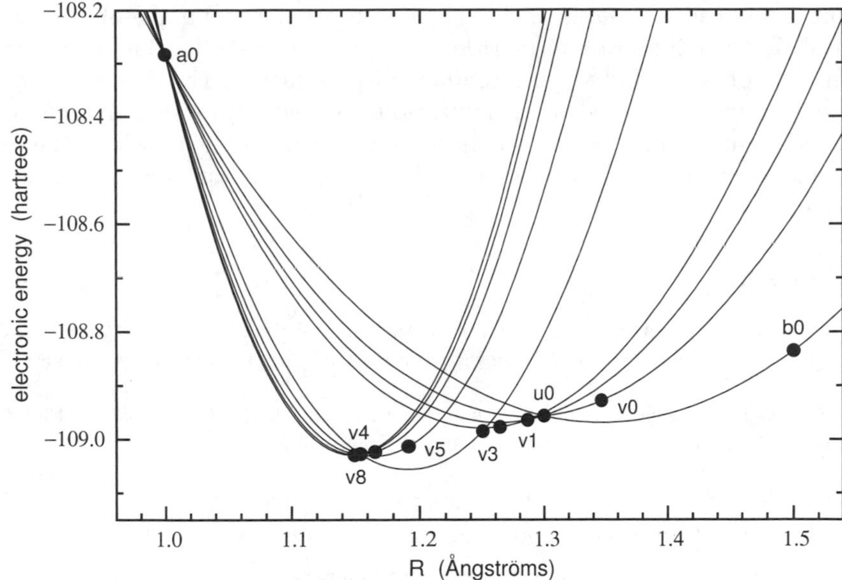

Figure 5.8: Using Brent's method to find the equilibrium bond length of the N_2 molecule. Each of the 11 energy values were computed using a time consuming full configuration-interaction computation.

In two dimensions, for a function $f(x, y)$, one could use Brent's method alternating between the dimensions, choosing some value of y, minimizing with respect to x with y held constant, then with x set at that minimum, minimize with respect to y, and then with the new y value minimize in x, and so on. However, this often proves inefficient in practice. Better methods are available. The most efficient make use of partial derivatives, if they can be computed directly at each point. Multidimensional minimization is a complicated problem, and there are various algorithms to choose from.[8]

There is no known algorithm that directly heads for the global minimum without the possibility of getting stuck at a local minimum. Especially for functions of high dimension this can be a problem, as they likely have many local minima. However, there is an indirect approach, called ***simulated annealing***, inspired by an example from chemistry, that can usually at least find a local minimum that is close to the global minimum.

If a liquid metal is cooled quickly, the resulting crystal will likely have defects, because many of the atoms will not have time to "find" their lowest-energy lattice positions. This metastable configuration is a local minimum of the crystal's potential energy. If the crystal is subjected to a series of heating and cooling cycles, with the temperature at the top of the heating cycle gradually decreased, the number of crystal defects can be greatly reduced. This

[8]The Nelder-Mead method, which will be discussed in in Section 12.4, is a popular choice. For other methods, see Box *et al.*, *Statistics for Experimenters* (Wiley, New York, 1978) or Press *et al.*, *ibid.* and references therein.

technique is called *annealing*. In the computational technique of *simulated* annealing, the minimization algorithm is allowed to find a local minimum and then that point is shifted by a random displacement. The shifted point is the initial guess for a second minimization. A new local minimum is found, it is randomly displaced, and so on, with the average magnitude of the displacements gradually decreased, according to some parameter analogous to temperature.

Exercises

5.1 Calculate (by hand) the Lagrange interpolant between the points $(1, 1)$, $(2, 3)$, $(4, 3.5)$. Write it as a polynomial and plot it showing the interpolating function and the points.

5.2* Calculate (by hand) cubic splines between the points $(1, 1)$, $(2, 3)$, $(4, 3.5)$. Plot the result, showing the interpolating function and the points.

5.3 Calculate the derivative of $e^{(1+x^4)^{1/3}}$ at $x = 1$ as follows:

(a) Using Eq. (5.6) for some arbitrary h. (b) Using Eq. (5.8). (c) Exactly.

Compare the accuracy of the two approximate results for given h. How does the accuracy depend on h? (Carry out a numerical experiment using different values of h.) What is the smallest h that is feasible to use on your computing device?

5.4 Test Eq. (5.7) for various functions for which you can calculate the derivative analytically.

5.5 Simulate a spectrum of overlapping peaks as in Example 5.2 and use the numerical second derivative to resolve the peaks. Experiment with different orders for the Lagrange interpolant. What are the advantages and disadvantages of increasing the order?

5.6 Calculate $\int_0^2 \ln(2 - 2x/3)e^{(1+x^4)^{1/3}} dx$ numerically using:

(a) Five quadrature rectangles.

(b) Five quadrature trapezoids.

(c) Five quadrature trapezoids, choosing unevenly spaced quadrature points to improve the accuracy.

(d*) Six Monte Carlo points. (Repeat the Monte Carlo computation several times to see how much the result varies.)

5.7* For the integral in the previous Exercise, carry out a numerical experiment to compare the relative efficacy of the following two strategies: (1) Taking the average of n Monte Carlo integrations each with N points. (2) A single Monte Carlo integral using nN points. Discuss your conclusion.

5.8 Estimate the locations of the roots of $f(x) = 7 - x + x^3 - 2x^4$ using (a) four bisections for each root, and (b) four Newton-Raphson steps for each root.

5.9* Consider the function $f(x) = x^{-12} - x^{-6} + x^{-1}/3$, which has a local minimum somewhere between 1.0 and 1.5. (a) Plot the function. (b) Use Brent's method to find the local minimum with a precision of 0.01. Compare the rate of convergence using various different initial guesses ranging from 0.8 to 2.0.

Chapter 6

Complex Numbers

6.1 Complex Arithmetic

The ***real numbers*** comprise an infinite set with a one-to-one correspondence between each number and a point on a line (for example, the x-axis). This set is designated $\mathbb{R}$. The notation $x \in \mathbb{R}$ is a concise way of saying, "x is a real number." ($\in$ means "is an element of.") $\mathbb{R}$ includes integers, 0, ± 1, $, \pm 2, \ldots$; ***rational numbers***,[1] such as $1/2$ or $7/4$; and ***irrational numbers***.

The real numbers usually are adequate for describing positions and trajectories of objects, but from a theoretical perspective they have a drawback. Consider the equation

$$x^2 = -1. \tag{6.1}$$

There is no real number x such that this simple equation has a solution. This was a cause of discomfort among mathematicians of the 16th and 17th centuries as they tried to develop a coherent theoretical foundation for algebra.

This difficulty was eliminated, in a sense, by *defining* a number, called i, such that

$$i^2 = -1. \tag{6.2}$$

One can then define a system of ***complex numbers***, $\mathbb{C}$, consisting of all

$$z = u + vi \tag{6.3}$$

where u and v are real numbers. Note that $\mathbb{R}$ is the set of all z in $\mathbb{C}$ such that $v = 0$. Thus, $\mathbb{R} \subset \mathbb{C}$; that is, the real numbers are a subset of the complex numbers. We can define another subset of $\mathbb{C}$ called the ***imaginary numbers***, which correspond to all z such that $u = 0$. In Eq. (6.3), u is called the ***real part*** of z and v is called the ***imaginary part*** of z. These are indicated by the notation Re z and Im z, respectively.

Example 6.1. *Real and imaginary parts.* Consider $2 + 3i$. $\operatorname{Re}(2 + 3i) = 2$ and $\operatorname{Im}(2 + 3i) = 3$. The "imaginary part" is the *real* number that multiplies i.

Arithmetic in $\mathbb{C}$ follows the same rules as in $\mathbb{R}$ except that i satisfies $i^2 = -1$. Thus, for example, addition follows the rule

$$(a + bi) + (c + di) = (a + c) + (b + d)i \tag{6.4}$$

and multiplication follows the rule

$$(a + bi)(c + di) = ac + adi + bci + bdi^2 = (ac - bd) + (ad + bc)i. \tag{6.5}$$

Example 6.2. *Calculate* $(2 + 3i)^3$:

$$(2 + 3i)^3 = (2 + 3i)(2 + 3i)(2 + 3i) = 8 + 36i + 54i^2 + 27i^3 = -46 + 9i,$$

using the fact that $i^2 = -1$ and $i^3 = i^2 i = -i$.

[1] A rational number is any number that can be expressed as a ratio of integers. Note that the integers are a subset of the rationals. For example, 2 can be expressed as 2/1. An irrational number is any real number that is not rational, such as π or $\sqrt{2}$.

The reciprocal $1/i$ is equal to $-i$. This follows immediately from the equation $i^2 = -1$, dividing both sides by i.

There is a new operation in $\mathbb{C}$ that is not needed in $\mathbb{R}$:

Definition: ***Complex conjugation*** is the operation that consists of substituting $-i$ for i wherever it appears.

For example, the ***complex conjugate*** of $a + bi$ is $a - bi$. The symbol for complex conjugation is a superscript asterisk,[2] for example,

$$(a + bi)^* = a - bi. \tag{6.6}$$

Complex conjugation is particularly useful because multiplication of a complex number by *its own* complex conjugate always converts it into a positive real number,

$$(a - bi)(a + bi) = a^2 + (ba - ab)i - b^2 i^2 = a^2 + b^2. \tag{6.7}$$

The absolute value of a complex number is defined as

$$|a + bi| = \sqrt{a^2 + b^2}. \tag{6.8}$$

In other words, for any $z \in \mathbb{C}$,

$$|z| = \sqrt{z^* z}\,. \tag{6.9}$$

Example 6.3. *Absolute value of complex numbers.*

$$|4 + 7i| = \sqrt{(4 - 7i)(4 + 7i)} = \sqrt{16 - 28i + 28i - 49i^2} = \sqrt{16 + 49} = \sqrt{65}\,.$$

$$\begin{aligned} |(4 + 7i)^{1/3}| &= \left[(4 - 7i)^{1/3}(4 + 7i)^{1/3}\right]^{1/2} \\ &= \left\{\left[(4 - 7i)(4 + 7i)\right]^{1/3}\right\}^{1/2} = \left(65^{1/3}\right)^{1/2} = 65^{1/6}. \end{aligned}$$

$$|3^{2i}| = \left(3^{-2i}\,3^{2i}\right)^{1/2} = \left(3^{-2i+2i}\right)^{1/2} = \left(3^0\right)^{1/2} = 1^{1/2} = 1.$$

Division of complex numbers can be carried out with the help of complex conjugation. Let $\alpha + \beta i$ be the quotient of $a + bi$ and $c + di$,

$$\alpha + \beta i = \frac{a + bi}{c + di}\,. \tag{6.10}$$

Suppose we know a, b, c, and d and want to determine α and β. A simple trick is to multiply and divide by the complex conjugate of the denominator,

$$\frac{(a + bi)}{(c + di)}\frac{(c - di)}{(c - di)} = \frac{(a + bi)(c - di)}{c^2 + d^2} = \frac{ac + bd}{c^2 + d^2} + \frac{bc - ad}{c^2 + d^2}\,i\,. \tag{6.11}$$

It follows that

$$\alpha = \frac{ac + bd}{c^2 + d^2}, \qquad \beta = \frac{bc - ad}{c^2 + d^2}\,. \tag{6.12}$$

---

[2]The asterisk is used in the scientific literature. Mathematicians often prefer to use a bar, $\overline{a + bi} = a - bi$.

6.2 Fundamental Theorem of Algebra

In $\mathbb{C}$ we can easily solve any equation of the form $z^2 = a$, where a is any real number. Thus, for example, $z^2 = -4$ has the solution $z = 2i$. Consider a less trivial equation,

$$z^2 - 2z + 4 = 0.$$

The roots (i.e., the values of z at which $z^2 - 2z + 4$ pass through zero) are given by the quadratic formula, Eq. (2.12), which in this case gives $z = 1 \pm \sqrt{-3}$. This number does not exist in $\mathbb{R}$, but in $\mathbb{C}$ it can be written $z = 1 \pm i\sqrt{3}$. In general, for the equation $az^2 + bz + c = 0$ with real parameters a, b, and c, the quantity

$$D = b^2 - 4ac, \tag{6.13}$$

which appears in the square root of the quadratic formula, is what determines whether or not the equation has solutions in $\mathbb{R}$. If $D > 0$ then there will be two real roots, specifically, $(-b \pm D^{1/2})/(2a)$, at which the polynomial passes through zero. If $D < 0$, then the polynomial has two roots with nonzero imaginary parts, and they are complex conjugates of each other. In that case, a plot of the polynomial vs. real x will not pass through zero. D is called the ***discriminant***, because it discriminates between these two cases.

The polynomials $g(x) = x^2 - 2x - 4$ and $f(x) = x^2 - 2x + 4$ are compared in Fig. 6.1. The roots of g are $x = 1 \pm \sqrt{5} = -1.236$ or $+3.236$. Because

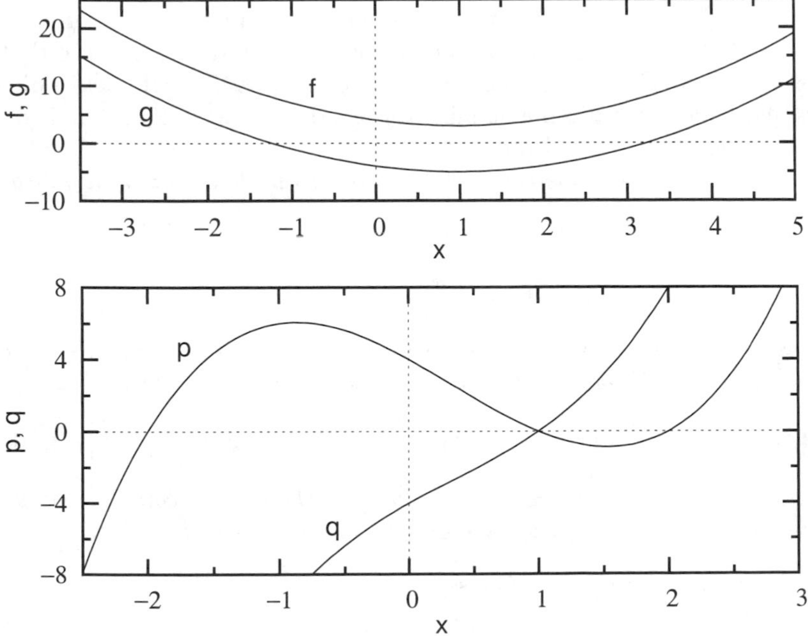

Figure 6.1: $f(x) = x^2 - 2x + 4$, $g(x) = x^2 - 2x - 4$, $p(x) = (x^2 - 4)(x - 1)$, and $q(x) = (x^2 + 4)(x - 1)$.

these are real numbers, we can see from the graph of $g(x)$ vs. real x that the function does indeed pass through the x-axis as expected. $f(x)$ has no roots in $\mathbb{R}$; it never passes through the x-axis. The lower panel of Fig. 6.1 compares *cubic* polynomials

$$p(x) = x^3 - x^2 - 4x + 4 = (x-1)(x-2)(x+2),$$

$$q(x) = x^3 - x^2 + 4x - 4 = (x-1)(x-2i)(x+2i).$$

It is clear from the graph that $p(x)$ passes through zero at each of its three roots, $x = 1$, 2, and -2, all real. $q(x)$ is zero only at its one real root $x = 1$.

In the limit of large $|x|$ the term with the largest exponent overwhelms all other terms in a polynomial. p and q behave as x^3 in the limit of large $|x|$. Consider a polynomial in which the term with the largest exponent is cx^n where n is an odd positive integer. Then

$$\lim_{x\to-\infty} cx^n = -\infty \quad \text{and} \quad \lim_{x\to+\infty} cx^n = +\infty, \qquad \text{for } n \text{ odd, } c \text{ positive}$$

$$\lim_{x\to-\infty} cx^n = +\infty \quad \text{and} \quad \lim_{x\to+\infty} cx^n = -\infty, \qquad \text{for } n \text{ odd, } c \text{ negative.}$$

In either case, the polynomial changes sign somewhere between $-\infty$ and $+\infty$, which means that it must pass through zero at at least one point. Therefore, any polynomial of odd degree must have at least one real root. A polynomial of even degree has the same sign in the limits $x \to \pm\infty$. It need not have any real roots. If it does have real roots, there must be an even number of them, because what goes down (or up) must come up (or down). Note also that for the examples considered here any complex roots appear in complex-conjugate pairs. For the quadratic polynomial f we had roots at $1 \pm i\sqrt{3}$ and for the cubic polynomial q we had roots at $\pm 2i$.

These observations illustrate the ***fundamental theorem of algebra***:

Theorem 6.2.1. *Consider a polynomial of degree n,*

$$f(z) = c_n z^n + c_{n-1} z^{n-1} + c_{n-2} z^{n-2} + \cdots + c_0\,,$$

in which all of the coefficients c_j are in $\mathbb{R}$. Then f has exactly n roots, $z_1, z_2, \ldots, z_n$, in $\mathbb{C}$ such that it can be factored into the form

$$f(z) = c_n (z - z_1)(z - z_2) \cdots (z - z_n)\,. \tag{6.14}$$

If any given root z_j has nonzero imaginary part, then its complex conjugate, z_j^*, is also a root. If n is odd then at least one root is real.

This theorem was proved by Gauss[3] in 1816.

[3]German mathematician Carl Friedrich Gauss (1777-1855) made very significant contributions to algebra, number theory, geometry, and probability theory, to astronomy, and to topics in physics such as optics and magnetism.

Example 6.4. *Real roots.* Given that the polynomial plotted at right is of degree 8, how many of its roots have nonzero imaginary part? It is evident that the function passes through zero at only two points, which implies it has only two real roots. According to Theorem 6.2.1, an 8th-degree polynomial has eight roots. Here, six of them have nonzero imaginary part.

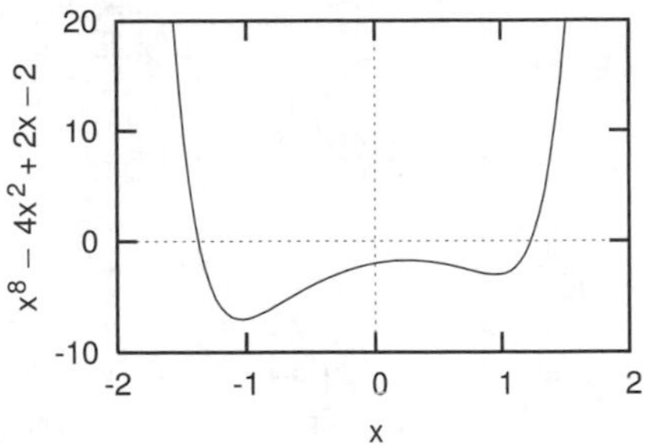

The theorem does not say that all the z_j must have different values. For example, $f(z) = (z-2)^6$ has six roots and they all have the value 2. A real multiple root of a polynomial will be a local extremum if the multiplicity of the root is even.

6.3 The Argand Diagram

We now develop a convenient representation of complex numbers as points in a two-dimensional plane. Let $z = x + iy$ where x and y are real numbers. There is a one-to-one correspondence between any complex number z and a point (x, y) in a two-dimensional Cartesian coordinate system. Thus, z can be plotted as a point in a plane. This is known as an ***Argand diagram.***[4] The plane defined by these two coordinates is called the ***complex plane***. An example is shown in Fig. 6.2.

The number 1 in an Argand diagram can be represented by the line segment between the points $(0, 0)$ and $(1, 0)$. The number i is represented by the line segment from $(0, 0)$ to $(0, 1)$. These two line segments are perpendicular to each other. Therefore, i can be obtained from 1 by rotation through 90°.

Using polar coordinates, Eqs. (3.1), any complex number $z = x + iy$ can be written

$$z = r(\cos\phi + i\sin\phi), \qquad r = |z| = \sqrt{z^*z} = \sqrt{x^2+y^2}. \tag{6.15}$$

r, the length of the line segment corresponding to the complex number z, is the absolute value of z. Note in particular that $z = 1$ corresponds to $\{r = 1,\ \phi = 0\}$ while $z = i$ corresponds to $\{r = 1,\ \phi = \pi/2\}$.

The angle ϕ in Eq. (6.15) is called the ***argument*** of the complex number z, and is designated $\arg z$. To calculate $\arg z$, we take advantage of the relations

$$\text{Re } z = r\cos\phi \qquad \text{Im } z = r\sin\phi, \tag{6.16}$$

$$\frac{\text{Im } z}{\text{Re } z} = \frac{\sin\phi}{\cos\phi} = \tan\phi. \tag{6.17}$$

[4]Named after the French mathematician Jean Argand who proposed this approach in 1806. Actually, the idea was developed earlier by Gauss, who never bothered to publish it, and by Danish mathematician Caspar Wessel, who published in Danish, a language few mathematicians could understand.

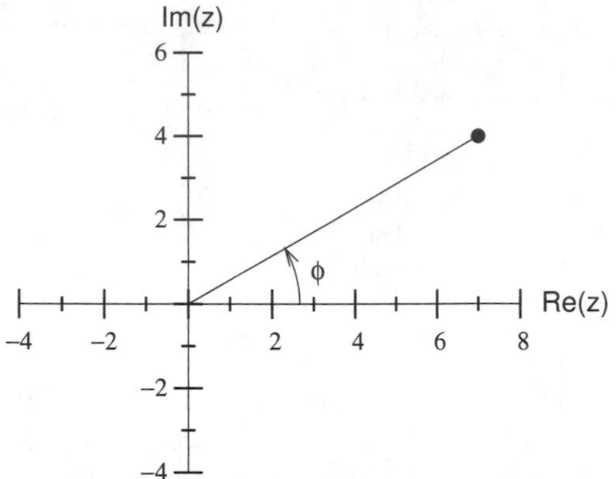

Figure 6.2: The complex number $z = 7 + 4i$ shown in an Argand diagram. The length of the line segment connecting the origin to the point is $|z| = (7^2 + 4^2)^{1/2} = \sqrt{65}$. The angle is $\phi = \arctan(4/7) = 29.7°$.

It follows that

$$\arg z = \arctan \frac{\text{Im } z}{\text{Re } z} . \tag{6.18}$$

To further explore the connection between trigonometry and complex arithmetic, note that

$$z^2 = r^2(\cos\phi + i\sin\phi)^2 = r^2(\cos^2\phi - \sin^2\phi + i\,2\sin\phi\,\cos\phi). \tag{6.19}$$

The identities for angle addition, Eqs. (2.39) and (2.40) with $\alpha = \beta = \phi$, give

$$\sin(2\phi) = 2\sin\phi\,\cos\phi, \qquad \cos(2\phi) = \cos^2\phi - \sin^2\phi. \tag{6.20}$$

Therefore,

$$(\cos\phi + i\sin\phi)^2 = \cos(2\phi) + i\sin(2\phi) . \tag{6.21}$$

Similarly, using Eqs. (2.39) and (2.40), one finds that

$$(\cos\phi + i\sin\phi)^3 = (\cos\phi + i\sin\phi)[\cos(2\phi) + i\sin(2\phi)] = \cos(3\phi) + i\sin(3\phi) .$$

Repeating this analysis $n - 1$ times gives ***de Moivre's formula***,[5]

$$(\cos\phi + i\sin\phi)^n = \cos(n\phi) + i\sin(n\phi) \tag{6.22}$$

for any positive integer n. Let $n = j + k$. It follows that

$$\begin{aligned}\cos[(j+k)\phi] + i\sin[(j+k)\phi] &= (\cos\phi + i\sin\phi)^{j+k} \\ &= (\cos\phi + i\sin\phi)^j(\cos\phi + i\sin\phi)^k \\ &= [\cos(j\phi) + i\sin(j\phi)][\cos(k\phi) + i\sin(k\phi)] .\end{aligned} \tag{6.23}$$

[5]This beautiful relationship was discovered by the French-born English mathematician Abraham de Moivre (1667-1754). He fled France to escape persecution for his Protestant religious beliefs. He arrived in England as an impoverished refugee and apparently was able to support himself by beating people at chess in a London coffee house.

Let $f(n) = \cos(n\phi) + i\sin(n\phi)$ for any given ϕ. Then Eq. (6.23) can be written as $f(j+k) = f(j)f(k)$. This is identical in form to Eq. (2.21), the characteristic property of the exponential function. Does this mean that $f(n)$ can be expressed as an exponential? Euler showed that $f(n)$ is in fact an exponential of an imaginary number. Specifically, he proved that

$$e^{i\phi} = \cos\phi + i\sin\phi. \tag{6.24}$$

This remarkable expression is known as ***Euler's formula***. (A more formal derivation will be given in Chapter 7.) Using the facts that $\cos(-\phi) = \cos\phi$ and $\sin(-\phi) = -\sin\phi$, we can easily demonstrate the following:

$$\cos\phi = \tfrac{1}{2}\left(e^{i\phi} + e^{-i\phi}\right), \qquad \sin\phi = \tfrac{1}{2i}\left(e^{i\phi} - e^{-i\phi}\right). \tag{6.25}$$

Substituting Eq. (6.24) into Eq. (6.15) gives

$$z = re^{i\phi} = |z|e^{i\arg z}, \tag{6.26}$$

which is an especially convenient way to represent complex numbers.

Example 6.5. *Complex numbers of unit length.* Note the following special values on the unit circle of the Argand diagram:

$$
\begin{aligned}
e^0 &= e^{i2\pi} = 1,\\
e^{i\pi} &= e^{-i\pi} = -1,\\
e^{i\pi/2} &= e^{-i3\pi/2} = i,\\
e^{i3\pi/2} &= e^{-i\pi/2} = -i,\\
e^{i\pi/4} &= e^{-i7\pi/4} = \frac{1}{\sqrt{2}}(1+i),\\
e^{i3\pi/4}&=e^{-i7\pi/4} = \frac{1}{\sqrt{2}}(-1+i),\\
e^{i5\pi/4}&=e^{-i3\pi/4} = \frac{1}{\sqrt{2}}(-1-i),\\
e^{i7\pi/4}&=e^{-i\pi/4} = \frac{1}{\sqrt{2}}(1-i).
\end{aligned}
$$

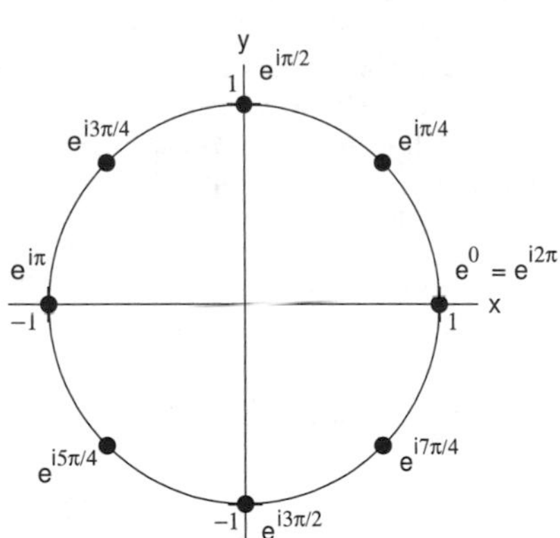

For any integer n,

$$e^{i2n\pi} = \left(e^{i2\pi}\right)^n = 1^n = 1. \tag{6.27}$$

Therefore, multiplying any complex number z by $e^{i2n\pi}$ leaves it unchanged:

$$z = |z|e^{i\arg z} = |z|e^{i\arg z}e^{i2n\pi} = |z|e^{i\arg z + i2n\pi}. \tag{6.28}$$

Euler's formula greatly simplifies algebra involving complex numbers:

Example 6.6. *Calculating a noninteger power of z.* Using the usual rules of exponentiation,

$$z^a = (re^{i\phi})^a = r^a e^{ia\phi} = r^a[\cos(a\phi) + i\sin(a\phi)], \tag{6.29}$$

with $\phi = \arg z$ given by Eq. (6.18) and with $r = |z|$. In particular, the nth root of z corresponds to the choice $a = 1/n$,

$$z^{1/n} = r^{1/n} e^{i\phi/n} = r^{1/n}[\cos(\phi/n) + i\sin(\phi/n)] \tag{6.30}$$

Let us calculate the fifth root of 32. In $\mathbb{R}$ the answer is simply 2, because $2\cdot2\cdot2\cdot2\cdot2 = 32$. In $\mathbb{C}$ the answer is more interesting. $32 = 32e^{i2\pi m}$ for any integer m. Therefore,

$$32^{1/5} = \left(32e^{i2\pi m}\right)^{1/5} = |32|^{1/5} e^{i2\pi m/5} = 2e^{i2\pi m/5}.$$

Thus, $\arg 32^{1/5} = 2\pi m/5$, as follows:

m	0	1	2	3	4	5
$\arg 32^{1/5}$	0	$\frac{2}{5}\pi$	$\frac{4}{5}\pi$	$\frac{6}{5}\pi$	$\frac{8}{5}\pi$	2π
$2e^{i2\pi m/5}$	1	$0.62 + i1.90$	$-1.62 + i1.18$	$-1.62 - i1.18$	$0.62 - i1.90$	1

The pattern repeats with a period of 5, because, for example,

$$2e^{i2\pi 6/5} = 2e^{i2\pi(1+5)/5} = 2e^{i2\pi/5}e^{i2\pi},$$

and $e^{i2\pi} = 1$. The five different solutions are shown at right in an Argand diagram. What we really have is a multiple valued function $z^{1/5}$, with five branches. On each branch, $\sqrt[5]{32}$ has a different value.

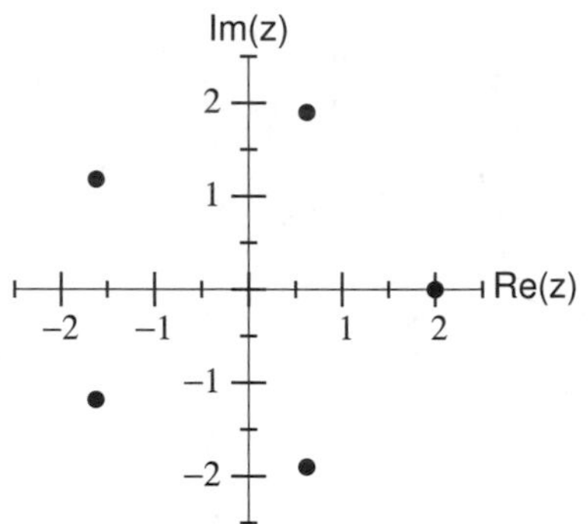

Euler's formula can also sometimes be used to simplify calculations that do not involve complex numbers:

Example 6.7. *Integrals involving circular functions.* Consider

$$I = \int e^{-ax}\cos(bx)\,dx. \tag{6.31}$$

According to Eqs. (6.25), $\cos(bx) = (e^{ibx} + e^{-ibx})/2$. Substituting into the integrand gives

$$I = \frac{1}{2}\int \left[e^{(-a+ib)x} + e^{(-a-ib)x}\right] dx$$

$$= -\frac{1}{2(a-ib)}\,e^{(-a+ib)x} - \frac{1}{2(a+ib)}\,e^{(-a-ib)x} + c. \tag{6.32}$$

Finally, let us derive the natural logarithm of a complex number. Note that any $z \in \mathbb{C}$ can be expressed as

$$z = |z|e^{i\arg z} = e^{\ln|z|}e^{i\arg z} = e^{\ln|z| + i\arg z}.$$

Taking the logarithm of both sides of the equation gives

$$\ln z = \ln|z| + i \arg z. \tag{6.33}$$

Example 6.8. *Calculate the logarithm of* $7 + 4i$.

$$\ln(7 + 4i) = \ln\sqrt{7^2 + 4^2} + i \arctan(4/7) = 2.0872 + 0.5191i$$

Check:

$$\begin{aligned} e^{2.0872+0.5191i} &= e^{2.0872} e^{0.5191i} = (8.0623)(\cos 0.5191 + i \sin 0.5191) \\ &= (8.0623)(0.8683 + 0.4961i) = 7 + 4i. \end{aligned}$$

6.4 Functions of a Complex Variable*

The theoretical framework for functions of a complex variable, a field of mathematics known as *complex analysis*, was developed almost single-handedly by Cauchy in the 1820's and 1830's.[6] Consider a function $f(z)$ where $z \in \mathbb{C}$ and $f(z) \in \mathbb{C}$. Let $z = x + iy$ and $f(z) = P(z) + iQ(z)$, where x, y, P, and Q are pure real. We can define the derivative at a point z_0 as

$$f'(z_0) = \lim_{z \to z_0} \frac{f(z_0 + \Delta z) - f(z_0)}{\Delta z}, \qquad \Delta z = z - z_0 . \tag{6.34}$$

But x and y are independent variables. To define a limit $z \to z_0$ we must specify a path of approach $(x, y) \to (x_0, y_0)$ in the Argand diagram. Similarly, we can define an integral over z as a line integral in the Argand diagram over some specified path $C(u) = \big(x(u), y(u)\big)$,

$$\begin{aligned} \int_C f(z)dz &= \int_C [P(x,y) + iQ(x,y)]\, d(x+iy) \\ &= \int \left[P(x,y)\frac{dx}{du} - Q(x,y)\frac{dy}{du}\right] du + i \int \left[Q(x,y)\frac{dx}{du} + P(x,y)\frac{dy}{du}\right] du. \end{aligned} \tag{6.35}$$

In order for the derivative to be well defined, its value $f'(z_0)$ must depend only on the value of z_0, not on the path of approach, and in order for $f'(z_0)$ to be independent of the path, we must have[7]

$$\frac{\partial P}{\partial x} = \frac{\partial Q}{\partial y} \quad \text{and} \quad \frac{\partial P}{\partial y} = -\frac{\partial Q}{\partial x} . \tag{6.36}$$

These are called the ***Cauchy-Riemann conditions***. They impose a severe restriction on the properties of $f(z)$, and as a result lead to a variety of powerful theorems.

[6] Many of Cauchy's results were first derived by Gauss, but he neglected to publish them.

[7] See Exercise 6.12. These conditions closely resemble Euler's test for exactness, Eq. (4.45), except for the minus sign, which results from the presence of i.

Theorem 6.4.1 (***Cauchy's fundamental theorem of complex integration***). *If a closed path C in the complex plane encloses a region in which f has no singularities, then*

$$\oint_C f(z)\,dz = 0. \tag{6.37}$$

The following is an important consequence:

Theorem 6.4.2. *The value of $\int_{z_1}^{z_2} f(z)\,dz$ is independent of path, as long as the path avoids any singular points.*

A surprising result is the following:

Theorem 6.4.3 (***Cauchy's fundamental formula***). *If a closed path in the complex plane encloses a region in which f has no singularities, and z_0 is a point within the region, then*

$$f(z_0) = \frac{1}{2\pi i}\oint_C \frac{f(z)}{z - z_0}\,dz. \tag{6.38}$$

Recall that a *pole* was defined in Chapter 1 as a singular point at which a function becomes infinite in proportion to $1/(x - x_0)^n$ where n is a positive integer. We can state this more precisely as follows:

Definition: The point $z = z_0$ is a ***pole of order n*** of a function $f(z)$, where n is an integer, if the limit

$$\rho = \lim_{z\to z_0} (z - z_0)^n f(z) \tag{6.39}$$

exists and is not zero or infinity. The value ρ is called the ***residue*** of the pole.

Example 6.9. *Residues of poles.* Consider $f(z) = (1 + 2z)/(3 - 2z^2)$. This is infinite at the points $z = \pm\sqrt{3/2}$. Using the fact that $3 - 2z^2 = 2(\sqrt{3/2} - z)(\sqrt{3/2} + z)$, we find that the pole is of first order and its residue at $+\sqrt{3/2}$ is

$$\rho = \lim_{z\to\sqrt{3/2}} \left(z - \sqrt{3/2}\right)\frac{1 + 2z}{3 - 2z^2} = \lim_{z\to\sqrt{3/2}} -\frac{1 + 2z}{2(\sqrt{3/2} + z)} = -\frac{1 + 2\sqrt{3/2}}{4\sqrt{3/2}}.$$

Consider $g(z) = (1 + 2z)/(3 - 2z)^2$. This is infinite at $z = 3/2$. After multiplying by $(z - 3/2)$ we still get a limit of ∞. We must multiply by $(z - 3/2)^2$ to get a finite nonzero residue, $\rho = \lim_{z\to 3/2} \frac{1}{4}(1 + 2z) = 1$. This is a second-order pole.

From Eq. (6.38) one can derive the following useful theorem:

Theorem 6.4.4 (***Cauchy's residue theorem***). *If $f(z)$ has one or more isolated*[8] *first-order poles within a closed region C but is otherwise nonsingular, then*

$$\oint_C f(z)\,dz = 2\pi i \sum_k \rho_k\,, \tag{6.40}$$

summing over the various poles, with ρ_k the residue of the kth pole.

[8] An pole is *isolated* if it is surrounded by a neighborhood in which the function is nonsingular.

The residue theorem is a valuable technique for solving definite integrals.

Example 6.10. *Applying the residue theorem.* Let us evaluate

$$I = \int_{-\infty}^{\infty} \frac{1+3x}{x^4+5x^2+4}\,dx.$$

Consider the cyclic integral $I_R + J_R$, where I_R is the integral from $-R$ to R along the real axis while J_R is the integral on the contour that closes a semicircular region:

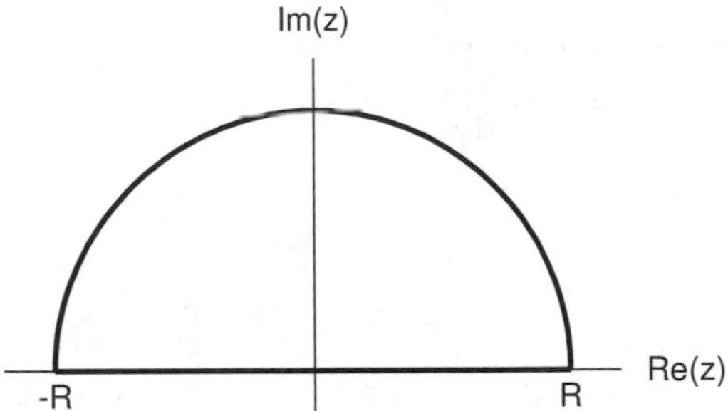

I is equal to $\lim_{R\to\infty} I_R$. The denominator of the integrand is

$$(x^2+1)(x^2+4) = (x+i)(x-i)(x+2i)(x-2i).$$

Within the contour we have a pole at i with residue $\rho_1 = (1+3i)/6i$ and a pole at $2i$ with residue $\rho_2 = -(1+6i)/12i$. $I_R + J_R$ (for large enough R) is, according to the residue theorem, equal to $2\pi i(\rho_1+\rho_2) = \pi/6$. Note that

$$J_R = \int \frac{1+3Re^{i\theta}}{R^4e^{4i\theta}+5R^2e^{2i\theta}+4} iRe^{i\theta}\,d\theta = \frac{1}{R^3}\int \frac{1/R+3e^{i\theta}}{e^{4i\theta}+5e^{2i\theta}/R^2+4/R^4} iRe^{i\theta}\,d\theta,$$

which goes to zero for $R\to\infty$. Therefore, $I = \pi/6$.

6.5 Branch Cuts*

Any complex number z can be specified with a value of its argument in the range

$$0 \le \arg z < 2\pi.$$

This is because $\cos(\phi+2\pi) = \cos\phi$ and $\sin(\phi+2\pi) = \sin\phi$, which imply that

$$\cos(\phi+2\pi) + i\sin(\phi+2\pi) = \cos\phi + i\sin\phi, \tag{6.41}$$

and $e^{i(\phi+2\pi)} = e^{i\phi}$, and, for that matter,

$$e^{i(\phi+2\pi n)} = e^{i\phi} \tag{6.42}$$

for any integer n. For example, the number $2+2i$ can be expressed as $2e^{i\pi/4}$, or $2e^{i(2\pi+\pi/4)}$, or as $2e^{i(2n\pi+\pi/4)}$. We can add any integer multiple of 2π to ϕ without changing the value of z. A better formulation of Euler's representation of z, Eq. (6.26), is therefore

$$z = |z|e^{i(2n\pi+\phi)}, \tag{6.43}$$

with

$$\arg z = \phi + 2n\pi. \tag{6.44}$$

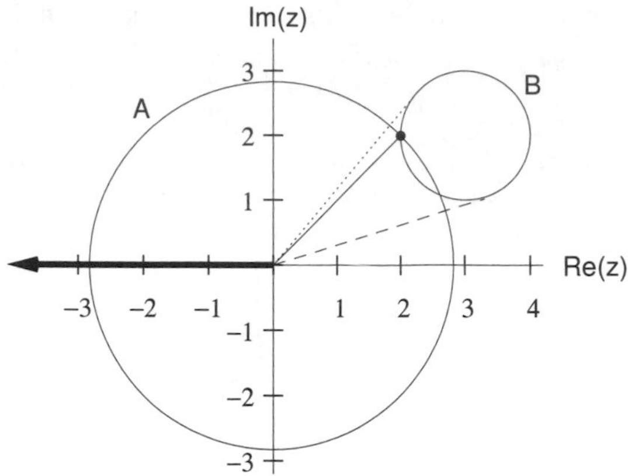

Figure 6.3: Two paths (labeled "A" and "B") in the complex plane that begin and end at $z = 2 + 2i$, with a branch cut corresponding to $-\pi < \arg z < \pi$. The dashed and dotted lines, respectively, connect the origin to the points on path B with the minimum and maximum values of $\arg z$.

While the addition of $2n\pi$ to the argument has no effect on the value of z, it does affect the value of $z^{1/5}$, as we saw in Example 6.6. It also affects the value of $\ln z$. Deriving the logarithm from Eq. (6.43) instead of from Eq. (6.26) gives

$$\ln z = \ln |z| + i(2n\pi + \phi)\,. \tag{6.45}$$

For $n = 0$ we have $\ln(2 + 2i) = \ln 2 + i\pi/4$ but for $n = 1$ we have $\ln(2 + 2i) = \ln 2 + i9\pi/4$, which are different numbers.

We can avoid this ambiguity by restricting $\arg z$ to some interval of length 2π. Suppose, for example, that we choose the interval $-\pi < \arg z < \pi$. We can indicate this interval on the Argand diagram by drawing a thick line from 0 to $-\infty$, as in Fig. 6.3, which will serve as a barrier. As long as we agree to avoid any paths that cross the barrier, then the function will be single valued. This is an example of a ***branch cut***, an artifical barrier in the complex plane that keeps a function single valued. For the function $\ln z$, any continuous curve between the singular point $z = 0$ and any point with infinite absolute value would serve.

Let us follow the function $\ln z$ as z moves from $z = 2 + 2i$ counterclockwise along the circle (path A) centered at the origin, returning to the original point. If we started out with $\arg z = \pi/4$ and $\ln z = \ln 2 + i\pi/4$, then when we return to the original point we have $\arg z = 9\pi/4$ and $\ln z = \ln 2 + i9\pi/4$. Now compare this with path B in Fig. 6.3, which is a circle of unit radius centered at $3 + 2i$. Moving counterclockwise, $\arg z$ at first decreases, reaching a minimum of 18° at the point indicated by the dashed line, then increases to a maximum of 50° at the point indicated by the dotted line, and then decreases back to the original argument of 45°. At the completion of path B the value of $\arg z$ has the same value as at the beginning, which means that the value of $\ln z$ has not changed.

Whenever a path crosses the branch cut we consider the function to have moved onto another ***branch***. The point $z = 0$, where the branch cut begins, is a ***branch point*** of $\ln z$. It is a point at which different branches come together. This is a kind of singular point because the derivative at a branch point is undefined. As a path in the complex plane passes through the branch point the function can switch between any of its branches. The derivative cannot then uniquely predict df from dz. The branch of the logarithm that corresponds to $n = 0$ is called the ***principal value***[9] of $\ln z$. The logarithm has an infinite number of branches, with a different branch for each integer n.

In contrast, the square root $z^{1/2}$ of a complex number has just two branches. To see this, express z in exponential notation. Then

$$\begin{aligned} z^{1/2} &= |z|^{1/2} e^{i(2n\pi+\phi)/2} \\ &= e^{in\pi} |z|^{1/2} e^{i\phi/2}. \end{aligned} \tag{6.46}$$

But $e^{in\pi}$ is equal to ± 1, with "+" if n is an even integer and "−" if n is an odd integer. If we start with the positive square root as the principal value and then circle about the branch point on path A of Fig. 6.3, then we end up with the negative of the original value. Another circuit about the branch point returns the function to the positive square root.

Exercises

6.1 Express $\sqrt{2} + \sqrt{2}\, i$ as an exponential.

6.2 Calculate the absolute value and the argument of $(28.2) - (37.6)i$.

6.3 Calculate the imaginary part of $(3 + 4i)/(1 - 2i)$.

6.4 Calculate all of the cube roots of $6.0 + 5.0i$. Verify that they are indeed cube roots by multiplying each solution by itself three times.

6.5 Calculate $z = \log_{10}(7 + 4i)$, and show that your answer is correct by calculating 10^z.

6.6 Calculate $z = \log_{16}(7 + 4i)$, and show that your answer is correct by calculating 16^z.

6.7 (a) Use Euler's formula to derive the identity $\cos(a + ib) = \cos a \cosh b - i \sin a \sinh b$. (b) Derive a similar identity for $\sin(a + ib)$.

6.8 Use Eqs. (6.25) to verify the identity $\sin(\alpha + \beta) = \sin\alpha\, \cos\beta + \cos\alpha\, \sin\beta$.

6.9 Suppose that $(c + id)^{1/3} = e^{i\phi}$. Solve for ϕ as a function of d.

[9]Not to be confused with the Cauchy principal value of an improper integral, which is something rather different.

6.10 The result of the integration in Example 6.7 is pure real. (It involves the sum of two terms that are complex conjugates of each other.) Express the result in a form that does not contain i.

6.11 Evaluate the integral $\int e^{-ax}\sin(bx)\,dx$.

6.12* Derive the Cauchy-Riemann conditions by requiring that a pure real path of approach to the limit give the same result as a pure imaginary path of approach.

6.13* Prove that

$$\frac{\partial^2 f}{\partial x^2} + \frac{\partial^2 f}{\partial y^2} = 0$$

for $f(x+iy) \in \mathbb{C}$ and x, $y \in \mathbb{R}$.

6.14* Derive the residue theorem from Cauchy's fundamental formula.

6.15* Starting at the point $z = 2$ in the Argand diagram, follow the function $z^{2/3}$ along a circle of radius 2 centered at the origin and determine the value of the function after a single full circuit, returning to $z = 2$.

6.16* Consider the function $f(z) = (2 + z^2)^{1/2}$.

(a) f has two branch points. Where are they?

(b) Demonstrate that a straight line, of finite length, connecting the two branch points will serve as a branch cut. You can prove this analytically, or you can show it graphically, with a parametric plot of the points $(\text{Re}f, \text{Im}f)$ where $f = (2 + r^2 e^{2i\phi})^{1/2}$, for various paths in $z = re^{i\phi}$.

6.17* In Section 21.2 we will need to evaluate the integral

$$I = \frac{1}{2\pi}\int_{-\infty}^{\infty} \frac{1}{1/\tau^2 + (\omega - \omega_0)^2}\,d\omega.$$

where τ and ω_0 are real constants. Evaluate this integral using the residue theorem. *(Hint: Start with the change of variable $z = \omega - \omega_0$.)*

6.18* Use the residue theorem to evaluate the integral $\int_{-\infty}^{\infty}(x^4 + 5x^2 + 4)^{-1}dx$.

Chapter 7

Extrapolation

In Section 5.1 we considered the problem of interpolating between known values of a function at a sequence of different points. Here we treat a situation in which we know the function's value and the value of its derivatives at some *single* point, x_0. Given this information, we can *extrapolate* to other points. The result is an expression that is exact at x_0, accurate in the neighborhood of x_0, and gradually less accurate as the distance from x_0 increases.

7.1 Taylor Series

The most commonly used extrapolant has the form

$$f(x) \approx f(x_0) + c_1 (x - x_0) + c_2 (x - x_0)^2 + c_3 (x - x_0)^3 + \cdots + c_n (x - x_0)^n. \quad (7.1)$$

This is called a ***power series***, as it expresses the function in terms of increasing powers of $(x - x_0)$. By construction, the correction terms all disappear in the limit $x \to x_0$. The arbitrary parameter n, the highest power, is called the ***order*** of the series.

We will determine the c_j values by requiring that the derivatives of the function at x_0 all agree with the derivatives of the power series. Taking the first derivatives of both sides of Eq. (7.1) and evaluating at x_0 gives

$$f'(x_0) = \lim_{x \to x_0} \left[c_1 + 2c_2(x - x_0) + 3c_3(x - x_0)^2 + 4c_4(x - x_0)^3 + \cdots\right] = c_1 .$$

Taking higher derivatives of Eq. (7.1) gives

$$\begin{aligned} f''(x_0) &= \lim_{x \to x_0} \big[2c_2 + (3 \cdot 2)c_3(x - x_0) \\ &\qquad + (4 \cdot 3)c_4(x - x_0)^2 + (5 \cdot 4)c_5(x - x_0)^3 + \cdots\big] = 2c_2 , \\ f^{(3)}(x_0) &= \lim_{x \to x_0} \big[(3 \cdot 2)c_3 + (4 \cdot 3 \cdot 2)c_4(x - x_0) \\ &\qquad + (5 \cdot 4 \cdot 3)c_5(x - x_0)^2 + (6 \cdot 5 \cdot 4)c_6(x - x_0)^3 + \cdots\big] = 6c_3 , \end{aligned}$$

and in general, $f^{(j)}(x_0) = j!\, c_j$. Thus, we arrive at a formula known as ***Taylor's theorem***:[1]

$$c_j = \frac{1}{j!} \left. \frac{d^j f}{dx^j} \right|_{x = x_0} . \quad (7.2)$$

A power series with these coefficients is called a ***Taylor series***.

[1] The English mathematician Brook Taylor (1685-1731) presented this result in his book *Methodus incrementorum directa et inversa*, published (in Latin) in 1715. Due to Taylor's terse writing style, the book was largely unappreciated by his contemporaries. Its significance was not fully realized until many decades after his death. Taylor also wrote articles on topics in physics, physical chemistry, and religious studies.

Table 7.1: Some common Taylor series, expanding about zero.

$$\frac{1}{1-x} \sim \sum_{j=0} x^j = 1 + x + x^2 + x^3 + \cdots. \tag{7.3a}$$

$$e^x \sim \sum_{j=0} \frac{1}{j!} x^j = 1 + x + \tfrac{1}{2!}x^2 + \tfrac{1}{3!}x^3 + \tfrac{1}{4!}x^4 \cdots. \tag{7.3b}$$

$$\ln(1+x) \sim \sum_{j=1} (-1)^{j+1} \tfrac{1}{j}\, x^j = x - \tfrac{1}{2}x^2 + \tfrac{1}{3}x^3 - \tfrac{1}{4}x^4 + \cdots. \tag{7.3c}$$

$$\cos x \sim \sum_{j=0} \frac{(-1)^j}{(2j)!}\, x^{2j} = 1 - \tfrac{1}{2!}x^2 + \tfrac{1}{4!}x^4 - \cdots. \tag{7.3d}$$

$$\sin x \sim \sum_{j=1} \frac{(-1)^j}{(2j+1)!}\, x^{2j+1} = x - \tfrac{1}{3!}x^3 + \tfrac{1}{5!}x^5 - \cdots. \tag{7.3e}$$

$$(1+x)^p \sim \sum_{j=0} \binom{p}{j} x^j, \qquad \binom{p}{j} = \frac{p(p-1)\cdots(p-j+1)}{j!} \tag{7.3f}$$

(This holds for noninteger p as well as well as for integer p.)

Table 7.1 shows some frequently encountered Taylor series. They can be derived using Taylor's theorem to expand about $x_0 = 0$. (The expansion variable is $x - x_0 = x$.) Other series can be derived through substitutions, and this is usually easier than using Taylor's theorem directly.

Example 7.1. *Taylor series related to* $(1-x)^{-1}$. Replacing x with $-x$ in Eq. (7.3a) gives[2]

$$\frac{1}{1+x} \sim \sum_{j=0} (-1)^j x^j = 1 - x + x^2 - x^3 + \cdots. \tag{7.4}$$

The function $(2-x)^{-1}$ can be expanded about $x_0 = 0$ as follows:

$$\frac{1}{2-x} = \frac{1}{2(1-\frac{x}{2})} \sim \frac{1}{2}\sum_{j=0} \left(\frac{x}{2}\right)^j, \tag{7.5}$$

using Eq. (7.3a) with $x/2$ substituted for x. To expand $(1-x)^{-1}$ about an arbitrary x_0, we can use a similar procedure. Let us expand about $x_0 = 0.5$:

$$\frac{1}{1-x} = \frac{1}{1-0.5-(x-0.5)} = \frac{1}{0.5-\delta} = \frac{2}{1-2\delta} \sim 2\sum_{j=0}(2\delta)^j \sim 2\sum_{j=0} 2^j (x-0.5)^j, \tag{7.6}$$

with $\delta = x - x_0$.

Example 7.2. *Taylor series of* $\sqrt{1+x}$ *about* $x_0 = 0$. Use Eq. (7.3f) with $p = 1/2$,

$$(1+x)^{1/2} \sim 1 + \tfrac{1}{2}x - \tfrac{1}{8}x^2 + \tfrac{1}{16}x^3 - \tfrac{5}{128}x^4 + \cdots, \tag{7.7}$$

using

$$\binom{1/2}{1} = \tfrac{1}{2}, \quad \binom{1/2}{2} = \frac{\frac{1}{2}\left(-\frac{1}{2}\right)}{2} = -\frac{1}{8}, \quad \binom{1/2}{3} = \frac{\frac{1}{2}\left(-\frac{1}{2}\right)\left(-\frac{3}{2}\right)}{3\cdot 2} = \frac{1}{16}, \quad \text{etc.}$$

[2]The symbol "$\sim$" is used instead of "$\approx$" to indicate a Taylor series.

Example 7.3. *Taylor series related to* e^x. A power of 10, expanded about $x_0 = 0$:

$$10^x = \left(e^{\ln 10}\right)^x = e^{(\ln 10)x} \sim \sum_{j=0} \frac{1}{j!}(\ln 10)^j x^j = 1 + (\ln 10)x + \tfrac{1}{2!}(\ln 10)^2 x^2 + \cdots . \tag{7.8}$$

Expanding about an arbitrary point, for example, $x = 0.5$:

$$e^x = e^{0.5+(x-0.5)} = e^{0.5+\delta} = e^{0.5}e^{\delta} \sim e^{0.5}\sum_{j=0}\frac{1}{j!}\delta^j = e^{0.5}\sum_{j=0}\frac{1}{j!}(x-0.5)^j .$$

Hyperbolic functions, expanded about $x_0 = 0$:

$$\cosh x = \tfrac{1}{2}(e^x + e^{-x}) \sim \frac{1}{2}\left(\sum_{j=0}\frac{1}{j!}x^j\right) + \frac{1}{2}\left[\sum_{j=0}\frac{1}{j!}(-x)^j\right]$$

$$= \frac{1}{2}\sum_{j=0}\frac{1}{j!}\left[x^j + (-x)^j\right] = 1 + \tfrac{1}{2!}x^2 + \tfrac{1}{4!}x^4 + \cdots = \sum_{k=0}\frac{1}{(2k)!}x^{2k}. \tag{7.9}$$

Taylor series can be added, subtracted, and multiplied, as long as the final result is expanded in powers of $\delta = x - x_0$ and care is taken to omit terms of degree higher than justified by the orders of the original series.

Example 7.4. *Multiplication of Taylor series.* Consider the series about $x_0 = 0$ of the function $e^x/(1-x)$. The second-order series can be calculated as follows:

$$\frac{e^x}{1-x} = e^x\frac{1}{1-x} \sim [1 + x + \tfrac{1}{2}x^2 + \mathcal{O}(x^3)][1 + x + x^2 + \mathcal{O}(x^3)]$$

$$\sim 1 + 2x + \tfrac{5}{2}x^2 + [\tfrac{3}{2}x^3 + 2\mathcal{O}(x^3)] + [\tfrac{1}{2}x^4 + 2x\,\mathcal{O}(x^3)] + \cdots \tag{7.10}$$

$$= 1 + 2x + \tfrac{5}{2}x^2 + \mathcal{O}(x^3)\,. \tag{7.11}$$

Terms of order x^3 and higher, for example, from multiplying $\frac{1}{2}x^2$ by x, are omitted from the final result, because x^3 is the order of the error. This can be seen from Eq. (7.10). The symbol $\mathcal{O}(x^3)$ represents a polynomial, $c_3x^3 + c_4x^4 + c_5x^5 + \cdots$, in which the coefficients are *unknown*. Adding a known quantity such as $\frac{3}{2}x^3$ to an unknown quantity yields an unknown quantity. Therefore, the terms in brackets in Eq. (7.10) represent some unknown error and must be omitted from the final expansion. Eq. (7.11) can also be calculated from Taylor's theorem applied directly to $e^x/(1-x)$, but that is much more tedious than multiplying the two separate series.

Example 7.5. *Expanding the expansion variable.* Consider $\sec^2\theta$. Let us expand it to fourth order about $\theta_0 = 0$. Use the fact that $\sec^2\theta = (\cos^2\theta)^{-1} = (1-\sin^2\theta)^{-1}$. We know that $\sin^2\theta$ goes to zero for $\theta \to 0$, so let us use $\xi = \sin^2\theta$ as an expansion variable:

$$\sec^2\theta = (1-\xi)^{-1} \sim 1 + \sin^2\theta + \sin^4\theta + \cdots .$$

Now substitute in the Taylor series about $\theta_0 = 0$ for $\sin\theta$:

$$\sec^2\theta \sim 1 + (\theta - \tfrac{1}{6}\theta^3 + \cdots)^2 + (\theta - \tfrac{1}{6}\theta^3 + \cdots)^4 + \cdots$$

$$\sim 1 + \theta^2[1 - \tfrac{1}{6}\theta^2 + \mathcal{O}(\theta^4)]^2 + \theta^4[1 + \mathcal{O}(\theta^2)]^4$$

$$\sim 1 + \theta^2[1 - \tfrac{2}{6}\theta^2 + \mathcal{O}(\theta^4)] + \theta^4[1 + \mathcal{O}(\theta^2)]$$

$$\sim 1 + \theta^2 + \tfrac{2}{3}\theta^4 + \mathcal{O}(\theta^6)\,. \tag{7.12}$$

Example 7.6. *Multivariate Taylor series.* Consider $f(x, y)$. Let $x = x_0 + \delta_x$ and $y = y_0 + \delta_y$. The Taylor series in powers of δ_y is

$$f(x_0 + \delta_x, y_0) + \left.\frac{\partial f}{\partial y}\right|_{x_0+\delta_x, y_0} \delta_y + \frac{1}{2!}\left.\frac{\partial^2 f}{\partial y^2}\right|_{x_0+\delta_x, y_0} \delta_y^2 + \frac{1}{3!}\left.\frac{\partial^3 f}{\partial y^3}\right|_{x_0+\delta_x, y_0} \delta_y^3 + \cdots .$$

Taking the Taylor series of this in powers of δ_x leads to

$$f(x, y) \sim f(x_0, y_0) + f^{(1,0)}\delta_x + f^{(0,1)}\delta_y + \frac{1}{2!}\left(f^{(2,0)}\delta_x^2 + 2f^{(1,1)}\delta_x\delta_y + f^{(0,2)}\delta_y^2\right)$$
$$+ \frac{1}{3!}\left(f^{(3,0)}\delta_x^2 + 3f^{(2,1)}\delta_x^2\delta_y + 3f^{(1,2)}\delta_x\delta_y^2 + f^{(0,3)}\delta_y^3\right) + \cdots , \quad (7.13)$$

where

$$f^{(j,k)} = \left.\frac{\partial^{j+k} f}{\partial x^j \partial y^k}\right|_{x=x_0, y=y_0} . \quad (7.14)$$

Instead of using Eq. (7.1) for the extrapolant, we could use the more general expression

$$f(x) = \sum_{j=-m}^{n} c_j\,(x - x_0)^j , \quad (7.15)$$

which includes negative powers of $(x - x_0)$ as well as positive powers. This is called a ***Laurent series***.[3] Suppose, for example, that f has a second-order pole at x_0. Then the appropriate extrapolant would be the Laurent series

$$f(x) \sim \frac{c_{-2}}{(x - x_0)^2} + \frac{c_{-1}}{x - x_0} + c_0 + c_1\,(x - x_0) + c_2\,(x - x_0)^2 + \cdots . \quad (7.16)$$

Example 7.7. *A Laurent series.* Let us expand $e^x/(1 - x)$ about $x_0 = 1$. We cannot use a Taylor series—the first term would be the value of the function at $x = x_0$, which is infinite. The function has a first-order pole at 1. We use a Laurent series of the form

$$\frac{e^x}{1 - x} \sim \frac{c_{-1}}{x - 1} + c_0 + c_1\,(x - 1) + c_2\,(x - 1)^2 + c_3\,(x - 1)^3 + \cdots .$$

Remove the pole by multiplying by $(x - 1)$. This gives a nonsingular function that can be expanded in a Taylor series,

$$(x - 1)\frac{e^x}{1 - x} = -e^x = -e^{1+(x-1)} = -e\,e^{x-1} \sim -e\sum_{j=0} \frac{1}{j!}(x - 1)^j .$$

We can then divide the Taylor series by $(x - 1)$ to obtain the Laurent series,

$$\frac{e^x}{1 - x} \sim \frac{-e}{x - 1}\sum_{j=0} \frac{1}{j!}(x - 1)^j = \frac{-e}{x - 1} - e - \frac{e}{2}(x - 1) - \frac{e}{6}(x - 1)^2 - \cdots . \quad (7.17)$$

If $f(x)$ is nonsingular in the limit $x \to \infty$, then it can be expanded about infinity. Make the change of variable $\xi = 1/x$, define $g(\xi) = f(1/\xi)$, and

[3] Named after a French military engineer, Pierre Laurent (1813-1854). Despite a favorable review by Cauchy, his paper describing this series approximation was rejected for publication, prompting him to give up mathematical research.

expand about $\xi_0 = 0$ in a Taylor series, $g(\xi) \sim g(0) + c_1\xi + c_2\xi^2 + \cdots$, so that

$$f(x) = g(1/x) \sim g(0) + c_1 x^{-1} + c_2 x^{-2} + c_3 x^{-3} + \cdots . \tag{7.18}$$

Example 7.8. *Expansion about infinity.* Consider $f(x) = (1 + 2x)/(1 + x)$.

$$\frac{1+2x}{1+x} = \frac{\frac{1}{x}+2}{\frac{1}{x}+1} = \frac{2+\xi}{1+\xi} \sim (2+\xi)(1-\xi+\xi^2-\xi^3+\cdots)$$

$$\sim 2 - \xi + \xi^2 - \xi^3 + \cdots = 2 + \sum_{j=1}(-1)^j \left(\frac{1}{x}\right)^j . \tag{7.19}$$

Example 7.9. *Stirling's formula as an expansion about infinity.* The gamma function has an ***essential*** singularity at infinity—it goes to infinity faster than any power. However, the function $x^{-(x+1/2)}e^x\Gamma(x+1)$ turns out to approach the finite value $\sqrt{2\pi}$ in the limit $x \to \infty$. Its Taylor series in powers of $1/x$ turns out to be

$$x^{-(x+1/2)}e^x\Gamma(x+1) \sim \sqrt{2\pi}\left(1 + \frac{1}{12x} + \frac{1}{288x^2} - \frac{139}{51840x^3} + \cdots\right) . \tag{7.20}$$

This leads to Stirling's formula, which was introduced in Section 2.2.

7.2 Partial Sums

The usual motivation for using a Taylor series is the hope that each term in the series will be smaller than its preceding term. This could allow us to truncate the series, ignoring any additional terms, once the desired level of accuracy is achieved. Let us define the ***partial sum*** of the power series as

$$S_n(x) = f(x_0) + \sum_{j=1}^{n} c_j\,(x - x_0)^j , \tag{7.21}$$

which truncates the infinite series at some specified finite order n. Each $c_{n+1}\,(x - x_0)^{n+1}$ is expected to be only a small correction to S_n. Thus, $S_2 = S_1 + c_2\,(x - x_0)^2$ is expected to be more accurate than S_1, $S_3 = S_2 + c_3\,(x - x_0)^3$ is expected to be more accurate than S_2, and so on. However, this expectation is not always realized in practice. If the S_n in the limit $n \to \infty$ approach a finite value, then the Taylor series is said to be ***convergent.*** Otherwise it is said to be ***divergent.***[4]

It is often the case that successive c_j values become harder and harder to calculate, so that it is important to use the smallest number of terms that can give the desired accuracy. The accuracy of $S_n(x)$ strongly depends on the value of x. If $|x - x_0|$ is small then $|(x - x_0)^{n+1}|$ is much smaller than $|(x - x_0)^n|$. For example, if $x_0 = 2$ and $x = 2.1$, then $(x - x_0)^3 = (0.1)^3 = 0.001$ is much smaller than $(0.1)^2 = 0.01$, and $(0.1)^4 = 0.0001$ is an order of magnitude smaller still. The c_j are constants, independent of x. We expect that $S_n(x)$

[4]Divergent series are frequently encountered in scientific applications. However, this is not cause for despair. If handled with care they can be very useful. This topic is treated in detail in Section 7.4.

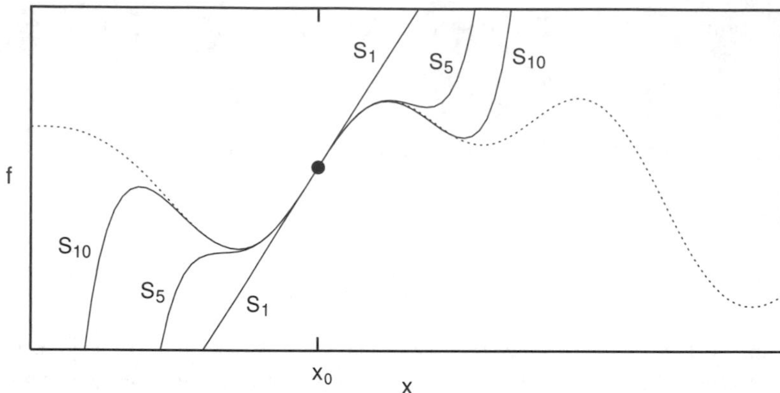

Figure 7.1: Accuracy of partial sums $S_n(x)$ (solid curves), $n = 1, 5, 10$, of a Taylor series about x_0, approximating a function $f(x)$ (dotted curve).

will be most accurate in the vicinity of x_0 and gradually less accurate as the distance from x_0 increases. Increasing the order n should increase the range of x in which $S_n(x)$ is accurate. This is illustrated by Fig. 7.1.

Example 7.10. *Comparison of extrapolation and interpolation.* The Taylor series for e^x at the point 0.5, truncated at second order, is

$$e^{0.5} + e^{0.5}(x - 0.5) + \tfrac{1}{2}e^{0.5}(x - 0.5)^2 = 1.64872 + 1.64872x + 0.824361x^2.$$

Compare this to the Lagrange interpolating polynomial through the three points $(0, e^0)$, $(0.5, e^{0.5})$, $(1, e^1)$,

$$1 + 0.876603x + 0.841679x^2.$$

These two approximations are compared in the following figures:

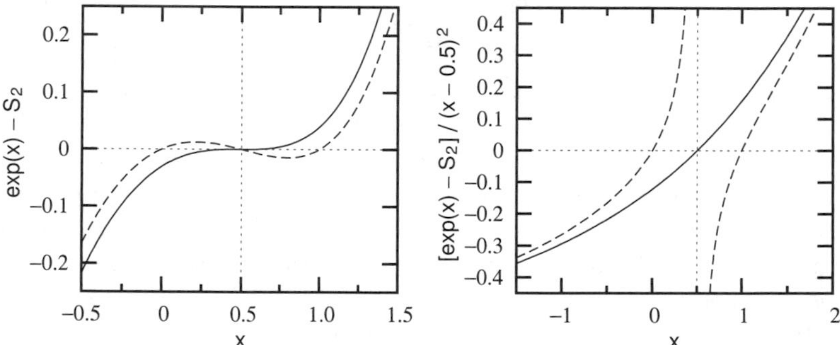

The left-hand figure shows the difference between the actual function and the approximations. In the right-hand figure those differences have been divided by $(x - 0.5)^2$, in order to see more clearly how the approximations behave in the immediate vicinity of 0.5. The solid curves correspond to the Taylor series while the dashed curves correspond to the interpolation. Over most of the range the difference is closer to zero for the interpolation than for the extrapolation. However, in the immediate vicinity of 0.5 the extrapolation is the more accurate. At 0.5 the extrapolation approaches the correct value fast enough that dividing the difference by $(x - 0.5)^2$ still gives zero. The interpolation does not; dividing by $(x - 0.5)^2$ gives infinity.

We can make these notions more precise. To simplify the notation, write

$$f(x) \sim c_0 + c_1\,\delta + c_2\,\delta^2 + c_3\,\delta^3 + \cdots + c_n\delta^n + \mathcal{O}(\delta^{n+1}), \tag{7.22}$$

where $c_0 = f(x_0)$ and $\delta = x - x_0$. The symbol $\mathcal{O}(\delta^{n+1})$ represents the error from truncating the series after the term proportional to δ^n,

$$\mathcal{O}(\delta^{n+1}) = f - S_n\,. \tag{7.23}$$

We can rearrange Eq. (7.22) as

$$\delta^{-n}\big[f(x) - c_0 - c_1\,\delta - c_2\,\delta^2 - c_3\,\delta^3 - \cdots - c_n\delta^n\big] = \delta^{-n}\mathcal{O}(\delta^{n+1}). \tag{7.24}$$

This suggests that we introduce a new relation, ***asymptotic equality*** ($\sim$), as follows:

Definition. Consider a function $f(x)$ and a power series $g = \sum_{j=-m}^{\infty} c_j\,\delta^j$, where $\delta = x - x_0$. Let $S_n = \sum_{j=-m}^{n} c_j\,\delta^j$. If, for all n,

$$\lim_{\delta\to 0} \frac{f(x_0+\delta) - S_n}{\delta^n} = 0, \tag{7.25}$$

then f is ***asymptotic*** to g as δ approaches zero: $\quad f \sim g, \quad \delta \to 0.$

In other words, if we use S_n as an approximation for a function f, the error $\mathcal{O}(\delta^{n+1})$ goes to zero faster than δ^n. It is possible to prove that the Taylor series is the unique power series asymptotic to a given function and that $\mathcal{O}(\delta^{n+1})$ is proportional to δ^{n+1} as δ approaches zero. Taylor and Laurent series are often called ***asymptotic series***.

7.3 Applications of Taylor Series

A common use of Taylor series is to simplify an analytical expression when it is known that some term will be small.

Example 7.11. *Simplifying a functional form.* Suppose that we have derived an equation $f(z) = e^{-z/c} - e^{z/c}$ and we will be applying it only for cases in which z is much smaller than c. Using $e^{\xi} \sim 1 + \xi + \frac{1}{2}\xi^2 + \mathcal{O}(\xi^3)$ with $\xi = z/c$, we find that

$$f(z) \sim [1-\xi+\tfrac{1}{2}\xi^2+\mathcal{O}(\xi^3)]-[1+\xi+\tfrac{1}{2}\xi^2+\mathcal{O}(\xi^3)] \sim -2\xi+\mathcal{O}(\xi^3) \sim -2z/c+\mathcal{O}\big((z/c)^3\big)\,.$$

It might be appropriate for this application to replace $f(z)$ with the simple approximate expression $-2z/c$. The error would be proportional to $(z/c)^3$.

Example 7.12. *Harmonic approximation for diatomic potential energy.* An important example of a Taylor series occurs in the analysis of molecular vibration. The potential energy for the force between atoms in a diatomic molecule is some function $V(R)$ of the distance R between them, but the exact expression for V is unknown. If we assume that the amplitude of the vibration is small, so that R remains close to an "equilibrium bond distance" R_e, then $V(R)$ can be replaced by a Taylor series

$$V(R) \sim V_0 + V_1\,(R - R_e) + V_2\,(R - R_e)^2 + \mathcal{O}\big((R - R_e)^3\big)\,,$$

where V_0, V_1, and V_2 are constants. Define R_e as a minimum of the function $V(R)$.

Taylor's theorem tells us that $V_1 = V'(R_e)$, but because R_e is a minimum, we know that $V'(R_e) = 0$. Thus we arrive at the ***harmonic oscillator***[5] approximation,

$$V(R) \sim V_0 + V_2 (R - R_e)^2, \tag{7.26}$$

which is simple enough that the quantum mechanical problem can be solved exactly for the allowed energy levels. We can treat V_0 and V_2 as empirical constants.

Example 7.13. *Buffers.* Consider an equimolar mixture of a weak acid HA and its conjugate base A^- in aqueous solution. Let K_a be the equilibrium constant for $HA \rightleftharpoons H^+ + A^-$,

$$K_a = \frac{[H^+][A^-]}{[HA]}, \tag{7.27}$$

where [X] is the equilibrium activity of substance X (approximately equal to its concentration divided by 1 mol/L). Taking the base-10 logarithm of both sides and then rearranging, we obtain the ***Henderson-Hasselbalch equation***,

$$\text{pH} = \text{p}K_a + \log_{10}([A^-]/[HA]). \tag{7.28}$$

where $\text{pH}=-\log_{10}[H^+]$ and $\text{p}K_a = -\log_{10} K_a$. Suppose that the initial activities are $[HA]_0 = [A^-]_0 = c_0/(1 \text{ mol/L})$. If we add x moles of a strong base per liter of the buffer solution, then

$$\text{pH} = \text{p}K_a + \log_{10}\frac{c_0 + x}{c_0 - x} = \text{p}K_a + \log_{10}\frac{1+\xi}{1-\xi} = \text{p}K_a + \frac{1}{\ln 10}\ln\frac{1+\xi}{1-\xi}, \tag{7.29}$$

where $\xi = x/c_0$. If x is significantly less than c_0, then $\xi \ll 1$ and Eq. (7.29) can be expanded in a Taylor series,

$$\begin{aligned} \text{pH} &\sim \text{p}K_a + \frac{1}{\ln 10}\ln\{(1+\xi)[1+\xi+\mathcal{O}(\xi^2)]\} \sim \text{p}K_a + \frac{1}{\ln 10}\ln[1+2\xi+\mathcal{O}(\xi^2)] \\ &\sim \text{p}K_a + \frac{2}{\ln 10}\xi + \mathcal{O}(\xi^2). \end{aligned} \tag{7.30}$$

Thus, we predict that pH increases linearly with the rather shallow slope of $2/\ln 10 = 0.87$. As shown below (by the dashed curve), this first-order approximation is quite accurate for $|x/c_0|$ up to around 0.6, with pH varying by less than about ± 0.5 for $-0.6 < x/c_0 < 0.6$. (Negative x corresponds to adding strong acid.)

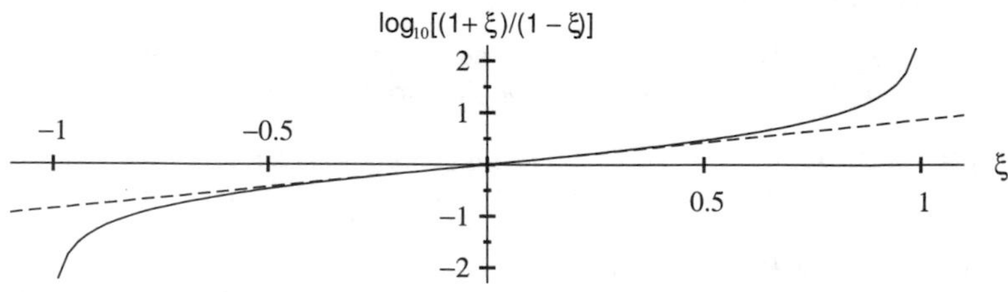

[5]This name comes from the fact that this model was developed, by the French mathematician and theoretical physicist Jean d'Alembert in 1747, to describe the pitches produced by a vibrating string of a musical instrument.

Sometimes a theoretical derivation results in an infinite series that can be identified as the Taylor series of a function and thus immediately summed.

Example 7.14. *Harmonic-oscillator partition function.* The partition function for the quantum-mechanical harmonic oscillator is given by $Q = \sum_{j=0}^{\infty} e^{-\beta(E_j - E_0)}$ with $E_j = \left(j + \frac{1}{2}\right)\hbar\omega$, where β, $\hbar$, and ω are constants. (See Section 8.2.5.) We can write this as

$$Q = \sum_{j=0}^{\infty} e^{-(j+1/2)\beta\hbar\omega + \beta\hbar\omega/2} = \sum_{j=0}^{\infty} e^{-j\beta\hbar\omega} = \sum_{j=0}^{\infty} \xi^j, \qquad \xi = e^{-\beta\hbar\omega},$$

which gives us the Taylor series of $(1-\xi)^{-1}$, according to Eq. (7.3a). Thus we obtain a closed-form solution

$$Q = \frac{1}{1 - e^{-\beta\hbar\omega}}. \tag{7.31}$$

We can use of this kind of reasoning to derive Euler's formula for the exponential of an imaginary number, $e^{i\phi} = \cos\phi + i\sin\phi$. In Section 2.2.1 the exponential function $\exp(x)$ was defined as the inverse of the logarithm, which in turn was defined in terms of an integral, Eq. (2.13). This definition is fine if x is real. However, a more general approach is to define the exponential as the inifinite power series of Eq. (7.3b). Substituting $x = i\phi$ gives

$$e^{i\phi} = \sum_{j=0}^{\infty} \frac{1}{j!}\, i^j \phi^j. \tag{7.32}$$

But i^j can have only four different values, i, $-i$, 1, or -1, according to

$$i^{2j} = (i^2)^j = (-1)^j, \qquad i^{2j+1} = (i^2)^j\, i = (-1)^j\, i.$$

Therefore,

$$e^{i\phi} = \sum_{j=0}^{\infty} \frac{1}{(2j)!}\, i^{2j}\phi^{2j} + \sum_{j=0}^{\infty} \frac{1}{(2j+1)!}\, i^{2j+1}\phi^{2j+1}$$

$$= \sum_{j=0}^{\infty} \frac{(-1)^j}{(2j)!}\,\phi^{2j} + i\sum_{j=1}^{\infty} \frac{(-1)^j}{(2j+1)!}\,\phi^{2j+1} = \cos\phi + i\sin\phi, \tag{7.33}$$

according to Eqs. (7.3d) and (7.3e). Thus, the derivation of one of the most famous formulas of mathematics is almost trivial using Taylor series.[6]

The Taylor series of the exponential can be used to define the exponential of anything,

$$e^A = \sum_{j=0}^{\infty} \frac{1}{j!} A^j, \tag{7.34}$$

where A is anything for which A^j has meaning. In particular, it is sometimes useful to take the exponential of an operator. Suppose that A represents the

---

[6] *Almost.* See Section 7.4.

first derivative operator, $A = \frac{d}{dx}$. We can define $A^2 = A\,A = \frac{d^2}{dx^2}$, $A^3 = \frac{d^3}{dx^3}$, and so on. Then

$$e^{d/dx} = \sum_{j=0}^{\infty} \frac{1}{j!} \frac{d^j}{dx^j}. \tag{7.35}$$

Example 7.15. *Exponential of the first-derivative operator.* We calculate $\exp\left(\frac{d}{dx}\right) f$ for $f = x^2 + 2x + 3$ as follows:

$$\begin{aligned} e^{d/dx}\,(x^2+2x+3) &= \left(1 + \frac{d}{dx} + \frac{1}{2}\frac{d^2}{dx^2}\right)(x^2+2x+3) \\ &= (x^2+2x+3) + (2x+2) + \tfrac{1}{2}\,(2) = x^2 + 4x + 6. \end{aligned}$$

$\exp\left(\frac{d}{dx}\right) f$ is *not* the same as $\exp\left(\frac{df}{dx}\right)$. In this particular case $\exp\left(\frac{df}{dx}\right) = e^{2x+2}$.

7.4 Convergence

Will increasing the order of the series always improve the accuracy? Consider $f(x) = (1+x)^{-1}$ expanded about $x_0 = 0$ and evaluated at three different x values,

$$\begin{aligned} f(0.25) &\sim 1 - 0.25 + 0.0625 - 0.015625 + 0.00390625 - \cdots, \\ f(0.5) &\sim 1 - 0.5 + 0.25 - 0.125 + 0.0625 - \cdots, \\ f(1.1) &\sim 1 - 1.1 + 1.21 - 1.331 + 1.4641 - \cdots. \end{aligned}$$

The partial sums are compared in Table 7.2. At 0.25 the series converges rapidly to the exact result. At 0.5 the series converges, but much more slowly; at third order the error at $x = 0.5$ is 4×10^{-2} compared with just 3×10^{-3} at $x = 0.25$. To reduce the error at 0.5 down to 3×10^{-3} would require going up to order 7. At $x = 1.1$ the partial sums bounce about wildly. This is a divergent series. Going to higher order makes the result less accurate.

There is an important theorem that tells us whether or not a Taylor series will converge at some given point:

Theorem 7.4.1. *Consider a function $f(z)$ for $z \in \mathbb{C}$, and its Taylor series about z_0. Let z_s be the singular point of f closest to z_0 in the complex plane. Let r be the distance between z_0 and z_s, $r = |z_s - z_0|$. The series will converge for any z within the disk $|z - z_0| < r$ and will diverge if $|z - z_0| > r$.*

r is called the ***radius of convergence*** of the series. This explains what went wrong for $f(x) = (1+x)^{-1}$ at $x = 1.1$. $f(z) = (1+z)^{-1}$ has a singularity at $z = -1$, which means the radius of convergence for expansion about $z_0 = 0$ is $r = 1$. The series diverges for any point farther from the origin than this.

The theorem is useful if one knows where the singular points are, but often in scientific applications all one knows about f is its series coefficients. It is possible in principle to estimate the radius of convergence with only this information. The idea is that to have a large radius of convergence the c_n must become steadily smaller with increasing n. The ***d'Alembert ratio test*** states that

$$r \approx |c_n / c_{n+1}|. \tag{7.36}$$

Table 7.2: Partial sums $S_n(x)$ of the Taylor series of $(1+x)^{-1}$ evaluated at different values of x. The last row shows the exact value of the function.

n	$S_n(0.25)$	$S_n(0.5)$	$S_n(1.1)$
0	1	1	1
1	0.75	0.5	−0.1
2	0.8125	0.75	1.11
3	0.79688	0.625	−0.221
4	0.80078	0.6875	1.2431
5	0.79980	0.65625	−0.36741
6	0.80005	0.67188	1.40415
exact	0.8	0.66667	0.47619

Table 7.3: Ratio test for Taylor series of $(1+x)^{1/2}$ about $x=0$.

n	5	10	15	20	25	30	40	50
$\lvert c_n/c_{n+1}\rvert$	1.333	1.158	1.103	1.077	1.061	1.051	1.038	1.030

For the example $(1+x)^{-1} \sim 1 - x + x^2 - x^3 + \cdots$, with $c_n = \pm 1$ for all n, it gives the exact radius of convergence, $r = 1$. However, the ratio test does not always work so well. Consider

$$(1+x)^{1/2} \sim 1 + \tfrac{1}{2}x - \tfrac{1}{8}x^2 + \tfrac{1}{16}x^3 - \tfrac{5}{128}x^4 + \tfrac{7}{256}x^5 - \cdots, \tag{7.37}$$

which also has a singular point at $x = -1$. Its coefficient ratios are shown in Table 7.3. They gradually approach $r = 1$ but at a slow rate. In fact, the precise statement of the ratio test is the following:

$$r = \lim_{n\to\infty} |c_n/c_{n+1}|. \tag{7.38}$$

It always works for very large n, but there is no guarantee it will be accurate at low to moderate n values. Better estimates of singularity locations from low-order series can sometimes be obtained from methods that explicitly model the function's singularity structure, as described in Section 7.5.

The ratio test is most useful when an explicit expression for c_n as a function of n is available. Consider e^x, for which $c_n = 1/n!$. Then

$$r = \lim_{n\to\infty} \frac{(n+1)!}{n!} = \lim_{n\to\infty} (n+1) = \infty. \tag{7.39}$$

Thus we prove that the exponential function has an infinite radius of convergence. e^z has no singularity at any finite value of z in the complex plane.[7]

[7]This fact makes the derivation of Euler's formula given in Section 7.3 a rigorous proof. It implies that $e^{i\theta} = \cos\theta + i\sin\theta$ is valid for any finite value of θ in $\mathbb{C}$.

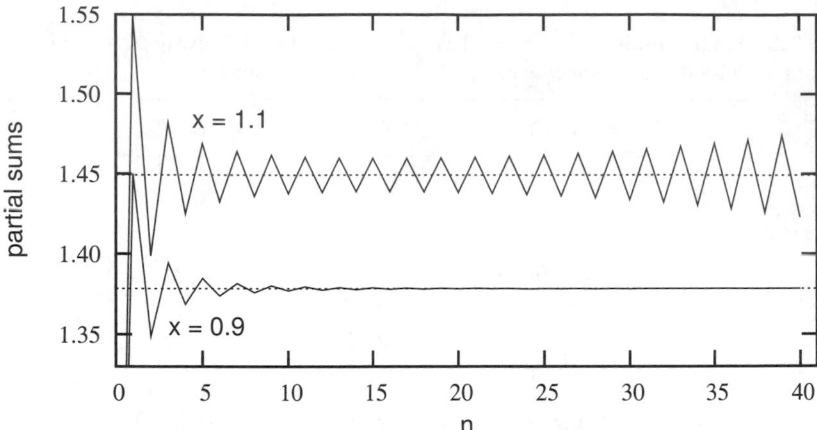

Figure 7.2: Partial sums vs. order n of the Taylor series of the function $(1+x)^{1/2}$ evaluated at $x = 0.9$ and at $x = 1.1$.

Return now to the function $(1+x)^{1/2}$ and consider it at the point $x = 1.1$, which is beyond the radius of convergence. The first few terms are

$$f(1.1) \sim 1 + 0.55 - 0.15 + 0.0831875 - \cdots, \tag{7.40}$$

which gives 1.483 at third order. Comparing this with the correct answer, $\sqrt{2.1} = 1.449$, shows that the series is giving a reasonably accurate result, even though we know from Theorem 7.4.1 that this must be a divergent series, on account of the singularity at $x = -1$. Fig. 7.2 shows the behavior of the partial sum of $(1+x)^{1/2}$ at $x = 1.1$ as a function of order. At low orders it seems to converge but eventually it diverges at high order.

Such behavior is often seen in practice. Divergent partial sums can be useful approximations, but the series order must be kept low enough that the divergence is not yet manifest. A commonly used approach is to take the term of order $n+1$ as an estimate of the error from truncating the series at order n. Here the error estimate $c_{n+1}(1.1)^{n+1}$ reaches a minimum at $n+1 = 16$, and in fact the partial sum is most accurate at 15th order for $x = 1.1$, with divergence beginning at 16th order. The partial sum truncated according to this criterion is called the ***optimal asymptotic approximation***. Despite the name, this "optimal" approximation is not necessarily the optimal summation strategy. Methods described in the following section very often give a more accurate summation.

7.5 Summation Approximants*

The partial sum S_n is a polynomial. A characteristic feature of polynomials is the fact that they have no singular points at any finite value of the variable. It is therefore not surprising that the partial sum has difficulty describing a function that has singularities. A possible strategy for dealing with this situation is to replace S_n with a functional form that has singular points, to more accurately model the singularity structure of the function of interest. Such an extrapolant is called a ***summation approximant***.

The most commonly used summation approximant is the ***Padé approximant***, proposed in 1892 in the doctoral thesis of French mathematician Henri Padé. His idea was to replace S_n with a ratio of polynomials (a *rational* function). For the sake of simplicity, let us just consider a series about $x_0 = 0$. Then

$$S_{[M/N]}(x) = \frac{p_0 + p_1 x + p_2 x^2 + \cdots + p_M x^M}{1 + q_1 x + q_2 x^2 + \cdots + q_N x^N}. \tag{7.41}$$

The index "$[M/N]$" indicates the degrees of the numerator and denominator polynomials, respectively. For example, the [4/2] approximant would be the ratio of a 4th-degree polynomial and a 2nd-degree polynomial.[8] The coefficients $p_0, p_1, p_2, \ldots$ and $q_1, q_2, \ldots$ of the two polynomials are constants, their values determined by requiring that the Taylor series of $S_{[M/N]}(x)$ agree with the Taylor series of the actual function $f(x)$. Note that the function $S_{[M/N]}(x)$ has a pole at each root of the denominator polynomial. If the Taylor series of f is diverging because of a pole, then it should be possible to fit that pole with one of the roots of the denominator and thereby obtain a convergent result. The generalization to a series about arbitrary x_0 is straightforward. On the right-hand side of Eq. (7.41) x would be replaced with $\delta = x - x_0$.

Let us introduce the notation P and Q for the numerator and denominator polynomials, and let F be the truncated nth-order Taylor series of f expanded about the origin,

$$f \sim F = \sum_{j=0}^{n} F_j x^j, \tag{7.42}$$

with coefficients F_j given by Taylor's theorem. The number of parameters in the Padé approximant is $M+N+1$. This means that we will need $M+N+1$ coefficients F_0, F_1,..., F_{M+N} to specify them. For given n, we can choose any M and N such that

$$M + N = n\,. \tag{7.43}$$

A given Taylor series has more than one Padé approximant. For example, a sixth-order series has the six approximants [0/6], [1/5], [2/4], [3/3], [4/2], [5/1]. Typically they give only slightly different values, but the approximants with $M \approx N$ tend to give the best results.

We require that $F \sim P/Q$. Multiplying both sides by Q gives

$$QF - P = \mathcal{O}(x^{M+N+1})\,. \tag{7.44}$$

The left-hand side can be multiplied out to give a polynomial. However, the equality with the right-hand side implies that this polynomial has no terms

[8] The "/" does not represent division! It is in the index to indicate that M is the degree of the polynomial above while N is the degree of the polynomial below. Note that the partial sum S_n is a special case of the Padé approximant, corresponding to index $[n/0]$.

of degree less than x^{M+N+1}. Setting each of the coefficients of the x^j equal to zero for $0 \le j \le M+N$ gives $M+N+1$ linear equations in the $M+N+1$ unknowns, which can be solved using standard methods. (See Section 18.7.)

Example 7.16. [1/1] *Padé approximant.* Start with the second-order Taylor series $F \sim F_0 + F_1 x + F_2 x^2$, with the values of F_j assumed to be known. The [1/1] approximant has the form $(p_0 + p_1 x)/(1 + q_1 x)$.

$$\begin{aligned}\mathcal{O}(x^3) &\sim (1 + q_1 x)(F_0 + F_1 x + F_2 x^2) - (p_0 + p_1 x) \\ &= F_0 - p_0 + (F_1 + F_0\, q_1 - p_1)x + (F_2 + F_1\, q_1)x^2 + \mathcal{O}(x^3)\,.\end{aligned}$$

Therefore, $p_0 = F_0$, $q_1 = -F_2/F_1$, and $p_1 = F_1 - F_0 F_2/F_1$.

Padé approximants provide a simple, easily computed method for improving the results from Taylor series. If the singularity of f that is determining the radius of convergence is indeed a pole, then the improvement in accuracy over partial summation can be quite significant.

But suppose the function has only branch-point singularities, with no poles. In practice, Padé approximants in such cases map out a branch cut with a sequence of nearby poles and zeros. (Zeros of the approximant correspond to roots of the numerator polynomial.) This can work fine as long as the point where the function is to be evaluated is not close to any branch point. However, a more efficient strategy is to design an approximant that actually has branch points. Let us generalize the linear equation, Eq. (7.44), to a quadratic equation,

$$QF^2 - PF + R \sim \mathcal{O}(x^{L+M+N+2})\,, \tag{7.45}$$

where R is a polynomial of degree L. Solving for F with the quadratic formula, we find it is asymptotically equal to a ***quadratic approximant***,

$$S_{[L,M/N]} = \frac{1}{2Q}\left(P \pm D^{1/2}\right), \tag{7.46}$$

where

$$D = P^2 - 4QR\,. \tag{7.47}$$

D is a polynomial. Its degree n_D is the maximum of $2M$ and $N + L$, and at each of the n_D roots of D, the approximant $S_{[L,M/N]}$ can have a square-root branch point.[9] It was also Padé who invented quadratic approximants and, in general, ***algebraic approximants*** based on polynomial equations in F of arbitrary degree, but the rational approximant, P/Q, was the only kind he studied in detail. Therefore, the term "Padé approximant" is usually reserved for the rational case.

[9]This is true for unique roots of D. If a root x_1 occurs as a factor $(x - x_1)^2$ then on taking the square root in Eq. (7.46) it becomes $(x - x_1)$, which is nonsingular at x_1.

Exercises

7.1 For each of the following calculate the second-order Taylor series about 0.

(a) e^{x+2} (b) e^{-x^2} (c) $\frac{\sin \pi x}{x}$ (d) $\frac{\sin \pi x}{x-3}$

7.2 Repeat Exercise 7.1 but expand instead about $x_0 = 3$.

7.3 Prove that Eq. (7.3a) is a special case of Eq. (7.3f).

7.4 Calculate the fourth-order Taylor series of $(1+x)^{-1/2}$ about $x_0 = 0$.

7.5 Prove that $2 \sin\theta \cos\theta = \sin(2\theta)$ by expanding both sides in infinite Taylor series.

7.6 (a) Use the fact that $\arctan x = \int_0^x \frac{dt}{1+t^2}$ to derive the Taylor series for the arctangent about $x_0 = 0$ by expanding the integrand. (b) Use the ratio test to determine the radius of convergence. (c) Where are the singular points that determine the radius of convergence?

7.7 (a) Expand $(1-x-y)^{-1}$ in a multivariate Taylor series about $x = y = 0$. *[Hint: Let $\delta = x + y$ and then use Eq. (7.3a). In this case, this is much easier than using the more general approach of Example 7.6.]* (b) Give the series coefficient of $x^m y^n$ for arbitrary m and n.

7.8 Calculate the Laurent series of $\cot^2\theta$ about $\theta_0 = 0$ through order θ^2.

7.9 The Morse potential, $V(R) = D[1 - e^{-a(R-R_e)}]^2$, is often used as an approximate model for the potential energy of a diatomic molecule. It is characterized by three parameters: the equilibrium bond distance R_e, the well depth D, and the parameter a that determines the width of the well.

(a) Calculate the Taylor series to second order in $(R - R_e)$.

(b) Plot V and its second-order Taylor series (the harmonic oscillator approximation) on a single graph. Why might you expect the Morse potential to give a more accurate description of a molecule than the harmonic oscillator?

(c) Show that D is the difference in potential energy between the bottom of the well and the potential energy at infinite separation.

(d) Values for D and for $V_2 = V''(R_e)$, the second derivative of the potential energy at the minimum, can in principle be estimated from an analysis of the molecule's absorption spectrum. Derive an expression for a in terms of these two quantities.

7.10 Expand $(1+x^2)/(1+x-x^2)$ about infinity to $\mathcal{O}(x^{-4})$.

7.11 Sum exactly the infinite series $e^{-ax}e^{d/dx}e^{ax}$ for arbitrary constant a.

7.12 Use Taylor's theorem and L'Hospital's rule to prove that $f(x) + e^{-1/x}$ has the same series about $x_0 = 0$ as does $f(x)$.

7.13* (a) Calculate the Taylor series of $\ln x$ about the point $x_0 = 2$ through fourth order. (b) What is the radius of convergence of this series? (c) Sum the series for various different values of x using a rational Padé approximant and using a quadratic approximant, and compare with the result from partial summation and with the exact result. (d) Estimate the radius of convergence using the ratio test, the nth-root test, and singularity analysis of the two approximants.

7.14* In Section 4.4 the claim was made that

$$g(x,a) = \frac{\sin(x/a)}{\pi x}$$

in the limit $a \to 0$ was a representation of the Dirac delta function. Show that

$$\int_{-\infty}^{\infty} g(x,a)dx = 1$$

for any nonzero positive a, which is one of the requirements for a representation of $\delta(x)$.

(Hint: Make the change of variable $u = x/a$. Then use the residue theorem with a small semicircle of radius r to divert the integration contour around the singular point $u = 0$. Approximate the integrand on this semicircle with a Taylor series for the sine, take the limit $r \to 0$, and then perform the integration.)

Part II

Statistics

Chapter 8

Estimation

A basic goal of science is to infer mathematical equations from physical measurements. This can be complicated by the fact that measurements usually have some amount of upredictable error. To analyze the effects of this we must introduce the concept of probability. This chapter presents an introduction to probability theory and addresses the question of how to extract a reliable estimate of a physical property from repeated measurements.

8.1 Error and Estimation

Error can be classified in two categories: ***random*** and ***systematic***.[1] If there is very little random error, we say the measurements are ***precise***. If there is very little systematic error, we say the measurements are ***accurate***. Random error can be reduced by repeating the measurement and taking an average. The assumption is that deviations in one direction are as likely as in any other, and therefore cancel out if the measurement is repeated many times. In contrast, systematic errors cause a bias in one particular direction that will not disappear with averaging.

Random error is usually assumed to be the result of a huge number of small deviations, such as thermal motions of atoms and molecules. It may be that these deviations are not truly random. Individual molecules follow trajectories that could in principle be predictable; however, there are so many of them that the net effect of the deviations, when observed on a macroscopic scale, is in practice unpredictable and without particular bias.

Given N measurements x_1, x_2, x_3, …, x_N of a physical property x, the most commonly used estimator is the ***mean***,

$$\bar{x} = (x_1 + x_2 + x_3 + \cdots + x_N)/N = \frac{1}{N}\sum_{j=1}^{N} x_j \,. \tag{8.1}$$

The mean is also called the ***average***. This is not the only possible estimator. Often the ***median*** is a more reliable estimator. It is defined as the middle value of the set—in other words, the value for which an equal number of measurements lie above as below it. If N is even, one takes the mean of the two middle values. We will use the notation $\operatorname{med} x$ to indicate the median.

[1] Also called ***indeterminate*** and ***determinate*** errors, respectively. The cause of a systematic error and the direction in which it skews the result can, at least in principle, be *determined* by tracing its source to a particular step in the experimental procedure. Although the existence of random error also can be attributed to specific causes, the direction of the resulting error in any given single measurement is impossible to predict.

Example 8.1. *Mean and median.* A property x is measured six times, with the following results:

$$\{237, 229, 323, 231, 233, 235\}$$

Note that the third result is much larger than any of the others. $\bar{x}$ is 248, even though five of the six measurements gave results within the range $229 \leq x \leq 237$. It is reasonable to suspect that 323 is anomolously large (perhaps it should have been 233 but the experimenter inadvertantly interchanged a 2 and a 3) and its presence in the data set significantly skews the mean. $\operatorname{med} x$ is $(233 + 235)/2 = 234$ (the average of the two middle values), which seems a more reasonable estimate of the true value.

Another estimator is the ***mode***, which is defined as the most probable value. One could construct a bar graph of the x_j values by dividing the x-axis into evenly spaced intervals, counting the number of values found in each interval, and then plotting the counts as a function of x. The x value at the center of the interval with the highest count is the mode. The mode is less convenient than the mean and median. It is harder to compute and its value depends on how one has defined the counting intervals. Its primary use is in the case of *bimodal* data distributions, in which the data show two distinct peaks. The two local maxima in the bar graph would give a most probable value and a second-most probable value, while the mean and median would only give a compromise value between the two peaks, without physical significance. This is shown in Fig. 8.1.

There are various ways to measure the amount of variation in a set of measurements. An obvious choice is the average of the magnitudes of the deviations,

$$|\hat{D}| = \frac{1}{N} \sum_{j=1}^{N} |x_j - \hat{\mu}|. \tag{8.2}$$

$\hat{\mu}$ is the mean or the median. $|\hat{D}|$ is called the ***mean absolute deviation***. A more commonly used error estimator is the ***standard deviation***, $\hat{\sigma}$. It is the square root of the ***variance***, $\hat{\sigma}^2$, which is given by[2]

$$\hat{\sigma}^2 = \frac{1}{N-1} \sum_{j=1}^{N} (x_j - \bar{x}\,)^2. \tag{8.3}$$

The use of $\hat{\sigma}$ instead of $|\hat{D}|$ can often be justified theoretically.

Estimators of data *variability*, such as $\hat{\sigma}$ and $|\hat{D}|$, are called ***scale estimators*** while estimators of the *value* of a property, such as $\bar{x}$ and $\operatorname{med} x$, are called ***location estimators***. We will see that the simple estimators that have been presented in this section are not necessarily the most dependable ones for practical applications. This topic will be revisited in Section 8.4.

[2]If the prefactor in Eq. (8.3) were $1/N$ instead of $1/(N-1)$, then $\hat{\sigma}^2$ would just be the average of the sum of the squares of the individual deviations. The reason for using $N-1$ in the prefactor instead of N is subtle. If the true value of the property x is known, then one should replace $\bar{x}$ in Eq. (8.3) with that value and use N instead of $N-1$ in the prefactor. This would apply to a situation in which the goal of the experiment is to study the measurement process itself rather than the property being measured.

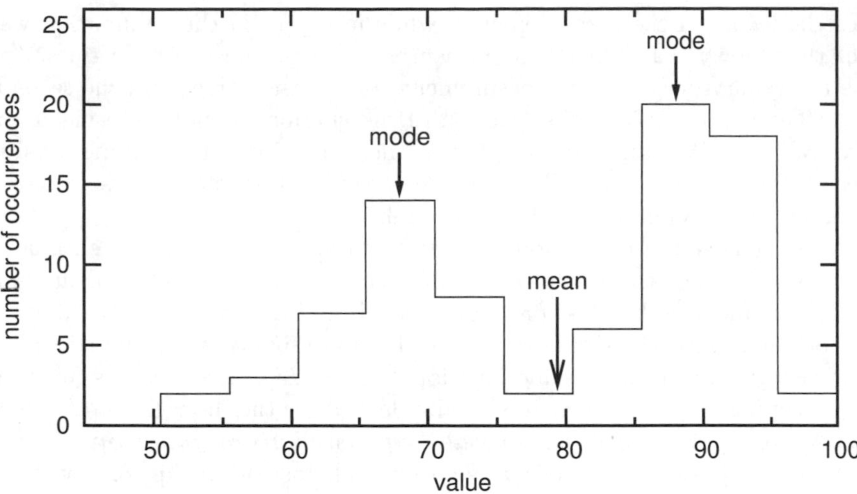

Figure 8.1: Bimodal distribution. The mean is a very poor estimator in this case.

A ***weighted average*** is an estimator of the form

$$\bar{x} = W^{-1} \sum_{j=1}^{N} w_j x_j \,, \qquad W = \sum_{j=1}^{N} w_j \,. \tag{8.4}$$

This allows for more dependable measurements to have a relatively stronger influence. For example, if the precisions of the measurements vary, those that are more precise can be given larger weights. $w_j = \hat{\sigma}_j^{-2}$ is a common choice, as it can often be justified theoretically. If all the w_j are equal, the weighted average reduces to the conventional average.

8.2 Probability Distributions

8.2.1 Probability Distribution Functions

Let us reconsider Eq. (8.1), the definition of the mean. To simplify the analysis, assume for now that x_j can only have discrete values. For example, x_j might be the scores on an examination for a class of N students. The set of data would consist of a collection of integers in the range $[0, 100]$. Let N_x be the number of occurrences of some given value x in the data set. For example, if 12 out of 100 students received a score of 90, then $N_{90} = 12$. One could predict that if the same examination is given the following year, the probability of someone scoring 90 will be $p(90) = 12/100 = 0.12$. Let us write Eq. (8.1) as

$$\bar{x} = \sum_{x=0}^{100} x\, p(x), \qquad p(x) = N_x/N. \tag{8.5}$$

The summation is over all the possible values of x. The two equations for $\bar{x}$ are numerically equivalent. The difference is that now the summation is over the *values* of the measurements, while in Eq. (8.1) the summation was over the index j that labeled the measurements regardless of the values. (x_1 was the value from the first measurement, x_2 was the value from the second, etc.) The ratio N_x/N is the fraction of measurements that have yielded a given value x. We can interpret $p(x)$ as the probability of obtaining a value x in future measurements. Thus, the mean can be thought of as a sum of the observed values weighted by their probabilities.

Let us remove the restriction to discrete data. x will now be a continuous variable. Consider the range of possible values in the infinitesimal interval between x and $x + dx$. The probability of obtaining a value in this range is proportional to dx, the size of the range. Let us designate the proportionality factor for given value x as some function $P(x)$. This is a continuous function of x, as opposed to $p(x)$, which was only defined for the discrete set of values x_1, x_2, etc. $P(x)$ is called a ***probability distribution function***. If x is continuous, the summation becomes an integral. Instead of Eq. (8.5) we have

$$\langle x \rangle = \int_{-\infty}^{\infty} x P(x) dx, \tag{8.6}$$

where $P(x)dx$ replaces $p(x)$. $\langle x \rangle$ is called the ***expectation value*** of x.[3]

Example 8.2. *Expectation value of a function of x.* Let us determine the expectation value of x^2 using a probability distribution of x. The correct way to calculate this is

$$\langle x^2 \rangle = \int_{-\infty}^{\infty} x^2 P(x) dx.$$

It would *not* be correct to compute $\langle x \rangle$ and then square the value; that is, $\langle x^2 \rangle \neq \langle x \rangle^2$. In general, the expectation of any function $f(x)$ of x is

$$\langle f(x) \rangle = \int_{-\infty}^{\infty} f(x)\, P(x) dx. \tag{8.7}$$

If the number N of x_j's is large, then $N - 1 \approx N$, and

$$\hat{\sigma}^2 \approx \frac{1}{N} \sum_{j=1}^{N} (x_j - \bar{x})^2, \tag{8.8}$$

which is the mean of the $(x_j - \bar{x})^2$. For continuous x, the number of x values is certainly large. (It is infinite!) Replacing mean values with expectation values gives

$$\sigma^2 = \left\langle \, (x - \langle x \rangle)^2 \, \right\rangle, \tag{8.9}$$

which defines the variance for a continuous probability distribution.

The range of all possible x values is $(-\infty, \infty)$. The fraction of values in the range of all possible values is of course unity. Therefore the integral of

[3] In this chapter the symbol $\bar{x}$ is used for the discrete average, Eq. (8.1), and $\langle x \rangle$ for the continuous average, Eq. (8.6). However, these symbols are often used interchangeably.

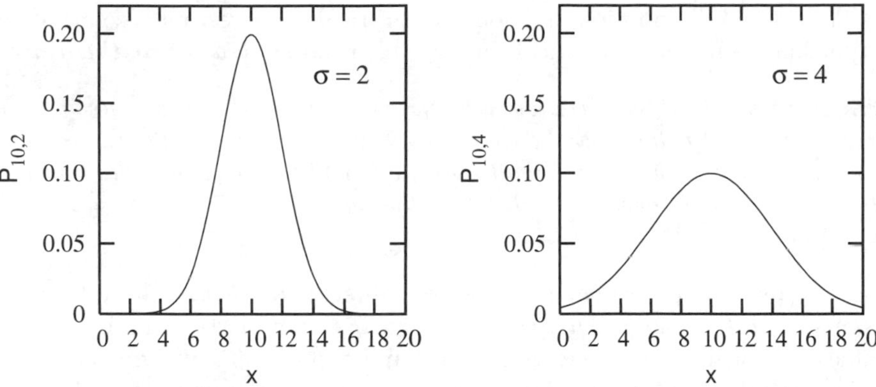

Figure 8.2: Normal distribution functions for two values of the scale σ.

the probability $P(x)dx$ over this range must be unity,

$$\int_{-\infty}^{\infty} P(x)dx = 1, \tag{8.10}$$

and $P(x)$ is said to be ***normalized.***

An important application of a distribution function is to calculate the probability of measuring a value for x within some portion $[x_{\text{min}}, x_{\text{max}}]$ of the full range of all possible values. This probability is obtained by replacing $(-\infty, \infty)$ with $[x_{\text{min}}, x_{\text{max}}]$ in the normalization integral. For example, the probability of a result for x between 46 and 48 is $\int_{46}^{48} P(x)dx$.

8.2.2 The Normal Distribution

The selection of an appropriate distribution function is usually the first step in a statistical analysis. The most common choice is

$$P_{\mu,\sigma}(x) = \frac{1}{\sqrt{2\pi\sigma^2}}\, e^{-(x-\mu)^2/2\sigma^2}. \tag{8.11}$$

μ and σ are parameters. Statisticians call this a ***normal distribution.***[4] Scientists often refer to it as a ***Gaussian distribution.***[5] $P_{\mu,\sigma}(x)$ gives a symmetric bell-shaped curve peaked at $x = \mu$, as shown in Fig. 8.2. The parameter σ is the ***scale*** of the distribution. It can be seen from the graphs that the width of the curve at half the peak height is approximately 2σ. The prefactor in Eq. (8.11) normalizes the distribution. Because the area under each curve in Fig. 8.2 is unity, the peak is higher if the scale is smaller.

[4]We now have two meanings for the word "normal." They should not be confused. What they have in common is the concept of "usual" or "ordinary." It is *usual* to require that a probability distribution be *normalized*, according to Eq. (8.10). This is not necessary but is usually convenient. The *normal distribution* is the *usual* probability distribution, appropriate for most applications.

[5]This usage dates from the early 1800's when Gauss used normal distributions to analyze astronomical data, but this distribution function was originally introduced by de Moivre.

The normal distribution function is of particular interest on account of an important result from statistical theory, called the ***central limit theorem***:

Theorem 8.2.1. *If the deviation in a series of repeated measurements of some given property x is the sum of deviations in many other mutually independent random variables, then in the limit that the number of such variables becomes infinite, the overall deviation will follow the normal distribution $P_{\mu,\sigma}(x)$ with $\mu = \bar{x}$ and $\sigma = \hat{\sigma}$.*[6]

Note the use of diacritical marks above the letter to indicate a value estimated from a *finite* data sample. μ is the true value of the property and σ is the true value of the scale. $\bar{x}$, pronounced "x bar," and $\hat{\sigma}$, pronounced "sigma hat," are estimates calculated from a finite number of measured values. The theorem implies $\bar{x} \to \mu$ and $\hat{\sigma} \to \sigma$ in limit $N \to \infty$.

Example 8.3. *An illustration of the central limit theorem.* Consider an examination consisting of only a single true/false question. Suppose that this question is a "random variable." (The student has not yet taken the course!) Then there will be a 50% probability of a perfect score of 100 points and a 50% chance of a score of 0. Now suppose there are two questions, worth 50 points each. The possible scores are $50 + 50 = 100$, $50 + 0 = 50$, $0 + 50 = 50$, and $0 + 0 = 0$; thus, there is a $1/4 = 25\%$ of scoring 100 or 0 and a $2/4 = 50\%$ chance of a score of 50. The effect on the probability distribution from increasing the number of question, N, is shown in the figures below, which show the relative probability for each possible total score.

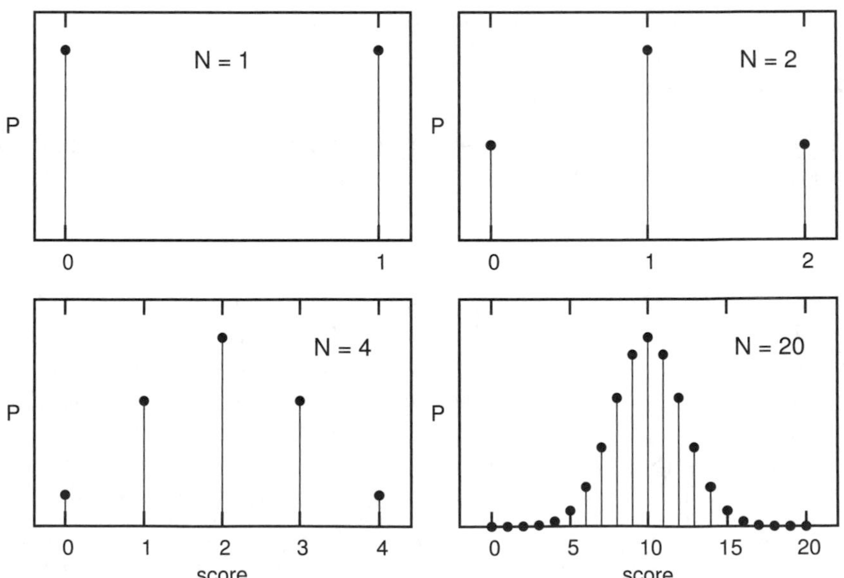

As N increases, P takes on the characteristic bell shape of the normal distribution.

[6]This is not a precise statement of the theorem. The terms "deviation" and "independent" need to be carefully defined, but that is beyond the scope of the present treatment.

Let us use the normal distribution function to calculate the probability of measuring a value of x within a specified range $[a, b]$. Actually, it is impossible to carry out this calculation using elementary functions. We will need a special function, called the ***cumulative normal distribution function***,[7]

$$\Phi(x) = \frac{1}{\sqrt{2\pi}} \int_{-\infty}^{x} e^{-u^2/2} du. \tag{8.12}$$

The probability in question is

$$\int_a^b P_{\mu,\sigma}(x)dx = \int_{-\infty}^{b} P_{\mu,\sigma}(x)dx - \int_{-\infty}^{a} P_{\mu,\sigma}(x)dx.$$

Making the change of integration variable $u = (x-\mu)/\sigma$, $dx = \sigma du$, gives

$$\begin{aligned} \int_{-\infty}^{b} P_{\mu,\sigma}(x)dx &= \frac{1}{\sqrt{2\pi\sigma^2}} \int_{-\infty}^{b} e^{-(x-\mu)^2/2\sigma^2} dx \\ &= \frac{1}{\sqrt{2\pi}} \int_{-\infty}^{(b-\mu)/\sigma} e^{-u^2/2} du = \Phi\left(\frac{b-\mu}{\sigma}\right) \end{aligned} \tag{8.13}$$

and therefore

$$\int_a^b P_{\mu,\sigma}(x)dx = \Phi\left(\frac{b-\mu}{\sigma}\right) - \Phi\left(\frac{a-\mu}{\sigma}\right). \tag{8.14}$$

The central limit theorem is based on the assumption that the deviation of the measured property is the *sum* of the deviations of other unmeasured properties. This will be the case if the measured property x is related to the unmeasured properties ξ_j according to

$$x = \xi_1 + \xi_2 + \xi_3 + \cdots . \tag{8.15}$$

But suppose that the true relationship between x and the ξ_j were

$$x = \xi_1 \xi_2 \xi_3 \cdots . \tag{8.16}$$

This would invalidate the theorem and in principle would make the normal distribution inappropriate for analyzing the data.

Let us examine the consequences of Eq. (8.16). The product can be converted to a sum by taking the logarithm of both sides:

$$\ln x = \ln \xi_1 + \ln \xi_2 + \ln \xi_3 + \cdots ,$$

or simply

$$z = \zeta_1 + \zeta_2 + \zeta_3 + \cdots , \qquad z = \ln x, \quad \zeta_j = \ln \xi_j .$$

[7] $\Phi(x)$ can be expressed in terms of the error function or the incomplete gamma function, as discussed in Section 2.2.3.

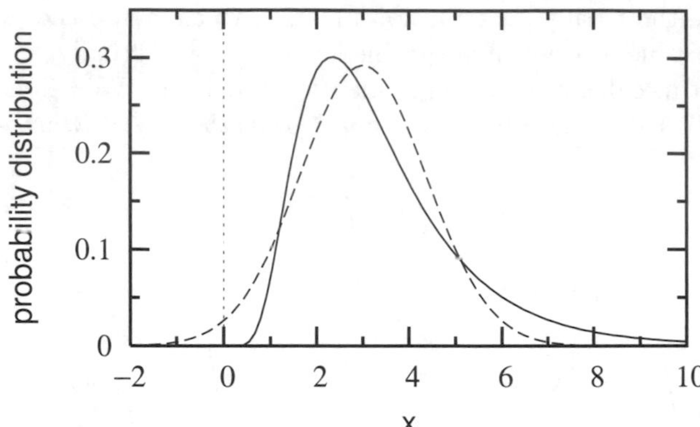

Figure 8.3: Comparison log-normal (solid curve) and normal (dashed curve) probability distribution functions.

z is a sum of the unmeasured properties ζ_j and therefore *is* described by a normal distribution. The distribution for x is obtained by changing the variable from z to x in the probability factor $P(z)dz$,

$$P_{\mu_z,\sigma_z}(z)dz = P_{\mu_z,\sigma_z}(\ln x)\,\frac{1}{x}dx, \tag{8.17}$$

which implies that the probability distribution for x is

$$P^{(LN)}_{\mu_z,\sigma_z}(x) = \frac{1}{x\sqrt{2\pi\sigma_z^2}}e^{(\ln x-\mu_z)^2/2\sigma_z^2}. \tag{8.18}$$

This is called the ***log-normal distribution***. In practice, the log-normal analysis is accomplished by taking logarithms of the measured data and then analyzing the $z_k = \ln x_k$ using methods based on the normal distribution. At the end of the analysis, the estimations for z are transformed back to x

Figure 8.3 compares shapes of the normal and the log-normal distributions. The most obvious difference is that the log-normal distribution is asymmetric, skewed toward lower values. Another difference is that the log-normal distribution begins at $x = 0$ while the normal distribution begins at $x = -\infty$, because $\ln 0 = -\infty$. A chemical concentration of course cannot be negative. This would seem to suggest that the normal distribution should not be used for concentration measurements, because it gives a nonzero probability for concentrations less than zero. This is not usually a problem in practice. As long as the concentrations are well above the detection limit of the measuring instrument and the standard deviation is not too large, then the normal probability of a negative value is vanishingly small. The choice of the probability distribution can be based on theoretical considerations or on an empirical analysis of the data (see Section 9.5). The log-normal distribution is very often used in trace analysis of environmental samples.[8]

[8]This is based largely on empirical experience. See, however, W. Ott, "A Physical Explanation of the Lognormality of Pollutant Concentrations," *J. Air Waste Manage. Assoc.* **40**, 1378-1383 (1990).

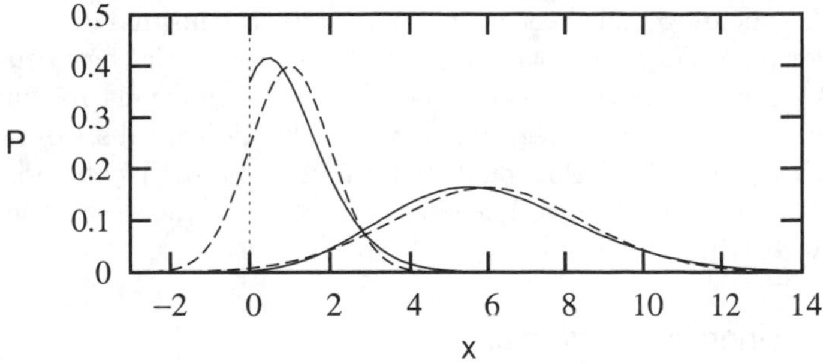

Figure 8.4: Comparison of Poisson distributions (solid curves) and normal distributions (dashed curves) with mean of 1 or 6.

8.2.3 The Poisson Distribution

Some processes, such as radioactive decay, which involve a counting of discrete events that are very rare, are not accurately described by a continuous distribution function. Consider the following question: Given a sample of radioactive material, what is the probability that n nuclear decays will occur in a specified time period? Suppose we know that the *average* number of decays expected in the time period is $\bar{n}$. Then the observed decays will follow a ***Poisson distribution***,[9]

$$p_{\bar{n}}(n) = \frac{1}{n!}\, e^{-\bar{n}}\, \bar{n}^{\,n}. \tag{8.19}$$

Example 8.4. *Radioactive decay probability.* While disassembling a nuclear warhead you accidentally inhale dust containing plutonium oxide, and 200,000 atoms of ^{239}Pu lodge in your lungs. The average number of decays expected from 200,000 ^{239}Pu atoms is $\bar{n} = 0.479$ per month with each decay emitting an α particle. What is the probability that no α particles will be emitted into your lungs during the next month?

The probability of no decays is $p_{0.479}(0) = e^{-0.479} = 0.619$, or 61.9%. This means there is a 38.1% chance that at least one α particle *will* be emitted.

What is the probability no more than three α particles will be emitted in one month?

$$p_{0.479}(0) + p_{0.479}(1) + p_{0.479}(2) + p_{0.479}(3)$$
$$= e^{-0.479}\left[1 + 0.479 + \tfrac{1}{2}(0.479)^2 + \tfrac{1}{6}(0.479)^3\right] = 0.998.$$

The Poisson distribution also describes such phenomena as the rate of mutation in a DNA sequence or the probability of observing a star in some given region of the sky. The distinguishing features are that these phenomena are inherently discrete (a nuclear decay either happens or it doesn't—there is no such thing as a fractional decay) and that the average n value is small.

Figure 8.4 compares Poisson distributions with normal distributions, with $\bar{n} = 1$ or 6. (It is possible to prove that the variance of a Poisson distribution

[9]The French mathematician Siméon-Denis Poisson (1781-1840) was a leading practitioner of applied mathematics, making contributions to theoretical chemistry, physics, and sociology. This probability distribution was presented in his book, *Research on the Probability of Judgments in Criminal and Civil Matters*, published in 1838.

is equal to its mean. σ^2 has been set equal to $\bar{n}$ for the normal distributions in the figure for purposes of comparison.) An obvious difference between the normal and Poisson distributions is that the normal distributions are symmetric, extending into the range of negative n, while the Poisson distributions stop at $n = 0$. The peaks of the Poisson distributions come slightly before $\bar{n}$, while the normal distribution is exactly centered at $\bar{n}$. As $\bar{n}$ increases, the Poisson and normal distributions become quite similar.

8.2.4 The Binomial Distribution*

The Poisson distribution is an example of a *discrete* distribution, in that the variable being described by the distribution can take on only discrete values. We now discuss such distributions in more detail. Consider an experiment with only two possible outcomes. Examples would be the flip of a coin or the collecting of particles into two bins. Let us refer to the two outcomes as "success" and "failure." Let p be the probability of success for any single particle or single flip of the coin. Then the probability of success twice in a row is p^2, three times in a row, p^3, and m times in a row, p^m. Probabilities of successive events are multiplicative.

In general, we would like to know the probability of n_1 successes and n_2 failures out of $N = n_1 + n_2$ experimental outcomes. Consider the partitioning of N particles into two bins. Let us count how many different arrangements there can be with n_1 particles in bin 1 and n_1 particles in bin 2. First we fill bin 1. We can choose any of N particles to be the first in the bin, $N-1$, to be second in the bin, $N-2$, to be third, down to $N - n_1 + 1$ for the n_1th. Therefore, the total number of arrangements for bin 1 is

$$N(N-1)(N-2)\cdots(N-n_1+1) = N!/(N-n_1)!\,.$$

However, the order in which we chose the particles does not matter to us. In how many ways could we reorder them? Any of n_1 could be chosen as "first," any of $n_1 - 1$, as "second," $n_1 - 2$, as "third," and so on, down to a single choice for the n_1th. The total number of possible orderings is

$$n_1(n_1-1)(n_1-2)\cdots 1 = n_1!\,.$$

Therefore, the total number of unordered arrangements is the first quantity divided by the second,

$$W_1(n_1) = \frac{N!}{(N-n_1)!\,n_1!} = \binom{N}{n_1}. \tag{8.20}$$

This is the binomial coefficient,[10] introduced in Section 2.1. Because there are only two possible outcomes, any particle not in bin 1 is in bin 2. Therefore, $W_1(n_1)$ tells us the total number of possible arrangements.

[10] The word *binomial* comes from the Latin prefix *bi-* (two) and the Greek word νόμος (portions, or arrangements).

The probability of a given value of n_1 occuring randomly is equal to the number of arrangements with that value of n_1, times the probability $p_1^{n_1}$ of n_1 particles in bin 1, times the probability $(1-p_1)^{n_2}$ of n_2 particles in bin 2. Thus we obtain

$$p_{N,p_1}^{(\text{binom})}(n_1) = \binom{N}{n_1} p_1^{n_1}(1-p_1)^{N-n_1}, \tag{8.21}$$

which is called the ***binomial distribution***. The mean of this distribution is the sum over possible n_1 values weighted by $p_{N,p_1}^{(\text{binom})}$,

$$\mu = \sum_{n_1=0}^{N} n_1\, p_{N,p_1}^{(\text{binom})}(n_1). \tag{8.22}$$

We can sum this quite easily by making a clever substitution (see Exercise 8.8). The result is

$$\mu = N p_1 \tag{8.23}$$

Example 8.3 described a binomial phenomenon (a set of true/false questions) with $p_1 = 1/2$. The center of the distribution, in the $N = 20$ case, for example, was $\mu = 10 = (20)(0.5)$, and for large N it was seen to resemble a normal distribution. It can be proved that the binomial distribution for specified value of p_1 approaches a normal distribution in the limit of large N. A rule of thumb[11] is that a binomial phenomenon can be adequately described using a normal distribution if both of the following conditions hold true:

$$N > 5 \qquad \text{and} \qquad \frac{1}{\sqrt{N}}\left(\sqrt{\frac{1-p_1}{p_1}} - \sqrt{\frac{p_1}{1-p_1}}\right) < 0.3\,. \tag{8.24}$$

The Poisson distribution can be derived from the binomial distribution by taking the limit of Eq. (8.21) as N becomes large and p_1 becomes small such that $\mu = Np_1$ remains nonzero and finite.

8.2.5 The Boltzmann Distribution*

Consider now the partitioning of N particles into M bins. The number of arrangements with n_1 particles in bin 1 is, as before, $W_1(n_1) = \binom{N}{n_1}$. After choosing the n_1 particles for bin 1, we then choose n_2 particles from among the remaining $N - n_1$ for bin 2. The number of arrangements for these is $W_2 = \binom{N-n_1}{n_2}$. For bin 3 we have $W_3 = \binom{N-n_1-n_2}{n_3}$, and so on. The total number of possible arrangements of the particles, for given $(n_1, n_2, \ldots, n_{M-1})$,

[11] G. E. P. Box, W. G. Hunter, and J. Stuart Hunter, *Statistics for Experimenters* (Wiley, New York, 1978), p. 130.

is the product of the W_j, for which we use the symbol W with no subscript,

$$\begin{aligned} W(n_1, n_2, n_3, \ldots, n_{M-1}) &= \prod_{j=1}^{M-1} W_j(n_j) \\ &= \frac{N(N-1)\cdots(N-n_1+1)}{n_1!} \times \frac{(N-n_1)\cdots(N-n_1-n_2+1)}{n_2!} \\ &\quad \times \frac{(N-n_1-n_2)\cdots(N-n_1-n_2-n_3+1)}{n_3!} \times \cdots \\ &= \frac{N!}{n_1!\, n_2!\, n_3! \cdots n_{M-1}!} \,. \end{aligned} \tag{8.25}$$

W is called the ***multiplicity*** of the system of particles. The ordered set of values $(n_1, n_2, \ldots, n_{M-1})$ is called the ***configuration*** of the system.

Boltzmann[12] conjectured that for a system with an extremely large number of possible configurations, the configuration observed in practice will almost always be one that has a value of W close to the maximum possible value. Let us apply this idea to a chemical process. Because W will be extremely large, we will consider $\ln W$ instead of W itself. Instead of bins, we will group the particles according to the possible values E_j of their energies. Let n_j be the total number of particles with a particular energy E_j, and consider a process in which the configuration changes. The effect on $\ln W$ is described by the differential

$$d(\ln W) = \sum_j \frac{\partial \ln W}{\partial n_j}\, dn_j \,.$$

The equilibrium condition will be that $\ln W$ is at a maximum, which requires that the differential equal zero. However, we have to impose some restrictions on the dn_j. Assume no particles are created or destroyed. Then the change in the total number of particles is zero, $0 = dN = \sum_j dn_j$. Let us assume that the system is closed, so that the total energy $E = \sum_j E_j$ is conserved. Then

$$0 = dE = \sum_j dE_j = \sum_j E_j dn_j \,. \tag{8.26}$$

This is now a problem of finding an extremum in the presence of constraints, well suited for the method of undetermined multipliers (described in Section 3.4.2). Let us write the extremum condition as

$$0 = d(\ln W) = \sum_j \left(\frac{\partial \ln W}{\partial n_j} - \lambda_n - \lambda_E E_j \right) dn_j \,, \tag{8.27}$$

[12]Austrian physicist Ludwig Boltzmann (1844-1906), who established the theoretical foundation for the field of statistical thermodynamics. He was a forceful advocate for the atomic theory of matter at a time when many physicists (and even some chemists!) still doubted it.

with two undetermined multipliers, λ_n and λ_E. The multipliers are new degrees of freedom. Their presence allows us to treat the n_j as independent coordinates, so that

$$\frac{\partial \ln W}{\partial n_j} - \lambda_n - \lambda_E E_j = 0 \tag{8.28}$$

for each j. Using Stirling's formula, in the form $\ln(x!) \sim x \ln x - x$, to approximate the factorials in $\ln W$, we find that

$$\frac{\partial \ln W}{\partial n_j} = -\ln(n_j/N). \tag{8.29}$$

But n_j/N is the probability $p(E_j)$ of a particle having energy E_j. It follows that

$$\ln p(E_j) = -\lambda_a - \lambda_E E_j\,, \qquad p(E_j) = e^{-\lambda_n} e^{-\lambda_E E_j}.$$

Finally, we need to assign values to the multipliers. The constraint $N = \sum_j n_j$ allows us to eliminate one degree of freedom. Thus,

$$N = \sum_j n_j = \sum_j N p(E_j) = N e^{-\lambda_n} \sum_j e^{-\lambda_E E_j},$$

which implies that

$$e^{-\lambda_n} = \left(\sum_j e^{-\lambda_E E_j} \right)^{-1}. \tag{8.30}$$

To assign a value to λ_E, we compare with experiment. It turns out that $\lambda_E = 1/(k_B T)$, where T is temperature in kelvins and k_B is Boltzmann's constant, 1.38065×10^{-23} J/K. Thus, using probability theory and two reasonable constraints, we derive one of the most important equations of physical chemsitry, the ***Boltzmann distribution*** for energies,

$$p(E_j) = e^{-E_j/(k_B T)} \Big/ \sum_j e^{-E_j/(k_B T)}. \tag{8.31}$$

Boltzmann was also able to show that $\ln W$, which plays such a key role in this derivation, is proportional to the entropy of the system. Incidentally, the denominator in Eq. (8.31),

$$Q = \sum_j e^{-E_j/(k_B T)}, \tag{8.32}$$

called the ***canonical partition function***, is of fundamental importance in statistical thermodynamics. It turns out that the various thermodynamic state functions can be expressed in terms of it, and this provides a connection between macroscopic thermodynamics and microscopic statistical mechanics.

8.3 Outliers

Outlier is the statisticians' term for a measurement in the data set that should not be there, presumably on account of a systematic error in that particular measurement. The assumption is that the data set ought to have random variability in accord with an underlying probability distribution. Outliers do not follow the underlying distribution. Therefore they interfere with attempts to infer the underlying distribution's mean and variance.[13]

Example 8.5. *Computer simulation of data samples.* A convenient technique for comparing estimation methods is to test them on computer generated "random" data sets. Computers can generate series of pseudorandom numbers. The numbers are not truly random, because they result from a reproduceable computational algorithm, but the relationship between one number and the next in the series is intentionally made to be so complicated that the series gives a realistic simulation of a random data set. A computer's random number generation software will output numbers between 0 and 1 according to a flat distribution. These numbers can then be scaled to simulate any desired distribution. This is easily accomplished using built-in algorithms that come with computer algebra software packages. For example, in *Mathematica*

```
Table[ RandomReal[ NormalDistribution[0, 1] ], {j, 1, 20} ]
```

will generate a list of 20 random numbers according to a normal distribution with $\mu = 0$ and $\sigma = 1$. Outliers can be simulated by choosing some numbers according to a distribution with a much larger σ. For example,

```
dataset1 = Table[ RandomReal[ NormalDistribution[0, 1] ], {j, 1, 18} ];
outliers = Table[ RandomReal[ NormalDistribution[0, 100] ], {j, 1, 2} ];
dataset = Join[dataset1, outliers]
```

would generate a set with two outliers, using $\sigma = 100$ for the outliers.

Given only a data set, and no other information about the measurement process, is it possible to identify an outlier? Consider the seven points along the bottom of Fig. 8.5. It is tempting to declare the far right point an outlier and throw it out, but is this justified? The normal distribution gives nonzero probability at any value of x. With only a small number of points, the apparent asymmetry could conceivably be consistent with a normal distribution.

One strategy is to use a ***rejection test***. This is any kind of objective test that states the probability that the observed data set would be generated by a random selection of elements governed by a specified distribution. The first step is to choose some property of the set to characterize its degree of scatter. Then one specifies a desired confidence level, typically 90% or 95%, and determines the probability that a randomly chosen set subject to the probability distribution would yield more scatter than observed. In other words, by choosing a 90% confidence level one is deciding to tolerate a 10%

[13] One should distinguish between a situation in which most of the data set has systematic error and a situation in which just a minority does. Only in the latter case do we have outliers. In the former case it is impossible to identify the existence of systematic error from the data set alone, without having a reference value with which to compare.

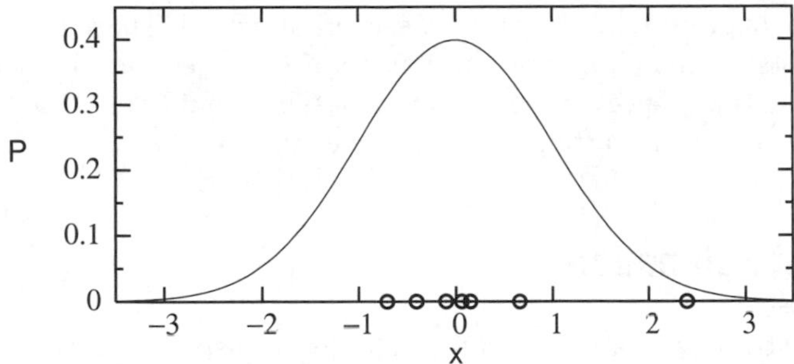

Figure 8.5: Normal distribution function with $\mu = 0$, $\sigma = 1$. The points on the x-axis are a small data set that may (or may not) be consistent with the distribution.

chance that the rejected value should not have been rejected. If the amount of scatter is more than predicted, the outlying data element is rejected.

Among analytical chemists the most popular rejection test has been Dixon's ***Q*-test**,[14] which is based on the normal distribution.

Example 8.6. *The Q-test.* Arrange a set of N observations in ascending order, $x_{(1)}, x_{(2)}, \ldots x_{(N)}$. For suspected outliers $x_{(1)}$ and $x_{(N)}$ calculate the test quantities

$$Q_1 = \frac{x_{(2)} - x_{(1)}}{x_{(N)} - x_{(1)}}, \qquad Q_N = \frac{x_{(N)} - x_{(N-1)}}{x_{(N)} - x_{(1)}}. \tag{8.33}$$

These are just the ratios of the nearest-neighbor distances to the full range. The largest of these Q's is compared with a tabulated value $Q_{\rm crit}^{(N)}$. The superscript indicates that the critical value depends on the sample size N. Values are typically tabulated for 90% confidence and 95% confidence. If the largest of Q_1 and Q_N is larger than $Q_{\rm crit}$, then the corresponding x value is rejected. It is usually recommended that no more than one element be rejected in this way, and that the test be applied only for $N \leq 10$.[15] The critical Q values are:[16]

$P \setminus N$	3	4	5	6	7	8	9	10
90%	0.941	0.765	0.642	0.560	0.507	0.468	0.437	0.412
95%	0.970	0.829	0.710	0.625	0.568	0.526	0.493	0.466

For the data shown in Fig. 8.5, the point on the far right would be rejected at the 90% confidence level but retained at the 95% level. If we increase the confidence level for rejection, then the non-normal behavior must be more extreme to merit rejection.

The Q-test has two significant drawbacks. The first is that it is easily confounded by the presence of *more than one* outlier. Consider the set

$$A = \{5.34, 5.38, 5.39, 5.40, 5.41, 5.43, 5.44, 77.24, 79.62\}. \tag{8.34}$$

This easily passes the Q-test at 90%, with $Q_9 = 0.032$, far below the 90% critical value 0.437, yet it is very likely that the two last elements both suffer

[14]Proposed by the statistician W. J. Dixon in 1950, and commonly described in analytical chemistry textbooks. See D. B. Rorabacher, *Anal. Chem.* **63**, 139 (1991).

[15]Dixon himself recommended this test only for $N \leq 7$, and proposed a series of other tests for higher ranges of N. See W. J. Dixon, *Ann. Math. Stat.* **21**, 488-506 (1950).

[16]Reproduced from D. B. Rorabacher, *Anal. Chem.* **63**, 139 (1991).

from a similarly huge systematic error. A second drawback, shared in general by rejection tests, is the all or nothing nature of the rejection. If there is a 10% chance that the suspect value is *not* an outlier, would it not be wise to include some effect, even if attenuated, on the estimator?

8.4 Robust Estimation

An alternative strategy to a rejection test is to leave suspected outliers in the data set but replace the mean and standard deviation with estimators that are less sensitive to the presence of outliers. This is called *robust* estimation. A ***robust estimator*** is an estimator whose value will not be significantly skewed by outliers. An example is the median. Consider the set A, Eq. (8.34), with and without the last two elements. The mean is 21.628 with the last elements included and 5.399 without them. The median is virtually unaffected by the outliers, going from 5.41 to 5.40.

The effectiveness of an estimation method is judged by its ***efficiency***, ϵ, and its ***breakdown point***, ϵ^*. Efficiency is determined by generating a large number of independent random data sets and analyzing the random scatter of the estimator values from the various sets. An efficient estimator gives consistent results, with little variability from one data set to the next, and the amount of variability decreases quickly as the size of the sets is increased. ϵ is usually expressed as a percentage of the efficiency of the mean (for a location estimator) and the standard deviation (for a scale estimator). It is determined using computer generated normally distributed samples.[17] For a normal distribution (with no outliers) $\bar{x}$ and $\hat{\sigma}$ are the most efficient estimators. For other estimators we expect $0 \leq \epsilon < 1$.

The efficiency of the estimator is expected to decrease as the fraction of the sample that consists of outliers increases. For any method, there exists some critical fraction at which the estimator will no longer perform properly. This fraction is the breakdown point of the method, given the symbol ϵ^*. If the fraction of outliers is greater than ϵ^*, then increasing the size of the data set will not improve the accuracy of the estimation.

Example 8.7. *Determining the breakdown point.* An easy way to determine ϵ^* is to see whether or not replacing M out of N numbers in a data set with ∞ will make the estimator infinite. We could replace any four of the nine elements of the set A with ∞ and still have a finite median. For the mean, setting even just one number to ∞ makes $\bar{x}$ infinite. ϵ^* is the smallest ratio M/N in the limit $N \to \infty$ that makes the estimator go to infinity. ϵ^* for med x is 50%, which is the maximum breakdown point possible. For $\bar{x}$, the breakdown point is zero. Just one outlier per sample is enough to prevent $\bar{x}$ from converging! *This strongly suggests that it is not safe to use the mean (at least without a rejection test) for experimental data analysis.*

---

[17]See P. J. Rousseeuw and C. Croux, *J. Amer. Stat. Assoc.* **88**, 424 (1993), for an example of how to obtain numerical values for ϵ.

Typically, there is a tradeoff between efficiency and breakdown point. med x is only 64% as efficient as the $\bar{x}$ for a pure normal distribution, but it is extremely resistent to outliers, showing little loss in efficiency until outliers make up a very large fraction. Furthermore, the 100% efficiency of $\bar{x}$ is for a pure normal distribution. If the distribution is only moderately non-normal (as is commonly the case with physical data sets) the efficiency of $\bar{x}$ drops to about that of med x.

Now consider scale estimators. The standard deviation, $\hat{\sigma}$, from Eq. (8.3), has $\epsilon^* = 0$ and is even more sensitive to outliers than is the mean, because the differences in the summation are squared. Even mild outliers can easily overshadow points that follow the underlying probability distribution. A robust replacement for $\hat{\sigma}$ was proposed by Gauss. It is the ***median absolute deviation*** (MAD),[18] defined as

$$\mathrm{MAD}\, x = b\, \mathrm{med}\, |\, x - \mathrm{med}\, x\, |\,, \qquad b = 1.4826\,. \tag{8.35}$$

The notation "med x" means "calculate the median of a set $\{x_1, x_2, \ldots, x_N\}$." In Eq. (8.35) the median of the set is calculated and then subtracted from each element of the set, to generate the set of deviations from the median

$$\{\, |x_1 - \mathrm{med}\, x|, |x_2 - \mathrm{med}\, x|, \ldots, |x_N - \mathrm{med}\, x|\, \}.$$

The median of this new set is taken as the measure of the data variability. The value given for the parameter b ensures that the MAD will converge to σ for a pure normal distribution in the limit of increasing sample size. Usually it is sufficient to truncate b to the easily remembered value 1.5.

Example 8.8. *Median absolute deviation.* For the set A given above, med $x = 5.41$ and

$$|x - \mathrm{med}\, x| = \{0.07,\ 0.03,\ 0.02,\ 0.01,\ 0,\ 0.02,\ 0.03,\ 71.83,\ 74.21\}.$$

Rearrange in these in increasing order:

$$\{0,\ 0.01,\ 0.02,\ 0.02,\ \mathbf{0.03},\ 0.03,\ 0.07,\ 71.83,\ 74.21\}.$$

The median is 0.03. Multiplying by b gives 0.044 as the value of the MAD. In contrast, the standard deviation gives a scale estimate for set A of 32.21, which is meaningless.

The MAD is quite robust, with $\epsilon^* = 50\%$, but its efficiency is rather low—only 37%. Consider a modification, proposed by Rousseeuw and Croux[19]:

$$S_{\mathrm{RC}} = c\, \mathrm{med}_i\{\, \mathrm{med}_j |x_i - x_j|\, \}, \qquad c = 1.1926\,. \tag{8.36}$$

This estimator has the same breakdown point as the MAD but is about twice as efficient. The notation "$\mathrm{med}_j |x_i - x_j|$" means that for a given element x_i, the set $\{|x_i - x_1|, |x_i - x_2|, \ldots, |x_i - x_N|\}$ is constructed and the median is

[18] Beware that some authors use "MAD" to indicate the *mean* absolute deviation, Eq. (8.2), for which we used the symbol $|\hat{D}|$.

[19] P. J. Rousseeuw and C. Croux, *ibid.*

taken. This yields a set of N different medians, one for each i. The median of this set is taken and multiplied by c. The value given for c ensures that S_{RC} converges to σ for a pure normal distribution. For the set A, we obtain $S_{\text{RC}} = 0.048$. A *Mathematica* routine for S_{RC} is given in Appendix A.1.

The development of estimation methods that are both efficient and stable in the presence of outliers is an active area of current research. It is impossible at present to definitively recommend a "best" method. A method that seems to offer a reasonable balance of efficiency and stability is ***Huber estimation***,[20] which is an example of a ***Winsorization*** method.[21] "Winsorization" refers to any method in which outliers are systematically moved in toward the center of the data sample. Some vestige of each outlier is included in the final estimation, just in case it might actually be an unusually large random fluctuation, but in such a way that it will not seriously damage the estimator accuracy if the outlier is the result of a systematic error.

Let us begin with initial estimates μ_0 and s_0 of location and scale. These could be $\bar{x}$ and $\hat{\sigma}$ or, probably better, $\text{med}\, x$ and S_{RC}. We then Winsorize any data values that lie outside the range $\mu_0 \pm 1.5 s_0$ by moving them in to the edges of the range. Values larger than $\mu_0 + 1.5 s_0$ are replaced by $\mu_0 + 1.5 s_0$ and values smaller than $\mu_0 - 1.5 s_0$ are replaced by $\mu_0 - 1.5 s_0$. The mean and standard deviation of the resulting set, $\bar{x}_1$ and $\hat{\sigma}_1$, are computed and then new estimates $\mu_1 = \bar{x}_1$ and $s_1 = 1.134\hat{\sigma}_1$ are used to adjust the previous Winsorization, using the range $\mu_1 \pm 1.5 s_1$. (The factor 1.134 is included to ensure the correct scale estimate for a pure normal ditribution.) The mean and standard deviation are once again calculated and we obtain improved estimates μ_2, s_2. The procedure is repeated until the estimates converge. We will call the converged estimates μ_{Hub} and S_{Hub}.

Example 8.9. *Huber estimators.* Let us calculate the Huber estimators for the set A starting with initial estimates $\mu_0 = \text{med}\, x = 5.41$ and $s_0 = S_{\text{RC}} = 0.048$.

Step 1. The Winsorization range is $[5.338, 5.482]$, which gives the Winsorized set $\{5.34, 5.38, 5.39, 5.40, 5.41, 5.43, 5.44, \mathbf{5.482}, \mathbf{5.482}\}$. The new location estimator is $\mu_1 = 5.417$ and the new scale estimator is $s_1 = (1.134)(0.0466) = 0.0529$.

Step 2. The corrected Winsorization range, $\mu_1 \pm 1.5 s_1$, is $[5.338, 5.496]$. The Winsorized set is adjusted to $\{5.34, 5.38, 5.39, 5.40, 5.41, 5.43, 5.44, \mathbf{5.496}, \mathbf{5.496}\}$. This gives $\mu_2 = 5.420$ and $s_2 = (1.134)(0.0579) = 0.0589$.

Step 3. The range is now $[5.332, 5.508]$, which leads to $\mu_3 = 5.423$, $s_3 = 0.0640$.

The values of μ_i and s_i tend to converge rather slowly, which makes it tedious to calculate them by hand, but it is straightforward to carry out the calculation with a computer. For this data set the scale estimator requires 84 iterations to converge to three significant figures, with the result $\mu_{\text{Hub}} = 5.451$ and $S_{\text{Hub}} = 0.124$.

---

[20] Peter J. Huber, a statistics professor at U. C. Berkeley and then at Harvard, laid the foundations of the mathematical theory of robust estimation, beginning in the 1960's. In 2001 the Analytical Methods Committee of the British Royal Society of Chemistry recommended Huber estimators as a reliable replacement for $\bar{x}$ and $\hat{\sigma}$.

[21] Named after American biostatistician Charles P. Winsor (1895-1951).

A *Mathematica* function for carrying this out is given in the Appendix A.1. Huber's method has very high efficiency, both for a pure normal distribution and for a distribution with only a small number of outliers. Its main drawback is a breakdown point of only 26%.

Example 8.10. *Breakdown of Huber estimation.* For the data set A, the fraction of outliers is 2/9, or 22%. Suppose we remove one element of the set and consider

$$B = \{5.34,\ 5.38,\ 5.39,\ 5.40,\ 5.43,\ 5.44,\ 77.24,\ 79.62\}\ .$$

This set has 25% outliers, which is very close to the breakdown point. In this case we get the *not* very robust values $\mu_{\text{Hub}} = 23.655$ and $S_{\text{Hub}} = 38.345$.

Thus, Huber estimation can fail even before its rather low breakdown point is reached. This last example demonstrates that Huber estimation is not a panacea—while it is more efficient than the median for a normal distribution, it is significantly less robust. Furthermore, the advantage in efficiency disappears if the distribution is non-normal. The use of med x and S_{RC} would appear to be the more prudent choice for a general method. Huber estimation should be reserved for cases in which you have reason to believe that the true underlying distribution is reasonably described as normal and you are confident that outliers make up no more than about one-fifth of the data.

The sample data sets A and B considered here were extreme examples, for the purpose of illustration. Any scientist faced with such a set would immediately remove the two outliers, based on common sense. In practice, however, outliers are often impossible to identify by superficial inspection.

The most satisfying way to deal with a suspected outlier is to reexamine the experimental procedure and show that the outlier resulted from a specific systematic error. Then the outlier can be rejected with a clear conscience, and perhaps replaced with a corrected value. (Quite often the error turns out to be a number that was written or read incorrectly.) If one suspects the presence of outliers but cannot clearly identify a cause, then a good course of action is to carry out repetitions of the measurement. A set of 10 measurements is considered by statisticians to be a "small" sample. Statistical methods (within their breakdown limits) become more reliable as the sample size increases. Finally, it is important to keep in mind that outliers may represent an unexpected discovery. Indeed, many important scientific advances have come from attempts to explain outliers in a data set.[22]

[22] A famous example is the discovery of argon by the physicist Lord Rayleigh (J. W. Strutt) in 1894. Rayleigh was comparing values for the density of nitrogen gas obtained from various measurement techniques. Some of the results seemed to be outliers, and these all happened to be from experiments in which the nitrogen was isolated from air rather than synthesized by chemical reaction. Rayleigh suggested that air contained a previously unknown inert chemical element. He then collaborated with William Ramsay, a chemist, to isolate the element, which they named "argon." Ramsay went on to isolate helium, neon, krypton, and xenon, adding a new column to the periodic table.

Exercises

8.1 Three laboratories, using different experimental techniques, report the following results for x ($\hat{\sigma}$): 7.31 (0.42), 7.630 (0.094), 6.9 (1.2). Estimate x.

8.2 Demonstrate for the data set A of Eq. (8.34) that $\overline{x^2} \neq \bar{x}^2$.

8.3 For a normal distribution with arbitrary parameters μ and σ, calculate $\langle x^4 \rangle$.

8.4 For a normal distribution with arbitrary parameters μ and σ, derive an explicit formula for the difference between $\langle x^2 \rangle$ and $\langle x \rangle^2$.

8.5 Prove that the normal distribution function satisfies the equation $\langle (x - \langle x \rangle)^2 \rangle = \sigma^2$.

8.6 Calculate the probability of measuring a result between 5 and 12 for a normal distribution with $\mu = 10$ and $\sigma = 3$. Express your answer in terms of error functions.

8.7 Prove that the Poisson distribution is normalized.

8.8 Derive the mean of the binomial distribution by making the substitutions $n_1 = k + 1$, $N = M + 1$ and showing that $\mu = (M+1)p_1 \sum_{k=0}^{M} p_{M,p_1}^{(\mathrm{binom})}(k)$. The sum of the distribution over the full range of k values must be unity.

8.9 For the set A of Eq. (8.34), calculate (by hand) (a) the standard deviation and (b) the scale estimator S_{RC} of Rousseeuw and Croux.

8.10 Carry out a computer study of how the choice of μ_0 and s_0 affects Huber estimation.

8.11 What is the breakdown point of the mean absolute deviation $|\hat{D}|$?

8.12 Consider the following set of concentration measurements, in mg/dL.

0.209 0.098 0.032 0.103 0.172 0.106 0.354 0.065 0.122 0.080

(a) Perform a Q-test on these data at the 95% confidence level under the assumption that they come from a normal distribution.

(b) Perform a Q-test on these data at the 95% confidence level under the assumption that they come from a log-normal distribution.

(c) Assuming a normal distribution, calculate the mean, the median, the standard deviation, the MAD, the Rousseeuw-Croux scale estimator, and the Huber estimators.

(d) Assuming a log-normal distribution, calculate the mean, the median, the standard deviation, the MAD, the Rousseeuw-Croux scale estimator, and the Huber estimators.

(e) Discuss your results. In your opinion, what is the best estimate for the value of the concentration and for its 95% confidence interval.

8.13 Using computer generated data samples with one or more outliers, study the robustness of the mean, the median, the mean with Q-test, and the Huber location estimator.

Chapter 9

Analysis of Significance

Statistical methods allow us not just to estimate the most probable value of a physical property but also to determine how much confidence we should have in the estimation. This chapter presents methods for determining the most probable *range* for the value of the property. Knowing this range can be even more useful than knowing the most probable value itself. It can tell us whether or not the result of an experiment is significant.

9.1 Confidence Intervals

Suppose that the random variability in measurements of some property x is governed by a normal probability distribution with scale σ. Let μ be the true value of x. In this section we will answer the following questions:

1. What is the probability that a single measurement x_j will yield a value outside of the interval
$$\mu - \delta < x_j < \mu + \delta$$
for some arbitrary specified value δ?

2. What is the probability that the true value μ lies outside the interval
$$x_j - \delta < \mu < x_j + \delta$$
for specified δ? (Actually, this is the same probability as before.)

3. What is the probability that μ lies outside the interval
$$\bar{x} - \delta < \mu < \bar{x} + \delta,$$
where $\bar{x}$ is the mean of N measurements? (This probability is lower, as the mean damps down random error.)

4. For specified probability, how can we estimate δ?

What, then, is the probability that a single measurement will give a result that deviates from the true value μ by more than some specified δ? Let α be the probability[1] of the measurement deviating by more than δ. We can calculate α by integrating the probability distribution function,

$$\alpha = \int_{-\infty}^{\mu-\delta} P_{\mu,\sigma}(\xi) d\xi + \int_{\mu+\delta}^{\infty} P_{\mu,\sigma}(\xi) d\xi \,. \tag{9.1}$$

This is illustrated in Fig. 9.1. The normal distribution function, $P_{\mu,\sigma}(\xi)$, is symmetric about $\xi = \mu$. The two integrals in Eq. (9.1) then are of equal value

[1] We will express probabilities as fractions, with values between 0 and 1. For example, a 15% probability would be expressed as 0.15.

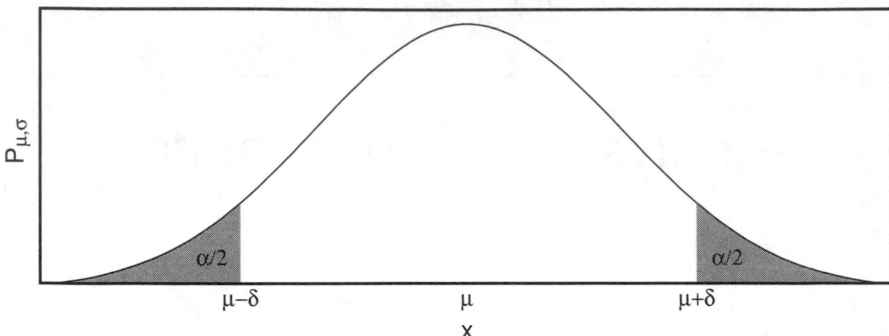

Figure 9.1: Probability α from integrating a probability distribution.

and we can use the simpler equation

$$\alpha/2 = \int_{\mu+\delta}^{\infty} P_{\mu,\sigma}(\xi)d\xi = \frac{1}{\sqrt{2\pi\sigma^2}} \int_{\mu+\delta}^{\infty} e^{-(\xi-\mu)^2/2\sigma^2} d\xi \,. \tag{9.2}$$

Because σ is a measure of the variability of the measurement process, we can expect that δ will depend on the of σ, with larger σ giving larger δ. The common practice is to consider a quantity

$$z_{\alpha/2} = \delta/\sigma. \tag{9.3}$$

The range expected to contain the fraction $1 - \alpha$ of the measurements is then

$$\mu - z_{\alpha/2}\,\sigma < x_j < \mu + z_{\alpha/2}\,\sigma, \tag{9.4}$$

with

$$\frac{\alpha}{2} = \frac{1}{\sqrt{2\pi\sigma^2}} \int_{\mu+z_{\alpha/2}\sigma}^{\infty} e^{-(\xi-\mu)^2/2\sigma^2} d\xi = \frac{1}{\sqrt{2\pi}} \int_{z_{\alpha/2}}^{\infty} e^{-u^2/2} du.$$

This expresses $\alpha/2$ in terms of $z_{\alpha/2}$. Notice that the result is independent of the values of μ and σ. It can be expressed in terms of the cumulative normal distribution function (which was defined in Section 8.2.1):

$$\frac{\alpha}{2} = \frac{1}{\sqrt{2\pi}} \int_{-\infty}^{\infty} e^{-u^2/2} du - \frac{1}{\sqrt{2\pi}} \int_{-\infty}^{z_{\alpha/2}} e^{-u^2/2} du = 1 - \Phi(z_{\alpha/2}). \tag{9.5}$$

Example 9.1. *Solving for* $z_{\alpha/2}$. What is the value of $z_{0.025}$? In other words, what z gives 95% confidence? We want to solve the equation $\Phi(z) = 1 - 0.025 = 0.975$. In *Mathematica*, $\Phi(z)$ is given by:

```
CDF[NormalDistribution[0, 1], z]
```

This is plotted at right. We can see that $z_{0.025}$ is slightly less than 2. To get a precise solution, we can use the following *Mathematica* statement:

```
InverseCDF[NormalDistribution[0, 1], 0.975]
```

The result is 1.95996.

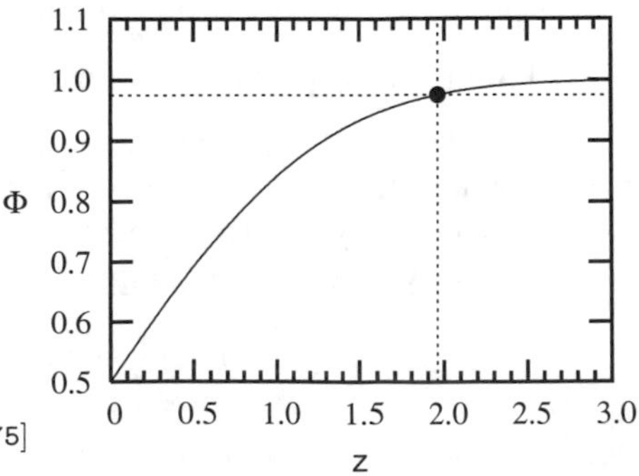

Example 9.2. *Solving for α.* Setting $z_{\alpha/2} = 1$ in Eq. (9.5) gives $\alpha = 0.3173$ while $z_{\alpha/2} = 2$ gives $\alpha = 0.0455$. Thus, the normal distribution predicts a result between $\mu - \sigma$ and $\mu + \sigma$ approximately two-thirds ($100\% - 31.73\% = 68.27\%$) of the time and a result between $\mu - 2\sigma$ and $\mu + 2\sigma$ approximately 95% of the time.

Eq. (9.4) makes a prediction about the result x_j of a measurement that is about to be performed. But suppose we do not know the true value μ and want to use a single measurement x_j as an estimator for μ. How reliable is it? The inequalities in Eq. (9.4) imply that $x_j - z_{\alpha/2}\,\sigma < \mu$ and that $\mu < x_j + z_{\alpha/2}\,\sigma$. Therefore, we have

$$x_j - z_{\alpha/2}\,\sigma < \mu < x_j + z_{\alpha/2}\,\sigma\,, \tag{9.6}$$

which tells us the reliability of the estimate. This is called a ***confidence interval***. For example, if $\alpha = 0.05$, it is a "95% confidence interval"; given a measured value x_j, the probability is 95% that μ lies within this interval.

What is the confidence interval if we use the mean of *two* measurements instead of just a single measurement? We expect the interval to be smaller, because the mean is more reliable than a single value. In fact, it turns out that σ^2 for the mean of two measurements is one-half that for a single measurement, which means σ needs to be divided by $\sqrt{2}$. For the mean of N measurements, the scale is $\sigma/\sqrt{N}$. (See Exercise 9.11.) It follows that the distribution for $\bar{x}$ is

$$P_{\mu,\,\sigma/\sqrt{N}}\,(\bar{x}) = \frac{1}{\sqrt{2\pi\sigma^2/N}}\, e^{-(\bar{x}-\mu)^2/(2\sigma^2/N)} \tag{9.7}$$

and

$$\bar{x} - z_{\alpha/2}\,\sigma/\sqrt{N} < \mu < \bar{x} + z_{\alpha/2}\,\sigma/\sqrt{N}. \tag{9.8}$$

This is still not quite what we need. It requires knowledge of the *exact* value σ of the normal distribution, which typically we do not know. In practice we use $\hat{\sigma}$, defined by Eq. (8.3), as an estimator for σ.

The standard deviation $\hat{\sigma}$ will agree with the true scale σ in the limit of infinite sample size. For a finite sample, however, $\hat{\sigma}$ will usually be slightly less than σ. This is because a finite sample will usually not contain any very large fluctuations from the mean. The probability of a very large fluctuation is small but it is not zero. As a consequence, the value of $\hat{\sigma}$ calculated from a *finite* sample will usually be smaller than its $N \to \infty$ limit. This was pointed out by a chemist, William S. Gosset, who then proceeded to derive an alternative to Eq. (9.7) appropriate for use with $\bar{\sigma}$ instead of σ. His analysis was presented in a paper published in 1908 under the pseudonym "Student."[2] The distribution function is called ***Student's t-distribution***. Gosset found that the distribution function to be used with $\hat{\sigma}$ calculated from a finite number

[2]Gosset was an English analytical chemist and mathematician who worked for the Guinness Brewery in Dublin. He used the pseudonym because the company had a policy that its scientists could not publish any of their research, so as not to reveal trade secrets.

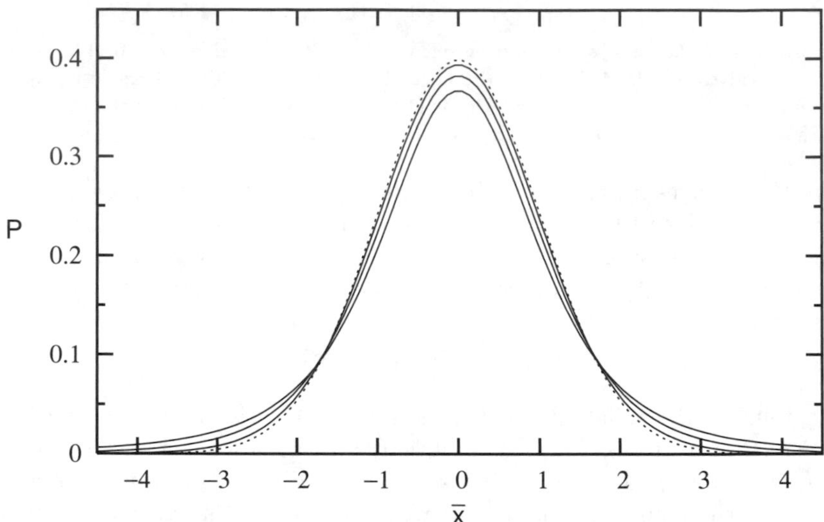

Figure 9.2: Student's t-distribution function $P_{0,1,\nu}(\bar{x})$ for $\nu = 3$ (the shallowest of the curves), $\nu = 6$, and $\nu = 20$ (the most sharply peaked). The dotted curve shows the normal distribution $P_{0,1}$.

of measurements is not the normal distribution but rather

$$P_{\mu,\hat{\sigma}/\sqrt{N},\nu}(\bar{x}) = \frac{\Gamma\left(\frac{\nu+1}{2}\right)}{\sqrt{\pi\nu\hat{\sigma}^2/N}\,\Gamma(\nu/2)}\left[1 + \frac{(\bar{x}-\mu)^2}{\nu\hat{\sigma}^2/N}\right]^{-(\nu+1)/2}, \tag{9.9}$$

with a parameter ν that is called the ***number of degrees of freedom.*** (ν is the Greek letter *nu*.) If we are using $\bar{x}$, the mean of N measurements, as our estimator for μ, then[3]

$$\nu = N - 1. \tag{9.10}$$

The Student distribution has higher tails than the normal distribution, dropping to zero more slowly with increasing $|\bar{x} - \mu|$, as shown in Fig. 9.2. In the limit of large ν, the Student and normal distributions become identical.

The confidence interval resulting from the Student distribution is usually written

$$\bar{x} - t_{\alpha/2,\,\nu}\,\hat{\sigma}/\sqrt{N} < \mu < \bar{x} + t_{\alpha/2,\,\nu}\,\hat{\sigma}/\sqrt{N}. \tag{9.11}$$

The scale factor is now designated t instead of z. $\hat{\sigma}$ is the standard deviation, calculated from the N data points. The quantity $\hat{\sigma}/\sqrt{N}$ is called the ***standard error*** or the ***standard deviation of the mean.*** It is, after all, the scale of the probability distribution for $\bar{x}$. The symbol $\hat{\sigma}_{\bar{x}}$ is often used for it:

$$\hat{\sigma}_{\bar{x}} = \hat{\sigma}/\sqrt{N}. \tag{9.12}$$

[3] As discussed in Chapter 3, ν is the minimum number of values needed to fully describe the state of a system. In this case, the "system" being described is the outcome of the experiment, which can be described with the set of N measurement results. If we assume that the mean $\bar{x}$ is an accurate estimate of the true value of the property x, then we have the equation $(x_1 + x_2 + \cdots + x_N)/N \approx x$, which imposes a constraint on the x_j values. Therefore, the number of degrees of freedom is $N - 1$.

The value $t_{\alpha/2,\,\nu}$ that yields a confidence level α for ν degrees of freedom is determined by the equation

$$\frac{\alpha}{2} = \int_{\bar{x}+t_{\alpha/2,\nu}\hat{\sigma}}^{\infty} P_{\bar{x},\hat{\sigma},\nu}(\xi)d\xi = \frac{\Gamma\left(\frac{\nu+1}{2}\right)}{\sqrt{\pi\nu}\,\Gamma(\nu/2)} \int_{t_{\alpha/2,\nu}}^{\infty} \left(1+t^2/\nu\right)^{-(\nu+1)/2} dt, \quad (9.13)$$

with the change of variable

$$t = \frac{\xi-\mu}{\hat{\sigma}/\sqrt{N}}. \quad (9.14)$$

Eq. (9.13) can be solved numerically for given α and ν. "t-tables" are widely available, listing $t_{\alpha/2,\,\nu}$ for various ν at select values of $\alpha/2$. To calculate the confidence interval using the Student distribution is to carry out a ***t*-test**. Note that the result in Eq. (9.13) does not depend on $\bar{x}$ or $\hat{\sigma}$. The usual practice is to express the distribution function in terms of t with the single subscript ν,

$$P_\nu(t) = \frac{\Gamma\left(\frac{\nu+1}{2}\right)}{\sqrt{\pi\nu}\,\Gamma(\nu/2)} \left(1+t^2/\nu\right)^{-(\nu+1)/2}. \quad (9.15)$$

Example 9.3. *A 95% confidence interval.* The concentration c of a solution is measured repeatedly, with the following results (in mol/L):

$$\{0.5091, 0.5540, 0.4813, 0.5485, 0.5239, 0.4712, 0.5010\}$$

The mean is $\bar{c} = 0.51271$ mol/L. The standard deviation is $\hat{\sigma} = 0.0315$ mol/L and the standard error is $\hat{\sigma}/\sqrt{7} = 0.0119$ mol/L. To determine t for 95% confidence, we solve (numerically) the equation

$$0.025 = \int_{t_{0.025,\,6}}^{\infty} P_6(t)\,dt\,.$$

In *Mathematica* the command would be:

```
InverseCDF[StudentTDistribution[6], 0.975]
```

(Or you could look up the answer in a table.) We find $t_{0.025,6} = 2.447$ and

$$\delta = t_{0.025,6}\,\hat{\sigma}/\sqrt{7} = 0.0292 \text{ mol/L}\,.$$

The 95% confidence interval is $c = 0.513 \pm 0.029$ mol/L. (It is customary to report δ to two significant figures and then round off the mean at the same decimal place as the last digit in δ.) We conclude there is a 5% chance that the true concentration is outside of the range 0.484 mol/L to 0.542 mol/L.

$\hat{\sigma}$ is defined as $\nu^{-1}\sum_j(x_j-\bar{x})^2$, but this is not the most prudent way to estimate its value. To avoid the effects of outliers, it is safer to use a robust estimator, such as $S_{\rm RC}$ or $S_{\rm Hub}$, as discussed in Section 8.3, to estimate $\hat{\sigma}$.

The t-test as given in Eq. (9.11) is the confidence interval most useful in practice. Suppose, however, that the exact value μ is known and measurements are carried out just to estimate the confidence interval. The goal would be to determine the precision of the measuring instrument using a sample of known composition. Then the precision of the instrument for a set of N measurements is $\pm t_{\alpha/2,\,N}\,\hat{\sigma}/\sqrt{N}$, with $\nu = N$ in the equations for P_ν and $\hat{\sigma}$.

Consider one additional question: Having determined a value for $\hat{\sigma}$ from a data set of N measurements of x, what is our best estimate for σ, the true scale parameter of the underlying normal probability distribution? First let us define a quantity

$$\chi^2 = \nu\,\hat{\sigma}^2/\sigma^2 \tag{9.16}$$

called "chi-square." χ is the Greek letter *chi.*[4] ν is the number of degrees of freedom, $N-1$. χ^2 is described by a rather simple distribution function,

$$P_\nu^{(\chi^2)}(u) = \frac{1}{2^{\nu/2}\Gamma(\nu/2)}\,u^{(\nu-2)/2}e^{-u/2}, \tag{9.17}$$

called the ***chi-square distribution***. The cumulative chi-square probability, that is, the probability of obtaining χ^2 less than some specified value χ_1^2, is

$$P_\nu^{(\chi^2<\chi_1^2)} = \int_0^{\chi_1^2} P_\nu^{(\chi^2)}(u)\,du. \tag{9.18}$$

This can be used to calculate a confidence interval for σ. (See Exercise 9.6.)

Example 9.4. *Using $\hat{\sigma}$ to estimate σ.* The best estimate of σ is a value corresponding to an optimal χ_1^2 (let us call it χ^2_{opt}) such that $P_\nu^{(\chi^2<\chi^2_{\mathrm{opt}})} = 0.5$, because there will then be a 50% chance χ^2_{opt} will be too large and a 50% chance it will be too small. χ^2_{opt} can be determined by solving Eq. (9.18). σ can then be estimated from

$$\sigma = \left(\frac{\nu}{\chi^2_{\mathrm{opt}}}\right)^{1/2}\hat{\sigma}. \tag{9.19}$$

Suppose $\hat{\sigma}$ was determined from 10 measurements. For $\nu = 9$, χ^2_{opt} is[5] 8.34283, and $\sigma = \sqrt{9/8.34283}\,\hat{\sigma} = 1.03864\,\hat{\sigma}$. In general, $\hat{\sigma}$ is expected to slightly underestimate σ.

The chi-square distribution was derived by Helmert[6] in 1876. It is quite fundamental to statistical theory. For example, the Student distribution is obtained from an average of the normal distribution weighted by the chi-square distribution. The average is performed by integrating over variable σ for specified $\hat{\sigma}$

9.2 Propagation of Error

Suppose that some physical property f depends on a measured property x according to a function $f(x)$. Let δ_x be a measure of the random error in x such that the empirical result can be described as $x \pm \delta_x$. For example, δ_x might be the result of a statistical analysis such as a 95% confidence interval from a t-test. If we only have one or two measurements, which is too few for applying statistical methods, we could try to predict a confidence interval from an analysis of the experimental procedure. For example, the precision with which you can read the instrument's signal can in some cases be used as a crude estimate of δ_x. If the instrument has a predicted measurement precision stated by the manufacturer, then that could be used. What we would like to know is the magnitude of error δ_f in f that results from the error in δ_x in x.

---

[4]Pronounced "kī" (rhymes with "pie") in English.

[5]In *Mathematica*, `InverseCDF[ChiSquareDistribution[9],0.5]` gives χ^2_{opt} for $\nu = 9$.

[6]German geodesist and statistician Friedrich R. Helmert (1842-1917). He also derived the "Student" distribution decades before Gosset independently rediscovered it.

Let us start with

$$f + df = f + \frac{df}{dx}dx. \tag{9.20}$$

df and dx are infinitesimally small. Suppose, however, that we approximate them with finite values

$$|df| \approx \delta_f\,, \qquad |dx| \approx \delta_x\,. \tag{9.21}$$

If δ_x is small, this is a reasonable approximation. Then

$$\delta_f \approx \left|\frac{df}{dx}\right| \delta_x\,. \tag{9.22}$$

The empirical value for f would be reported as $f \pm \delta_f$.

If f depends on two measured properties, x and y with error estimates δ_x and δ_y, then we can write

$$df = \frac{\partial f}{\partial x}\,dx + \frac{\partial f}{\partial y}\,dy \tag{9.23}$$

and

$$\delta_f \approx \frac{\partial f}{\partial x}\,\delta_x + \frac{\partial f}{\partial y}\,\delta_y\,, \tag{9.24}$$

but this is appropriate only for propagating *systematic* errors. If the errors in δ_x and δ_y represent *random* variation, then this equation overestimates the resulting error in f. For random variation, δ_x and δ_y are, by definition, positive quantities, appearing in confidence intervals with a plus or minus in front. Eq. (9.24) does not take into account the fact that half the time the errors in x and y have opposite sign. This must be the case if x and y are independent random variables.

It is easy to derive a more appropriate error estimate for δ_f for the case of random errors. Let us square Eq. (9.23),

$$(df)^2 = \left(\frac{\partial f}{\partial x}\right)^2 (dx)^2 + \left(\frac{\partial f}{\partial y}\right)^2 (dy)^2 + 2\frac{\partial f}{\partial x}\frac{\partial f}{\partial y}\,dxdy,$$

and then average over many repetitions of the experiment,[7]

$$\langle (df)^2\rangle = \left(\frac{\partial f}{\partial x}\right)^2 \langle (dx)^2\rangle + \left(\frac{\partial f}{\partial y}\right)^2 \langle (dy)^2\rangle + 2\frac{\partial f}{\partial x}\frac{\partial f}{\partial y}\langle dxdy\rangle. \tag{9.25}$$

The value of $dxdy$ for random fluctuations is as likely negative as positive. Therefore, $\langle dxdy\rangle = 0$. Identifying δ_f^2, δ_x^2, and δ_y^2, with $\langle (df)^2\rangle$, $\langle (dx)^2\rangle$, and $\langle (dy)^2\rangle$, we obtain

$$\delta_f^2 \approx \left(\frac{\partial f}{\partial x}\right)^2 \delta_x^2 + \left(\frac{\partial f}{\partial y}\right)^2 \delta_y^2\,. \tag{9.26}$$

With two or more independent variables with random error, an expression such as Eq. (9.26) should be used. If the error is systematic, then Eq. (9.24)

[7] The expectation value notation, $\langle \cdots \rangle$, defined in Section 8.2, is being used here rather carelessly. Strictly speaking, the argument is as follows: Replace dx with $(x - \mu_x)$ and dy with $(y - \mu_y)$, where μ_x and μ_y are the true values, and then calculate the expectation value with a probability distribution function $P(x, y)$ that is symmetric about μ_x with respect to x and symmetric about μ_y with respect to y.

should be used. A systematic error is a consistent bias in one direction. It that case it is necessary to take into account the signs of the errors.

Example 9.5. *Standard error.* We can use a propagation of error analysis to derive $\hat{\sigma}/\sqrt{N}$ as an estimate for the standard deviation of the mean of N measurements. Let $\bar{x} = (x_1 + x_2 + \cdots + x_N)/N$ be our function $f(x_1, x_2, \ldots, x_N)$ and assume that each x_N has the same error measure δ_x. Then,

$$\delta_{\bar{x}}^2 = \sum_{n=1}^{N} \left(\frac{\partial \bar{x}}{\partial x_n}\right)^2 \delta_x^2 = \sum_{n=1}^{N} \left(\frac{1}{N}\right)^2 \delta_x^2 = \frac{1}{N^2}\delta_x^2 \sum_{n=1}^{N} 1 = \frac{1}{N^2}\delta_x^2 N = \delta_x^2/N.$$

With $\delta_x = \hat{\sigma}$, we obtain the standard error as the standard deviation of the mean,

$$\hat{\sigma}_{\bar{x}} = \hat{\sigma}/\sqrt{N}\,. \tag{9.27}$$

Eq. (9.26) can be a useful aid in designing an experimental procedure. Suppose, for example, that $(\partial f/\partial x)^2$ is much larger than $(\partial f/\partial y)^2$. This would mean that δ_f is extremely sensitive to δ_x. The experiment should then be designed to make the x measurement as precise and accurate as possible. The precision and accuracy of the y measurement would be of less concern.

The rules for significant figures, given in first-year chemistry textbooks, are a convenient way to keep track of error propagation from simple arithmetic operations. They can be derived from the equations for propagation of error.

Example 9.6. *Rules for significant figures.* Let $f = x + y$ and suppose that the first uncertain digit to the right of the decimal place is for x the n_xth digit and for y the n_yth digit. Let $\delta_x = 10^{-n_x}$ and $\delta_y = 10^{-n_y}$. It follows from Eq. (9.24) that

$$\delta_{x+y}^2 = \delta_x^2 + \delta_y^2 = 10^{-2n_x} + 10^{-2n_y} = 10^{-2n_x}\left[1 + 10^{-2(n_y - n_x)}\right]. \tag{9.28}$$

Suppose that $n_x < n_y$; in other words, x has fewer significant digits to the right of the decimal than does y. Then $10^{-2(n_y-n_x)} \ll 1$, which implies that $\delta_{x+y} \approx 10^{-n_x}$. Thus, we obtain the rule for addition (and subtraction): The number of significant digits to the right of the decimal in $x + y$ is the same as in x. What if $n_x = n_y$? Then $\delta_{x+y} = \sqrt{2}\,\delta = 1.4 \times 10^{-n}$, which is slightly larger than what the "rule" suggests.

Now consider multiplication. The rule in this case is that the number with the smallest number of total significant figures determines the number of total significant figures in the product. Let us define the total number of significant figures as

$$\mathrm{sf}_x = 1 - \log_{10}\left(\delta_x/|x|\right) = 1 - \tfrac{1}{2}\log_{10}\left(\delta_x^2/x^2\right). \tag{9.29}$$

(You can verify that this definition is reasonable using numerical examples.) Propagation of error gives us

$$\delta_{xy}^2 = y^2\delta_x^2 + x^2\delta_y^2, \tag{9.30}$$

$$\begin{aligned}\mathrm{sf}_{xy} &= 1 - \tfrac{1}{2}\log_{10}\left(\delta_{xy}^2/x^2y^2\right) = 1 - \frac{1}{2}\log_{10}\left[\frac{\delta_x^2}{x^2}\left(1 + \frac{\delta_y^2 x^2}{\delta_x^2 y^2}\right)\right] \\ &= \mathrm{sf}_x - \tfrac{1}{2}\log_{10}\left(1 + \frac{\delta_y^2/y^2}{\delta_x^2/x^2}\right).\end{aligned} \tag{9.31}$$
$$\tag{9.32}$$

If $\mathrm{sf}_x < \mathrm{sf}_y$, then $\delta_y^2/y^2 < \delta_x^2/x^2$, the logarithm is about equal to zero, and $\mathrm{sf}_{xy} \approx \mathrm{sf}_x$. If $\mathrm{sf}_x \approx \mathrm{sf}_y$, then $\delta_y^2/y^2 \approx \delta_x^2/x^2$, $\mathrm{sf}_{xy} \approx \mathrm{sf}_x - (\log_{10} 2)/2 = \mathrm{sf}_x - 0.15$, and the rule very slightly overestimates the number of significant figures.

9.3 Monte Carlo Simulation of Error

Another way to obtain a confidence interval for a set of N measurements is to simulate the measurement process on a computer. This can be useful as an alternative to the t-test if the underlying distribution is not normal or log-normal. It also provides an alternative to Eqs. (9.22) and (9.26) for propagation of random error, which can be useful for estimating a value that depends on the measured property in a complicated way. The idea is to simulate the results of many repetitions of the experiment with computer-generated random numbers according to a specified probability distribution. The value of a property of interest is calculated for each simulated data sample and then the values are arranged in increasing order. A centrally positioned range containing 95% of the values is taken as an estimate for the 95% confidence interval. For example, for a collection of 1000 simulated samples, the 25th and 975th ordered values are the limits of the 95% confidence interval.

Example 9.7. *Monte Carlo determination of 95% confidence interval of the mean.* To illustrate the procedure, let us use a Monte Carlo simulation of N measurements of a property x for which the variability is described by a normal distribution. This ought to reproduce the confidence interval given by the t-test, Eq. (9.11).

Suppose that our measured data set is the following:

$$\{4.866, 4.797, 4.993, 4.847, 4.880, 4.816, 4.953, 4.909, 4.730, 4.962\}.$$

First we must characterize the distribution function. Let us use the mean, $\bar{x} = 4.8753$, as the center of the distribution, μ. The standard deviation, $\hat{\sigma} = 0.0818$, underestimates the true scale, σ, of the distribution. For a sample of 10 elements, we need to increase it by a factor of 1.039 (see Example 9.4), which gives us $\sigma = 0.0850$.

Next, we generate 1000 simulated data sets, each with 10 elements, compute the mean of each set, and arrange the resulting values in increasing order. This can be accomplished in *Mathematica* with the following statement:

```
mcmeanlist = Sort[ Table[
  Mean[ RandomReal[NormalDistribution[4.8753, 0.0850], 10] ], {1000}] ]
```

The 25th element in this list is 4.823 while the 975th element is 4.928. This agrees reasonably well with the result from the t-test, $4.817 < \mu < 4.934$. Alternatively, we can use the *median* of each sample instead of the mean:

```
mcmedlist = Sort[ Table[
 Median[ RandomReal[NormalDistribution[4.8753, 0.08505], 10] ], {1000}] ]
```

The 25th and 975th elements give the confidence interval $4.811 < \mu < 4.937$, which also agrees well with the t-test.

Monte Carlo simulation can only be performed if the distribution function is known. There is a related technique, called ***bootstrap resampling***, that constructs a set of simulated data samples using only the original sample as input.[8] No assumption is made about the underlying distribution function.

[8]The word "bootstrap" here refers to the technique of lifting oneself up from a prone position on the floor by pulling on one's bootstraps. Bootstrap resampling was invented by American statistician Bradley Efron in 1979. For a detailed discussion see B. Efron and R. J. Tibshirani, *An Introduction to the Bootstrap* (Chapman & Hall, New York, 1993).

Given a measured set of N data, a new sample is created by replacing each element of the set with a randomly chosen element of the set. This is done repeatedly until one has enough samples to determine a confidence interval.

Example 9.8. *Bootstrap resampling.* Let us start with the initial data sample of Example 9.7. The following *Mathematica* function bootstraps a new sample:

```
bootstrapresample[sample_] :=
  Table[sample[[ RandomInteger[{1, Length[sample]}] ]], {Length[sample]}]
```

The following are three examples of bootstraps of the original sample:

{ 4.866, 4.866, 4.847, 4.962, 4.993, 4.962, 4.880, 4.962, 4.847, 4.953 }
{ 4.953, 4.962, 4.909, 4.816, 4.797, 4.847, 4.962, 4.953, 4.953, 4.866 }
{ 4.993, 4.797, 4.953, 4.953, 4.866, 4.797, 4.847, 4.847, 4.816, 4.953 }

We can generate 1000 such samples and analyze them as in the Monte Carlo example. Because the bootstrap is a very general method, applicable for any kind of underlying probability distribution, we will use the median rather than the mean. (The mean is only applicable to symmetric distributions.) The 25th and 975th ordered medians in this case give the confidence interval $4.816 < \mu < 4.952$.

9.4 Significance of Difference

Another use of Student's t is to test whether or not two sets of data are measuring the same sample population. For example, suppose that you measure the concentration in river water of some toxic substance, carrying out N_u measurements upstream from a factory and N_d measurements downstream from it. The individual measurements will give a range of results on account of random error, but we would like to know if there is a significant difference in concentrations that can be attributed to the factory.

Suppose the downstream mean is $\bar{x}_\mathrm{d}$ while the upstream mean is $\bar{x}_\mathrm{u}$, with standard errors $\hat{\sigma}_{\bar{x}_\mathrm{d}} = \hat{\sigma}_\mathrm{d}/\sqrt{N_\mathrm{d}}$ and $\hat{\sigma}_{\bar{x}_\mathrm{u}} = \hat{\sigma}_\mathrm{u}/\sqrt{N_\mathrm{u}}$. What we need in order to compare $\bar{x}_\mathrm{d}$ and $\bar{x}_\mathrm{u}$ is the standard error for the quantity $(\bar{x}_\mathrm{d} - \bar{x}_\mathrm{u})$. Propagation of error implies that $\sigma^2_{\bar{x}_\mathrm{d}-\bar{x}_\mathrm{u}} = \sigma^2_{\bar{x}_\mathrm{d}} + \sigma^2_{\bar{x}_\mathrm{u}}$. Assume the error distribution has the same scale σ for each sample. Then, proceeding as in Example 9.5, we obtain

$$\sigma^2_{\bar{x}_\mathrm{d}-\bar{x}_\mathrm{u}} = \sigma^2 \left(\frac{1}{N_\mathrm{d}} + \frac{1}{N_\mathrm{u}} \right). \tag{9.33}$$

Our experiment has given us two independent estimates of σ^2,

$$\hat{\sigma}^2_\mathrm{d} = \frac{1}{\nu_\mathrm{d}} \sum_{j=1}^{N_d} (x_{\mathrm{d},j} - \bar{x}_\mathrm{d})^2, \qquad \hat{\sigma}^2_\mathrm{u} = \frac{1}{\nu_\mathrm{u}} \sum_{j=1}^{N_u} (x_{\mathrm{u},j} - \bar{x}_\mathrm{u})^2. \tag{9.34}$$

Let us average $\hat{\sigma}^2_\mathrm{d}$ and $\hat{\sigma}^2_\mathrm{u}$, weighting each of them by the number of degrees of freedom,

$$\hat{\sigma}^2 = (\nu_\mathrm{d}\hat{\sigma}^2_\mathrm{d} + \nu_\mathrm{u}\hat{\sigma}^2_\mathrm{u})/\nu, \tag{9.35}$$

with $\nu_\mathrm{d} = N_\mathrm{d} - 1$, $\nu_\mathrm{u} = N_\mathrm{u} - 1$, and $\nu = \nu_\mathrm{d} + \nu_\mathrm{u}$, and use this to estimate σ^2 in Eq. (9.33),

$$\hat{\sigma}^2_{\bar{x}_\mathrm{d}-\bar{x}_\mathrm{u}} = \frac{1}{\nu} \left(\nu_\mathrm{d}\hat{\sigma}^2_\mathrm{d} + \nu_\mathrm{u}\hat{\sigma}^2_\mathrm{u} \right) \left(\frac{1}{N_\mathrm{d}} + \frac{1}{N_\mathrm{u}} \right). \tag{9.36}$$

We can use this to construct a t-test for $\mu_{\rm d} - \mu_{\rm u}$,

$$\bar{x}_{\rm d} - \bar{x}_{\rm u} - t_{\alpha/2,\,\nu}\,\hat{\sigma}_{\bar{x}_{\rm d}-\bar{x}_{\rm u}} < \mu_{\rm d} - \mu_{\rm u} < \bar{x}_{\rm d} - \bar{x}_{\rm u} + t_{\alpha/2,\nu}\,\hat{\sigma}_{\bar{x}_{\rm d}-\bar{x}_{\rm u}}. \tag{9.37}$$

The hypothesis we wish to test is "$\mu_{\rm d}$ is equal to $\mu_{\rm u}$," which would imply that $\bar{x}_{\rm d} - \bar{x}_{\rm u} - t_{\alpha/2,\,\nu}\,\hat{\sigma}_{\bar{x}_{\rm d}-\bar{x}_{\rm u}} < 0$ and $0 < \bar{x}_{\rm d} - \bar{x}_{\rm u} + t_{\alpha/2,\,\nu}\,\hat{\sigma}_{\bar{x}_{\rm d}-\bar{x}_{\rm u}}$. These are equivalent to

$$-t_{\alpha/2,\,\nu}\,\hat{\sigma}_{\bar{x}_{\rm d}-\bar{x}_{\rm u}} < \bar{x}_{\rm d} - \bar{x}_{\rm u} < t_{\alpha/2,\,\nu}\,\hat{\sigma}_{\bar{x}_{\rm d}-\bar{x}_{\rm u}}. \tag{9.38}$$

If this condition is satisfied, we conclude that our measurements are consistent with the hypothesis $\mu_{\rm d} = \mu_{\rm u}$ at confidence level $1 - \alpha$. Suppose we choose $\alpha = 0.05$. Then if $\bar{x}_{\rm d} - \bar{x}_{\rm u}$ lies within the confidence interval, we conclude there is a 95% chance the factory has no effect on the concentration of the toxic substance. (α is the probability of falsely accepting the hypothesis $\mu_{\rm d} \neq \mu_{\rm u}$. This analysis does not consider the probability of falsely rejecting it. Please read Section 12.1 before applying this test to real-life problems!)

In fact we expect the true concentration will never be *lower* downstream than upstream. Either the factory increases the concentration or it has no effect, but it is unlikely to decrease it. It makes more sense then to use a ***one-way t-test***,[9]

$$\bar{x}_{\rm d} < \bar{x}_{\rm u} + t_{\alpha,\,\nu}\,\hat{\sigma}_{\bar{x}_{\rm u}-\bar{x}_{\rm d}}, \tag{9.39}$$

with $t_{\alpha,\,\nu}$ instead of $t_{\alpha/2,\,\nu}$. For a confidence level of 95%, we use $t_{0.05,\nu}$.

This use of the t-test is based on the assumption that the two different data sets are governed by normal probability distributions with the same scale σ, with any differences between the measured $\hat{\sigma}_{\rm u}$ and $\hat{\sigma}_{\rm d}$ due to random fluctuations. Let us now ask another question: Given two sets of measurements, can we conclude that they correspond to different values of σ? A procedure for answering this question was developed by Fisher in the 1920's.[10] Consider two samples, one with N_1 elements and another with N_2 elements, and compute the variances $\hat{\sigma}_1^2$ and $\hat{\sigma}_2^2$ of each sample and the ratio

$$F = \hat{\sigma}_1^2 / \hat{\sigma}_2^2\,. \tag{9.40}$$

This is called the ***F-ratio*** ("F" in honor of Fisher). Suppose that both samples are described by a normal distribution (with no outliers) and both have the *same* value for the scale σ. Fisher showed that the probability distribution function for observing any given value of F is then

$$P^{(F)}_{\nu_1,\nu_2}(F) = \frac{1}{B(\nu_1/2, \nu_2/2)} \left(\frac{\nu_1}{\nu_2}\right)^{\nu_1/2} F^{(\nu_1-2)/2} \left(1 + F\nu_1/\nu_2\right)^{(\nu_1+\nu_2)/2}, \tag{9.41}$$

where B is the beta function. Fig. 9.3 shows an example. The area under the curve for F less than some value $F_{\rm low}$ is given by the cumulative distribution

[9] Also called a *one-sided* test.

[10] Ronald A. Fisher (1890-1962) was an English biologist and statistician. He developed a mathematically rigorous but computationally tractable statistical methodology to aid his work on evolutionary genetics. It was he who realized the importance of the work of Gosset (and Helmert) and developed it as a practical tool.

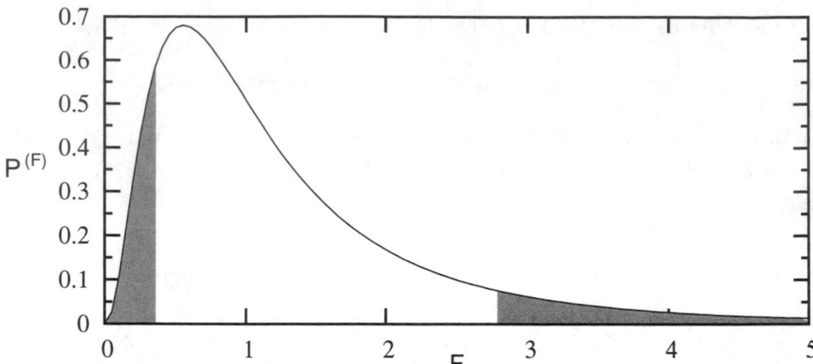

Figure 9.3: Probability distribution for the F-ratio with $\nu_1 = \nu_2 = 7$. The shaded regions each have area 0.1. The range of F values corresponding to the unshaded region gives a confidence interval within which the observed $F = \hat{\sigma}_1^2/\hat{\sigma}_2^2$ values would be expected to fall 80% of the time.

function

$$Q^{(F)}_{\nu_1,\nu_2}(F_{\rm low}) = \int_0^{F_{\rm low}} P^{(F)}_{\nu_1,\nu_2}(f)df = I_{F_{\rm low}\nu_1/(F_{\rm low}\nu_1+\nu_2)}\left(\frac{\nu_1}{2}, \frac{\nu_2}{2}\right), \qquad (9.42)$$

where $I_b(p,q)$ is the regularized incomplete beta function, according to Eq. (2.56). The probability the observed F will be greater than some value $F_{\rm high}$ is

$$1 - Q^{(F)}_{\nu_1,\nu_2}(F_{\rm high}) = 1 - \int_0^{F_{\rm high}} P^{(F)}_{\nu_1,\nu_2}(f)df = 1 - I_{F_{\rm high}\nu_1/(F_{\rm high}\nu_1+\nu_2)}\left(\frac{\nu_1}{2}, \frac{\nu_2}{2}\right).$$

For specified confidence level α, we can determine the confidence interval $F_{\rm low} < F < F_{\rm high}$ by solving numerically the equations $Q^{(F)}_{\nu_1,\nu_2}(F_{\rm low}) = \alpha/2$ and $Q^{(F)}_{\nu_1,\nu_2}(F_{\rm high}) = \alpha/2$ for $F_{\rm low}$ and $F_{\rm high}$, respectively. The total area of the shaded regions in Fig. 9.3 is equal to 0.20, which implies that $0.36 < F < 2.78$ is an 80% confidence interval.

A common practice, however, is to define

$$F = \hat{\sigma}^2_{\rm large}/\hat{\sigma}^2_{\rm small} , \qquad (9.43)$$

where $\hat{\sigma}^2_{\rm large}$ is the larger of the two variances and $\hat{\sigma}^2_{\rm small}$ is the smaller. With this definition, any instances of $F = \hat{\sigma}_1^2/\hat{\sigma}_2^2 < 1$ will be converted to $F > 1$ by taking the reciprocal. $P^{(F)}$ is then zero for $F < 1$, and for $F > 1$ it is given by the expression in Eq. (9.41) multiplied by 2, so that the area under curve for $F > 1$ is normalized to unity. In Fig. 9.3 the shaded region on the right-hand side, when multiplied by 2, has area $\alpha = 0.10$, instead of $\alpha/2 = 0.10$, for a 90% confidence interval $F < 2.78$. In general, we obtain the confidence interval[11]

$$F < F_{\alpha,\, \nu_{\rm large},\, \nu_{\rm small}} \qquad (9.44)$$

[11]The subscripts "large" and "small" refer to the variances, not the degrees of freedom. The number of measurements in the small-variance sample could very well be larger than that in the large-variance sample, in which case $\nu_{\rm small}$ would be larger than $\nu_{\rm large}$.

for F as defined in Eq. (9.43), with the limiting F value obtained by solving the equation

$$I_{F\nu_{\rm large}/(F\nu_{\rm large}+\nu_{\rm small})}\left(\frac{\nu_{\rm large}}{2}, \frac{\nu_{\rm small}}{2}\right) = 1 - \alpha. \tag{9.45}$$

This confidence interval allows us to carry out a test, called the ***F*-test**, of the following hypothesis:

> *For a given observed value of F, both samples come from distributions with the same scale σ.*

This is an example of a ***null hypothesis***, a statement making the claim that the phenomenon we seem to be observing is not actually occurring. Even though the observed variances $\hat{\sigma}_1^2$ and $\hat{\sigma}_2^2$ are not exactly the same, we nevertheless hypothesize that the true scales σ_1 and σ_2 (without hats) of the underlying distributions are equal.

Example 9.9. *Testing significance of difference.* Six measurements of a toxic substance were carried out upstream from a factory and ten measurements were carried out downstream. The values $\bar{x}_{\rm u} = 0.262$ mg/L, $\hat{\sigma}_{\rm u}^2 = 0.00548$ mg^2/L^2 upstream and $\bar{x}_{\rm d} = 0.303$ mg/L, $\hat{\sigma}_{\rm u}^2 = 0.00219$ mg^2/L^2 downstream were obtained using Huber estimators. Let us use an F-test at the 95% confidence level to compare the variances. $\nu_{\rm large} = 6 - 1 = 5$ and $\nu_{\rm small} = 9$. $F_{0.05,5,9}$ for the confidence limit is determined by solving the equation

$$I_{5F/(5F+9)}\left(\tfrac{5}{2}, \tfrac{9}{2}\right) = 0.95.$$

for F. The solution[12] is 3.48. The observed F, from Eq. (9.43), is 0.00548/0.00219 = 2.50, which is less. We accept the null hypothesis that $\sigma_{\rm d} = \sigma_{\rm u}$.

We can then proceed with a one-way t-test of the $\hat{\mu}$ values. Eq. (9.36) gives

$$\hat{\sigma}^2_{\bar{x}_{\rm d}-\bar{x}_{\rm u}} = \frac{9 \cdot 0.00219 \text{ mg}^2/\text{L}^2 + 5 \cdot 0.00548 \text{ mg}^2/\text{L}^2}{14}\left(\tfrac{1}{10} + \tfrac{1}{6}\right) = 0.000897 \text{ mg}^2/\text{L}^2.$$

The $t_{\alpha,\nu}$ value for $\alpha = 0.05$ and $\nu = 14$ is 1.761. The t-test is

$$\bar{x}_{\rm d} = 0.303 \text{ mg/L} \overset{?}{<} \bar{x}_{\rm u} + t_{0.05,14}\,\hat{\sigma}_{\hat{x}_{\rm d}-\hat{x}_{\rm u}} = 0.315 \text{ mg/L}\,,$$

which is true. Our experiment cannot detect a difference between $\mu_{\rm d}$ and $\mu_{\rm u}$ with 95% confidence. It is a good idea to also report the confidence level p at which $\bar{x}_{\rm d} = \bar{x}_{\rm u} + t_{p,14}\hat{\sigma}_{\bar{x}_{\rm d}-\bar{x}_{\rm u}}$. In this case, $p = 0.096$, corresponding to confidence level of $1 - p = 0.904$, or 90.4%. An F-test with α as large as 0.096 would be consistent with the null hypothesis. We conclude that there is a 9.6 % chance that the apparent concentration increase is just due to random error in the measurements.

If the samples "pass" the F-test, then we conclude that the variances are the same, with level of significance $1 - \alpha$, and then perform the t-test for significance of difference of the means. If the samples fail the F-test then the analysis is more complicated. Various alternatives to the t-test have been proposed for the case of unequal variances but they involve approximations that make them less rigorous than the t-test. It is probably better to instead carry out a Monte Carlo simulation.

[12]Using *Mathematica*, we could obtain this from the following:

```
Solve[BetaRegularized[5 * f/(5 * f + 9), 5/2, 9/2] == 0.95, f]
```

Example 9.10. *Monte Carlo test of significance of difference.* Consider two samples, with means $\bar{x}_1$ and $\bar{x}_2$, $\bar{x}_1 > \bar{x}_2$, and variances $\hat{\sigma}_1^2$, $\hat{\sigma}_2^2$ that the F-test has declared to be significantly different from each other. Let us test the hypothesis that the underlying probability distributions of both samples are centered at the same value μ. Use the pooled mean

$$\mu \approx \bar{x} = (N_1\bar{x}_1 + N_2\bar{x}_2)/(N_1 + N_2) \tag{9.46}$$

as the center of two normal distributions, one with scale σ_1 and the other with scale σ_2. Following the procedure in Example 9.7, generate N_1 random values for sample 1 and N_2 random values for sample 2, and then calculate $\Delta = \bar{x}_1 - \bar{x}_2$. This can be repeated very many times. Counting the number of simulated samples for which Δ is greater than that for the observed samples gives the confidence limit.

9.5 Distribution Testing*

Given only a measured data sample, how does one determine which kind of probability distribution describes the random variation? A simple approach is to prepare a bar graph, by dividing the x-axis into evenly spaced intervals and plotting the number of data points in each interval. This is called a ***histogram***. If the data sample size is sufficiently large, the shape of the histogram might suggest the appropriate probability distribution function.

Example 9.11. *Histogram of a normally distributed data set.* The left-hand panel of Fig. 9.4 shows a histogram of 100 numbers chosen randomly from a normal distribution. It uses a bin width of 2. For example, we can see from the figure that there are 13 data elements with values between 26 and 28. Superimposed on the histogram is a plot of the distribution function multiplied by the total number of data elements. Although the histogram is slightly asymmetric, it shows the expected bell shape. The other two panels show histograms of subsets of points randomly chosen from the initial set. With 20 points the histogram is sufficiently asymmetric that one might doubt that the underlying distribution was normal. With just five points the histogram bears little resemblance to the actual distribution.

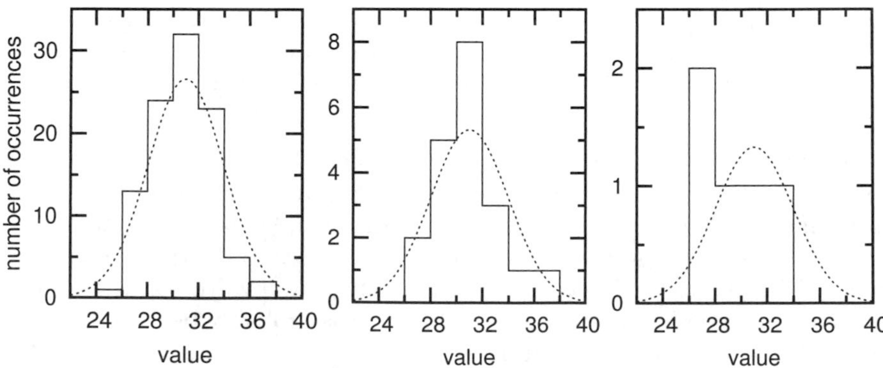

Figure 9.4: Histograms of 100 (left-hand panel), 20 (center panel), and 5 (right-hand panel) normally distributed random numbers with $\mu = 31$ and $\sigma = 3$. Dotted curves show the distribution function normalized to the same area as the histogram.

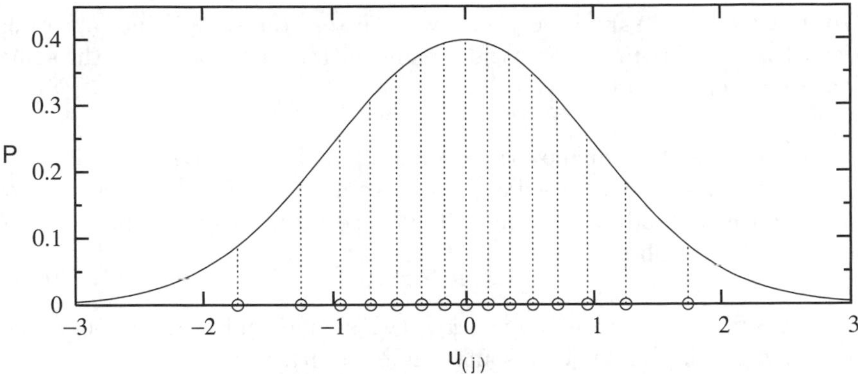

Figure 9.5: Plot of the distribution function for the standard normal deviate showing the order statistic for a sample size of 15.

In general, small data samples will tend to look asymmetric even if they come from a symmetric distribution, and the histrogram will not be very informative. A more sensitive method for testing a sample for normality is an ***ordered probability plot***. This a special kind of plot constructed so that it will give a straight line if the data come from a normal distribution. It is based on the idea that when the data sample is arranged in ascending order, the spacing of the points will show a distinct pattern if the distribution is normal, with bunching near the center.

A simple way to obtain the expected spacing pattern is a Monte Carlo simulation. The following *Mathematica* functions accomplish this:

```
ostatnormal[n_] := Sort[RandomReal[NormalDistribution[0, 1], n] ]

mcostatnormal[n_, m_] :=
  Module[{sum, avg}, sum = Table[0, {n}];
    Do[sum = sum + ostatnormal[n], {j, 1, m}]; avg = sum/m]
```

The first function generates a random set of `n` elements from a normal distribution centered at zero with $\sigma = 1$. The second averages `m` such sets. Averaging 2,000,000 simulated samples of 15 elements, we obtain

$$\{-1.736,\ -1.248,\ -0.948,\ -0.715,\ -0.516,\ -0.335,\ -0.165,\ 0.000,\ 0.165,\ 0.335,\ 0.516,\ 0.715,\ 0.948,\ 1.248,\ 1.736\}\,.$$

This set is called the ***standard deviate order statistic*** for a normal distribution with a sample size of 15. (An *order* statistic is something obtained from a statistical sample that has been arranged in ascending order. A *standard deviate* is a random variable from a distribution centered at $\mu = 0$ with $\sigma = 1$.) We will use the notation $u_{(1)}, u_{(2)}, u_{(3)}, \ldots,\ u_{(15)}$ for the elements of this set, with subscripts in parentheses to indicate that the elements are arranged in ascending order. As can be seen in Fig. 9.5, the values are indeed somewhat bunched up at the center and spread out at the tails.

Given a measured sample set $\{x_j\}$, we can sort the sample into an order statistic $\{x_{(j)}\}$ and then compare the spacing with that of the $u_{(j)}$ of the same sample size. Let us write

$$x = \mu + \sigma u,$$

where μ and σ are the (unknown) parameters of the underlying distribution of the physical property x. Solving for u we obtain $u = (x - \mu)/\sigma$, which by construction is a standard deviate.[13] If we use the measured values $x_{(j)}$ as estimates for x, we obtain

$$x_{(j)} \approx \mu + \sigma u_{(j)}\,. \tag{9.47}$$

This implies that if we plot $x_{(j)}$ vs. $u_{(j)}$, we should find that the points lie along a straight line, albeit with some random scatter.

Example 9.12. *Probability plots.* Consider the following set of 15 values randomly generated from a normal distribution with $\mu = 2$ and $\sigma = 0.5$, arranged in ascending order:

$\{0.77, 1.09, 1.68, 1.95, 1.98, 2.07, 2.26,$
$2.28, 2.35, 2.42, 2.56, 2.59, 2.82, 2.88, 3.23\}$

These are plotted vs. $u_{(j)}$ at right. They more or less lie on a straight line, as they should.

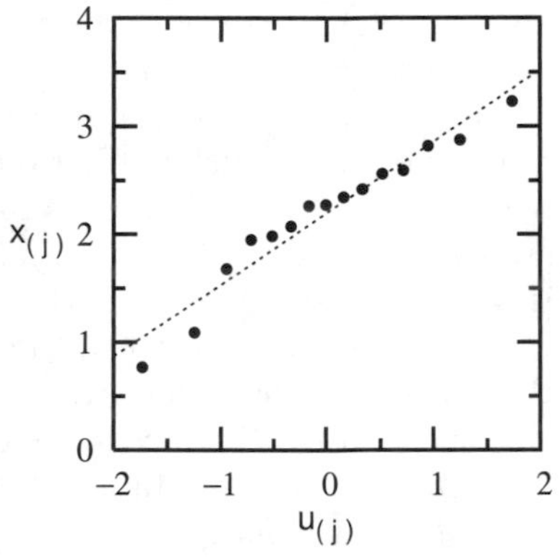

Consider another sample, this one from a log-normal distribution, with $\mu = \ln 2$, $\sigma = 0.5$:

$\{1.15, 1.34, 1.38, 1.47, 1.58, 1.59, 1.65,$
$1.93, 1.97, 2.54, 2.77, 2.84, 2.87, 4.16, 5.18\}$

The plot below on the left shows a marked curvature, which suggests that the sample is not from normal distribution. The log plot on the right is much straighter.

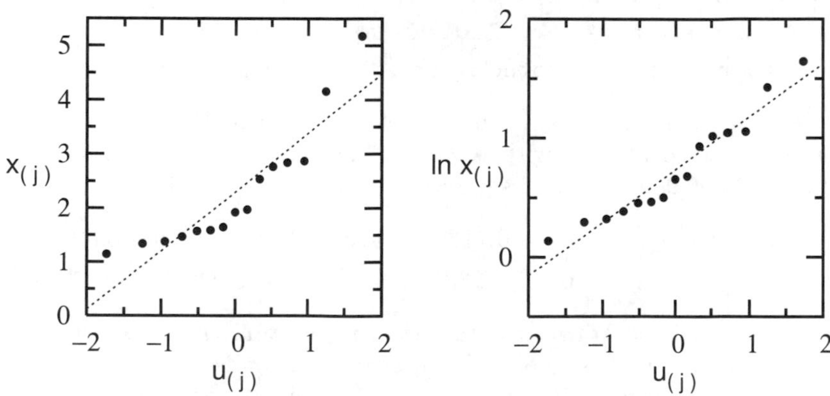

Plots of this sort can be useful but they are not definitive. We can see from Example 9.12 that even for a simulated sample from a normal distribution with no outliers, there is some curvature. Ambiguous cases are common.

[13] To prove this, make the change of variable $x = \mu + \sigma u$, $dx = \sigma du$ in $P_{\mu,\sigma}(x)dx$.

Table 9.1: Shapiro-Wilk parameters a_{N-k+1} for sample size N. [S. S. Shapiro and M. L. Wilk, "An Analysis of Variance Test for Normality (Complete Samples)," *Biometrika* **52**(3-4), 603 (1965), by permission of Oxford University Press.]

$k \backslash N$	3	4	5	6	7	8	9	10	11
1	0.7071	0.6872	0.6646	0.6431	0.6233	0.6052	0.5888	0.5739	0.5601
2		0.1677	0.2413	0.2806	0.3031	0.3164	0.3244	0.3291	0.3315
3				0.0875	0.1401	0.1743	0.1976	0.2141	0.2260
4						0.0561	0.0947	0.1224	0.1429
5								0.0399	0.0695

$k \backslash N$	12	13	14	15	16	17	18	19	20
1	0.5475	0.5359	0.5251	0.5150	0.5056	0.4968	0.4886	0.4808	0.4734
2	0.3325	0.3325	0.3318	0.3306	0.3290	0.3273	0.3253	0.3232	0.3211
3	0.2347	0.2412	0.2460	0.2495	0.2521	0.2540	0.2553	0.2561	0.2565
4	0.1586	0.1707	0.1802	0.1878	0.1939	0.1988	0.2027	0.2059	0.2085
5	0.0922	0.1099	0.1240	0.1353	0.1447	0.1524	0.1587	0.1641	0.1686
6	0.0303	0.0539	0.0727	0.0880	0.1005	0.1109	0.1197	0.1271	0.1334
7			0.0240	0.0433	0.0593	0.0725	0.0837	0.0932	0.1013
8					0.0196	0.0359	0.0496	0.0612	0.0711
9							0.0163	0.0303	0.0422
10									0.0140

Another approach is a test developed by Shapiro and Wilk.[14] It gives a test statistic to be compared with an expected value at a given confidence level, in a manner analogous to the t- and F-tests. The test statistic, called "W," is the ratio of the square of the slope of the best fit line through the probability plot to the sum of squares of deviations of the actual points from this line. If the points all lie close to the line then the denominator is small and W is large. Thus, a large value of W suggests a normal distribution. The slope of the probability plot line is an estimate for the scale of the distribution. A large slope predicts large deviations even if the distribution is normal. For this reason, W is taken as the ratio of these two quantities.

The slope of the best fit line can be computed using methods described in the next chapter. However, the usual procedure is simply to use the following formula, which gives an explicit solution:

$$W = b^2/SS\,, \quad SS = \sum_{j=1}^{N}(x_j - \bar{x})^2, \quad b = \sum_{j=1}^{[N/2]} a_{N-j+1}(x_{(N-j+1)} - x_{(j)})\,, \tag{9.48}$$

where "$[N/2]$" means the largest integer less than or equal to $N/2$, and the a_k are tabulated parameters. The set $\{a_k\}$ depends on the sample size. Table 9.1 shows the values for $N \leq 20$. (For larger sample sizes, see the original paper.) Critical values of W_α for $\alpha = 0.05$ (5% confidence) are are given in Table 9.2. These should be interpreted as follows: The probability that a sample from a

[14] S. S. Shapiro and M. L. Wilk, *Biometrika* **52**, 591-611 (1965).

Table 9.2: Critical values of the Shapiro-Wilk test statistic at 95% confidence for sample size N. [S. S. Shapiro and M. L. Wilk, *ibid.*, by permission of Oxford University Press.]

N	3	4	5	6	7	8	9	10	11
W_{crit}	0.767	0.748	0.762	0.788	0.803	0.818	0.829	0.842	0.850
N	12	13	14	15	16	17	18	19	20
W_{crit}	0.859	0.866	0.874	0.881	0.887	0.892	0.897	0.901	0.905

normal distribution would give a W value less than $W_{0.05}$ is only 5%. If the actual W value for the sample is less than the $W_{0.05}$ for the sample size, then one can conclude with 95% confidence that the sample does *not* come from a normal distribution.

Example 9.13. *Shapiro-Wilk test.* Let us test the sample of five values that were used to compute the histogram in the right-hand panel of Fig. 9.4.

j	1	2	3	4	5
$x_{(j)}$	27.262	27.781	29.828	30.784	32.595
$x_{(6-j)}$	32.595	30.784	29.828	27.781	27.262
$x_{(6-j)} - x_{(j)}$	5.333	3.003			
a_{6-j}	0.6646	0.2413			

The $x_{(N-j+1)}$ in are just the order statistic $x_{(j)}$ in reverse (descending) order.

$$b = 4.269, \quad \bar{x} = 29.650, \quad SS = 19.186, \quad W = 0.9498.$$

The critical value at 95% is $W_{0.05} = 0.762$. We conclude that this sample *is* consistent with a normal distribution, despite the fact that the histogram seems quite asymmetric. However, this is not the end of the analysis. This test does not rule out the possibility that the sample could also be consistent with some other distribution.

To see if the sample could have come from a log-normal distribution, we replace $x_{(j)}$ with $\ln x_{(j)}$:

j	1	2	3	4	5
$x_{(j)}$	3.30549	3.32435	3.39545	3.42700	3.48416
$x_{(6-j)} - x_{(j)}$	0.17867	0.10264			

$$b = 0.14351, \quad \bar{x} = 3.38729, \quad SS = 0.021678, \quad W = 0.9500.$$

Thus, we find that a log-normal distribution is also consistent with the sample. To distinguish between these two possible distributions one would have to increase the sample size by carrying out additional measurements.

Example 9.13 demonstrates that the Shapiro-Wilk test can *reject* the hypothesis that the underlying distribution is normal but it cannot *confirm* it. In practice, if the sample passes the Shapiro-Wilk test then it is considered safe to use analytical methods based on the assumption of normality. Even if the distribution is non-normal, it will sufficiently resemble a normal distribution that methods such as the t- and F-tests will give reasonable results.

The Shapiro-Wilk test is not robust. It can be seriously skewed by outliers. It is a good idea to use the test only in conjuction with the probability plot. If a point in the plot lies far from a line that closely fits the other points, then the corresponding value may be an outlier. Removing the suspected outlier and recalculating W might allow the sample to pass the test.

So what does one do with a phenomenon that cannot be described by a normal (or log-normal) distribution? If the distribution is known then a Monte-Carlo simulation can be used. However, it is not uncommon that the distribution is non-normal and unknown. In that case one can resort to a ***nonparametric*** statistical analysis. A *parametric* statistical analysis is one based on a known distribution function characterized by parameters such as μ and σ. Nonparametric methods make no such assumption. Bootstrap resampling is an example and many other nonparametric methods are also available. For example, there is a nonparametric alternative to the Student t-test for significance of difference between means, called the Mann-Whitney U-test.[15] If the probability distribution is normal, nonparametric tests are still valid but can be expected to be less sensitive than the corresponding parametric test.

Exercises

9.1 Repeat the analysis in Example 9.3 to calculate the 95% confidence interval assuming that the measurements were performed on a standard of known concentration 0.500 mol/L.

9.2 Repeat the analysis in Example 9.3 to calculate the 95% confidence interval but use the MAD to estimate the standard deviation. Repeat the analysis again with S_{RC} instead the MAD. Repeat the analysis yet again, but this time use S_{Hub}. Discuss your results. Which method is most appropriate for this data set?

9.3 Consider again Example 9.3. Suppose that you return to the laboratory the next day and conduct one more measurement, and, to your surprise, the result is 0.3983 mol/L, which is lower than you expected it would be. Do your best to provide a realistic estimate of the 95% confidence interval and discuss your conclusion.

9.4 Calculate the $t_{\alpha/2,\,5}$ value and the $t_{\alpha,\,5}$ value for an 82% confidence interval.

9.5 Express the cumulative chi-square distribution in terms of an incomplete gamma function.

9.6 Suppose $\hat{\sigma}$ of a set of 10 measurements is 7.2. Calculate the 95% confidence interval for the scale of the presumably normal distribution. Plot the appropriate chi-square distribution. Indicate on the plot the χ^2 values corresponding to the 95% confidence interval.

9.7 (a) Plot an example of the chi-square distribution. (b) Determine the midpoint of the distribution (50% of area on each side). (c) Determine the most probable χ^2 value (the maximum). (d) Give a qualitative explanation for why these are, in general, not the same.

[15]See E. A. McBean and F. A. Rovers, *Statistical Procedures for Analysis of Environmental Monitoring Data and Risk Assessment* (Prentice-Hall, Upper Saddle River, NJ, 1998), for an overview of nonparametric statistical tests.

9.8 For Example 9.9, (a) verify that $p = 0.096$ and (b) check if the data pass the F-test at $\alpha = 0.096$.

9.9 Use a Taylor's series to derive an additive correction to Eq. (9.26) for error propagation.

9.10 (a) Prove that $\langle (x - \langle x \rangle)^2 \rangle = \langle x^2 \rangle - \langle x \rangle^2$. (b) Starting from the definition $\sigma_{\bar{x}}^2 = \langle (\bar{x} - \langle \bar{x} \rangle)^2 \rangle$, derive the expression for the standard error, $\sigma_{\bar{x}} = \sigma / \sqrt{N}$ for the mean of N measurements.

9.11 Demonstrate by computer simulation of random data samples that the standard deviation of the mean of N measurements is smaller than the standard deviation of single measurements by a factor of $\sqrt{N}$, as follows:

(a) Generate a set of 25 random numbers drawn from a normal distribution of standard deviation $\hat{\sigma} = 1$ and centered at $\hat{\mu} = 0$, and compute the standard deviation of the set.

(b) Generate a set of 25 *pairs* of random numbers using the normal distribution of part (a). Take the mean of each pair and construct a new set of the 25 means. Is the standard deviation of the new list smaller by a factor of approximately $\sqrt{2}$?

(c) Repeat part (b) but with groups of n numbers each instead of pairs, for $1 \leq n \leq 30$.

9.12 Three different laboratories report the following results for a property x (and its corresponding N and $\hat{\sigma}$): 11.2 (5, 1.7), 12.3 (12, 1.0) 10.4 (7, 2.9). Use a Monte Carlo simulation to determine a 95% confidence interval for the weighted average.

9.13* The Shapiro-Wilk test can give an erroneous conclusion (claiming a data set is non-normal when in fact it is normal) if the data set contains an outlier. Therefore, it is prudent to try to remove outliers from the data before applying the test. Why would it *not* be a good idea to use the Q-test for this purpose?

9.14* Carry out Shapiro-Wilk tests on the two data samples in Example 9.12.

9.15 Prove that the normal distribution is the $\nu \to \infty$ limit of the Student distribution, Eq. (9.15).

9.16* Derive the Student distribution $P_\nu(\bar{x})$ as an average of the normal distribution weighted by the chi-square distribution.

9.17 Consider error propagation in 10^x.

(a) Derive the following expression for the number of significant figures in 10^x:

$$\mathrm{sf}_{10^x} = \mathrm{sf}_x - \log_{10}|x| - 0.36.$$

(b) Demonstrate that this is consistent with the following rule: The number of significant figures in 10^x is equal to the number of digits to the right of the decimal in x.

9.18 Carry out Monte Carlo test of significance of difference, as described in Example 9.10, using simulated data samples.

Chapter 10

Fitting

Suppose that we have a set of N data points,

$$\{(x_1, f_1), (x_2, f_2), (x_3, f_3), \ldots, (x_N, f_N)\},$$

and that we wish to describe them with some function $f(x)$. Typically the f_k would be experimental measurements of some physical property for different values of a coordinate x_k, and the function f would contain adjustable parameters. Let us accept that the f_k values, and perhaps also the x_k values, contain random error. The presence of error makes this problem different from that of interpolation. If the points (x_k, f_k) themselves contain error, it is not so important for $f(x)$ to reproduce the points exactly. What we want is a ***fit*** rather than an interpolation; that is, we want a function $f(x)$ that is the most probable guess for the true underlying functional relationship, rather than an equation that simply interpolates between the measured values.

10.1 Method of Least Squares

10.1.1 Polynomial Fitting

Let us express f as a polynomial,

$$f(x) = \sum_{j=0}^{M} \hat{c}_j x^j , \tag{10.1}$$

with as yet undetermined parameters $\hat{c}_0$, $\hat{c}_1, \ldots,$ $\hat{c}_M$. The set of monomials $\{1, x, x^2, \ldots, x^M\}$ is called the *basis set* and the parameters $\{\hat{c}_0, \hat{c}_1, \hat{c}_2, \ldots, \hat{c}_M\}$ are called the *coefficients*. We want the set[1] $\{\hat{c}_j\}$ that gives the best fit of the data. First, however, we must define what we mean by "best."

It seems reasonable to choose $\{\hat{c}_j\}$ as the set that minimizes the deviations between the predicted values of $f(x_k)$ and the measured data points f_k. We saw in Chapter 8 that there is no unique definition of "deviation." We will take our inspiration from Eq. (8.3), the definition of the variance, and consider the sum of squares of the differences,[2]

$$SS = \sum_{k=1}^{N} [\, f_k - f(x_k)\,]^2, \tag{10.2}$$

as our measure of deviation from the data for any given choice of the set $\{\hat{c}_j\}$.

[1] $\{\hat{c}_j\}$ is used here as a concise notation for $\{\hat{c}_0, \hat{c}_1, \hat{c}_2, \ldots, \hat{c}_M\}$. The hats indicate that the values of the $\hat{c}_j$ will be determined by an empirical data set.

[2] The prefactor $(N-1)^{-1}$ in Eq. (8.3) is omitted here because it would have no effect on this analysis.

Determining the $\hat{c}_j$ by minimizing SS is called the ***method of least squares*** (abbreviated "LS"). The differences $f_k - f(x_k)$ are called ***residuals***. Note that SS is not a function of the variable x, but rather a functional $SS[f]$ of the function f, in which the "variable" is the choice of definition for f.

A necessary condition for f to minimize SS is that the partial derivatives $\partial SS/\partial \hat{c}_j$ each be zero,

$$0 = \frac{\partial}{\partial \hat{c}_j} \sum_{k=1}^{N} \left(f_k - \sum_{i=0}^{M} \hat{c}_i x_k^i \right)^2 = -2 \sum_{k=1}^{N} x_k^j \left(f_k - \sum_{i=0}^{M} \hat{c}_i x_k^i \right) . \tag{10.3}$$

Example 10.1. *Fitting with a straight line.* The most commonly used fit is a straight line. Let us write the fitting function as $y(x) = \hat{a} + \hat{b}x$ and designate the data set as $\{(x_1, y_1), (x_2, y_2), \ldots, (x_N, y_N)\}$. Then

$$0 = \frac{\partial}{\partial \hat{a}} \sum_{k=1}^{N} (y_k - \hat{a} - \hat{b}x_k)^2, \qquad 0 = \frac{\partial}{\partial \hat{b}} \sum_{k=1}^{N} (y_k - \hat{a} - \hat{b}x_k)^2.$$

But $(\partial/\partial\hat{a})\,(y_k - \hat{a} - \hat{b}x_k)^2 = -2(y_k - \hat{a} - \hat{b}x_k)$. Therefore, $0 = -2\sum_{k=1}^{N}(y_k - \hat{a} - \hat{b}x_k)$, which implies that

$$\sum_{k=1}^{N} y_k - N\hat{a} - \hat{b} \sum_{k=1}^{N} x_k = 0 .$$

We have used the fact that $\sum_{k=1}^{N} \hat{a} = \hat{a} + \hat{a} + \cdots + \hat{a}$ (N times), which equals $N\hat{a}$. Dividing through by N gives

$$\hat{a} = \bar{y} - \hat{b}\,\bar{x}. \tag{10.4}$$

For the derivative with respect to $\hat{b}$, we have $0 = -2\sum_{k=1}^{N} x_k(y_k - \hat{a} - \hat{b}\,x_k)$, which implies that

$$\hat{a} \sum_{k=1}^{N} x_k + \hat{b} \sum_{k=1}^{N} x_k^2 = \sum_{k=1}^{N} x_k y_k .$$

Substituting Eq.(10.4) for $\hat{a}$ into this equation and dividing by N gives

$$\bar{y}\bar{x} - \hat{b}\,\bar{x}^2 + \hat{b}\,\overline{x^2} = \overline{xy} ,$$

where

$$\overline{x^2} = \frac{1}{N} \sum_{k=1}^{N} x_k^2 , \qquad \overline{xy} = \frac{1}{N} \sum_{k=1}^{N} x_k y_k .$$

Solving for $\hat{b}$ and substituting the result into Eq. (10.4) gives the solution

$$\hat{b} = \frac{\overline{xy} - \bar{x}\bar{y}}{\overline{x^2} - \bar{x}^2} , \qquad \hat{a} = \frac{\bar{y}\,\overline{x^2} - \bar{x}\,\overline{xy}}{\overline{x^2} - \bar{x}^2} . \tag{10.5}$$

For arbitrary value of the polynomial degree M, Eqs. (10.3) can be rearranged into the following set of equations, one for each value of $j = 0, 1, 2, \ldots, M$:

$$\begin{aligned}
\hat{c}_0\overline{x^0} + \hat{c}_1\overline{x^1} + \hat{c}_2\overline{x^2} + \cdots + \hat{c}_M\overline{x^M} &= \overline{x^0 f} , \\
\hat{c}_0\overline{x^1} + \hat{c}_1\overline{x^2} + \hat{c}_2\overline{x^3} + \cdots + \hat{c}_M\overline{x^{M+1}} &= \overline{x^1 f} , \\
\hat{c}_0\overline{x^2} + \hat{c}_1\overline{x^3} + \hat{c}_2\overline{x^4} + \cdots + \hat{c}_M\overline{x^{M+2}} &= \overline{x^2 f} , \\
&\vdots \\
\hat{c}_0\overline{x^M} + \hat{c}_1\overline{x^{M+1}} + \hat{c}_2\overline{x^{M+2}} + \cdots + \hat{c}_M\overline{x^{2M}} &= \overline{x^M f} ,
\end{aligned} \tag{10.6}$$

where
$$\overline{x^m f} = \frac{1}{N} \sum_{k=1}^{N} x_k^m f_k \, .$$

Eqs. (10.6) are called the ***normal equations***. They are $M+1$ linear equations in the $M+1$ unknowns $\hat{c}_0, \hat{c}_1, \hat{c}_2, \ldots \hat{c}_M$. If M is small then the solution is easy to obtain analytically, as in Example 10.1. For higher M one usually uses a computer to solve for the $\hat{c}_j$. Efficient computer algorithms are available.[3]

The use of *SS* as the quality-of-fit functional can be justified from theoretical considerations. If the x_k are known exactly and the uncertainties in the f_k follow a normal distribution, then it is possible to prove that the $\{\hat{c}_j\}$ given by the LS method is the set of values that gives the highest probability of being consistent with the observed data set.[4] This kind of analysis, in which statistical methods are used to infer a fitting function from data containing random scatter, is called a ***regression***.[5] The fitting function is itself often called a "regression."

It is important to distinguish between least-squares polynomial fitting and Lagrange interpolation. Both methods model a data set using a polynomial, but interpolation gives *exact* agreement between the polynomial and the data at each point while the LS fit is not expected to give exact agreement with any of the data points. A justification for using a fit instead of an interpolation is the assumption that the data contain uncertainty. It would be artificial to force exact agreement. The number of coefficients in the polynomial is $M+1$, where M is the polynomial degree. In the case of interpolation, this number must equal the number of data points, N. With the method of least squares, the number of coefficients is usually chosen to be much smaller than the number of data points. The difference,

$$\nu = N - M - 1 \, , \tag{10.7}$$

is the number of degrees of freedom in the problem.

We can also derive normal equations to fit a function of more than one variable. This is called a ***multivariate least-squares fit***, or a ***multiple regression***. For a function $f(x, y)$, the data set would have the form $\{(x_k, y_k, f_k)\}$. Multiple straight-line regression, such as $f = \hat{a} + \hat{b}x + \hat{c}y$, is often used by economists, social scientists, and medical researchers, who routinely deal with phenomena that depend on many independent factors. The method has occasional chemistry applications.

---

[3]This is a standard problem of linear algebra. (See Section 18.7.) *Mathematica* has various different functions, such as `NSolve`, `Fit`, `FindFit`, and `LinearModelFit`, that can be used to carry out the computation. They differ in their input and output formats.

[4]This important result was proved by Gauss. It is one of the fundamental ideas of statistical analysis.

[5]The reason for this terminology is somewhat obscure. Statistical fitting methods were popularized by biologists in the late 19th century, and an early application was to try to explain the fact that the heights of offspring of individuals who are much taller or much shorter than average will in subsequent generations tend to "regress" toward the average height of the population.

Example 10.2. *Experimental determination of a reaction rate law.* The rate of a chemical reaction

$$\mathrm{A + B + C \rightarrow products}$$

will often (but not always!) be described by a "rate law" of the form

$$v = k[\mathrm{A}]^\alpha[\mathrm{B}]^\beta[\mathrm{C}]^\gamma. \tag{10.8}$$

[A], [B], and [C] are concentrations while k, α, β, and γ are constants to be determined from empirical analysis. The rate of the reaction is $v = -d[\mathrm{A}]/dt$, where t is time. The exponents α, β, γ typically are integers or half-integers. To convert the rate law to a form amenable to multiple regression, we will take the logarithm of the rate law. However, the function $\ln x$ was defined for unitless x. Let us define unitless concentrations, $\widetilde{[\mathrm{A}]} = [\mathrm{A}]/(\mathrm{mol\ L^{-1}})$, etc., unitless rate $\tilde{v} = v/(\mathrm{mol\ L^{-1}s^{-1}})$, and unitless rate constant $\tilde{k}$. (The units of k will be determined by the values we find for α, β, and γ.) Then,

$$\ln \tilde{v} = \ln \tilde{k} + \alpha \ln \widetilde{[\mathrm{A}]} + \beta \ln \widetilde{[\mathrm{B}]} + \gamma \ln \widetilde{[\mathrm{C}]}\,. \tag{10.9}$$

We can write this as

$$f(x, y, z) = \kappa + \alpha x + \beta y + \gamma z\,, \tag{10.10}$$

where $f = \ln \tilde{v}$, $\kappa = \ln \tilde{k}$, $x = \ln \widetilde{[\mathrm{A}]}$, $y = \ln \widetilde{[\mathrm{B}]}$, and $z = \ln [\mathrm{C}]$. Collect data points (x, y, z, f) by measuring the reaction rate for various different values of [A], [B], and [C]. Then solve the normal equations to determine the parameter values. In *Mathematica* this can be done with the following statement:

```
FindFit[data, kappa + alpha * x+beta * y + gamma * z,
                    {kappa, alpha, beta, gamma}, {x, y, z}]
```

`data` is the list of data points.

10.1.2 Weighted Least Squares

Suppose that some of the values f_k are known with greater certainty than others. It would seem reasonable to give greater emphasis to the values that have less error. This is easily done by modifying Eq. (10.2) as follows:

$$SS = \sum_{k=1}^{N} w_k\,[\,f_k - f(x_k)\,]^2\,, \tag{10.11}$$

where w_k is a weight factor. A relatively large w_k will give a greater emphasis to the corresponding f_k. This formulation leads to solvable linear equations. If the errors in f_k follow normal distributions

$$P^{(k)}(y) \propto e^{-[y-f(x_k)]^2/\sigma_k^2}\,, \tag{10.12}$$

with a different scale parameter σ_k for each f_k, then an obvious choice for the weighted SS is[6]

$$\chi^2 = \sum_{k=1}^{N} \left[\frac{f_k - f(x_k)}{\sigma_k} \right]^2, \tag{10.13}$$

with $w_k = 1/\sigma_k^2$. Points with higher precision (smaller σ_k) are weighted more heavily than points with lower precision.

To use χ^2 instead of the unweighted SS for a least-squares analysis, one needs values for the σ_k. If the property f is measured repeatedly for a given value x_k, then σ at that x_k can be estimated using methods of Chapters 8 and 9. Alternatively, relative values of the w_k might be determined from *a priori*

[6]This is a generalization of the definition of χ^2 ("chi-square") used in Section 9.1.

estimates of the precision of the measuring device as a function of x or from an analysis of the propagation of errors in the experiment.

A common situation that requires weighted fitting is when a nonlinear function is transformed into a linear one to carry out a straight-line fit.

Example 10.3. *Exponential fit.* Suppose a theoretical analysis predicts the relationship $y = \alpha e^{bx}$, and we want to estimate α and b from measurements of y for various x. Assuming y is positive, we take the logarithm of both sides and get $\ln y = \ln \alpha + bx$. We fit this using $f(x) = a + bx$ with $f_k = \ln y_k$ and $a = \ln \alpha$. Suppose that the x_k are known exactly but that the variability of the y_k follows a normal distribution with scale σ_y, with the same scale for each point. Then the scale for f given by propagation of error is

$$\sigma_f = \left|\frac{df}{dy}\right| \sigma_y = \frac{1}{y}\sigma_y = e^{-f}\sigma_y \,. \tag{10.14}$$

Using $\sigma_y e^{-f_k}$ for σ_k gives

$$\chi^2 = \sigma_y^{-2} \sum_{k=1}^{N} e^{2f_k} [f_k - f(x_k)]^2 \,. \tag{10.15}$$

Normal equations are obtained by setting the derivatives $\partial\chi^2/\partial\hat{a}$ and $\partial\chi^2/\partial\hat{b}$ to zero. Because σ_y^{-2} here is just an overall multiplicative factor, it will not affect the normal equations and can be omitted.

To see why the weighting might be needed, consider a large value $y_1 = 2000.0$ and a small value $y_2 = 0.2$, each with a confidence interval of ± 0.1. Then the confidence interval for f_1 ranges from $\ln 1999.9 = 7.60085$ to $\ln 2000.1 = 7.600095$, or ± 0.00005, while the interval for f_2 is from $\ln 0.1 = -2.3$ to $\ln 0.3 = -1.2$, or ± 0.5. The value f_1 deserves a larger weight because it is much more precise.

10.1.3 Generalizations of the Least-Squares Method*

A *moving* fit is a fit in which the parameters are functions of the coordinates. The fit is optimized for the region of space in which it will be used. This is conceptually similar to the idea behind spline interpolation, in which a different interpolating function is used in each interval.

The weighted least-squares method can be modified to allow the weights to be functions of x. To see why this might be useful, we first consider a ***moving weighted average***

$$\bar{f}(x) = \sum_{k=1}^{N} \widetilde{w}_k(x) f_k \tag{10.16}$$

of measured values f_k, The $\widetilde{w}_k(x)$ are normalized weight functions, normalized in the sense that $\sum_{j=1}^{N} \widetilde{w}_k(x) = 1$. The idea is to construct the weight functions such that they are peaked in the vicinity of a data point and drop toward zero as the distance of x from a data point increases. For example, we could use Gaussian functions,

$$w_k(x) = e^{-(x-x_k)^2/2\beta^2}, \qquad \widetilde{w}_k(x) = \frac{w_k(x)}{w_1(x) + w_2(x) + \cdots + w_N(x)} \,. \tag{10.17}$$

The parameter β determines how sharply the weight functions are peaked. With a very small β value we get a very close fit to each data point, but between data points $\bar{f}(x)$ drops almost to zero. A very large β gives a smooth curve but does not ensure that $\bar{f}(x_k)$ will be close to f_k. However, an intermediate value of β can give a reasonable fit.

The weights are "moving" in the sense that they depend on x. The rationale is that we should give greater weight in the fitting function to ranges of x in the vicinities of the data points, because those are where we have information about the true function. This method is not often used for curve fitting, because the resulting fit is usually not much more accurate than a conventional LS fit and the functional form of $\bar{f}(x)$ is rather more complicated.

However, it is interesting to note that Eq. (10.16) can be derived from a weighted least-squares analysis. Consider a weighted least-squares fit with a polynomial of degree $M = 0$. In other words, $f(x) = \hat{a}_0$. The LS condition is

$$0 = \frac{\partial}{\partial \hat{a}_0} \sum_{k=1}^{N} w_k(x)(f_k - \hat{a}_0)^2 , \tag{10.18}$$

giving the normal equation $\left[\hat{a}_0 \sum_{k=1}^{N} w_k(x)\right] - \left[\sum_{k=1}^{N} w_k(x) f_k\right] = 0$. Solving for $\hat{a}_0$ leads immediately to Eq. (10.16). This analysis suggests that we could improve the accuracy of the fit by using a nonzero value of M with moving weight functions, and thus take advantage of both the moving weight strategy and the least-squares strategy. This method, with $M = 1$ or 2, can be useful but the resulting functional form of $f(x)$ is quite complicated.

An important application of the moving weighted LS formalism is to derive an interpolation method called the ***method of interpolating moving least squares***. The idea is to choose the $w_k(x)$ so that they are *infinitely* peaked at the corresponding x_k. A common choice is

$$w_k(x) = (x - x_k)^{-2}. \tag{10.19}$$

At each data point x_k the weight $w_k(x_k)$ is infinite. This forces the fit to pass exactly through each data point. In contrast to spline interpolation, this method gives a single function valid for all coordinate ranges. In contrast to Lagrange interpolation, it gives reasonable behavior between data points even for very large data sets.[7]

The method of least squares is easily extended to basis sets other than monomials. For example, Gaussian functions centered at data points,

$$\left\{ e^{-(x-x_1)^2/2\beta^2}, e^{-(x-x_2)^2/2\beta^2}, e^{-(x-x_3)^2/2\beta^2}, \ldots, e^{-(x-x_m)^2/2\beta^2} \right\}, \tag{10.20}$$

can be a useful basis set. The fitting function is then

$$f(x) = \sum_{k=1}^{m} \hat{a}_k e^{-(x-x_k)^2/2\beta^2}. \tag{10.21}$$

[7] Although presented here for one variable, the method can be used with multiple variables. It seems a good technique for constructing molecular potential energy hypersurfaces. See, for example, T. Ishida and G. C. Schatz, *Chem. Phys. Lett.* **324**, 369 (1999).

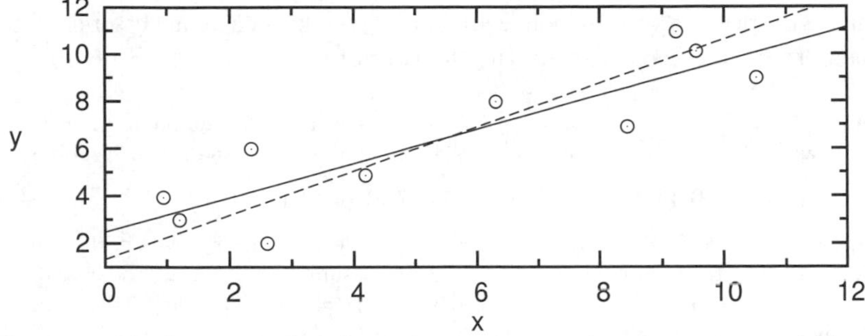

Figure 10.1: A straight-line fit with error assumed to be present only in y (solid line), and with error assumed to be only in x (dashed line).

The arbitrary parameter β determines the width of each basis function. A very small β value (narrow functions) leads to a spiked curve that closely fits each of the m data points but goes almost to zero in between, while a large β (wide functions) gives a very smooth fitting function but less close agreement with the individual data points. A basis function can be placed at each data point (for an interpolation) or at only a subset of them.

A common problem for physical chemists is to construct a fit of a curved surface.[8] In that case a function $f(x, y)$ that is linear in each of the coordinates is not likely to be sufficient. An obvious generalization is to use a basis set $\{1, x, y, x^2, y^2, xy, x^3, x^2y, xy^2, y^3, \dots\}$, but this can lead to computational difficulties. A reasonable alternative is to use a Gaussian basis,

$$\{e^{-[(x-x_1)^2+(y-y_1)^2]/2\beta^2}, e^{-[(x-x_2)^2+(y-y_2)^2]/2\beta^2}, \dots, e^{-[(x-x_m)^2+(y-y_m)^2]/2\beta^2}\}.$$

10.2 Fitting with Error in Both Variables

10.2.1 Uncontrolled Error in x

The conventional least-squares functional SS of Eq. (10.2) includes the residuals $y_k - y$ but does not take into account $x_k - x$. This means an implicit assumption is being made that there is no error in the x values. We can use instead

$$SS = \sum_{k=1}^{N} [\, x_k - x(y_k)\,]^2 \tag{10.22}$$

to fit $x(y)$. Thus we can obtain two different fits for a given set of (x_k, y_k) values, as illustrated in Fig. 10.1. We get different fits but all that changed

[8]Surface fitting is an important topic with many applications in science and engineering, but from a mathematical standpoint it can be very complicated. For an accessible general introduction, see P. Lancaster and K. Šalkauskas, *Curve and Surface Fitting* (Academic Press, San Diego, 1986). For examples of techniques for molecular potential energy surfaces see D. W. Brenner, *Phys. Rev. B* **42**, 9458 (1990); **46**, 1948 (1992); D. W. Zhang, Y. M. Li, and J. Z. H. Zhang, *J. Theor. Comp. Chem.* **2**, 119 (2003); X. Huang, B. J. Braams, and J. M. Bowman, *J. Chem. Phys.* **122**, 044308 (2005); Z. Xie, B. J. Braams, and J. M. Bowman, *J. Chem. Phys.* **122**, 224307 (2005).

was our assumption about which coordinate was known exactly and which contained error. This is a discomfiting feature of conventional LS fitting.

Example 10.4. *Effect of error assumption on least-squares fit.* The points shown in Fig. 10.1 were obtained by adding normally distributed error to the set of 10 points

$$\{(1,2),(2,3),(3,4),(4,5),(5,6),(6,7),(7,8),(8,9),(9,10),(10,11)\},$$

corresponding to the function $y(x) = 1 + x$. The scale for the error was $\sigma = 0.1$ for y and 2.0 for x. To obtain the fit with the error assumed to be in x, the x and y coordinates of each point were interchanged and $x(y) = \hat{a} + \hat{b}y$ was fit to the points (y, x). Inverting this expression gave $y(x) = (-\hat{a}/\hat{b}) + (1/\hat{b})x = 1.32 + 0.93\,x$, which is closer to the true function $1 + x$ that is the conventional LS fit, $y(x) = 2.47 + 0.72\,x$.

Suppose that the x_k and the y_k both contain significant random error. Let

$$x_k = X_k + \xi_k, \qquad y_k = Y_k + \eta_k, \tag{10.23}$$

where X_k and Y_k are the true values, x_k and y_k are the measured values, and ξ_k and η_k are independent errors from normal distributions with scale parameters λ_k and σ_k, respectively. Then a reasonable choice for χ^2 is

$$\chi^2 = \sum_{k=1}^{N}\left(\frac{\xi_k}{\lambda_k}\right)^2 + \sum_{k=1}^{N}\left(\frac{\eta_k}{\sigma_k}\right)^2, \tag{10.24}$$

with errors ξ_k and η_k weighted by their respective scales. Use of this statistic is called ***maximum likelihood estimation.***[9] Consider $y(x) = \hat{a} + \hat{b}x$. Substituting Eqs. (10.23) into Eq. (10.24) and replacing Y_k with $y_k = y(X_k)$ gives

$$\chi^2 = \sum_{k=1}^{N}\frac{1}{\lambda_k^2}(x_k - X_k)^2 + \sum_{k=1}^{N}\frac{1}{\sigma_k^2}(y_k - \hat{a} - \hat{b}X_k)^2. \tag{10.25}$$

This equation includes the exact values X_k, which are unknown to us. Let us give them hats and treat them as parameters to be fit to the data. We will estimate the $\hat{X}_k$ values in the same way we estimate the parameters $\hat{a}$ and $\hat{b}$—by minimizing χ^2. Our measure of fit deviation is now

$$\chi^2 = \sum_{k=1}^{N}\frac{1}{\lambda_k^2}(x_k - \hat{X}_k)^2 + \sum_{k=1}^{N}\frac{1}{\sigma_k^2}(y_k - \hat{a} - \hat{b}\,\hat{X}_k)^2. \tag{10.26}$$

Obtaining a fit by minimizing Eq. (10.26) is known as *functional relationship estimation by maximum likelihood*, abreviated ***FREML***.[10]

[9] This approach was developed in the 1940's by American statistician and musician W. Edwards Deming (1900-1993).

[10] This method was developed by Canadian physicists D. York and J. H. Williamson. [See *Canadian J. Phys.* **46**, 1845 (1968).] It was rediscovered, and studied in more detail, by B. D. Ripley and M. Thompson [*Analyst*, **112**, 377 (1987)], who called it "maximum-likelihood fitting of a functional relationship," with the unpronounceable acronym MLFR. Note that the analysis here applies only to a straight-line fit.

Equations for the $\hat{X}_k$ can be obtained from $\partial\chi^2/\partial\hat{X}_k = 0$. This leads to the solutions

$$\hat{X}_k = w_k[\sigma_k^2\, x_k + \lambda_k^2\, \hat{b}\,(y_k - \hat{a})] \,, \tag{10.27}$$

where

$$w_k = (\sigma_k^2 + \hat{b}^2\lambda_k^2)^{-1} \,. \tag{10.28}$$

Substituting Eq. (10.27) into Eq. (10.26) we obtain, after some algebra,

$$\chi^2 = \sum_{k=1}^{N} w_k(\hat{b})\,(y_k - \hat{a} - \hat{b}\,x_k)^2 \,. \tag{10.29}$$

We are now writing $w_k(\hat{b})$ to emphasize that the w_k are functions of $\hat{b}$, according to Eq. (10.28). Next we set $\partial\chi^2/\partial\hat{a}$ to zero, which gives

$$\hat{a}(\hat{b}) = \frac{1}{W(\hat{b})}\sum_{k=1}^{N} w_k(\hat{b})\,(y_k - \hat{b}\,x_k) \,, \qquad W(\hat{b}) = \sum_{k=1}^{N} w_k(\hat{b}) \,. \tag{10.30}$$

This leaves us with the problem of minimizing

$$\chi^2(\hat{b}) = \sum_{k=1}^{N} w_k(\hat{b})\,[y_k - \hat{a}(\hat{b}) - \hat{b}\,x_k]^2 \tag{10.31}$$

with respect to $\hat{b}$. We have reduced the number of parameters over which we are minimizing from $N+2$ down to just one. However, this last minimization, in the parameter $\hat{b}$, proves to be complicated.

Consider first the special case in which x measurements all have the same scale value λ and y measurements all have the same scale value σ, and let γ be their ratio,

$$\sigma_k = \sigma, \quad \lambda_k = \gamma\sigma, \quad \gamma = \lambda/\sigma \,. \tag{10.32}$$

From Eqs. (10.30) we now have $W(\hat{b}) = Nw(\hat{b})$ (because the w_k are all the same) and

$$\hat{a}(\hat{b}) = \bar{y} - \hat{b}\,\bar{x} \,. \tag{10.33}$$

Substituting this into χ^2 gives, after some algebra,

$$\chi^2 = \frac{1}{\sigma^2}\,\frac{1}{1+\gamma^2\hat{b}^2}\,(s_{yy} - 2\hat{b}s_{xy} + \hat{b}^2 s_{xx}) \,, \tag{10.34}$$

where

$$s_{xx} = \sum_{k=1}^{N}(x_k - \bar{x})^2, \quad s_{yy} = \sum_{k=1}^{N}(y_k - \bar{y})^2, \quad s_{xy} = \sum_{k=1}^{N}(x_k - \bar{x})(y_k - \bar{y}) \,. \tag{10.35}$$

Expanding the squares and product in Eqs. (10.35) leads to the simple expressions

$$s_{xx}/N = \left(\overline{x^2} - \bar{x}^2\right), \quad s_{yy}/N = \left(\overline{y^2} - \bar{y}^2\right), \quad s_{xy}/N = (\overline{xy} - \bar{x}\bar{y}) \,. \tag{10.36}$$

Finally, we set $d\chi^2/d\hat{b}$ to zero, with the result

$$0 = (1 + \gamma^2 \hat{b}^2)(s_{xx} \hat{b} - s_{xy}) - \gamma^2 \hat{b} (s_{yy} - 2\hat{b} s_{xy} + \hat{b}^2 s_{xx}) .$$

Collecting powers of $\hat{b}$, we obtain

$$\gamma^2 s_{xy} \hat{b}^2 + (s_{xx} - \gamma^2 s_{yy}) \hat{b} - s_{xy} = 0 , \tag{10.37}$$

a quadratic equation in $\hat{b}$. The quadratic formula gives the two solutions,

$$\hat{b} = \frac{1}{2\gamma^2 s_{xy}} \left[\gamma^2 s_{yy} - s_{xx} \pm \sqrt{(\gamma^2 s_{yy} - s_{xx})^2 + 4\gamma^2 s_{xy}^2} \right] . \tag{10.38}$$

One of these minimizes χ^2 while the other maximizes it. It is usually clear which solution we want, as the "+" sign gives a positive slope while the "−" sign gives a negative slope. (It should be obvious from looking at a plot of the points whether the slope of the fit should be positive or negative.) We then substitute this solution for $\hat{b}$ into Eq. (10.33) to calculate $\hat{a}$.

The general problem of minimizing χ^2 of Eq. (10.31) with arbitrary values of σ_k and λ_k is a nonlinear problem, which cannot be solved simply by setting a derivative equal to zero. However, it is quite feasible to solve it using an iterative numerical procedure. Starting with an initial guess for $\hat{b}$ obtained from a conventional fit with error in only one variable, a computer can systematically search for a nearby $\hat{b}$ value that minimizes $\chi^2(\hat{b})$. A computer program to accomplish this is given in Appendix A.2.

10.2.2 Controlled Error in x

We now consider straight-line fits that might appear to require a FREML analysis but in fact should be treated with error assumed only in y. We will need to make a subtle but important distinction between two different classes of random error. In the case of the FREML analysis, the x_k values are determined from measurements of the true values X_k with random variation ξ_k such that $x_k = X_k + \xi_k$. The error in x in this case is said to be ***uncontrolled***. However, it is quite often the case in science experiments that the values of x_k are exactly determined by the experimenter. Then the uncertainty lies not in the datum x_k but in the experimenter's knowledge of the true value X_k, according to $X_k = x_k + \Xi_k$, where Ξ_k is a random variation. This error in x is said to be ***controlled***. It is perhaps best explained with examples.

Example 10.5. *Controlled vs. uncontrolled variables.* Consider an experiment to determine the rate constant k (with units divided out) of a chemical reaction taking place in a heat bath of temperature T (in kelvins). Let $y = \ln k$ and $x = 1/T$. We expect, from kinetics theory, that $y = a + bx$. In advance of the experiment, we calibrate the heat bath's thermostat dial by measuring T for several different dial settings. Here are two ways to determine the x_j values to be used in a straight-line fit:

(a) Let x_j be $1/T_j$ with T_j the T value previously determined to correspond to the jth dial setting. x_j is determined by the experiment design. Its value is *controlled*. Use a conventional LS fit, *not* FREML.

(b) During each run of the experiment, with the thermostat dial set at one of the calibration marks, make repeated measurements of T_j (which show significant variability). Let x_j be the average of $1/\tilde{T}_j$ over the measured T_j for the jth run. The x_j then vary from the true $1/T$ values according to a random distribution. They are *uncontrolled*. FREML can be used.

Now for another series of measurements let x be the molar concentration in the reaction mixture of Cu^{2+}, which is suspected of catalyzing the reaction such that k will depend linearly on x. Prepare a series of reaction mixtures of the same total volume but each with a different number of aliquots of a stock solution of Cu^{2+}.

(a) Determine the x_j values by calculating them, using the concentration value listed on the stock solution bottle, the precisely known volume of the volumetric pipette used for the aliquots, and the number of aliquots. These x_j values are *controlled.*

(b) Determine the x value for the jth mixture by measuring the true Cu^{2+} concentration with a visible-frequency absorption spectrometer. The x_j contain random error due to the spectrometric measurement process. x is *uncontrolled.*

Note that for both cases with controlled x, a full repetition of all the experimental measurements (other than the initial temperature calibration) would result in exactly the same x_j values being used in the calculation of the fit, while for both cases of uncontrolled x, a different set of x_j values would be obtained in each repetition.[11]

There is one difference between the case of controlled error in x as opposed to *no* error in x. The errors in the observed y values are

$$y_k = \hat{a} + \hat{b}x_k + \eta_k = \hat{a} + \hat{b}X_k + \eta_k - \hat{b}\,\Xi_k\,. \tag{10.39}$$

Let $\sigma_{y_k}^2$ be the variance due to the error η_k and let $\sigma_{x_k}^2$ be the variance due to Ξ_k. Then by propagation of error, the observed variance in y is

$$\sigma_{y_k,\mathrm{obs}}^2 = \sigma_{y_k}^2 + \hat{b}^2\sigma_{x_k}^2. \tag{10.40}$$

The fit calculation is identical to that in the case of no error in x. All that changes is our interpretation of the observed variance in y.

The following example shows that conventional least squares is also the appropriate approach if the error in x is correlated with the error in y:

Example 10.6. *Enzyme kinetics: The Eadie-Hofstee plot.* The Michaelis-Menton analysis of the rates of enzyme catalyzed reactions leads to the equation

$$v = k[\mathrm{E}]_0/(1 + K_\mathrm{M}/[\mathrm{S}]_0)\,, \tag{10.41}$$

where $[\mathrm{S}]_0$ is the initial concentration of "substrate" (i.e., reactant), and $[\mathrm{E}]_0$, of enzyme, v is the rate of the reaction, and k and K_M are unknown constants to be determined from measurements of v vs. $[\mathrm{S}]_0$. Because reaction rates can be difficult to measure precisely, it is reasonable to expect a large scale σ_v for the v values.

Eq. (10.41) is nonlinear in K_M. We could carry out a fit for v using a nonlinear function of $[\mathrm{S}]_0$ (see Section 10.3), but this is more computationally demanding than using a linear function. In the days before computers were widely available, the usual practice was to transform Eq. (10.41) into a linear equation. There were various ways to do

[11] This distinction between fitting with controlled as opposed to uncontrolled x was noted by American statistician Joseph Berkson (1899-1982). The controlled case is often called the *Berkson error model.* See J. Berkson, *J. Amer. Stat. Assoc.* **45**, 164-180 (1950).

this. Probably the best was the Eadie-Hofstee method, with the equation rearranged as

$$v = k[\mathrm{E}]_0 - K_\mathrm{M} x\,, \qquad x = v/[\mathrm{S}]_0\,, \tag{10.42}$$

so that a straight-line fit of v vs. x determines the constants.

Given that v has random error with scale σ_v, we know that x has random error with scale $\sigma_v/[\mathrm{S}]_0$, which could be large. This might seem like a case for FREML. However, the error ξ_k in x is predictable from the error η_k in v, according to $\xi_k = \eta_k/[\mathrm{S}]_0$. Therefore,

$$\chi^2 = \sum_{k=1}^{N}\left(\frac{\eta_k/[\mathrm{S}]_0}{\sigma_k/[\mathrm{S}]_0}\right)^2 + \sum_{k=1}^{N}\left(\frac{\eta_k}{\sigma_k}\right)^2 = 2\sum_{k=1}^{N}\left(\frac{\eta_k}{\sigma_k}\right)^2 .$$

The factor of 2 has no effect on the minimization. Therefore, this is equivalent to conventional weighted LS except that

$$\sigma_{v,\mathrm{obs}}^2 = \sigma_v^2 + K_\mathrm{M}^2\sigma_x^2 = (1 + K_\mathrm{M}^2/[\mathrm{S}]_0^2)\,\sigma_v^2\,. \tag{10.43}$$

In summary, FREML straight-line fitting should be used only if y and x are both uncontrolled variables (i.e., repeating the experiment will change the values of $\{y_k\}$ and of $\{x_k\}$) and the errors in x are independent of the errors in y. If the fitting function is not a straight line, even if x is controlled, the analysis is not so simple. Then the usual practice is simply to ignore error in x. A better practice, if feasible, is to design the experiment so that the error in x is small.

10.3 Nonlinear Fitting

A ***nonlinear fit*** is a fit using a function that is nonlinear in the *parameters.* In this context, "linear" means that each term in the expression for the fitting function is multiplied by a single parameter to the first power. The function

$$f(x) = a_0 + a_1 \ln x + a_2 x^{-3} e^{(x-1)^2} + a_3 \tanh^3(x^{1/5}) \tag{10.44}$$

is not linear in x but it *is* linear in the parameters a_j. An expression of the form $\sum_j a_j g_j(x)$, regardless of the complexity of the functions $g_j(x)$, will be linear in the a_j, as long as the $g_j(x)$ do not depend on the a_j. In contrast, $a_0 + (a_1 + a_2 x)^{-1}$ and $a_0 e^{a_1 x}$ are both *non*linear in the parameters.

One approach to nonlinear fitting is to "linearize" the functional relationship. We saw this in Examples 10.2, 10.3, and 10.6. The function is rearranged by algebraic manipulations to make it linear in the parameters.

Example 10.7. *Linearization.* We saw that in Example 10.3 that an exponential can be linearized by taking logarithms. This works for any exponential of a polynomial,

$$y = a \exp\left(\sum_{j=1}^{N} b_j x^j\right) \qquad \Longrightarrow \qquad \ln y = c + \sum_{j=1}^{N} b_j x^j ,$$

where $c = \ln a$. The set $\{(x_j, \ln y_j)\}$ is fit instead of $\{(x_j, y_j)\}$. Similarly, powers can be linearized:

$$y = ax^b = ae^{b\ln x} \qquad \Longrightarrow \qquad \ln y = c + b\ln x\,.$$

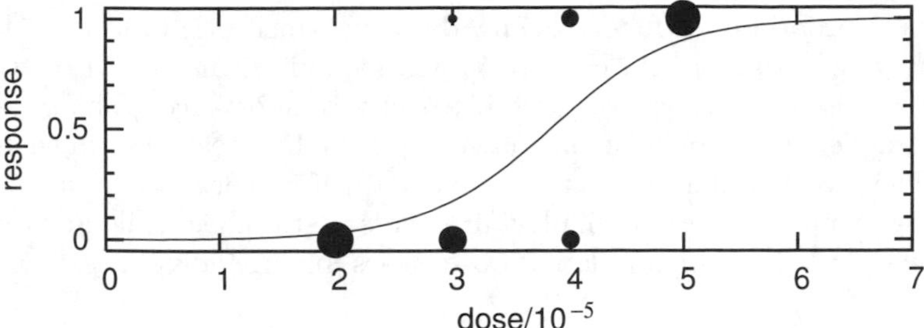

Figure 10.2: A logistic regression for dose vs. response. Dose is typically expressed as mass of substance divided by mass of the subject. The diameters of the points are proportional to the number of subjects with given response.

Ratios of polynomials are linearized by multiplying by the denominator. Consider

$$y = (a_0 + a_1 x)/(b_0 + b_1 x + b_2 x^2)\,.$$

This seems to have five parameters, but in fact one is redundant. Dividing numerator and denominator by b_0 gives a four-parameter equation that we then linearize:

$$y = (\tilde{a}_0 + \tilde{a}_1 x)/(1 + \tilde{b}_1 x + \tilde{b}_2 x^2), \quad \tilde{a}_j = a_j/b_0,\ \tilde{b}_j = b_j/b_0\,,$$
$$\Longrightarrow \quad y = \tilde{a}_0 + \tilde{a}_1 x - \tilde{b}_1 xy - \tilde{b}_2 x^2 y\,.$$

This can be treated as a multivariate fit $y = c_0 + c_1 x + c_2 p + c_3 q$ with $p = xy$, $q = x^2 y$.

Linearization is not always possible, and even when it is, it is often not the best approach. The alternative is to minimize the nonlinear quality-of-fit functional with an iterative numerical algorithm. There is however the possibility that the iterations will not converge. To increase the likelihood of convergence, it is prudent to start with a reasonable initial guess for the parameters. The rate of convergence slows with increasing number of parameters, and the more parameters, the greater the likelihood that the iterations will converge to a local minimum that has all first derivatives equal to zero but gives a worse fit than the parameters at some other minimum. However, nonlinear fitting with just a few parameters is usually straightforward. Computer algebra systems generally include a nonlinear fitting routine.[12]

A common application of nonlinear fitting is ***logistic regression***, with the fitting function

$$\pi(x) = e^{a+bx}/(1 + e^{a+bx})\,, \tag{10.45}$$

called a ***logistic curve***.[13] It has a characteristic *"S"* shape as in Fig. 10.2. Note its behavior in the limits of large $|x|$:

$$\pi(x) = \frac{e^{a+b|x|}}{1+e^{a+b|x|}} \approx \frac{e^{a+b|x|}}{e^{a+b|x|}} = 1 \quad \text{(large positive } x)\,,$$

$$\pi(x) = \frac{e^{a-b|x|}}{1+e^{a-b|x|}} \approx \frac{e^{a-b|x|}}{1} \to 0 \quad \text{(large negative } x)\,.$$

[12]In *Mathematica*, `NonlinearModelFit` and `FindFit` accept functions that are nonlinear in the parameters and allow for initial guesses of the parameter values.

[13]"π" is the symbol conventionally used for this function. It has nothing to do with the constant π from trigonometry.

This is appropriate for fitting *binary* data, in which y is 0 or 1. $\pi(x)$ is mostly very close to 0 or very close to 1, with a rapid transition in between. Consider a dose-response curve for the toxicity of substance. The x value is, for example, the amount of the substance that has been consumed by a mouse. The y value is 0 if the mouse is alive and 1 if it is dead. The substance is fed to a group of mice. At small x all mice are still alive while at large x all are dead, but the minimum lethal dose varies among individuals.

Example 10.8. *Dose-response curve.* Fig. 10.2 was prepared by fitting $\pi(x)$ to the data set

$$\{(2,0),(2,0),(2,0),(2,0),(3,0),(3,1),(3,0),(3,0),\\ (4,0),(4,1),(4,1),(4,0),(5,1),(5,1),(5,1),(5,1)\}.$$

The toxicity of the substance is characterized by its **LD₅₀** ("lethal dose, 50%"), the dose at which half of the population is predicted to have responded, which is determined by solving $\pi(\mathrm{LD}_{50}) = 0.5$. For this example, $\mathrm{LD}_{50} = 3.83 \times 10^{-5}$.

Exercises

10.1 Derive explicit expressions for the straight-line weighted fit for the exponential, Example 10.3, analogous to Eqs. (10.5).

10.2 Solve the normal equations for a multivariate fit using $f = \hat{a} + \hat{b}x + \hat{c}y$.

10.3 (a) Derive the identity $\sum_k \sum_i u_k v_{i,k} = \sum_k u_k \sum_i v_{i,k}$. (b) Derive Eqs. (10.6).

10.4 Derive Eq. (10.29).

10.5 Discuss the connection between χ^2 as defined in Eq. (10.13) and in Eq. (9.16).

10.6 Suppose that you have obtained the set of experimental points

$$\{\,(1.3,2.1),\ (2.0,4.7),\ (3.1,10.7),\ (3.9,12.4),\ (4.8,23.2)\,\}$$

and you are confident they should be fit with $y(x) = a + bx + cx^2$. (a) Assuming error only in y, determine the fit. (b) Assuming the error only in in x, determine the fit. *(Hint: Use a nonlinear fit of x as a function of y.)* (c) On a single graph, plot the points and fits.

10.7 Use the function in Eq. (10.44) to fit the points in Exercise 10.6, and compare the result with a cubic polynomial fit by plotting the points and the two fits on a single graph.

10.8 Linearize the following functions and state what quantities comprise the data set.

(*a*) $y = a_0 + (a_1 + a_2 x)/(b_0 + b_1 x + b_2 x^2)$ (*b*) $\ln(bx + y) = a$

Chapter 11

Quality of Fit

Determining the quality of a fit is less straightforward than determining the quality of a point estimation. As with point estimation, we need to determine the effect of random measurement error on the precision of the calculated result. Now, however, we must also consider whether or not the form of the fitting function (the "model") is an accurate description of the underlying physical phenomenon. Distinguishing between the effects of error in the data as opposed to error in the model is not always easy.

11.1 Confidence Intervals for Parameters

In this section we will assume that the model is correct and consider only the effects of error in the data. The normal equations for a least-squares fit yield expressions for the fitting parameters in terms of the x_k and y_k. If we have estimates for the errors in x_k and y_k, then we can propagate them to the parameters in the manner of Eq. (9.26).

Consider a straight-line fit $y = \hat{a} + \hat{b}x$ of a data set for which the x_k are known with much higher precision than are the y_k. (Assume in effect that $\sigma_{x_k} \approx 0$.) Then the scale estimate for the parameter b is given by

$$\hat{\sigma}_b^2 = \sum_{k=1}^{N} \left(\frac{\partial \hat{b}}{\partial y_k} \right)^2 \sigma_{y_k}^2 . \tag{11.1}$$

Suppose that all of the y_k have the same scale, σ_y. Then the solution for the fit, according to Eqs. (10.5) and (10.36), is

$$\hat{b} = s_{xy}/s_{xx}, \qquad \hat{a} = \bar{y} - \hat{b}\bar{x}, \tag{11.2}$$

with

$$s_{xx} = \sum_{k=1}^{N} (x_k - \bar{x})^2 = N\,(\overline{x^2} - \bar{x}^2), \qquad s_{xy} = \sum_{k=1}^{N} (x_k - \bar{x})(y_k - \bar{y}) = N\,(\overline{xy} - \bar{x}\bar{y}),$$

and

$$\frac{\partial \hat{b}}{\partial y_k} = \frac{1}{s_{xx}} \frac{\partial s_{xy}}{\partial y_k}, \tag{11.3}$$

because s_{xx} is independent of y_k. The derivatives of the residuals with respect to y_k are

$$\frac{\partial}{\partial y_k}(y_k - \bar{y}) = 1 - \frac{1}{N} \sum_{i=1}^{N} \frac{\partial y_i}{\partial y_k} = 1 - \frac{1}{N}$$

and, for $k \neq j$,

$$\frac{\partial}{\partial y_k}(y_j - \bar{y}) = -\frac{1}{N} .$$

Note also that

$$\sum_{j=1}^{N}(x_j - \bar{x}) = \sum_{j=1}^{N} x_j - \sum_{j=1}^{N} \bar{x} = N\frac{1}{N}\sum_{j=1}^{N} x_j - N\bar{x} = N(\bar{x} - \bar{x}) = 0\,.$$

It follows that

$$\frac{\partial s_{xy}}{\partial y_k} = \sum_{j=1}^{N}(x_j - \bar{x})\frac{\partial}{\partial y_k}(y_j - \bar{y}) = (x_k - \bar{x}) - \frac{1}{N}\sum_{j=1}^{N}(x_j - \bar{x}) = x_k - \bar{x}\,, \quad (11.4)$$

which implies that $\partial\hat{b}/\partial y_k = (x_k - \bar{x})/s_{xx}$ and, from Eqs. (11.1) and (11.3),

$$\hat{\sigma}_b^2 = \sigma_y^2 \frac{1}{s_{xx}^2}\sum_{j=1}^{N}(x_k - \bar{x})^2 = \frac{1}{s_{xx}}\sigma_y^2. \quad (11.5)$$

Using the fact that $\hat{b} = s_{xy}/s_{xx}$, this can also be written

$$\hat{\sigma}_b^2 = \frac{1}{s_{xy}}\hat{b}\,\sigma_y^2. \quad (11.6)$$

For $\hat{a}$ we have

$$\hat{\sigma}_a^2 = \sum_{k=1}^{N}\left(\frac{\partial\hat{a}}{\partial y_k}\right)^2 \sigma_{y_k}^2 = \sigma_y^2 \sum_{k=1}^{N}\left(\frac{\partial\hat{a}}{\partial y_k}\right)^2 \quad (11.7)$$

and

$$\frac{\partial\hat{a}}{\partial y_k} = \frac{\partial}{\partial y_k}(\bar{y} - \hat{b}\bar{x}) = \frac{\partial\bar{y}}{\partial y_k} - \bar{x}\frac{\partial\hat{b}}{\partial y_k} = \frac{1}{N} - \bar{x}\frac{\partial\hat{b}}{\partial y_k}\,. \quad (11.8)$$

After a little more algebra, we obtain

$$\hat{\sigma}_a^2 = \frac{1}{N}\sigma_y^2 + \bar{x}^2\hat{\sigma}_b^2\,. \quad (11.9)$$

This analysis requires a value for σ_y. If the error distribution is normal, then it is possible to show that the best estimate[1] of σ_y, given no information other than the data points, is

$$\hat{\sigma}_y^2 = \frac{1}{N-2}\sum_{k=1}^{N}(y_k - \hat{a} - \hat{b}x_k)^2\,. \quad (11.10)$$

This is just the sum of the squared residuals divided by ν (with $\nu = N - 2$ degrees of freedom because we have already used the data sample to estimate *two* other values, namely, a and b). Confidence intervals for the parameters are given by

$$\hat{a} \pm t_{\alpha/2,\,\nu}\,\hat{\sigma}_a\,, \qquad \hat{b} \pm t_{\alpha/2,\,\nu}\,\hat{\sigma}_b\,, \quad (11.11)$$

using the Student t-distribution with $\nu = N - 2$.

[1]This estimate is "best" in the sense of "most probable," based on the assumption that the random error in y follows a normal distribution.

If the y_k have different standard deviations, a similar analysis can be applied to the method of weighted least squares. This analysis can be applied to almost any least-squares fit, linear or nonlinear, regardless of the functional form and basis set. The resulting values for standard deviations of the parameters can be obtained from standard statistical software packages and computer algebra systems.

Now consider a straight-line fit with uncontrolled error in *both* variables. Standard deviations from propagation of error for this case are not usually provided in standard software packages. For the particular case of a straight-line fit with $\sigma_x = \gamma\sigma_y$, which was treated in detail in Section 10.2, it is simpler to start with the quadratic equation for $\hat{b}$,

$$\gamma^2 s_{xy}\hat{b}^2 + (s_{xx} - \gamma^2 s_{yy})\hat{b} - s_{xy} = 0\,,$$

rather than with the explicit solution for $\hat{b}$. Let us take the derivative of this equation with respect to y_k:

$$0 = \gamma^2\hat{b}^2\frac{\partial s_{xy}}{\partial y_k} + 2\gamma^2 s_{xy}\,\hat{b}\,\frac{\partial \hat{b}}{\partial y_k} - \gamma^2\hat{b}\,\frac{\partial s_{yy}}{\partial y_k} + (s_{xx} - \gamma^2 s_{yy})\frac{\partial \hat{b}}{\partial y_k} - \frac{\partial s_{xy}}{\partial y_k}\,.$$

Solving this for $\partial\hat{b}/\partial y_k$, and evaluating the derivatives of s_{xy} and s_{yy}, we obtain

$$\frac{\partial \hat{b}}{\partial y_k} = \frac{\hat{b}}{1+\gamma^2\hat{b}^2}\left[2\gamma^2\hat{b}\,\frac{y_k - \bar{y}}{s_{xy}} + (1-\gamma^2\hat{b}^2)\,\frac{x_k - \bar{x}}{s_{xy}}\right]. \tag{11.12}$$

An expression for $\partial\hat{b}/\partial x_k$ is found similarly. After a fair amount of additional algebra, we obtain a surprisingly simple result for the propagation of error:

$$\hat{\sigma}_b^2 = \hat{\sigma}_y^2\sum_{k=1}^{N}\left[\left(\frac{\partial \hat{b}}{\partial y_k}\right)^2 + \gamma^2\left(\frac{\partial \hat{b}}{\partial x_k}\right)^2\right] = \frac{s_{xx}+\gamma^2 s_{yy}}{s_{xy}^2}\,\hat{b}^2\,\hat{\sigma}_y^2. \tag{11.13}$$

For $\hat{a}$,

$$\frac{\partial \hat{a}}{\partial y_k} = \frac{1}{N} - \bar{x}\frac{\partial \hat{b}}{\partial y_k}, \qquad \frac{\partial \hat{a}}{\partial x_k} = -\frac{1}{N}\,\hat{b} - \bar{x}\frac{\partial \hat{b}}{\partial x_k}\,. \tag{11.14}$$

After still more algebra, we obtain

$$\hat{\sigma}_a^2 = \hat{\sigma}_y^2\sum_{k=1}^{N}\left[\left(\frac{\partial \hat{a}}{\partial y_k}\right)^2 + \gamma^2\left(\frac{\partial \hat{a}}{\partial x_k}\right)^2\right] = \frac{1}{N}\,(1+\gamma^2\hat{b}^2)\hat{\sigma}_y^2 + \bar{x}^2\hat{\sigma}_b^2. \tag{11.15}$$

Note that in the limit $\gamma \to 0$, in which the errors in the x_k disappear, these results for $\hat{\sigma}_b^2$ and $\hat{\sigma}_a^2$ reduce to the results for conventional least squares, Eqs. (11.6) and (11.9) with $\hat{b} = s_{xy}/s_{xx}$.

For the general FREML problem, with standard deviations $\hat{\sigma}_{x_k}$ and $\hat{\sigma}_{y_k}$, it is possible to derive explicit expressions for the variances even though $\hat{b}$ can only be computed numerically. The expressions are complicated.[2] Alternatively, because we know all the standard deviations, we could carry out a Monte Carlo analysis to determine confidence intervals.

[2]See K. I. Mahon, *Int. Geol. Rev.* **38**, 293 (1996).

11.2 Confidence Band for a Calibration Line

A common use for straight-line fitting is to calibrate a measuring instrument. Suppose we have reason to expect that the response of the instrument to some property should follow a straight line. Let x be the property, y, the response, and $y = a + bx$, the model. We then prepare a collection of samples, called ***standards***, with precisely known values of x. Let us designate their values $x_{\mathrm{std,\,j}}$. We measure the corresponding instrument responses $y_{\mathrm{std},j}$ and fit the coefficients $\hat{a}$ and $\hat{b}$. Next we perform measurements on a sample with unknown x, obtaining responses $y_{\mathrm{u},\,j}$. Finally, we estimate the unknown x value from the fitted line,

$$\hat{x}_u = (\bar{y}_{\mathrm{u}} - \hat{a})/\hat{b}. \tag{11.16}$$

What is the confidence interval for $\hat{x}_{\mathrm{u}}$? The answer is somewhat complicated. Fig. 11.1 shows a simple way to carry out a very approximate analysis. We get a rough estimate of the error in the fit by examining the variation obtained when the fit is performed on subsets of the data points. If there were no error, all the points would lie in a line and any subset of two of the three points would give the same fit, but because the points contain random error, lines through different subsets yield a range of slopes and y-intercepts. We could estimate the confidence bounds of x_{u} by measuring y_{u} and calculating the corresponding x first using the line with the largest slope and then using the line with the smallest slope. The distance between those two extreme lines is a rough estimate of the error in y_{u} as a function of x. The implication is that this error is smallest at the center of the distribution of the $x_{\mathrm{std},j}$ values and flares symmetrically away from the center.

Suppose that the variation in y follows a normal distribution. Then it is possible to derive an x-dependent function $\hat{\sigma}_{\bar{y}_{\mathrm{u}}}(x)$ for the standard error in y_{u}, such that the confidence interval for y_{u} is

$$\bar{y}_{\mathrm{u}} - t_{\alpha/2,\,\nu_{\mathrm{std}}}\hat{\sigma}_{\bar{y}_{\mathrm{u}}}(x) < y_{\mathrm{u}} < \bar{y}_{\mathrm{u}} + t_{\alpha/2,\,\nu_{\mathrm{std}}}\hat{\sigma}_{\bar{y}_{\mathrm{u}}}(x), \tag{11.17}$$

with the Student t value for $\nu_{\mathrm{std}} = N_{\mathrm{std}} - 2$ degrees of freedom. An estimate for the variance of y in the absence of x dependence was given by $\hat{\sigma}_y^2$ of

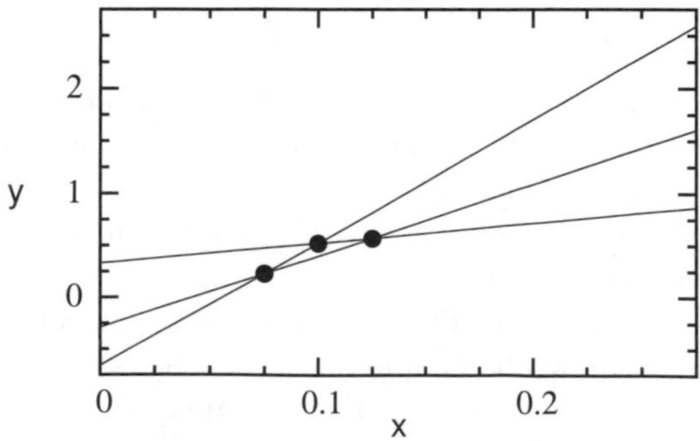

Figure 11.1: Straight lines through subsets of two points from a set of three points.

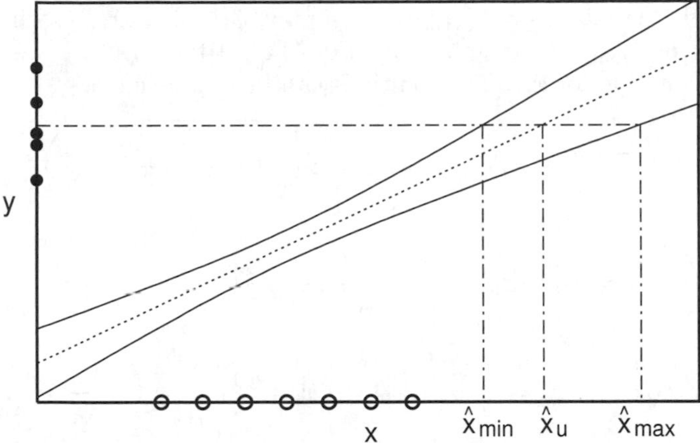

Figure 11.2: Confidence band of a calibration line, according to Eqs. (11.19). The dotted line is the least-squares fit for the standards.

Eq. (11.10). The x-dependent standard error, $\hat{\sigma}_{\bar{y}_\mathrm{u}}(x)$, will be given by $\hat{\sigma}_y$ multiplied by some function of x that will be smallest at the center of the $x_{\mathrm{std},j}$ distribution and larger away from it. The "center" is the average of the $x_{\mathrm{std},j}$ values, $\bar{x}_\mathrm{std} = \left(\sum_{k=1}^{N_\mathrm{std}} x_{\mathrm{std},k}\right)/N_\mathrm{std}$. We expect the thickness of the confidence band to decrease as the number of measurements increases. We saw before that standard error decreases as $1/\sqrt{N}$, but here we have two different N values, N_std and N_u. Taking all these effects into account, it turns out that the standard error in y_u as a function of x is

$$\hat{\sigma}_{\bar{y}_\mathrm{u}}(x) = \hat{\sigma}_y \left[\frac{1}{N_\mathrm{std}} + \frac{1}{N_\mathrm{u}} + \frac{(x - \bar{x}_\mathrm{std})^2}{\sum_{k=1}^{N_\mathrm{std}} (x_{\mathrm{std},k} - \bar{x}_\mathrm{std})^2} \right]^{1/2} . \tag{11.18}$$

The region in the plot of y vs. x bordered by the curves

$$\hat{a} + \hat{b}x + t_{\alpha/2,\,\nu_\mathrm{std}} \hat{\sigma}_{\bar{y}_\mathrm{u}}(x), \qquad \hat{a} + \hat{b}x - t_{\alpha/2,\,\nu_\mathrm{std}} \hat{\sigma}_{\bar{y}_\mathrm{u}}(x) \tag{11.19}$$

is called the ***Hotelling-Working confidence band.***[3] Fig. 11.2 shows that this region is narrow in the center and flares out at the edges, as expected.

Fig. 11.2 also illustrates how to use the calibration band to obtain a confidence interval for x_u. It shows a straight-line fit to y values corresponding to seven different values of x_std (open circles). Five measurements of y_u (filled circles) are averaged and a horizontal line is drawn from $\bar{y}_\mathrm{u}$. The estimate $\hat{x}_\mathrm{u}$ is obtained from the intersection with the fitted line while the confidence interval $\hat{x}_\mathrm{min} < x_\mathrm{u} < \hat{x}_\mathrm{max}$ is obtained from the intersections with the borders of the confidence band. Thus, $\hat{x}_\mathrm{min}$ and $\hat{x}_\mathrm{max}$ are determined by the equations

$$\hat{a} + \hat{b}\hat{x}_\mathrm{min} + t_{\alpha/2,\nu_\mathrm{std}} \hat{\sigma}_{\bar{y}_\mathrm{u}}(\hat{x}_\mathrm{min}) = \bar{y}_u \,, \tag{11.20a}$$

$$\hat{a} + \hat{b}\hat{x}_\mathrm{max} - t_{\alpha/2,\nu_\mathrm{std}} \hat{\sigma}_{\bar{y}_\mathrm{u}}(\hat{x}_\mathrm{max}) = \bar{y}_u \,. \tag{11.20b}$$

[3]This analysis was developed in the 1920's by the American economists Harold Hotelling and Holbrook Working.

These can be written as $[t_{\alpha/2,\nu_{\rm std}}\hat{\sigma}_{\bar{y}_{\rm u}}(\hat{x}_{\rm m})]^2 = (\bar{y}_{\rm u} - \hat{a} - \hat{b}\hat{x}_{\rm m})^2$, where $\hat{x}_{\rm m}$ is either $\hat{x}_{\rm min}$ or $\hat{x}_{\rm max}$. If we substitute Eq. (11.18) for $\hat{\sigma}_{\bar{y}_{\rm u}}(\hat{x}_{\rm m})$, we obtain a quadratic equation for $\hat{x}_{\rm m}$. The solutions can be expressed as

$$\hat{x}_{\rm m} = \hat{x}_{\rm u} - \left(\frac{\Delta_0}{\sigma_{\rm std}}\right)^2 \epsilon \pm \Delta_{\rm u}, \qquad \Delta_{\rm u} = \Delta_0\sqrt{1 + \frac{1}{1-\tau^2}\left(\frac{\epsilon}{\sigma_{\rm std}}\right)^2}\,, \qquad (11.21)$$

with

$$\hat{x}_{\rm u} = \frac{\bar{y}_{\rm u} - \hat{a}}{\hat{b}}\,, \quad \epsilon = \bar{x}_{\rm std} - \hat{x}_{\rm u}\,, \quad \Delta_0 = \frac{1}{\sqrt{1-\tau^2}}\frac{t_{\alpha/2,\nu_{\rm std}}\hat{\sigma}_{\bar{y}}}{\hat{b}}\,, \quad \tau = \frac{t_{\alpha/2,\nu_{\rm std}}\hat{\sigma}_{\bar{y}}}{\hat{b}\,\sigma_{\rm std}}\,,$$

$$\hat{\sigma}_{\bar{y}} = \hat{\sigma}_y/\sqrt{\mu}\,, \quad \sigma^2_{\rm std} = \frac{1}{\mu}\sum_{k=1}^{N_{\rm std}}(x_{{\rm std},k} - \bar{x}_{\rm std})^2, \quad \frac{1}{\mu} = \left(\frac{1}{N_{\rm std}} + \frac{1}{N_{\rm u}}\right).$$

Δ_0 is the half-width of the band at its narrowest, where $\hat{x}_{\rm u} = \bar{x}_{\rm std}$. The term in $\Delta_{\rm u}$ with $(\epsilon/\sigma_{\rm std})^2$ is responsible for the flaring of the band.

The confidence interval for $x_{\rm u}$ is not symmetric; the answer cannot be written $\hat{x}_{\rm u} \pm \delta$ because $\hat{x}_{\rm u}$ does not lie halfway between $\hat{x}_{\rm min}$ and $\hat{x}_{\rm max}$. One would report the result as "$\hat{x}_{\rm u}$ with a 95% confidence interval of $\hat{x}_{\rm min}$ to $\hat{x}_{\rm max}$" (if we used the t factor with $\alpha = 0.05$). Note that the confidence interval in Fig. 11.2 is quite large. This is because $x_{\rm u}$ in this case is well outside the range of x values of the standards. If possible, one should avoid using a calibration line beyond the region in which it has been calibrated.

Example 11.1. *Designing an optimal procedure for estimating* $x_{\rm u}$. Suppose that due to constraints of time or cost you can only perform a total of eight measurements. How many of these should be measurements of standards and how many should be of the unknown sample? The size of the confidence interval is $2\Delta_{\rm u}$, which is proportional to $\hat{\sigma}_{\bar{y}}$, which is proportional to $\sqrt{N_{\rm std}^{-1} + N_{\rm u}^{-1}}$. For given N, this is minimized by choosing $N_{\rm std} = N_{\rm u}$. For example, $1/3 + 1/5 = 0.533$ while $1/4 + 1/4 = 0.500$. We might conclude that the best choice is $N_{\rm std} = N_{\rm u} = N/2 = 4$, but this ignores the fact that the width is also proportional to $t_{\alpha/2,\,\nu_{\rm std}}$, which depends on $\nu_{\rm std} = N_{\rm std} - 2$. The t value decreases with increasing $N_{\rm std}$. Consider the following table:

$N_{\rm std}$	3	4	5	6	7
$t_{0.025,\,N_{\rm std}-2}$	12.706	4.303	3.182	2.776	2.571
$N_{\rm std}^{-1} + N_{\rm u}^{-1}$	0.533	0.500	0.533	0.667	1.143
$t_{0.025,\,N_{\rm std}-2}\sqrt{N_{\rm std}^{-1} + N_{\rm u}^{-1}}$	5.682	3.043	2.324	2.267	2.749

This suggests that the optimal choice is $N_{\rm std} = 6$, $N_{\rm u} = 2$. The spacing of the six $x_{{\rm std},\,k}$ values should be chosen so that $\bar{x}_{\rm std} \approx x_{\rm u}$, to minimize ϵ.

The upper boundary of the confidence band can be used to specify the lower ***limit of detection*** (LOD) of an instrument. There are two possible situations, illustrated in Fig. 11.3, distinguished by the sign of the y-intercept. With no error we would have a y-intercept of zero, because "no sample" would correspond to "no signal." We use the different definitions of the detection limit in the two cases shown because it is meaningless to have a negative concentration. The relevant confidence interval for this analysis is not Eq. (11.17)

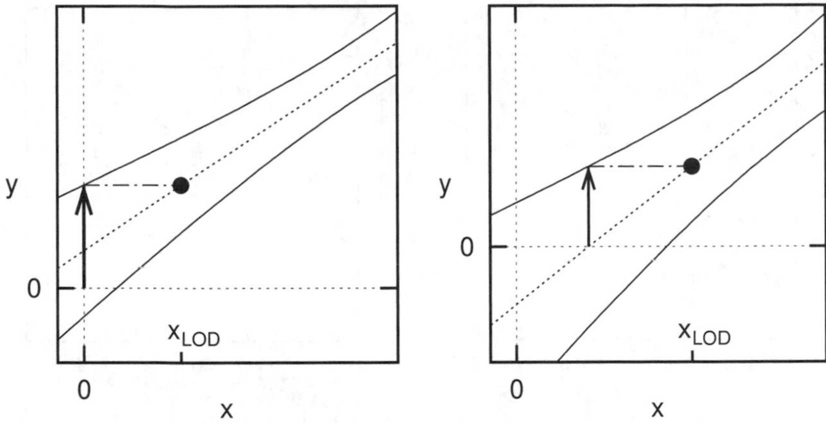

Figure 11.3: Limit of detection, as given by the upper limit of the confidence band. The y value at the top of the arrow is the lowest value that would be significant at the confidence level $1 - \alpha$. In the left panel, the y-intercept from the straight-line fit is positive while in the right panel it is negative.

but rather

$$y_{\mathrm{u}} < t_{\alpha,\nu_{\mathrm{std}}} \hat{\sigma}_{\bar{y}_{\mathrm{u}}}(x_0)\,, \tag{11.22}$$

where $x_0 = 0$ in the left-hand panel and $x_0 = -\hat{a}/\hat{b}$, the point of intersection with the x-axis, in the right-hand panel. If a measured y value is greater than $t_{\alpha,\nu_{\mathrm{std}}} \hat{\sigma}_{\bar{y}_{\mathrm{u}}}(x_0)$, then the probability is less than α that the true signal is zero.

Note that $\bar{y}_{\mathrm{u}}$ in Eq. (11.17) has been replaced with zero in Eq. (11.22). Note also that we are now carrying out a one-way t-test, with subscript α instead of $\alpha/2$; we only care if y is above the upper boundary of the confidence band, not if it is below the lower boundary. The corresponding x value,

$$x_{\mathrm{LOD}} = [t_{\alpha,\nu_{\mathrm{std}}} \hat{\sigma}_{\bar{y}_{\mathrm{u}}}(x_0) - \hat{a}]/\hat{b}\,, \tag{11.23}$$

is the lowest x value that is significantly distinguishable from zero at a confidence level of $1 - \alpha$. By convention, the LOD confidence level, unless stated otherwise, is chosen as 99%, that is, $\alpha = 0.01$.

11.3 Outliers and Leverage Points

The method of least squares is notoriously sensitive to outliers. Consider the straight-line fits shown in Fig. 11.4. It is quite obvious in both panels that all of the points but one lie very close to a straight line. However, the least-squares fit (the solid curve) is clearly the wrong straight line.

Note that the degree to which the fit is skewed by the outlier depends very strongly on the relative position of the outlier in the x coordinate. In the first case, the x value lies in the middle of set. In other words, if the coordinates of

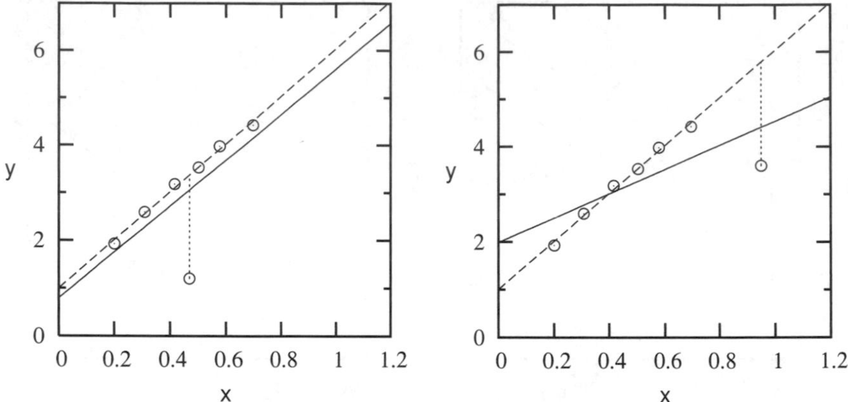

Figure 11.4: Conventional least-squares straight-line fit of data set with one outlier. The solid line is a fit to all the points while the dashed line is a fit to the set with the outlier removed. The dotted line shows the vertical distance between the outliers and the fit to the remaining points.

the outlier are $(x_{\text{out}}, y_{\text{out}})$, then, for the left panel, $x_{\text{out}} \approx \bar{x}$. In this case the entire line is shifted somewhat in the direction of the outlier but the slope of the least-squares fit is not much affected. In the second case x_{out} is far from $\bar{x}$, and the effect of the outlier is catastrophic.

The reason for this sensitivity is clear from the expression for the functional $SS[y]$, Eq. (10.2). To simplify the notation, we will write

$$\rho_k = y_k - y(x_k) \tag{11.24}$$

for the residuals, which are the vertical distances between the actual measured values y_k corresponding to a given x_k and the predicted values $y(x_k)$ from some candidate fitting function. If the residual ρ_{out} is much larger than any of the others, then

$$SS[y] = \sum_{k=1}^{N} \rho_k^2 \approx \rho_{\text{out}}^2. \tag{11.25}$$

This effect is exaggerated by the fact that the residuals are squared. A small ρ_k becomes even smaller when squared while a larger residual becomes relatively larger. In order to minimize SS, the function $y(x)$ reduces ρ_{out} while distancing itself from all of the "good" points. The result is a "fit" that fits none of the points. In the case of an outlier for which x_{out} is far from $\bar{x}$, the attempt to minimize ρ_{out}^2 skews the slope.

Points in the data set with x values far from $\bar{x}$ are called ***leverage points***, because they have the potential to exert an undue influence on the fit.[4] Clearly, it is a dangerous situation for a leverage point to be an outlier. However, leverage points are of significance even if they are not outliers. We expect random variation in all data points. Random fluctuations in leverage points

[4]The analogy here is to a mechanical lever. A force at the end of the lever has a much greater effect than does a force near the fulcrum.

have a larger effect on the fit than do random variations in central points. This should be taken into account when designing an experiment. If possible, efforts should be made to reduce the uncertainty at leverage points, for example, by performing extra measurements at the extremes of the x range.

11.4 Robust Fitting*

One way to make the least-squares fit robust is to make repeated measurements $y_{k,1}, y_{k,2}, \ldots,$ for each x_k. A robust estimator, such as the median or the Huber estimator could then give a best estimate $\hat{Y}_k$ to replace the individual $y_{k,j}$'s, and then the data set $\{(x_k, \hat{Y}_k)\}$ could be fit using the method of weighted least squares, with the weights w_k given by the reciprocal of the square of the scale estimator corresponding to each $\hat{Y}_k$. If the variability in x is significant, then repeated measurements of x could be used to robustly estimate a location and scale for each x_k, to be used in a FREML fit.

However, the most common situation is to have only a single value of y_k for each x_k. There are two strategies for dealing with this: try to identify and remove outliers with some kind of rejection test, or use some kind of robust fitting method less strongly skewed by outliers. A disadvantage of rejection tests is the difficulty in distinguishing between an outlier, *in*consistent with the underlying probability distribution, and a random but unusually large deviation consistent with the distribution. A disadvantage of robust methods is that they typically are less efficient than conventional least-squares fitting if the error follows a normal distribution with no outliers.

Let us reexamine the method of least squares, with the intention of modifying it to increase its robustness. For the sake of simplicity, we consider a linear fitting function, $y(x) = a + bx$, but the analysis that follows can also be applied to more complicated fitting functions. The method of least squares can be concisely described as follows:

$$\underset{a,\ b}{\text{minimize}} \sum_{k=1}^{N} \rho_k^2 \,, \tag{11.26}$$

in terms of the residuals $\rho_k = y_k - a - bx_k$. The values of a and b are chosen as those that minimize the sum of the squares of the ρ_k. A more accurate name for this approach would be the "method of least *sum of* squares." Note that

$$\frac{1}{N} \sum_{k=1}^{N} \rho_k^2 = \text{mean } \rho_k^2 \,. \tag{11.27}$$

Minimizing the sum is the same as minimizing the mean. Thus, the method of least squares is also the "method of least *mean* of squares."

Suppose that we replace the mean with the median:

$$\underset{a,b}{\text{minimize}} \ \underset{k}{\text{med}} \ \rho_k^2 \,. \tag{11.28}$$

This is the ***method of least median of squares***. We will use "LMS" as the acronym for this method and "LS" as the acronym for the conventional least sum of squares method. The LMS method was developed by Rousseeuw[5] in the 1980's.

Example 11.2. *LMS as a point estimation method.* To illustrate the method, consider the fitting function $y(x) = \hat{a}$. This would pertain to a situation in which repeated measurements are made of y_k with x_k set to zero. Consider the data sample $\{y_k\} = \{1, 3, 6, 7, 9\}$. Let us define a function

$$F(\hat{a}) = \operatorname*{med}_k \rho_k^2 .$$

For a given value of $\hat{a}$, this function computes the five values of $\rho_k^2 = (y_k - \hat{a})^2$, arranges them in ascending order, and chooses the median of them. To find the minimum of $F(\hat{a})$ we can plot it vs. $\hat{a}$:

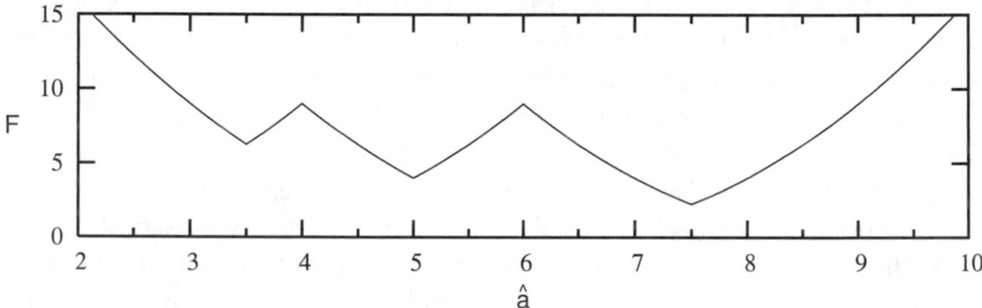

The minimum occurs at $\hat{a} = 7.5$.

Rousseeuw has developed a convenient algorithm for computing the LMS. It is based on the following theorem.

Theorem 11.4.1. *Consider a data set $\{(x_j, y_j)\}$ and a fitting function $y(x) = \hat{a} + \hat{b}x$. For any given value of $\hat{b}$, the value of $\hat{a}$ that minimizes the least median of squares of the residuals can be obtained as follows:*

1. *Calculate the set of $z_j = y_j - \hat{b}x_j$ and arrange them in ascending order, $z_{(1)}, z_{(2)}, \ldots, z_{(N)}$.*

2. *Let h be the largest integer less than or equal to $(N/2)+1$, and calculate the set of differences*

$$d_1 = z_{(h)} - z_{(1)},\ d_2 = z_{(h+1)} - z_{(2)}, \ldots,\ d_{N-h+1} = z_{(N)} - z_{(N-h+1)}.$$

3. *Let k be the index corresponding to the smallest of the d_j. Then the LMS solution for $\hat{a}$ (for given $\hat{b}$) is*

$$\hat{a} = \big(z_{(k)} + z_{(h+k-1)}\big)/2.$$

[5]Peter J. Rousseeuw, a Belgian statistician whose research has focused on ways to make classical statistical techniques robust to outliers.

Example 11.3. *Algorithm for LMS point estimation.* Consider again the data set in Example 11.2. Let us apply Theorem 11.4.1 using the fitting function $y(x) = \hat{a} + \hat{b}x$ with $\hat{b} = 0$. Then $\{z_{(j)}\} = \{1, 3, 6, 7, 9\}$, $h = 3$, and

$$\{d_j\} = \{6 - 1, 7 - 3, 9 - 6\} = \{5, 4, 3\},$$

which, according to the theorem, implies that $\hat{a} = (z_{(3)} + z_{(6)})/2 = 15/2 = 7.5$, which agrees with the result in the previous example.[6]

Let us define

$$G(\hat{b}) = \operatorname*{med}_k \, [\, y_k - \hat{a}(\hat{b}) - \hat{b}x_k \,]^2, \tag{11.29}$$

where $\hat{a}(\hat{b})$ is the LMS solution corresponding to $y(x) = \hat{a} + \hat{b}x$ with specified value for $\hat{b}$. The LMS solution for the slope, $\hat{b}_{\text{LMS}}$, is the value of $\hat{b}$ that minimizes $G(\hat{b})$. This can be computed numerically. (See Section 5.6.)

Example 11.4. *LMS straight-line fitting.* Consider yet again the data set in Example 11.2 but suppose now that that set just comprises the y_j values of the two-dimensional set

$$\{(x_j, y_j\} = \{(1.3, 1), (2.1, 3), (3.0, 6), (3.9, 7), (5.1, 9)\}.$$

The conventional least-squares fit is shown below in the left panel.

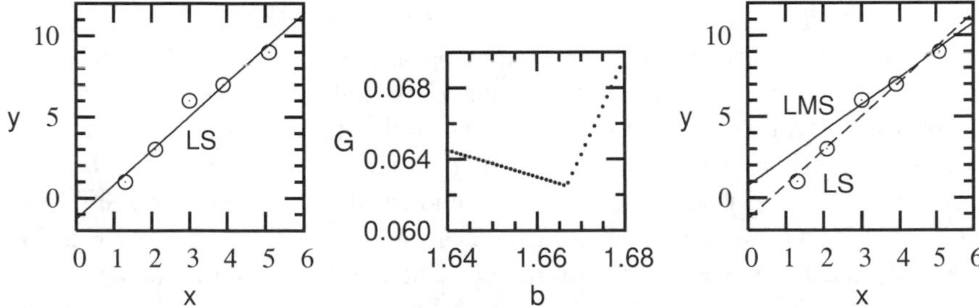

Is the middle point an outlier? Perhaps the LMS fit will tell us. $G(\hat{b})$ has a minimum at $b = 1.666$, which corresponds to $a = 0.7527$. The resulting line is the solid curve in the right panel. Note that the LMS fit essentially ignores the first two points. It does not treat the central point as an outlier—it treats the first two points as outliers!

Example 11.4 illustrates an important idea. For point estimation (i.e., estimation of a single value, the subject of Chapters 8 and 9) we were concerned with whether or not the observed deviation was within a range consistent with an underlying probability distribution of random error. When fitting a function we must also ask if the chosen function has the correct *form*. In Example 11.4 the answer seems to be that two of the points do not lie on the straight line that fits the other three points. This is probably due not to random error in the points, but to a failure of the assumption that the underlying relationship between x and y is a straight line. This is a failure of the *model*. A different kind of function might be able to fit all the points.

Because fitting, as compared with point estimation, has this added complication of distinguishing between random error and model error, the LMS

[6] This algorithm may seem like magic but the proof is reasonably straightforward. See P. J. Rousseeuw, *J. Amer. Stat. Assoc.* **79**, 871 (1984).

method is perhaps best used as a rejection test.[7] If outliers are suspected, it is a good idea to calculate the LMS fit and then plot the LMS residuals. If an isolated point has an unusually large residual ρ_k, then it would be reasonable to reject it if it lies outside the 99% confidence interval given by

$$|\rho_k| > t_{0.005,\nu}\,\hat{\sigma}_y\,, \tag{11.30}$$

using the $t_{\alpha/2,\nu}$ corresponding to 99% confidence for $\nu = N - 2$ degrees of freedom, with the standard deviation calculated from Eq. (11.10) but with the outlier omitted,

$$\hat{\sigma}_y^2 = \frac{1}{N-3} \sum_{k=1;\ k\neq k_{\text{out}}}^{N} (y_k - \hat{a}_{\text{LMS}} - \hat{b}_{\text{LMS}}\, x_k)^2. \tag{11.31}$$

If a group of points with consecutive x_k's at one end of the data set have large LMS residuals while the other points have small residuals, as in Example 11.4, then it is likely that the functional form being used is not appropriate.

11.5 Model Testing

The form of the function used in a fit is called the ***model***. The method of least squares is a way to set the values of parameters in a specified model. It does not tell us whether or not the choice of model was reasonable.

Analysis of variance (ANOVA), the method introduced in Section 9.4 to determine whether or not two sets of measurements are samples of the same population, can be used as a test of the choice of model. Before seeing how this works, let us carefully consider the question the test will answer. Fig. 11.5 shows two least-squares fits of a data set. The solid line is a fit using

$$y_{\text{r}}(x) = \hat{a}_{\text{r}} + \hat{b}_{\text{r}}x, \tag{11.32}$$

which we will call the "reduced" model, while the dashed curve uses

$$y_{\text{f}}(x) = \hat{a}_{\text{f}} + \hat{b}_{\text{f}}x + \hat{c}_{\text{f}}x^2, \tag{11.33}$$

which we will call the "full" model. Which is the better fit?

To answer this question, we will test the following hypothesis:

> *The reduced model describes the data at least as well as does the full model.*

This will be the null hypothesis. If our test confirms it, then we will reject the full model in favor of the reduced model.[8]

---

[7] See M. Cruz Ortiz, L. A. Sarabia, and A. Herrero, "Robust Regression Techniques: A Useful Alternative for the Detection of Outlier Data in Chemical Analysis," *Talanta* **70**, 499-512 (2006), for a comparative study of various methods.

[8] This is an example of a principle known as *Ockham's razor*, popularized by 14th-century English philosopher William of Ockham: Given a choice between two competing theories each of which explains the observed phenomena, the theory with the fewest assumptions should be preferred. (This is also called the *principle of parsimony*.) In this case, if the two-term expression $\hat{a} + \hat{b}x$ agrees with the data as well as does the three-term expression $\hat{a} + \hat{b}x + \hat{c}x^2$, then the third term $\hat{c}x^2$ is unnecessary and should be omitted.

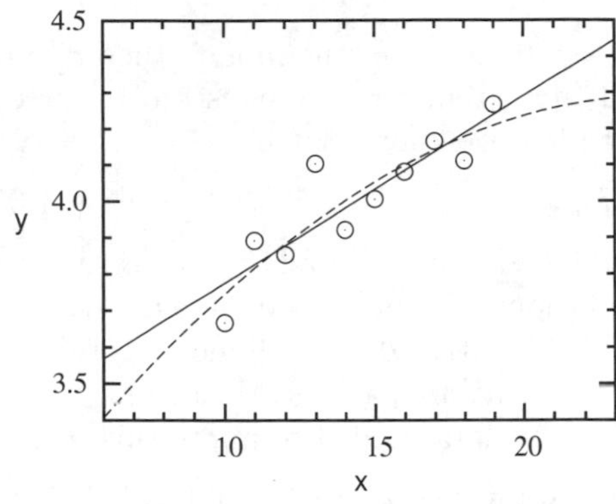

Figure 11.5: Comparison of two models for a least-squares fit: a linear polynomial (solid line) and a quadratic polynomial (dashed curve).

Let us compare the sums of squares for the two models,

$$SS_\mathrm{r} = \sum_{j=1}^{N} [y_j - y_\mathrm{r}(x_j)]^2, \qquad SS_\mathrm{f} = \sum_{j=1}^{N} [y_j - y_\mathrm{f}(x_j)]^2.$$

We will always have $SS_\mathrm{f} \leq SS_\mathrm{r}$, because the model y_f has more degrees of freedom than does y_r. This is the case even if y_r is the true model.[9] Let us write

$$SS_\mathrm{r} = SS_\mathrm{f} + SS_\Delta \,, \tag{11.34}$$

where SS_Δ is the increase in the sum of squares from reducing the number of degrees of freedom. SS_Δ will always be greater than or equal to zero, but if the null hypothesis is false (i.e., y_f describes the data better than does y_r), then we expect SS_f to be relatively smaller and SS_Δ relatively larger than would be the case if the null hypothesis is true. The question is, how large does SS_Δ/SS_f have to be in order for the null hypothesis to be rejected?

The sum of squared residuals from a fit, divided by the number of degrees of freedom, is an estimator of the variance σ_y^2 of y. Thus,

$$\hat{\sigma}_\mathrm{f}^2 = \frac{1}{N - p_\mathrm{f}} SS_\mathrm{f} \,, \tag{11.35}$$

where p_f is the number of parameters in y_f, gives an estimate for σ_y^2. It turns out that

$$\hat{\sigma}_\Delta^2 = \frac{1}{p_\mathrm{f} - p_\mathrm{r}} SS_\Delta \,, \tag{11.36}$$

where p_r is the number of parameters in the reduced model, will also be an estimator of σ_y^2, *but only if y_r is the true model.*[10] If y_r is not the true model, then $\hat{\sigma}_\Delta^2$ will tend to be larger than σ_y^2.

---

[9] For example, if the straight line y_r is the true model and we have three data points, then SS_r will be greater than zero due to random error but SS_f will *equal* zero, because y_f will be an interpolation, with the same number of degrees of freedom as data points.

[10] The most curious feature of Eq. (11.36) is the prefactor, which suggests the number of degrees of freedom is $p_\mathrm{f} - p_\mathrm{r}$. Consider the case $p_\mathrm{f} = 3$, $p_\mathrm{r} = 2$. Because of the assumption that y_r is the correct model, the coefficients $\hat{a}_\mathrm{f}$, $\hat{b}_\mathrm{f}$, and $\hat{c}_\mathrm{f}$ of the incorrect model y_f will exhibit random fluctuations about the true values a_r, b_r, and zero. The set $\{\hat{a}_\mathrm{f}, \hat{b}_\mathrm{f}, \hat{c}_\mathrm{f}\}$ of y_f can be viewed as the "result" of the experiment. The normal equations for the calculated values $\hat{a}_\mathrm{r}$ and $\hat{b}_\mathrm{r}$ are constrained to estimate the true values a_r and b_r. Given p_r terms in y_r, we get p_r constraints. Thus, the number of degrees of freedom is $p_\mathrm{f} - p_\mathrm{r}$.

We can view the situation as follows: If y_r is the true model, then $\hat{\sigma}_\mathrm{f}^2$ and $\hat{\sigma}_\Delta^2$ are two independent estimates of the variance σ_y^2. This is a case for the F-test. From the discussion in Section 9.4, we know that if

$$\hat{\sigma}_\Delta^2/\hat{\sigma}_\mathrm{f}^2 < F_{\alpha,\, p_\mathrm{f}-p_\mathrm{r},\, N-p_\mathrm{f}} \tag{11.37}$$

then the probability is greater than $1-\alpha$ that $\hat{\sigma}_\Delta^2$ and $\hat{\sigma}_\mathrm{f}^2$ are estimating the variance of the same sample population. This is a reasonable criterion for testing the null hypothesis. If Eq. (11.37) is satisfied, the full model is rejected in favor of the reduced model. Statistical software packages for computing fits routinely provide a table that contains the data needed to apply this test.

Example 11.5. *Choosing between models.* Let us analyze the data in Fig. 11.5. We begin by comparing a quadratic polynomial with a linear polynomial. Here is the output from the LS fitting routine in *Mathematica*:

```
sample = {{10, 3.667}, {11, 3.893}, {12, 3.853}, {13, 4.103},
   {14, 3.921}, {15, 4.006}, {16, 4.080}, {17, 4.165}, {18, 4.111}, {19, 4.267}};

linpolyfit = LinearModelFit[sample, {1, x}, x];
linpolyfn[x_] := linpolyfit["BestFit"];

linpolyfn[x]

3.25945 + 0.0515273 x

linpolyfit["ANOVATable"]
```

	DF	SS	MS	F Statistic	P-Value
x	1	0.219042	0.219042	29.9085	0.000595245
Error	8	0.05859	0.00732375		
Total	9	0.277632			

```
quadpolyfita = LinearModelFit[sample, {1, x, x^2}, x];
quadpolyfna[x_] := quadpolyfita["BestFit"];

quadpolyfna[x]
```

$2.75216 + 0.124357\,x - 0.00251136\,x^2$

```
quadpolyfita["ANOVATable"]
```

	DF	SS	MS	F Statistic	P-Value
x	1	0.219042	0.219042	27.747	0.00116373
x^2	1	0.00333007	0.00333007	0.421834	0.536745
Error	7	0.0552599	0.00789427		
Total	9	0.277632			

```
InverseCDF[FRatioDistribution[1, 7], 1 - 0.536745]

0.421833
```

The two models being compared here are

$$y_\mathrm{r} = 3.25945 + 0.0515273\,x, \qquad y_\mathrm{f} = 2.75216 + 0.124357\,x - 0.00251136\,x^2.$$

The value of $\hat{\sigma}_\mathrm{f}^2 = 0.00789427$ is listed in the ANOVA table of the quadratic polynomial, where it is referred to as the mean square ("MS") of the error. $\hat{\sigma}_\Delta^2$ is easily calculated from entries in the sum of squares ("SS") columns:

$$\hat{\sigma}_\Delta^2 = \frac{1}{3-2}(SS_\mathrm{r} - SS_\mathrm{f}) = 0.058590 - 0.0552599 = 0.003330.$$

This value is listed as the "MS" for the x^2 term. Thus, the empirical F-ratio is $0.003330/0.007894 = 0.4218$, as listed in the F column of the full model.

We focus on the "P-values." These are the values of α such that $F_{\alpha,p_{\rm f}-p_{\rm r},N-p_{\rm f}}$ equals the empirical F statistic. This is shown at right. The area α is the probability that an observed F equals 0.4218 and yet the null hypothesis is true. Here, α is 54% of the total area under the curve. We interpret this as the probability that $y_{\rm f}$ is no better than $y_{\rm r}$. We therefore reject $y_{\rm f}$. Let us consider alternative models. How about $y_{\rm f} = a+bx+c\ln x$? This gives $\alpha = 44\%$. Because α is now less than 50%, this $y_{\rm f}$ is perhaps a better fit than is $y_{\rm r}$, but not by much. If we accept this $y_{\rm f}$, we have to accept that there is a 44% chance we are making a mistake. Including x^{-1} in the basis set turns out to be more successful. The following table shows results from various comparisons:

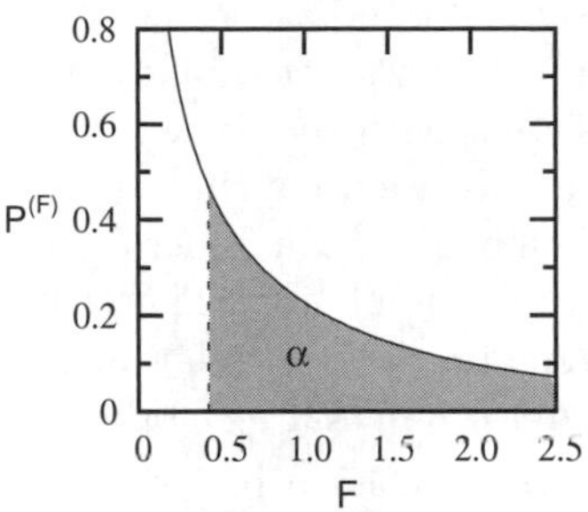

Models	F	α
$a+bx$ vs. $a+bx+cx^2$	0.4218	0.5367
$a+bx$ vs. $a+bx+c\ln x$	0.6664	0.4412
$a+bx$ vs. $a+bx+cx^{-1}$	0.8134	0.3971
$a+bx^{-1}$ vs. $a+bx^{-1}+cx$	0.006722	0.9370

For each comparison $y_{\rm f}$ must consist of $y_{\rm r}$ plus one or more additional terms. The last comparison shows that there is a 94% probability that the linear term is not needed. We conclude that the true model is most likely $y(x) = a + bx^{-1}$. The coefficients of this fit are $\hat{a} = 4.728$ and $\hat{b} = -10.042$. (In fact, the data points for this exercise were generated by adding random error to $y = 4.7 - 10\,x^{-1}$.)

Another way to judge the quality of fit of a linear fitting function is to calculate the ***linear correlation coefficient***,

$$r = \frac{\sum_{i=1}^N (x_i-\bar{x})(y_i-\bar{y})}{\left\{\left[\sum_{i=1}^N (x_i-\bar{x})^2\right]\left[\sum_{i=1}^N (y_i-\bar{y})^2\right]\right\}^{1/2}} = \frac{s_{xy}}{\sqrt{s_{xx}s_{yy}}}\,. \tag{11.38}$$

The closer $|r|$ is to 1, the greater the agreement between the data and a straight line. It is always the case that $|r| \le 1$. As an extreme example, suppose that the points lie exactly on a straight line $y(x) = a + bx$. Then $y_i = a + bx_i$, which implies that $\bar{y} = a + b\bar{x}$, $y_i - \bar{y} = b\,(x_i - \bar{x})$, and $r = b/\sqrt{b^2}$. If the slope is positive then $r = 1$ and if it is negative then $r = -1$. Now suppose that the x_i and y_i values are not correlated at all. In general, we expect $(x_i - \bar{x})$ to be positive as often as it is negative, and the same for $(y_i - \bar{y})$. In the absence of any correlation between x_i and y_i, the product $(x_i - \bar{x})(y_i - \bar{y})$ in the numerator of Eq. (11.38) will also be positive as often as it is negative. Then the sum in the numerator would be approximately equal to zero.

Assuming a normal distribution of error, it is possible to derive from the Student t-distribution the following expression for the probability that a set of N uncorrelated x_i, y_i values would give $|r|$ greater than or equal to some given $|r_0|$,

$$P_N^{(|r|\ge|r_0|)} = \frac{2\,\Gamma\left(\frac{N-1}{2}\right)}{\sqrt{\pi}\,\Gamma\left(\frac{N-2}{2}\right)}\int_{|r_0|}^1 (1-\rho^2)^{(N-4)/2}d\rho\,. \tag{11.39}$$

For example, suppose that you have calculated a correlation coefficient of

$r_0 = 0.70$ for a data set with $N = 10$. The value of $P_{10}^{(|r| \geq 0.70)}$ is 0.024. There is only a 2.4% probability that the hypothesis y *is not a function of* x is consistent with the data.

Can we then conclude that there is a 97.6% probability that y vs. x is a straight line? In general, no. Suppose that y is a function of x but the function is nonlinear. Then r might be less than 1 even with no random error in the data. The correlation coefficient useful if the only two alternatives are y *is not a function of* x or y *is linear in* x. If there is the possibility that the functional relationship is nonlinear, then the $P_N^{(|r| \geq |r_0|}$ value is hard to interpret, and the F ratio will probably be a more useful test of the model. Furthermore, there are more versatile tests of the hypothesis y *is not a function of* x, which will be described in Section 12.3.1.

Exercises

11.1 Consider the data set $\{(0.2053, 2.10), (0.4345, 3.09), (0.5920, 3.89), (0.7793, 4.92)\}$. Calculate by hand the straight-line least-squares fit and the 95% and 90% confidence intervals for the two coefficients. Compare with results from computer algebra.

11.2 For the data set in Exercise 11.1 calculate the the 90% and 95% confidence intervals for the coefficients under the assumption that the values of the y coordinates are known exactly but the x coordinates contain random error.

11.3* For the data in Exercise 11.1 calculate the LMS straight-line fit and plot the residuals.

11.4 For the data set in Exercise 11.1 carry out an F-test to see if a quadratic fit might be more appropriate than a straight-line fit.

11.5 Suppose that the data set in Exercise 11.1 are calibration points for a set of standards. (a) Plot the Hotelling-Working confidence band boundaries along with the calibration points and the best-fit line. (b) Calculate the 95% confidence interval for the x value corresponding to a single y measurement of 4.89. (c) Determine the detection limit of the instrument.

11.6 Derive Eq. (11.9).

11.7 In the case of the confidence intervals given by Expressions (11.11), why not divide $\hat{\sigma}_a$ and $\hat{\sigma}_b$ by $\sqrt{N}$ as was done for point estimation? Under what circumstances would it be appropriate to use $\hat{a} \pm t_{\alpha/2}\hat{\sigma}_a/\sqrt{M}$ and $\hat{b} \pm t_{\alpha/2}\hat{\sigma}_b/\sqrt{M}$? What would M represent?

11.8 Consider an experiment in which you will calibrate a measuring instrument and then measure an unknown. For each of the following values of α and $N_{\mathrm{tot}} = N_{\mathrm{std}} + N_{\mathrm{u}}$, determine the optimal number of these measurements that should be used for measuring the unknown. (a) $\alpha = 0.05$, $N_{\mathrm{tot}} = 6$. (b) $\alpha = 0.10$, $N_{\mathrm{tot}} = 6$. (c) $\alpha = 0.05$, $N_{\mathrm{tot}} = 10$. (d) $\alpha = 0.10$, $N_{\mathrm{tot}} = 10$. (e) $\alpha = 0.05$, $N_{\mathrm{tot}} = 20$. (f) $\alpha = 0.10$, $N_{\mathrm{tot}} = 20$.

11.9 Write a computer algebra function that generates a simulated measurement value $y(x) = 2x + \eta$, where η is normally distributed error with $\sigma = 0.1$.

11.10 Using your measurement simulation function from Exercise 11.9, simulate an experiment in which you calibrate the instrument and use it to determine the value of an unknown x_{u}. Assume that the true value of x_{u} is 0.4117 and that the instrument gives dependable results for $0.05 < x < 1.0$. Try different spacings of the $x_{\mathrm{std},\,k}$ and compare the results for $\hat{x}_{\mathrm{u}}$. Develop and test an optimal strategy for choosing the spacings.

Chapter 12

Experiment Design

Here we discuss various ways in which statistical considerations can be taken into account when designing an experimental study. First we examine in more detail the logic of hypothesis testing and see that an appropriate experiment design must take into account the degree to which we are willing to risk making an incorrect decision. Next we consider the important technique of *randomization*, which can improve the statistical significance of an experiment at no additional cost. Then we briefly survey the design and analysis of *multiple comparisons*, in which the effects of different experimental conditions are compared. This is often used as preliminary screening, to be followed by a more thorough examination of the identified conditions. Finally we consider the question of how to systematically optimize experimental conditions.

12.1 Risk Assessment

Let us review the procedure for the t-test to answer the question, Is x greater than some specified value μ_0? We carry out N measurements of the property x and use the results to estimate $\bar{x}$ and $\hat{\sigma}$. Let us call this observed value of the mean $\bar{x}_{\text{obs}}$. We then develop a hypothetical probability distribution function centered at μ_0 with scale $\hat{\sigma}/\sqrt{N}$ and use it to predict a maximum value $\mu_0 + \delta_\alpha$ below which $\bar{x}$ would be predicted to be found with a probability of $1-\alpha$. If $\bar{x}_{\text{obs}}$ lies below $\mu_0 + \delta_\alpha$ we conclude, at confidence level $1-\alpha$, that there is no significant difference between $\bar{x}_{\text{obs}}$ and μ_0.

Let us express this in statistical jargon. We will call the hypothetical population described by a normal probability distribution centered at μ_0 "population 0" and the actual population being sampled by the experiment, "population 1." Let μ_1 be the true value of the measured property for population 1. The t-test is testing the null hypothesis that population 1 is identical to population 0. We will abbreviate this with the notation

$$\mathrm{H}_0: \quad \mu_1 = \mu_0\,, \quad \sigma_1^2 = \sigma_0^2\,. \tag{12.1}$$

This kind of experiment is called a comparison with a *control*. The "control" is population 0, with a specified value for μ_0.

The way in which we carry out the analysis depends on how we choose to define the alternative to the null hypothesis, which is called the ***alternative hypothesis***, designated by the symbol "$\mathrm{H_A}$." Among the possible choices are the following:

$$\mathrm{H_A}: \quad \mu_1 > \mu_0\,, \quad \sigma_1^2 = \sigma_0^2\,, \tag{12.2a}$$

$$\mathrm{H_A}: \quad \mu_1 \neq \mu_0\,, \quad \sigma_1^2 = \sigma_0^2\,, \tag{12.2b}$$

$$\mathrm{H_A}: \quad \mu_1 < \mu_0\,, \quad \sigma_1^2 = \sigma_0^2\,. \tag{12.2c}$$

(We could also consider inequalities for the variances, as discussed in Chapter 9.) Each choice leads to a different confidence interval for the null hypothesis, as follows, respectively:

$$\bar{x} \leq \mu_0 + \delta_\alpha, \qquad \delta_\alpha = t_{\alpha,\nu}\,\hat{\sigma}_{\bar{x}}\,, \tag{12.3a}$$

$$\mu_0 - \delta_\alpha \leq \bar{x} \leq \mu_0 + \delta_\alpha, \qquad \delta_\alpha = t_{\alpha/2,\nu}\,\hat{\sigma}_{\bar{x}}\,, \tag{12.3b}$$

$$\mu_0 - \delta_\alpha \leq \bar{x}, \qquad \delta_\alpha = t_{\alpha,\nu}\,\hat{\sigma}_{\bar{x}}\,, \tag{12.3c}$$

where $\hat{\sigma}_{\bar{x}}$ is the standard error, $\hat{\sigma}/\sqrt{N}$. At first glance one might think that only $\mathrm{H_A} : \mu_1 \neq \mu_0$ is an appropriate alternative to $H_0 : \mu_1 = \mu_0$. In fact, the hypotheses $\mu_1 > \mu_0$ and $\mu_1 < \mu_0$ do also test $\mathrm{H_0}$ but they implicitly incorporate additional information about the situation. For example, μ_0 might be the known concentration of some solute in a prepared solution and μ_1, the concentration of the solute after a treatment to remove it. We would probably make the implicit assumption that the treatment cannot increase the solute concentration—either it will decrease it or have no effect. An experiment that yields $\bar{x}_{\mathrm{obs}} > \mu_0$ would be taken as evidence for the hypothesis $\mu_0 = \mu_1$, because $\mu_1 > \mu_0$ is by assumption impossible. The apparent concentration increase would be presumed to be due to error in the measurements.

While the t-test is useful, it is not the whole story. What it tests is $\mathrm{H_0}$; *it does not test* $\mathrm{H_A}$. This is an important distinction. α is the probability that the t-test will give an incorrect result: $\mathrm{H_0}$ will be rejected, and $\mathrm{H_A}$ accepted, while in fact $\mathrm{H_A}$ is false. In other words, α is the probability of a ***false positive***—we conclude the treatment has an effect, while in fact it does not. This is called a ***type I error***. The t-test does not, by itself, take into account the possibility of a ***false negative***, in which the alternative hypothesis is rejected when it is in fact true. This is a ***type II error***. Its probability is called β. The nomenclature is summarized in Table 12.1.

Type II error is important—often more important than type I error. In environmental risk assessment, for example, it is a type II error to conclude that a toxic pollutant is absent when in fact it is present. This kind of error could endanger public safety. The consequence of the type I error, which might be to force a factory to install unnecessary pollution control equipment, would likely be less consequential.

Table 12.1: Errors in inference.

Inference \ Reality:	$\mathrm{H_A}$ is false.	$\mathrm{H_A}$ is true.
Reject $\mathrm{H_A}$.	no error (true negative) prob. $1-\alpha$	type II error (false negative) prob. β
Accept $\mathrm{H_A}$.	type I error (false positive) prob. α	no error (true positive) prob. $1-\beta$

To calculate β it is necessary to specify our tolerance for accepting H_0. In practice, it will almost always be the case that we can tolerate some small difference between μ_1 and μ_0 and still consider H_0 to be true. We will use the symbol ϵ to designate this minimum acceptable difference. It is the smallest population difference for which we will reject H_0. In the case of a treatment to remove a solute, ϵ would be the minimum decrease deemed worth the cost of the treatment. In the case of a polluting factory, ϵ would be the minimum concentration of the pollutant deemed to be dangerous. In the following example a quantitative relationship between α, β, ϵ, σ^2, and N is developed.

Example 12.1. *Type II error for one-way comparison with a control.* $KClO_4$ is commonly used as a propellant in fireworks. The perchlorate ion is known to interfere with the uptake of iodide by the thyroid gland. Because perchlorate salts are extremely soluble, it can be expected that much of the perchlorate released into the air will end up in groundwater and surface waters. In Massachusetts many cities and towns sponsor fireworks displays on the fourth of July to commemorate Independence Day. Suppose that you are concerned that your town's fireworks will contaminate the drinking water that you and your neighbors get from private wells.

Throughout the months of May and June you collect weekly water samples from five different wells in order to determine the background ClO_4^- level. Some perchlorate exists in raindrops, on account of absorption from sea spray and perhaps from natural chlorine chemistry in the atmosphere. Another source may be agricultural fertilizer, which sometimes contains small amounts of perchlorate salts. Let x be the concentration of ClO_4^- in μg/L. You see no correlations in results with respect to time or location of measurement, so you make the assumption that your total data set of 45 measurements is sampling a population described at least approximately by a normal distribution. You find a mean background concentration of 0.39 with standard error $\sigma_{\bar{x}} = 0.058$. This implies a quite large scale estimate of $\hat{\sigma} = 0.42$. You propose to collect five more samples on the fifth of July.

Is this experiment design adequate? Let us set α at 5% and then determine β. The Massachusetts Department of Environmental Protection has set 1 μg/L as the "advisory level" for ClO_4^- in drinking water, at which there might be toxic effects on pregnant women and young children. Let us choose $\mu_0 + \epsilon - 1.00$ with $\mu_0 = 0.39$. We then plot the probability distribution expected for the mean $\bar{x}$ of our new measurements if the null hypothesis is true. This was given in Chapter 9. It is a Student distribution, Eq. (9.9), centered at $\mu_0 = 0.39$ with $\nu = 4$ degrees of freedom. The scale is the standard error, $\sigma_{\bar{x}} = 0.42/\sqrt{5}$. The distribution function is

$$P_{\mu_0,\,\hat{\sigma}/\sqrt{N},\,N-1}(\bar{x}) = \frac{\Gamma(N/2)}{\sqrt{\pi(N-1)\hat{\sigma}^2/N}\;\Gamma\big((N-1)/2\big)}\left[1 + \frac{(\bar{x}-\mu_0)^2}{(N-1)\hat{\sigma}^2/N}\right]^{-N/2}$$

$$= (1.996)\big[1 + (7.086)(\bar{x} - 0.39)^2\big]^{-5/2}.$$

The equation determining δ_α is

$$\alpha = \int_{\delta_\alpha}^{\infty} P_{\mu_0,\,\hat{\sigma}/\sqrt{N},\,N-1}(\bar{x})\,d\bar{x}\,. \tag{12.4}$$

Solving this numerically gives $\delta_\alpha = 0.79$.

The alternative hypothesis is $\mu_1 > \mu_0$ where μ_1 is the concentration on July 5. Consider a second distribution function, describing our expectation for $\bar{x}$ if the true value of μ_1 were equal to the minimum value we consider to be consequential, that is,

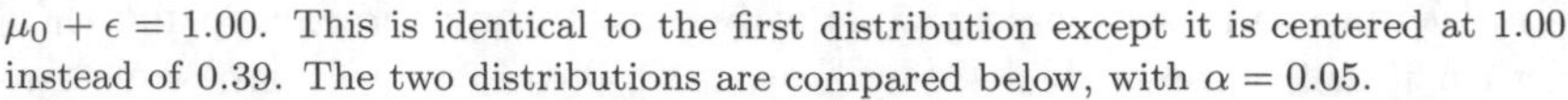
$\mu_0 + \epsilon = 1.00$. This is identical to the first distribution except it is centered at 1.00 instead of 0.39. The two distributions are compared below, with $\alpha = 0.05$.

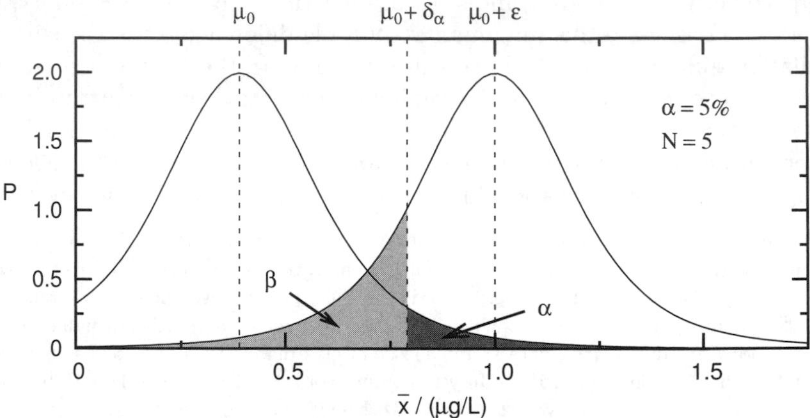

Clearly, our proposed experiment design gives a very large β. This is the probability of a type II error (false negative) for a population with μ_1 at the limit of toxicity, 1.00 μg/L. Integrating the distribution from $-\infty$ to $\mu + \delta_\alpha$ gives $\beta = 16.4\%$. Considering that the consequences of a false negative in this case (possible developmental problems in fetuses and young children) are more serious than the consequences of a false positive (the town having to switch to perchlorate-free fireworks), one might want β less than α, perhaps 1% or even smaller. What is the lowest level that the proposed experiment can detect with $\beta = 1\%$? β drops to this level if $\mu_0 + \epsilon$, the center of the μ_1 distribution in the graph, is increased to 1.38. If the samples collected on July 5 yield $\bar{x}$ greater than 1.38, we will be able to conclude with confidence greater than $1 - \beta = 99\%$ that μ_1 is above the limit of 1.00 μg/L. If $\bar{x}$ is less than 0.39, our estimate for μ_0, we conclude with confidence greater than $1 - \alpha = 95\%$ that the water is safe. However, if $0.39 \leq \bar{x} \leq 1.38$, the experiment will be inconclusive.

We might want to consider increasing the number of samples to be collected. The scale of the Student distributions is the standard error $\hat{\sigma}/\sqrt{N}$. Increasing N narrows the distributions, decreasing δ_α and decreasing β. The graph below shows the result with $N = 9$, the minimum needed for β to drop below 1% with α at 5%.

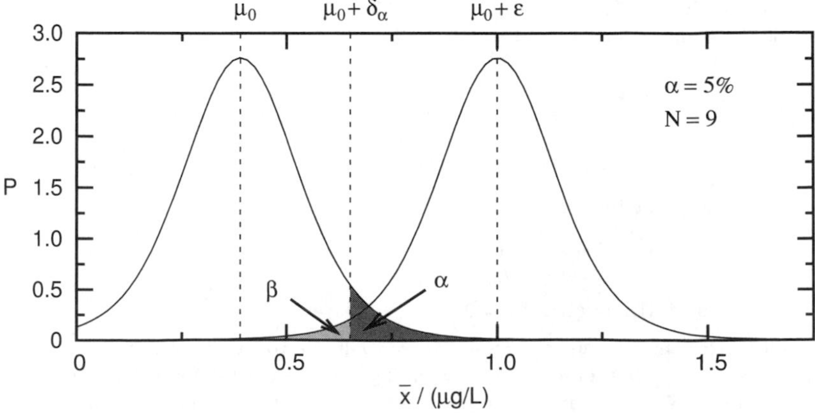

The amount of type II error is an indicator of the experiment's sensitivity—the smaller the β value, the lower the ϵ value and the minimum value $\mu_0+\epsilon$ the experiment can detect. The quantity $1-\beta$ is called the ***statistical power*** of the experiment. The closer it approaches 1, the more powerful (i.e., sensitive) the experiment. In general, the following strategies increase the power (i.e., decrease β):

- Increase the number of measurements, N.
- Decrease the scale σ of the probability distributions.
- Increase the tolerance ϵ, so that it is significantly larger than δ_α.
- Increase α, and thereby accept a larger probability of type I error.

12.2 Randomization

Here we develop a technique for dealing with *systematic* error. Consider the following experiment design, intended to determine whether or not a proposed catalyst increases the yield of a chemical synthesis:

> The synthesis will be repeated ten times. In five of the syntheses a predetermined amount (the same for each experiment) of catalyst will be added to the mixture of reactants. The yield of product will be measured after a predetermined time interval.

Consider the sources of variability. First of all, there is of course some random error in the measurement of the yield, which can presumably be described with a normal or log-normal distribution. However, for this experiment we will assume that the measurement error is insignificant compared with uncontrolled systematic errors. Some possible sources of systematic error are:

- The room temperature of the laboratory might vary.
- One of the reagents might spontaneously decompose, so that its concentration varies.
- The amount of and manner of stirring might vary.
- Some of the glassware might be contaminated with something that affects the reactions.

In a complicated multistep synthetic procedure there are quite likely many other possible sources of variation as well. It might be difficult to quantify these effects and, in any case, it is unlikely they would follow any of the standard probability distributions. In fact, it is unlikely they would even be random—they are systematic errors.

Nevertheless, we seek a statistical test of the null hypothesis

> H_0: *The treatment does not significantly increase the value of y.*

y is the yield and the "treatment" is the addition of catalyst. We will use

Table 12.2: Results of a randomized study of the effect of a catalyst on the yield of a chemical synthesis. The "+" and "−" labels indicate whether or not the catalyst is used. The numbers in parentheses are the yield of desired product as a percentage of the stoichiometric yield.

#1	#2	#3	#4	#5
+ (5.3)	− (5.4)	− (3.2)	+ (3.6)	− (1.4)
#6	**#7**	**#8**	**#9**	**#10**
+ (7.6)	− (2.8)	+ (7.3)	+ (5.6)	− (4.2)

a technique called ***randomization***[1] to test H_0 without the need to assume any particular probability distribution for the errors. What we will do is randomly select five of the syntheses to be assigned the label "−," with the rest assigned the label "+," and then add the catalyst only to the "+" syntheses. The selection is done in advance, before performing the first synthesis.[2] Suppose that the results of the experiment are as shown in Table 12.2. The observed difference in the means between the treated and the untreated cases is

$$\Delta = \bar{y}^{(+)} - \bar{y}^{(-)} = 2.48\,.$$

H_0 will be accepted if this value is not significantly greater than zero. The question then is how large must Δ be for us to declare it "significant."

If H_0 is true (the catalyst has no significant effect) then changing the labels has no significant effect on the results of the experiment. The number of different ways the labels can be assigned is[3]

$$W = \frac{N!}{n^{(+)}!\,n^{(-)}!} = \frac{10!}{5!\,5!} = 252\,. \tag{12.5}$$

Given the results in Fig. 12.2, let us recalculate Δ for each of the 251 other possible labelings with five cases labeled "−" and five labeled "+." Each different choice of labels gives a different Δ value. Each alternative labeling will cause one or more treated syntheses to be considered as untreated and one or more untreated syntheses to be considered treated. If the treatment has a significant effect then this should decrease $\bar{y}^{(+)}$ and increase $\bar{y}^{(-)}$, thereby decreasing Δ.

---

[1] This is another statistical technique developed by Fisher. The systematic error will "randomized" so that it can be analyzed as if it were random error.

[2] In *Mathematica*, `RandomSample[tlist]` randomizes a list of treatments. The original list could be `tlist={"-","-","-","-","-","+","+","+","+","+"}`, for example. `Permutations[tlist]` lists all possible alternative orderings.

[3] See Section 8.2.4.

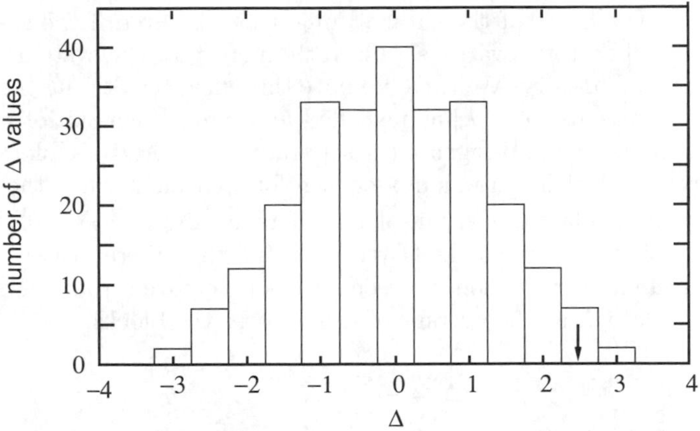

Figure 12.1: Histogram of $\Delta = \bar{y}^{(+)} - \bar{y}^{(-)}$ from all 252 different +/− labelings of the results in Table 12.2. The arrow marks the observed value.

The full collection of these Δ values is displayed as a histogram in Fig. 12.1. This is, in effect, a probability distribution for Δ, and it takes into account systematic as well as random variation. Let us arrange the values in ascending ordering, using the notation $\Delta_{(k)}$ with $1 \leq k \leq 252$, where $\Delta_{(1)}$ is the lowest value. $\Delta_{(1)}$ corresponds to a probability between 0 and 1/252; let us call it 0.5/252, in the center of the range. $\Delta_{(2)}$ corresponds to 1.5/252 and in general the probability $(1 - \alpha)$ corresponding to $\Delta_{(k)}$ is $(1 - \alpha) = (k - 0.5)/W)$. It follows that

$$k = 0.5 + (1 - \alpha)W. \tag{12.6}$$

The 95th percentile in this case corresponds to $0.5 + (0.95)(252) \approx 240$. The observed Δ value of 2.48 corresponds to $k = 247$, which is among the upper 5%. The null hypothesis is rejected at the 95% confidence level. The P value (i.e., the value of α corresponding to the observed Δ) is

$$P = 1 - \frac{247 - 0.5}{252} = 0.022\,.$$

Thus we reject H_0 with the expectation that this will be an incorrect decision only 2.2% of the time, and we accept instead the alternative hypothesis

H_A: *The treatment does significantly increase the value of* y.

Having convinced ourselves there *is* a statistically significant effect, we can plan a new series of syntheses in which each synthesis uses a *different amount* of catalyst. The results can then be modeled with a fit $y(x)$ using the methods of Chapters 10 and 11.

Now suppose the ten syntheses were carried out in five different laboratories. In each laboratory two syntheses are performed, one with catalyst and one without. The effect of the catalyst might be obscured if the systematic error *between* laboratories varies significantly more that the systematic error *within* a single laboratory. We can eliminate this effect by altering the way we analyze the results, using a technique called ***blocking***. Each pair of syntheses in a single laboratory is designated a separate *block*. With a block size of two, one from each pair is randomly chosen for treatment while the other is untreated. We calculate the within-block differences $\Delta_m = y_m^{(+)} - y_m^{(-)}$, where $m = 1, 2, 3, 4, 5$ labels the blocks. By calculating the differences within each block, the systematic variation between blocks is removed from the analysis. We then calculate Δ as the average of the δ's over the blocks,

$$\Delta = \frac{1}{5} \sum_{m=1}^{5} \Delta_m \,.$$

A randomization analysis can be carried out by calculating this Δ for all the different possible alternate $+/-$ labelings, exchanging $+$ and $-$ only within each block. The number of different labelings with five blocks of two samples each is

$$W = 2^5 = 32\,,$$

which is much smaller than 252 different labelings with the unblocked experiment. In general, with M blocks, with N_m members in the mth block and $n_m^{(+)}$ of them treated, $n_m^{(-)}$ untreated, the total multiplicity of labelings is

$$W = \prod_{m=1}^{M} W_m\,, \qquad W_m = \frac{N_m!}{n_m^{(+)}!\,n_m^{(-)}!}\,. \tag{12.7}$$

W_m is the multiplicity within the mth block.

The blocked multiplicity, W from Eq. (12.7), is always smaller than the unblocked multiplicity, and usually very much smaller. A smaller W implies less sensitivity. With a small W, the Δ value will have to be large in order to detect any effect of the treatment. Thus, there is a tradeoff involved with blocking. It should be used only when one has reason to suspect that there is significant systematic variation between the designated blocks.

12.3 Multiple Comparisons

In Section 12.2 we developed a test for whether a treatment has an effect on the property of interest. We did this by developing a probability distribution for the difference of the means of treated and untreated populations. Suppose now that there are multiple different treatments to consider. Then it turns out to be more efficient to analyze variances rather than means, using Fisher's ANOVA (analysis of variance) methodology.[4] The methods discussed here are best used in the preliminary stage of an investigation, as a tool for deciding

[4]ANOVA has a variety of applications. We used it in Section 9.4 to test if two different samples have different variances, and in Section 11.5, to compare fitting models.

which of many different possible factors are of importance for further study. We consider here experiment designs in which each sample is given at most one of the various treatments. However, it is also possible to analyze a situation in which multiple treatments are applied to each sample. Sophisticated experiment designs for facilitating such an analysis, called *factorial designs*, have been developed, but this is beyond the scope of the presentation here. Details can be found in references in the Bibliography.

12.3.1 ANOVA*

Consider an experiment to compare k different treatments on a response y. For example, we might have k different catalysts, and each of a series of chemical syntheses will use one of them or none at all. Let $y_{t,j}$ be the jth measurement with treatment t, let N_t be the number of samples subjected to treatment t, and let the "within-treatment" means be $\bar{y}_t = N_t^{-1} \sum_j y_{t,j}$.

We will test the null hypothesis

> H_0 : *The within-treatment means are not significantly different from each other.*

If H_0 is true, then all of the $y_{t,j}$ come from the same population. In that case there are various ways to estimate the scale σ of the probability distribution. Let us assume that the distribution is normal.[5] Then the obvious way to estimate σ is to use the sum of squared residuals of the full sample. However, we could also estimate σ as follows:

$$\sigma^2 \approx \hat{\sigma}_{\mathrm{R}}^2 = \frac{1}{\nu_{\mathrm{R}}} SS_{\mathrm{R}}, \qquad SS_{\mathrm{R}} = \sum_{t=1}^{k} \sum_{j=1}^{N_t} (y_{t,j} - \bar{y}_t)^2. \tag{12.8}$$

This is a *within-treatment* sum of squared residuals (the subscript "R" stands for "residuals") with the differences $(y_{t,j} - \bar{y}_t)$ calculated from the within-treatment mean $\bar{y}_t$ instead of from the "grand" mean $\bar{y}$. If H_0 is true, each block will give an estimate of the same true scale σ. The number of degrees of freedom is equal to the number of measured $y_{t,j}$ values, which is N, minus the number of other intermediate values that we have to calculate from the $y_{t,j}$. We have to calculate k different within-treatment means, $\bar{y}_t$. Therefore,

$$\nu_{\mathrm{R}} = N - k\,. \tag{12.9}$$

Another way to estimate σ is to use the variance of *between-treatment* means,

$$\sigma^2 \approx \hat{\sigma}_{\mathrm{T}}^2 = \frac{1}{\nu_{\mathrm{T}}} SS_{\mathrm{T}}, \qquad SS_{\mathrm{T}} = \sum_{t=1}^{k} N_t\,(\bar{y}_t - \bar{y})^2. \tag{12.10}$$

("T" stands for "treatments.") The difference is now between the grand mean $\bar{y}$ and the mean for a single kind of treatment, weighted by the number of

[5]If this assumption is not appropriate, a nonparametric alternative to ANOVA, the *Kruskal-Wallis test*, could be used.

measurements N_t corresponding to the particular treatment. Now let us determine $\nu_{\rm T}$. By substituting in the definitions of the SS's and then expanding the squares one can show that

$$SS_{\rm T} + SS_{\rm R} = \sum_{t=1}^{k} \sum_{j=1}^{N_t} y_{t,j}^2 - N\bar{y}^2. \tag{12.11}$$

The right-hand side has $N - 1$ degrees of freedom: N measured values $y_{t,j}$ minus one for the derived quantity $\bar{y}$. Therefore,

$$\nu_{\rm T} + \nu_{\rm R} = N - 1,$$

which implies that

$$\nu_{\rm T} = k - 1\,. \tag{12.12}$$

Thus we have the independent variance estimates

$$\hat{\sigma}_{\rm T}^2 = SS_{\rm T}/(k-1), \qquad \hat{\sigma}_{\rm R}^2 = SS_{\rm R}/(N-k)\,. \tag{12.13}$$

N is the total number of measurements and k is the number of treatments.

If any of the treatments do have significant effects on the within-treatment means, then the variance $\sigma_{\rm T}^2$ *between* treatments should be larger than the variance $\sigma_{\rm R}^2$ *within* treatments. In that case the ratio

$$F = \hat{\sigma}_{\rm T}^2/\hat{\sigma}_{\rm R}^2 \tag{12.14}$$

will be relatively larger than it would be if $\rm H_0$ were true. All that is left to do is to determine the critical value $F_{\alpha,\nu_{\rm T},\nu_{\rm R}}$ corresponding to the desired confidence level α, which can be looked up in tables or computed using Eq. (9.45).

In principle we could also determine this critical F value using randomization, randomly changing the treatment labels, recomputing F, and then sorting them in ascending order. Note that in Fig. 12.1 the histogram of the randomization distribution looks rather like a normal distribution. As N increases, the randomization distribution for F becomes equivalent to a normal distribution. The number of different randomization labelings is

$$W = N! \Big/ \prod_t N_t!\,.$$

This is usually an exceedingly large number. For example, if $N_t = 5$ for all t and $k = 4$, then $W = 20!/(5!)^4 = 1.2 \times 10^{10}$. This is many orders of magnitude larger than W for a sequence of separate experiments first comparing treatment 2 with treatment 1, then treatment 3 with treatment 1, then 4 to 1, and so on. A larger W for simultaneously comparing all the treatments implies that the simultaneous comparison will give higher sensitivity than will the separate comparisons for a given total number of experiments.

12.3.2 Post-Hoc Tests*

If $F < F_{\alpha,\nu_{\rm T},\nu_{\rm R}}$, then we accept the null hypothesis and conclude that the experiment shows no evidence that any of the treatments have an effect. The analysis is finished. However, if $F > F_{\alpha,\nu_{\rm T},\nu_{\rm R}}$, then we conclude that at least one of the treatments has an effect. What we probably then want to know is *which* of the treatments are responsible. To answer this question we need to directly test the difference between two treatment means, $\bar{y}_{\rm a} - \bar{y}_{\rm b}$, using an additional "post-hoc" test.[6]

We saw in Section 9.4 how to do this for a comparison of two particular treatments.[7] The result was a confidence interval $|\bar{x}_{\rm a} - \bar{x}_{\rm b}| < t_{\alpha/2}\,\hat{\sigma}_{\bar{x}_{\rm a}-\bar{x}_{\rm b}}$. To apply this to our present purpose we could calculate a standard deviation for the difference of means using propagation of error,

$$\sigma^2_{\bar{y}_{\rm a}-\bar{y}_{\rm b}} = \sum_{j=1}^{N_{\rm a}} \left(\frac{\partial \bar{y}_{\rm a}}{\partial y_{{\rm a},j}}\right)^2 \sigma^2_{{\rm a},j} + \sum_{j=1}^{N_{\rm b}} \left(\frac{\partial \bar{y}_{\rm b}}{\partial y_{{\rm b},j}}\right)^2 \sigma^2_{{\rm b},j}\,. \tag{12.15}$$

The derivatives are trivial to calculate:

$$\frac{\partial \bar{y}_{\rm a}}{\partial y_{{\rm a},j}} = \frac{\partial}{\partial y_{{\rm a},j}} \frac{1}{N_{\rm a}} \sum_{m=1}^{N_{\rm a}} y_{{\rm a},m} = \frac{1}{N_{\rm a}}, \qquad \frac{\partial \bar{y}_{\rm b}}{\partial y_{{\rm b},j}} = \frac{\partial}{\partial y_{{\rm b},j}} \frac{1}{N_{\rm b}} \sum_{m=1}^{N_{\rm b}} y_{{\rm b},m} = \frac{1}{N_{\rm b}}.$$

Our strategy is to test the null hypothesis (H_0) that the effect of treatment "a" is not significantly different from the effect of treatment "b." If H_0 is true, then $\hat{\sigma}^2_{\rm R}$ is a reasonable estimate both for $\sigma^2_{{\rm a},j}$ and for $\sigma^2_{{\rm b},j}$, which would mean that

$$\hat{\sigma}^2_{\bar{y}_{\rm a}-\bar{y}_{\rm b}} = \left(\frac{1}{N_{\rm a}} + \frac{1}{N_{\rm b}}\right)\hat{\sigma}^2_{\rm R}\,. \tag{12.16}$$

However, there is an added complication. The t distribution is appropriate only for comparing a single pair of treatments. For our present experiment, with k treatments, the Student t value would give too small a confidence interval. It would test the hypothesis "$\bar{y}_{\rm a}-\bar{y}_{\rm b}$ is not significantly different from zero" for a single specific pair of treatments "a" and "b," giving a probability $1-\alpha_1$ that $|\bar{y}_{\rm a} - \bar{y}_{\rm b}|$ is less than its confidence limit $t_{\alpha_1/2}\hat{\sigma}_{\bar{y}_{\rm a}-\bar{y}_{\rm b}}$. But suppose that we also make a comparison between treatments "c" and "d" such that there is a $1-\alpha_2$ probability that $|\bar{y}_{\rm c} - \bar{y}_{\rm d}|$ is less than its confidence limit $t_{\alpha_2/2}\hat{\sigma}_{\bar{y}_{\rm c}-\bar{y}_{\rm d}}$. The probability that *both* of the pairs are within these confidence limits is $(1-\alpha_1)(1-\alpha_2)$. We want this product to equal $1-\alpha$. To ensure a $1-\alpha$ probability for multiple comparisons we need an α value larger than the α_i's for individual comparisons.

We will consider two different post-hoc tests. The first is ***Dunnett's test*** for multiple comparisons with a control.[8] This test simultaneously compares

[6] *Post hoc* is Latin for "after this." These methods are performed only after ANOVA indicates that a significant effect is present.

[7] We are using the word *treatment* quite broadly. In Section 9.4 the different "treatments" consisted of collecting the samples at two different locations.

[8] Developed by Canadian statistician Charles W. Dunnett (1921-2007).

Table 12.3: Multivariate $t_{k,\alpha,\nu_\text{R}}$ statistic for a one-way Dunnett test at the 95% confidence level ($\alpha = 0.05$). k is the number of treatments (including the control). $\nu_\text{R} = N - k$ is the number of residual degrees of freedom. [From C. W. Dunnett, *J. Amer. Stat. Assoc.* **50**, 1096-1121 (1955). Reproduced with permission. Copyright by the American Statistical Association, all rights reserved. Tables for two-way tests are in C. W. Dunnett, *Biometrics* **20**, 482-491 (1964).]

$\nu_\text{R} \setminus k$	2	3	4	5	6	7	8	9	10
5	2.02	2.44	2.68	2.85	2.98	3.08	3.16	3.24	3.30
6	1.94	2.34	2.56	2.71	2.83	2.92	3.00	3.07	3.12
7	1.89	2.27	2.48	2.62	2.73	2.82	2.89	2.95	3.01
8	1.86	2.22	2.42	2.55	2.66	2.74	2.81	2.87	2.92
10	1.81	2.15	2.34	2.47	2.56	2.64	2.70	2.76	2.81
13	1.77	2.09	2.27	2.39	2.48	2.55	2.61	2.66	2.71
20	1.72	2.03	2.19	2.30	2.39	2.46	2.51	2.56	2.60
40	1.68	1.97	2.13	2.23	2.31	2.37	2.42	2.47	2.51
∞	1.64	1.92	2.06	2.16	2.23	2.29	2.34	2.38	2.42

all the treatments with treatment 1, which is designated the "control" or "standard" treatment. Typically, treatment 1 is "no treatment," in which case Dunnett's test will identify those treatments that have no significant effect on the result. The one-way test for no effect with probability $1 - \alpha$ is

$$\bar{y}_m - \bar{y}_1 < t_{k,\alpha,\nu_\text{R}} \left(\frac{1}{N_1} + \frac{1}{N_m} \right)^{1/2} \hat{\sigma}_\text{R} \,, \tag{12.17}$$

where $t_{k,\alpha,\nu_\text{R}}$ is the ***multivariate Student's t*** for k treatments. These are larger than the usual "single-variable" t_{α,ν_R}. Values at the 95% level are given in Table 12.3. Values for ν_R's not listed in the table can be estimated by interpolation. Dunnett's test is the most commonly used post-hoc test.

Another post-hoc test is ***Tukey's paired comparison test.***[9] If the number of measurements for each treatment is the same value n, then the confidence interval for no difference between any two particular treatments i and j, taking into account all possible comparisons, can be written in the form

$$-q_{k,\alpha/2,\nu_\text{R}} \hat{\sigma}_\text{R}/\sqrt{n} < \bar{y}_i - \bar{y}_j < q_{k,\alpha/2,\nu_\text{R}} \hat{\sigma}_\text{R}/\sqrt{n} \,. \tag{12.18}$$

The q statistic at 95% is given in Table 12.4. (Note that this is a two-way test.) This gives the maximum range of variation between treatment means that would be consistent with the null hypothesis. If a pair of treatments passes this test then we conclude that the effects of the two treatments are not statistically different.

[9] Often referred to simply as "Tukey's pairs." One of many statistical techniques developed by the American chemist and mathematician John W. Tukey (1915-2000). His work in statistics and in signal processing was extremely influential.

Table 12.4: Critical values of the studentized range statistic $q_{k,\alpha/2,\nu_R}$ for Tukey's paired comparison test. These values are for a two-way test at the 95% confidence level ($\alpha = 0.05$). k is the number of treatments and $\nu_R = N - k$ is the number of residual degrees of freedom. [From H. L. Harter, *Ann. Math. Stat.* **31**(4), 1122-1147 (1960). Reproduced with permission. Copyright by the Institute of Mathematical Statistics, all rights reserved.]

$\nu_R \setminus k$	2	3	4	5	6	7	8	9	10
5	3.64	4.60	5.22	5.67	6.03	6.33	6.58	6.80	6.99
6	3.46	4.34	4.90	5.30	5.63	5.90	6.12	6.32	6.49
7	3.34	4.16	4.68	5.06	5.36	5.61	5.82	6.00	6.16
8	3.26	4.04	4.53	4.89	5.17	5.40	5.60	5.77	5.92
9	3.20	3.95	4.41	4.76	5.02	5.24	5.43	5.59	5.74
10	3.15	3.88	4.33	4.65	4.91	5.12	5.30	5.46	5.60
12	3.08	3.77	4.20	4.51	4.75	4.95	5.12	5.27	5.39
15	3.01	3.67	4.08	4.37	4.59	4.78	4.94	5.08	5.20
20	2.95	3.58	3.96	4.23	4.45	4.62	4.77	4.90	5.01
40	2.86	3.44	3.79	4.04	4.23	4.39	4.52	4.63	4.73
60	2.83	3.40	3.74	3.98	4.16	4.31	4.44	4.55	4.65
∞	2.77	3.31	3.63	3.86	4.03	4.17	4.29	4.39	4.47

Although Tukey derived this only for the case in which all treatment blocks have the same number of measurements, numerical simulations indicate that it gives a reasonably accurate estimate even if the block sizes are slightly uneven. In that case, it takes the form

$$|\bar{y}_i - \bar{y}_j| < \frac{q_{k,\alpha/2,\nu_R}}{\sqrt{2}} \left(\frac{1}{N_i} + \frac{1}{N_j} \right)^{1/2} \hat{\sigma}_R \,. \tag{12.19}$$

Example 12.2. *Multiple comparisons.* Consider the following set of data from an experiment intended to determine the effect of four different treatments ("2" through "5") on a measured property y. "Treatment 1" was a control, with no treatment. Two replicates were done with each treatment.

1	2	3	4	5
$205^{(9)}$	$257^{(5)}$	$244^{(8)}$	$267^{(7)}$	$218^{(10)}$
$213^{(4)}$	$245^{(2)}$	$236^{(1)}$	$275^{(3)}$	$222^{(6)}$

The numbers in parentheses show the randomized assignments of treatments to the 10 different samples.

Let us first carry out a one-sided ANOVA analysis. This can be done manually, using Eqs. (12.13) and (12.14), or using any standard statistical software package. With

Mathematica, for example, we could do it as follows:

```
Needs["ANOVA`"];
dataset = {{1, 205}, {1, 213}, {2, 257}, {2, 245},
  {3, 244}, {3, 236}, {4, 267}, {4, 275}, {5, 218}, {5, 222}};
ANOVA[dataset]
```

```
                       DF   SumOfSq   MeanSq   FRatio    PValue
{ANOVA →      Model    4    4853.6    1213.4   34.4716   0.000781669
              Error    5    176.      35.2                            ,
              Total    9    5029.6

                All        238.2
                Model[1]   209.
                Model[2]   251.
  CellMeans →   Model[3]   240.   }
                Model[4]   271.
                Model[5]   220.
```

The F-ratio is quite large, giving a P value of only 0.078%, well below the 5% threshold for significance. Examining the data set, we see that all the treated samples have higher values of y than do the control runs. Are they *each* large enough to be significant? Let us apply Dunnett's test. We have $\nu_{\mathrm{R}} = 10 - 5 = 5$, and from the ANOVA table we see that the sum of squares for the residuals, SS_{R}, is 176. The multivariate t statistic from Table 12.3 is 2.85. The confidence interval for the null hypothesis is

$$\bar{y}_m - \bar{y}_1 < (2.85)\left(\frac{1}{2}+\frac{1}{2}\right)^{1/2}\left(\frac{176}{5}\right)^{1/2} = 17\,.$$

Any $\bar{y}_m$ less than $\bar{y}_1 + 17 = 226$ cannot be distinguished from the control. We conclude that treatment 5, with $\bar{y}_5 = 220$, has no significant effect.

Let us also apply the Tukey test. The q statistic is 5.67. The confidence interval for the null hypothesis is

$$|\bar{y}_i - \bar{y}_j| < (5.67)\,2^{-1/2}\left(\frac{176}{5}\right)^{1/2} = 24\,.$$

Any difference less than 24 is deemed insignificant. The actual values of the $\bar{y}_i - \bar{y}_j$ are:

j: 2	3	4	5	i	Conclusions
−42	**−31**	**−62**	−11	1	
	11	−20	**31**	2	"2"≈"3", "2"≈"4", "2">"5"
		−31	20	3	"4">"3", "3"≈"5"
			51	4	"4">"5"

Treatment differences larger than the significance threshold are in boldface. We conclude, for example, that treatment 4 has a significantly larger effect than treatment 3, but cannot conclude that 4 has a greater effect than 2 nor that 2 has a greater effect than 3. Tukey's test also gives a comparison with the control, treatment 1, but Dunnett's test, with a limit of 17 instead of 24 is a more sensitive test for this purpose.

The 5% confidence level is commonly chosen, but with computer algebra the choice is arbitrary. For example, to perform the tests at the 99% level we use the following:

```
ANOVA[dataset, PostTests->{Dunnett,Tukey}, SignificanceLevel->0.01]
```

With Dunnett's test, the best practice is to make N_1, the number of control measurements, larger than the number of measurements for each of the treatments by a factor of $\sqrt{k}$. For the experiment in Example 12.2, with two measurements for each treatment, we should have designed the experiment to have four measurements (i.e., $2\sqrt{5}$) with no treatment.

12.4 Optimization*

We now consider the problem of optimizing factors for a desired effect. (When speaking of different treatments simultaneously applied to a single sample, the term ***factor*** is commonly used instead of *treatment*.) An example is a chemical synthesis, in which the yield can depend on such factors as temperature, pH, concentrations of reactants, concentrations of catalysts, the rate of stirring, the amount of time before the product is separated, etc. If there is no interaction between factors then this is just a simple problem in one-dimensional maximization, which can be efficiently solved using a one-dimensional algorithm such as Brent's method. However, it is often the case that factors interact, either destructively, interfering with each other, or constructively, creating a stronger effect than either alone. Optimization is then a multidimensional maximization, which is a much greater challenge.

Example 12.3. *Contour plot of a chemical synthesis.* The stoichiometric equation for the conversion of reactants to products may actually be the result of a reaction mechanism consisting of many different elementary reactions (see Section 15.2). There may be competing reactions that produce undesired products or that destroy the desired product. This can cause the effects of experimental conditions to interact in complicated ways. A common example is the interaction of temperature T and time t. Suppose that there is a competing reaction that destroys the product,

$$\text{reactants} \xrightarrow{k_1} \text{desired product}$$
$$\text{desired product} \xrightarrow{k_2} \text{undesired product}$$

with rate constants k_1 and k_2. Given enough time, the desired product will be converted completely into undesired product, but if the rate of the second reaction is slower than that of the first, there will be an intermediate value of t at which the yield will be at a maximum. However, reaction rate constants are affected by temperature. Suppose that k_2 increases with T much more rapidly than does k_1. Then at higher T the maximum yield will be at smaller t. This is conveniently described with a ***contour plot***, shown here at two different levels of resolution:

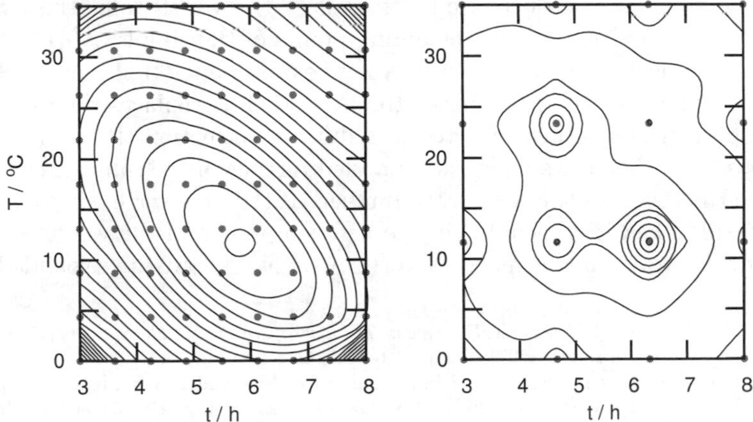

Each dot represents an experiment. The points are chosen to lie on a regular grid. The contours show evenly spaced levels of constant y, determined from cubic spline interpolations between neighboring points. The plot on the left-hand side, with a dense grid, gives an accuracte description of the function $y(t, T)$, showing the maximum at approximately $t = 5.8$ hours and $T = 12$ °C. The plot on the right-hand side, with a sparse grid, is useless; it does not give an accurate interpolation between data points. It simply places the maximum at the data point with the highest y value.

If the experiments are cheap and easy to perform, and only two factors are of interest, then a contour plot is an excellent method to use. It maps out a large region of function values. Graphics software packages or computer algebra systems can create the plot. However, the contour plot also has serious disadvantages. It is extremely inefficient. In Example 12.3, we needed 81 separate experiments to obtain an accurate plot. Furthermore, this approach is not very convenient with more than two factors.

The most efficient approach to optimization is to use an iterative numerical algorithm, such as we considered in Section 5.6. (There we treated the problem of minimization, but maximization is just minimization of the negative of the function.) For multidimensional optimization there are various algorithms to choose from. Because we do not know the derivative of our function y, we are restricted to algorithms that do not require derivatives.[10]

A particularly popular iterative method is the ***Nelder-Mead algorithm***, also called the ***downhill simplex method***.[11] A *simplex* is a geometrical figure in N-dimensional space with $N + 1$ vertices (i.e., points where sides meet). In two dimensions a simplex is called a "triangle." In three dimensions it is called a "tetrahedron." For higher dimensions we cannot visualize a simplex but we can specify it mathematically. The Nelder-Mead algorithm begins with a simplex defined by a set of $N + 1$ experimental results, each with a different combination of factor values. The simplex is then transformed by moving the vertex that gives the lowest response to a position that the algorithm predicts will give a higher response.

The various transformations are illustrated in Fig. 12.2 for the two-factor case. The primary transformation is to reflect the worst point through the centroid of the positions of the remaining points of the simplex.[12] The factor values of the new vertex are used in a new experiment. If the new value of y is better than the worst but is not the best, then it replaces the worst in a new simplex, which is then reflected in a different direction. If the new value is the best one, then the simplex is expanded further in the same direction, in the hope that the result will improve further. If the new vertex gives an even better result, then it is included in the new simplex, and the worst vertex is removed. However, if the reflected vertex is better than the expanded one,

---

[10]See W. H. Press *et al.*, *Numerical Recipes: The Art of Scientific Computing* (Cambridge University Press, Cambridge, 2007), Chap. 10, and references therein.

[11]It was developed by British statisticians John A. Nelder and Roger Mead, *Comput. J.* **7**, 308-313 (1965). It is called "downhill" because its usual formulation is as a method for minimization. What we will consider here is, strictly speaking, the *uphill* simplex method.

[12]The *centroid* of N points is defined as the average of their vectors, $N^{-1}\sum_j \vec{x}_j$. See Sections 13.2.1 and 16.1 for discussions of coordinate vectors.

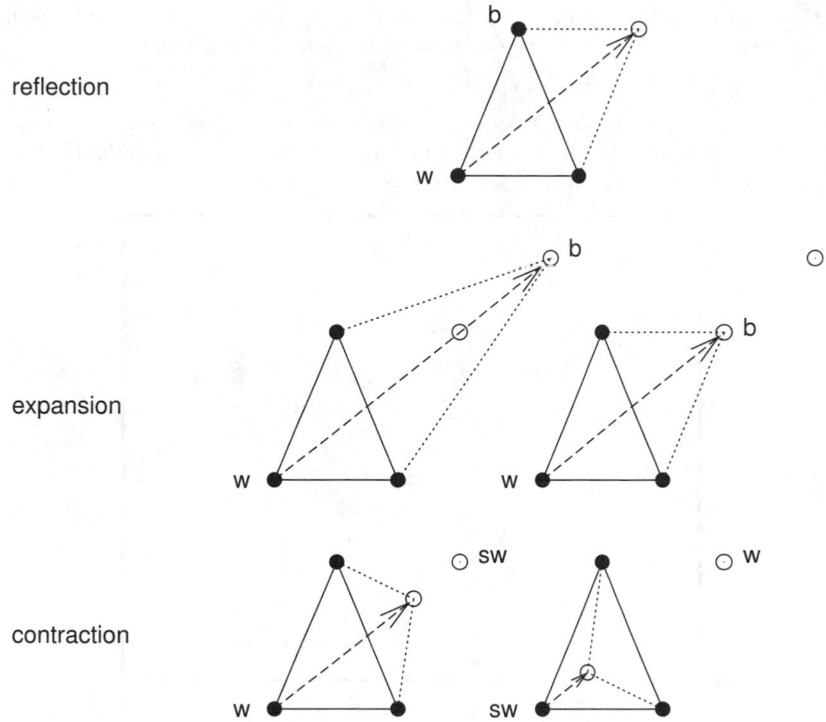

Figure 12.2: Nelder-Mead simplex transformations for a two-factor optimization. The labels "b," "w," and "sw" indicate the best (i.e. highest yield), worst, and second worst vertices, respectively.

the expanded vertex is ignored. If the reflected vertex gives the second worst result, better only than the vertex to be replaced, then a new vertex is used that is contracted halfway toward the center. If the reflected vertex is the worst of all, it is ignored and a new vertex that is a half-way contraction of the original unreflected vertex is used. The algorithm gives a sequence of simplices that scoot across the surface, with much zigzagging, converging toward the maximum. Typically, only a few simplices are needed to get reasonably close.

Example 12.4. *Optimization using the Nelder-Mead simplex algorithm.* Let us apply the algorithm to the problem in Example 12.3. To begin, we need to choose an initial simplex. Suppose that we have read that the optimum conditions for a similar reaction are 6.5 hours at 24 °C. We will make this the center of the initial simplex,

$$\mathbf{s}_0 = \{(6.0, 23), (6.5, 27), (7.0, 23)\}.$$

These conditions give yields of only 16.0%, 2.7%, and 3.7%, respectively. The worst vertex is $(6.5, 27)$. Reflection maps it to $(6.5, 19)$. Performing the synthesis with these conditions gives a 19.7% yield, which is now the best vertex. Expansion takes us to

(6.5, 15), which gives a 34.6% yield. According to the method of Nelder and Mead, we can perform no more than one expansion. Therefore, our new simplex is

$$\mathbf{s}_1 = \{(6.0, 23), (7.0, 23), (6.5, 15)\}.$$

The algorithm continues with a reflection, replacing (7.0, 23), now the worst vertex, with (5.5, 15), and so on. The result of the algorithm using ten repetitions of the synthesis is shown below, superimposed on the contour plot:

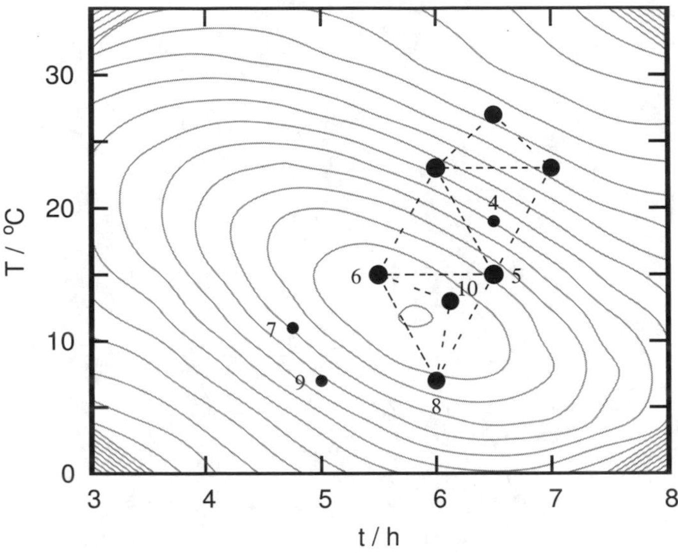

Six syntheses bring the yield up to 51.7%.

This example is typical; just two or three simplices are very often sufficient to get close to the maximum. After that the rate of convergence can be rather slow, and the criteria for when to terminate the algorithm are somewhat subjective. A reasonable practice is to polish off the optimization by switching to a local interpolation method once the simplices no longer seem to be significantly improving the yield.

Example 12.5. *Polishing the optimization with local modeling.* Near the maximum it is reasonable to express the response surface $y(t, T)$ in the form of a second-order two-dimensional Taylor series about the (as yet unknown) maximum:

$$y(t, T) \sim y_{\mathrm{max}} - a_{2,0}(t - t_{\mathrm{max}})^2 - a_{1,1}(t - t_{\mathrm{max}})(T - T_{\mathrm{max}}) - a_{0,2}(T - T_{\mathrm{max}})^2. \quad (12.20)$$

Expanding this as a polynomial gives

$$y(t, T) \approx b_{0,0} + b_{1,0}t + b_{0,1}T - a_{2,0}t^2 - a_{1,1}tT - a_{0,2}T^2,$$

where

$$b_{1,0} = 2a_{2,0}t_{\mathrm{max}} + a_{1,1}T_{\mathrm{max}}\,, \qquad b_{0,1} = 2a_{0,2}T_{\mathrm{max}} + a_{1,1}t_{\mathrm{max}}\,,$$
$$b_{0,0} = y_{\mathrm{max}} - a_{2,0}t_{\mathrm{max}}^2 - a_{1,1}t_{max}T_{max} - a_{2,0}T_{\mathrm{max}}^2.$$

The equations for $b_{1,0}$ and $b_{0,1}$ give

$$t_{\mathrm{max}} = \frac{2a_{0,2}b_{1,0} - a_{1,1}b_{0,1}}{4a_{0,2}a_{2,0} - a_{1,1}^2}\,, \qquad T_{\mathrm{max}} = \frac{2a_{2,0}b_{0,1} - a_{1,1}b_{1,0}}{4a_{0,2}a_{2,0} - a_{1,1}^2}\,. \quad (12.21)$$

Let us approximate the six coefficients $b_{j,k}$ by interpolation through the fifth through tenth (t, T, y) values used in the simplex algorithm,

$$(6.5, 15.0, 34.6), \quad (5.5, 15.0, 51.7), \quad (4.75, 11.0, 30.8),$$
$$(6.0, 7.0, 42.2), \quad (5.0, 7.0, 22.4), \quad (6.125, 13.0, 49.0).$$

The result is

$$y(t, T) \approx -799.9 + 251.0\, t - 2.173\, t\, T - 19.63\, t^2 + 21.65\, T - 0.3586\, T^2,$$

which, according to Eqs. (12.21), has its maximum at the point

$$t_{\max} = 5.69 \text{ h}, \qquad T_{\max} = 13.0\ ^\circ\text{C}.$$

Finally, let us carry out four replicates of the synthesis at these factor values. Suppose that the results are

$$(5.69, 13.0, 52.2), \quad (5.69, 13.0, 53.1), \quad (5.69, 13.0, 51.3), \quad (5.69, 13.0, 52.4)\,.$$

The mean is $y = 52.20\%$ with a standard error of 0.38%. Seeing as the polishing improved on the third-simplex maximum of 51.7% by only 0.5%, which is less than twice the standard error, it is likely that any further optimization would be fruitless, as the additional improvement would be comparable to the level of random error.

It is straightforward to generalize the simplex algorithm to higher dimensionality. For details, see Appendix A.3, which also gives *Mathematica* routines for implementing the algorithm. An advantage of this method is the fact that it scales quite favorably with increasing dimension. With N_f factors the initial simplex needs to have $N_\mathrm{f} + 1$ vertices, but each subsequent step requires just one more experiment, regardless of how large N_f is.

The disadvantages are the need for a separate polishing algorithm, the subjectivity of the termination criterion, and the fact that the algorithm can in rare instances get stuck rather far from the maximum. Industrial engineers tend to prefer a more laborious approach to optimization called ***response surface methodology***. This involves setting up symmetrical configurations of points in factor space, fitting them with a polynomial, determining the direction in which the polynomial increases most rapidly, and then finding the maximum along a line in that direction using a one-dimensional maximization algorithm. That one-dimensional maximum is then used as the center of a new multidimensional symmetrical configuration and the whole process is repeated, as many times as necessary.[13]

Exercises

12.1 Carry out a two-sided t-test of the null hypothesis $x = 50$ mg/L to determine the β value corresponding to $\alpha = 0.05$ with $N = 10$ and a tolerance of ± 5 mg/L, and determine the minimum N value such that β is also at most 0.05.

12.2 Consider a manufacturing process intended to produce a pill with 75 mmol of the active ingredient with a tolerance of ± 5 mmol. The amount in a pill can be measured very precisely but the measurement destroys the pill. Assume that the pill composition within

[13] See references in the Bibliography for detailed descriptions of this method.

a batch varies according to a normal distribution. Design a quality monitoring protocol in which N pills randomly selected from each production run are sacrificed. Suppose that α and β must both be no greater than 0.001 (i.e. 0.1%) or the entire production run is disposed of. How would you choose the value of N?

12.3 Considering only experiments 1 through 5 in Table 12.2, carry out a randomization analysis (by hand) and calculate the P value for $\bar{y}^{(+)} - \bar{y}^{(-)}$.

12.4 Suppose that the results in Table 12.2 had been obtained in five different laboratories, results 1 and 2 in the first, 3 and 4 in the second, 5 and 6 in the third, 7 and 8 in the fourth, 9 and 10 in the fifth. Use a blocked randomization analysis to calculate the P value for rejecting the null hypothesis.

12.5* For a comparison of k treatments each with N_t replicates, (a) prove that $N\bar{y} = \sum_{t=1}^{k} N_t \bar{y}_t$ and (b) derive Eq. (12.11).

12.6* A response y is measured for two different treatments in a randomized experiment, with three replicates. The results with the first treatment are 507, 519, and 531 while the results with the second treatment are 534, 532, and 527. Four replicates with no treatment give the results 513, 518, 505, and 503.

(a) Calculate $\hat{\sigma}_{\mathrm{T}}^2$, $\hat{\sigma}_{\mathrm{R}}^2$, and the F-ratio, and compare with the critical F value at $\alpha = 0.05$. Is there a significant effect from the treatments?

(b) Do either of the treatments significantly increase y, at a confidence level of 95%?

(c) Are the effects of the two treatments significantly different from each other, at a confidence level of 95%?

12.7* Using computer algebra, determine the P value for the comparison of treatment 3 with the control in Example 12.2.

12.8* Determine the 11th and 12th points (t, T) in the simplex optimization of Examples 12.4 and 12.5.

12.9* Simulate the optimization of a chemical synthesis:

(a) Write a computer algebra routine that simulates the yield, using the function

$$y(x_1, x_2) = (1.1 + 0.8x_1 + 0.3x_2 + 0.04x_1x_2)e^{-0.02(x_1-16)^2 + 0.01(x_1-16)(x_2-31) - 0.01(x_2-31)^2}$$

with random error added according to a log-normal distribution.

(b) Optimize the synthesis using the Nelder-Mead simplex algorithm followed by a local interpolation. Repeat this analysis for several different initial simplices and for different amounts of random error.

12.10* Repeat Exercise 12.9 but this time perform a three-factor optimization, using

$$y(x_1, x_2, x_3) = (1.1 + 0.8x_1 + 0.3x_2 + 0.4x_3 + 0.04x_1x_2) \times e^{-0.02(x_1-16)^2 + 0.01(x_1-16)(x_2-31) - 0.01(x_2-31)^2 - 0.02(x_3-21)^2}.$$

(The local interpolation will require more data points than in the two-dimensional case.)

Part III

Differential Equations

13. Examples of Differential Equations

14. Solving Differential Equations, I

15. Solving Differential Equations, II

Chapter 13

Examples of Differential Equations

A ***differential equation*** is an equation involving differentials or derivatives. Such equations are ubiquitous in science and engineering. This chapter presents various examples. Solution methods are treated in subsequent chapters.

13.1 Chemical Reaction Rates

Chemical processes are commonly described in terms of stoichiometric reaction equations,[1] which show that changes in amounts of reactants and products occur in integer ratios. Consider the stoichiometric equation $2\mathrm{A} \to \mathrm{X}$. If the molar concentration [X] of substance X increases by an amount $d[\mathrm{X}]$, then we know from the stoichiometry that [A] decreases by twice as much; that is, $d[\mathrm{A}] = -2d[\mathrm{X}]$, or $d[\mathrm{X}] = -\frac{1}{2}\,d[\mathrm{A}]$. We need the minus sign because a decrease in A implies an increase in X. If the stoichiometry is $2\mathrm{A} \to 3\mathrm{X}$ then we have $d[\mathrm{A}] = -\frac{2}{3}\,d[\mathrm{X}]$, or $\frac{1}{3}\,d[\mathrm{X}] = -\frac{1}{2}\,d[\mathrm{A}]$. Consider $\mathrm{A} + 3\mathrm{B} \to 4\mathrm{X} + 3\mathrm{Y}$. Following the same reasoning, we find that $-d[\mathrm{A}] = \frac{1}{4}\,d[\mathrm{X}] = \frac{1}{3}\,d[\mathrm{Y}] = -\frac{1}{3}\,d[\mathrm{B}]$.

We can streamline this analysis by labeling the substances according to a numerical index, $j = 1, 2, 3, \ldots$, with the molar concentrations indicated by c_j. Let us define the ***stoichiometric number*** of the jth substance, with symbol ν_j, as the number multiplying the substance in the stoichiometric reaction equation but with a minus sign for reactants. We can summarize the above analyses simply by stating that the $\nu_j^{-1} dc_j$ are equal for all j.

For a stoichiometric equation with M substances, the equations $\nu_j^{-1} dc_j = \nu_k^{-1} dc_k$ comprise a set of $M-1$ constraints. We have only one degree of freedom. Let us write

$$d\xi = \nu_j^{-1}\, dc_j \tag{13.1}$$

for the one independent quantity. The variable ξ defined by this differential is called the ***extent of reaction***. The ***reaction rate*** can be defined as

$$\dot{\xi} = d\xi/dt. \tag{13.2}$$

(A dot above a symbol indicates a time derivative.) By integrating Eq. (13.1),

$$\int_0^{\xi(t)} d\xi = \nu_j^{-1} \int_{c_j(0)}^{c_j(t)} dc_j\,,$$

[1] The word *stoichiometry*was coined by the German chemist Jeremias Richter (1762-1807) in the title of his book *Der Stochiometrie oder Messkunst chemischer Elemente* ("Stoichiometry, or the Art of Measuring Chemical Elements") in 1794. The word is a combination of the Greek στοιχεῖα ("element") and μέτρον ("measure"). Richter was also the inventor of acid-base titration. His published work is mostly devoted to attempts to develop a mathematical theory of chemistry. However, it was only with the development of atomic theory by John Dalton (1766-1844) in the following decade that stoichiometry was given a clear rational foundation.

we obtain the individual concentrations in terms of ξ:

$$c_j(t) = c_{j,0} + \nu_j \xi(t) \,. \tag{13.3}$$

$c_{j,0} = c_j(0)$ is the initial concentration of substance j, at time $t = 0$.

A ***rate law*** is a differential equation that expresses the reaction rate as a function of the concentrations,

$$\dot{\xi} = f(c_1, c_2, c_3 \dots) \,. \tag{13.4}$$

This is more complicated than it looks, because the c_j depend on ξ. Furthermore, the functional form of f usually has to be determined empirically.

The determination of rate laws is a major goal of physical chemistry and chemical engineering. A rate law allows one to predict the reaction rate for any given amounts of the reacting substances. They can be complicated expressions. For example, the rate law of $H_2(g)+Br_2(g) \to 2HBr(g)$, determined from experiment, is

$$\dot{\xi} \approx \frac{k_a c_1 c_2^{3/2}}{c_2 + k_b c_3} \,, \tag{13.5}$$

with $c_1 = [H_2]$, $c_2 = [Br_2]$, and $c_3 = [HBr]$, where k_a and k_b are empirical constants.[2] It is not possible in general to infer a rate law from the reaction stoichiometry. Additional information is needed. If one knows the reaction *mechanism*, that is, the sequence of of atomic rearrangements that leads from reactants to products, then it may be possible to derive an approximate analytical rate law. (This will be discussed in Section 15.2.)

Example 13.1. *Expressing the rate law in terms of the extent of reaction.* Consider a reaction with stoichiometry

$$2A + 3B \to X.$$

Then $-\frac{1}{2}\frac{dc_A}{dt} = -\frac{1}{3}\frac{dc_B}{dt} = \frac{dc_X}{dt} = \dot{\xi}$. Suppose experiments show the rate law to have the form

$$\dot{\xi} = k c_A c_B^2 \,. \tag{13.6}$$

Using $c_j(t) = c_{j,0} + \nu_j \xi(t)$ to express concentrations in terms of ξ, we obtain

$$\dot{\xi} = k(c_{A,0} - 2\xi)(c_{B,0} - 3\xi)^2 \,. \tag{13.7}$$

$c_{A,0}$ and $c_{B,0}$ are concentrations at an arbitrary initial time that we call $t = 0$. The solution of this equation will give ξ as a function of time.

For the reaction $H_2+Br_2 \to 2HBr$, we have $-\frac{dc_1}{dt} = -\frac{dc_2}{dt} = \frac{1}{2}\frac{dc_3}{dt} = \dot{\xi}$, and Eq. (13.5) gives

$$\dot{\xi} = \frac{k_a\,(c_{1,0} - \xi)(c_{2,0} - \xi)^{3/2}}{c_{2,0} - \xi + k_b\,(c_{3,0} + 2\xi)} \,. \tag{13.8}$$

Example 13.2. *Empirical determination of reaction rate.* $\dot{\xi}$ can be measured in the laboratory, as long as we have some way of measuring the concentration of one of the reactants or products. Consider $A + 2B \to X$, with "B" being something we can measure. According to Eq. (13.1), reaction rate is $\dot{\xi} = -\frac{1}{2}dc_B/dt$. This derivative can be estimated using methods described in Section 5.2. For example, we might measure $c_B(0)$, $c_B(h)$, and $c_B(2h)$ for a small time step h and use Eq. (5.10).

[2] They are "constant" in that they do not depend on concentrations and time. However, they do depend on temperature.

13.2 Classical Mechanics

Perhaps the most impressive achievement in the history of science was Newton's theory of *mechanics*, describing the motions of particles in response to forces.[3] Not only did this theory provide a rigorous foundation for much of physics and engineering, but it proved to be extremely influential in the development of mathematics. Indeed, Newton needed to create a new branch of mathematics (calculus) in order to formulate his theory. A significant focus of much of 18th- and 19th-century mathematics was the attempt to solve Newton's equations for practical applications.

Here we explore three formulations of classical mechanics. The first is the one presented by Newton, in terms of trajectories, forces, and velocities. We then consider two more abstract formulations, *Lagrangian mechanics* and *Hamiltonian mechanics*, which are more elegant than Newtonian mechanics and can be easier to generalize to complicated problems.

13.2.1 Newtonian Mechanics

Newton's equation for motion of a particle in response to a force is

$$F = ma, \tag{13.9}$$

where F is the force, m is the particle mass, and a is the acceleration,

$$a = \frac{dv}{dt} = \frac{d^2x}{dt^2}\,. \tag{13.10}$$

We can write this succinctly in the dot notation as $F = m\dot{v}$ in terms of velocity $v = \dot{x}$, or equivalently, $F = m\ddot{x}$. The units of acceleration are distance over time squared, $\mathrm{m/s^2}$ in the SI fundamental units, which implies that force has units $\mathrm{kg\ m/s^2}$. This unit is called the ***newton***, with symbol N.

With three dimensions, we need to solve a set of three differential equations,

$$F_x = m\ddot{x}, \qquad F_y = m\ddot{y}, \qquad F_z = m\ddot{z}, \tag{13.11}$$

in terms of forces in the x, y, and z dimensions, respectively. Today these equations are usually written in the vector notation developed by Gibbs and Heaviside in the 1880's.[4] We define a Cartesian coordinate position *vector*[5] as the ordered set (x, y, z) of the three coordinates. This can be visualized as an arrow reaching from the origin to the point (x, y, z). We will distinguish between a vector and a scalar (i.e., a single number) by using boldface type for vectors, for example, $\mathbf{q} = (x, y, z)$. The notation $\vec{q}$ is also often used.

---

[3] This theory was published in 1686 in his book *Philosophiæ Naturalis Principia Mathematica* (Mathematical Principles of Natural Philosophy). Today it is referred to as *classical* mechanics, to distinguish it from the more comprehensive theory called *quantum* mechanics developed in the 20th century.

[4] Physical chemist J. Willard Gibbs and the English engineer and physicist Oliver Heaviside (1850-1925) independently developed this very convenient notational system.

[5] A more systematic treatment of vectors will be given in Chapter 16.

Taking the derivative of a vector preserves the ordering. For example, $\dot{\mathbf{q}} = (\dot{x}, \dot{y}, \dot{z})$. The acceleration vector is $\mathbf{a} = \ddot{\mathbf{q}} = (\ddot{x}, \ddot{y}, \ddot{z})$. Eqs. (13.11) can be written

$$\mathbf{F} = m\ddot{\mathbf{q}} = m\mathbf{a}\,, \tag{13.12}$$

where $\mathbf{F} = (F_x, F_y, F_z)$ and $m\mathbf{a} = (m\ddot{x}, m\ddot{y}, m\ddot{z})$. The goal of classical mechanics is to solve Eq. (13.12) for the particle's ***trajectory***,

$$\mathbf{q}(t) = \big(x(t), y(t), z(t)\big)\,,$$

a vector function of time.

It is often possible to express the vector $\mathbf{F}$ in terms of a scalar function $V(x, y, z)$ according to

$$F_x = -\frac{\partial}{\partial x}V, \quad F_y = -\frac{\partial}{\partial y}V, \quad F_z = -\frac{\partial}{\partial z}V. \tag{13.13}$$

Then V is called the ***potential energy***. Let us introduce a vector operator

$$\vec{\nabla} = \left(\frac{\partial}{\partial x}, \frac{\partial}{\partial y}, \frac{\partial}{\partial z}\right), \tag{13.14}$$

called ***del***. This operator maps a scalar function $f(x, y, z)$ to a vector

$$\vec{\nabla} f = \left(\frac{\partial f}{\partial x}, \frac{\partial f}{\partial y}, \frac{\partial f}{\partial z}\right) \tag{13.15}$$

called the **gradient** of f.[6] Then we can write $\mathbf{F} = -\vec{\nabla} V$. In one dimension this reduces to $F = -dV/dx$.

Example 13.3. *A free particle.* A free particle is not subject to any forces. We have $\mathbf{F} = (0, 0, 0)$, which implies that V must be a constant, independent of x, y, and z. Force only depends on the *change* in V with position, not on the actual value of V. We can add any arbitrary constant to V (that is, change the zero of the potential energy scale) with no effect on the motion of the particle. Newton's equation for a free particle gives $0 = \mathbf{a} = d\mathbf{v}/dt$. Therefore, $\mathbf{v}$ is independent of t. This is Newton's Second Law: In the absence of force, velocity is constant.

Energy is a fundamental unifying concept of modern physical science. Potential energy, V, is just one example. From Eqs. (13.13) we see that V has units of force times distance, or $\mathrm{kg\,m^2/s^2}$ in the SI system. This unit is called the ***joule***, with symbol J. Another kind of energy is ***work***. Consider the motion of a particle during a time interval t_0 to t_1, from initial position $x_0 = x(t_0)$ to final position $x_1 = x(t_1)$. The work of this process is defined as

$$w = \int_{x_0}^{x_1} F dx\,, \tag{13.16}$$

which has units of force times distance, and therefore can be thought of as a

[6]Other common notations for the gradient are ∇f (without the arrow) and grad f.

kind of energy. Substituting Newton's equation for F gives

$$w = m \int_{x_0}^{x_1} \frac{dv}{dt}\, dx.$$

Let us change the integration variable from x to t, with $dx = \frac{dx}{dt}\, dt = vdt$:

$$w = m \int_{t_0}^{t_1} \frac{dv}{dt}\, vdt = m\, \frac{1}{2} \int_{t_0}^{t_1} \left[\frac{d}{dt}(v^2)\right] dt = \tfrac{1}{2} m v^2 \Big|_{t_0}^{t_1} = \tfrac{1}{2} m v_1^2 - \tfrac{1}{2} m v_0^2, \tag{13.17}$$

in terms of the initial and final velocities $v_0 = v(t_0)$ and $v_1 = v(t_1)$. Let us define ***kinetic energy*** as $T = \frac{1}{2}mv^2$. Eq. (13.17) tells us that work is the change in kinetic energy due to the application of a force.

With three dimensions, work is the sum of three components,

$$w_x = \int_{x_0}^{x_1} F_x dx, \quad w_y = \int_{y_0}^{y_1} F_y dy, \quad w_z = \int_{z_0}^{z_1} F_z dz.$$

We have $w = w_x + w_y + w_z = T_1 - T_0$, with

$$T_1 = \tfrac{1}{2} m (v_{1,x}^2 + v_{1,y}^2 + v_{1,z}^2), \quad T_0 = \tfrac{1}{2} m (v_{0,x}^2 + v_{0,y}^2 + v_{0,z}^2).$$

In vector notation these equations can be written as

$$w = \int_{(x_0,y_0,z_0)}^{(x_1,y_1,z_1)} (F_x, F_y, F_z) \cdot (dx, dy, dz) = \int_{\mathbf{q}_0}^{\mathbf{q}_1} \mathbf{F} \cdot d\mathbf{q}, \tag{13.18}$$

with

$$(F_x, F_y, F_z) \cdot (dx, dy, dz) = F_x dx + F_y dy + F_z dz, \tag{13.19}$$

and

$$T = \tfrac{1}{2} m \mathbf{v} \cdot \mathbf{v}, \tag{13.20}$$

with

$$\mathbf{v} \cdot \mathbf{v} = (v_x, v_y, v_z) \cdot (v_x, v_y, v_z) = v_x^2 + v_y^2 + v_z^2. \tag{13.21}$$

Newton formulated his theory not in terms of velocity but in terms of velocity times mass,

$$\mathbf{p} = m\mathbf{v} = (m\dot{x}, m\dot{y}, m\dot{z}). \tag{13.22}$$

Then $\mathbf{F} = m\mathbf{a}$ becomes simply

$$\mathbf{F} = \dot{\mathbf{p}}. \tag{13.23}$$

$\mathbf{p}$ is called the ***linear momentum***.[7] Kinetic energy in terms of $\mathbf{p}$ is

$$T = \frac{1}{2m}\left(p_x^2 + p_y^2 + p_z^2\right) = \frac{1}{2m}\, \mathbf{p} \cdot \mathbf{p}. \tag{13.24}$$

[7] *Momentum* means "movement" in Latin. The use of this word for the product of mass and velocity apparently was coined by English mathematician John Wallis (1616-1703), although the concept was developed much earlier, by the Persian chemist, physicist, physician, philosopher, music theorist, and poet Abu Ali Ibn Sina (980?-1037), who called it *mayl* ("inclination" in Arabic). Newton called it *quantitas motus* ("quantity of motion").

13.2.2 Lagrangian and Hamiltonian Mechanics

In Newtonian mechanics the first step typically is to specify a force vector. Then one solves for the trajectory. There are two alternative formulations of mechanics that derive the trajectory from the kinetic and potential energies instead of from forces. The first of these, developed by Lagrange,[8] starts with the *difference* between the kinetic energy and the potential energy,

$$\mathcal{L} = T - V. \tag{13.25}$$

This is called the ***Lagrangian***. Lagrange proved that the trajectory must satisfy the differential equation

$$\frac{d}{dt}\frac{\partial \mathcal{L}}{\partial \dot{x}} - \frac{\partial \mathcal{L}}{\partial x} = 0, \tag{13.26}$$

which is called ***Lagrange's equation***.

Example 13.4. *Lagrange's equation in one dimension.* The Lagrangian is

$$\mathcal{L} = \tfrac{1}{2}m\dot{x}^2 - V(x)\,. \tag{13.27}$$

For purposes of the partial derivatives, x and $\dot{x}$ are treated as independent coordinates. We have

$$\frac{\partial \mathcal{L}}{\partial \dot{x}} = m\dot{x}\,, \qquad \frac{\partial \mathcal{L}}{\partial x} = -\frac{\partial V}{\partial x}\,. \tag{13.28}$$

Lagrange's equation is

$$0 = m\frac{d\dot{x}}{dt} + \frac{\partial V}{\partial x} = ma - F\,,$$

which is equivalent to Newton's equation.

For problems in three dimensions, we have three equations,

$$\frac{d}{dt}\frac{\partial \mathcal{L}}{\partial \dot{x}} - \frac{\partial \mathcal{L}}{\partial x} = 0, \quad \frac{d}{dt}\frac{\partial \mathcal{L}}{\partial \dot{y}} - \frac{\partial \mathcal{L}}{\partial y} = 0, \quad \frac{d}{dt}\frac{\partial \mathcal{L}}{\partial \dot{z}} - \frac{\partial \mathcal{L}}{\partial z} = 0. \tag{13.29}$$

The main advantage of Lagrange's formulation is that it facilitates the use of non-Cartesian coordinates. It turns out that Lagrange's equations have the form of Eqs. (13.29) for *any* set of coordinates, as long as the coordinates are all independent variables and they exactly suffice to uniquely specify the configuration of the system and the value of the potential energy at any given point. Consider the spherical polar coordinate system. The kinetic energy in these coordinates was derived in Section 3.1:

$$T = \tfrac{1}{2}m\dot{r}^2 + \tfrac{1}{2}mr^2\dot{\theta}^2 + \tfrac{1}{2}mr^2(\sin^2\theta)\dot{\phi}^2. \tag{13.30}$$

[8] The French-Italian Joseph Louis Lagrange (1736-1813) (pronounced *la-grawnj*) was a leading figure of 18th-century theoretical physics. A largely self-taught child prodigy, he was appointed a university professor while still a teenager. He made many contributions to mathematics as well as to physics.

For a potential energy function $V(r,\phi,\theta)$ with no velocity dependence, we have the Lagrange equations of motion

$$\frac{d}{dt}\left(\frac{\partial T}{\partial \dot r}\right) - \frac{\partial T}{\partial r} + \frac{\partial V}{\partial r} = m\ddot r - mr(\dot\theta^2 + \dot\phi^2 \sin^2\theta) + \frac{\partial V}{\partial r} = 0,$$

$$\frac{d}{dt}\left(\frac{\partial T}{\partial \dot\theta}\right) - \frac{\partial T}{\partial \theta} + \frac{\partial V}{\partial \theta} = mr^2\ddot\theta + 2mr\dot r\dot\theta - mr^2\dot\phi^2 \sin\theta\cos\theta + \frac{\partial V}{\partial \theta} = 0,$$

$$\frac{d}{dt}\left(\frac{\partial T}{\partial \dot\phi}\right) - \frac{\partial T}{\partial \phi} + \frac{\partial V}{\partial \phi} = mr^2\ddot\phi\sin^2\theta + 2mr\dot r\dot\phi\sin^2\theta$$
$$+ 2mr^2\dot\phi\,\dot\theta\sin\theta\cos\theta + \frac{\partial V}{\partial \phi} = 0. \quad (13.31)$$

The Lagrangian formulation is particularly useful for describing constrained motion. Consider a particle constrained to a circle. r and θ are constants and therefore $\dot r$ and $\dot\theta$ are zero. The only free coordinate is ϕ. The kinetic energy reduces to

$$T = \tfrac{1}{2}mr^2(\sin^2\theta)\dot\phi^2. \quad (13.32)$$

The single Lagrange equation is

$$mr^2\ddot\phi\sin^2\theta = -\frac{\partial V}{\partial \phi}. \quad (13.33)$$

Some kinds of forces cannot be expressed in terms of a potential energy function $V(\mathbf{q})$, in which V depends only on the position vector $\mathbf{q}$. The most important such case is friction, which depends on the velocity vector, $\dot{\mathbf{q}}$. Lagrange's equation then needs to be modified as follows:

$$\frac{d}{dt}\frac{\partial \mathfrak{L}}{\partial \dot q_j} - \frac{\partial \mathfrak{L}}{\partial q_j} = \mathcal{F}_j\,, \quad (13.34)$$

where $\mathcal{F}_j$ is the sum of all forces acting on motion in the coordinate q_j that are not accounted for by $V(\mathbf{q})$. Friction is present only when an object is moving. If the magnitude of the friction is small, it is reasonable to describe it as proportional to the velocity, $\mathcal{F}_j = -\gamma\dot q_j$, where γ is an empirical constant. The negative sign expresses the fact that the frictional force vector points in the direction opposite the velocity vector.

We can use Lagrange's formulation to generalize the concept of momentum. Let us define a ***canonical momentum***[9] as

$$p_q = \frac{\partial \mathfrak{L}}{\partial \dot q}\,. \quad (13.35)$$

This is referred to as the momentum that is ***conjugate*** to coordinate q.[10] In Cartesian coordinates the canonical momentum is identical to the linear

[9] "Canonical" in the sense of "according to a rule." κανών means "rule."

[10] Not to be confused with "complex conjugate." The word *conjugate* simply means "joined together according to some relationship."

momentum,

$$\mathcal{L} = \tfrac{1}{2}m\dot{x}^2 + \tfrac{1}{2}m\dot{y}^2 + \tfrac{1}{2}m\dot{z}^2 - V(x,y,z), \qquad p_x = \frac{\partial \mathcal{L}}{\partial \dot{x}} = m\dot{x},$$

and similarly, $p_y = m\dot{y}$, $p_z = m\dot{z}$. In polar coordinates, the canonical momenta are

$$p_r = m\dot{r}\,, \quad p_\theta = mr^2\dot{\theta}, \quad p_\phi = mr^2(\sin^2\theta)\dot{\phi}\,. \tag{13.36}$$

Now we consider another formulation of mechanics, this one developed by Hamilton.[11] Let us write the *sum* of the kinetic and potential energies as

$$H = T + V. \tag{13.37}$$

This is called the ***Hamiltonian function***, or, simply, the "Hamiltonian." H is a function of the particle trajectory. Let us choose a set of coordinates and calculate their conjugate momenta, according to Eq. (13.35), and then express H as a function these coordinates and momenta. Hamilton proved that

$$\dot{q} = \frac{\partial H}{\partial p}\,, \qquad \dot{p} = -\frac{\partial H}{\partial q} \tag{13.38}$$

for each conjugate pair of coordinate and momentum.

Example 13.5. *Hamilton's equations in one dimension.* The Hamiltonian is

$$H = \frac{1}{2m}p^2 + V(x). \tag{13.39}$$

Hamilton's equations are

$$\dot{x} = p/m, \qquad \dot{p} = -dV/dx. \tag{13.40}$$

The first equation is just a restatement of the definition of linear momentum and the second is just another way to write $ma = F$.

Now let us formulate Hamilton's equations for a free particle in polar coordinates. The first step is to express T in terms of the canonical momenta. Substituting Eqs. (13.36) into Eq. (13.30), we obtain

$$T = \frac{1}{2m}p_r^2 + \frac{1}{2mr^2}p_\theta^2 + \frac{1}{2mr^2\sin^2\theta}p_\phi^2\,. \tag{13.41}$$

For a free particle, $V = 0$, implying $H = T$. Hamilton's equations are then

$$\dot{r} = \frac{1}{m}p_r\,, \qquad \dot{\theta} = \frac{1}{mr^2}p_\theta\,, \qquad \dot{\phi} = \frac{1}{mr^2\sin^2\theta}p_\phi\,,$$

$$\dot{p}_r = \frac{2}{mr^3}\left(p_\theta^2 + \frac{1}{\sin^2\theta}p_\phi^2\right), \qquad \dot{p}_\theta = \frac{2\cos\theta}{mr^2\sin^3\theta}p_\phi^2\,, \qquad \dot{p}_\phi = 0\,. \tag{13.42}$$

The equations for $\dot{r}$, $\dot{\theta}$, and $\dot{\phi}$ just restate the definitions of the momenta, but the last three equations are interesting. For a free particle in *Cartesian* coordinates, momenta $\dot{p}_x$, $\dot{p}_y$, and $\dot{p}_z$ are each equal to zero. In other words, in the absence of forces, the Cartesian momenta are constants. But now we

[11] Irish mathematician William Rowan Hamilton (1805-1865).

find that in polar coordinates two of the momenta are nonzero, even though no forces are acting on the particle. Only p_ϕ is a constant of the motion.

Hamilton's formulation doubles the number of differential equations. This is because the number of coordinates has been doubled; the "spatial" coordinates q_j and the momenta p_j are each treated as independent coordinates. Thus, three-dimensional motion is described by a six-dimensional coordinate system. At each value of time, the particle's motion is expressed as a "point" $(q_1, q_2, q_3, p_1, p_2, p_3)$. This coordinate system is called ***phase space***.

Hamilton's equations are more elegant than Lagrange's, but there is usually no advantage to using them for simple mechanical problems. The advantage is seen in such fields as quantum mechanics and statistical mechanics, which place a central focus on energy. Note that $H = T + V$ is the total energy. Thus, H has a direct physical interpretation while $\mathfrak{L} = T - V$ does not.

13.2.3 Angular Momentum

A property that remains constant during the course of motion is said to be ***conserved***. For a free particle, p_x, p_y, and p_z are each conserved. In spherical polar coordinates, the only conserved canonical momentum is, according to Eqs. (13.42),

$$p_\phi = mr^2(\sin^2\theta)\dot{\phi}\,, \tag{13.43}$$

which is the momentum for rotation about the z-axis. This momentum is called the ***angular momentum*** for rotation about the z-axis. More generally, we can define an ***angular momentum vector***, $\mathbf{L}$, with components

$$L_x = m(y\dot{z} - z\dot{y}), \quad L_y = m(-x\dot{z} + z\dot{x}), \quad L_z = m(x\dot{y} - y\dot{x}). \tag{13.44}$$

Now note what happens if we write the z component in spherical polar coordinates. The expressions for $\dot{x}$ and $\dot{y}$ were given in Eqs. (3.6). Substituting them into the expression for L_z gives

$$\begin{aligned} L_z &= mr\cos\phi\sin\theta(\dot{r}\sin\phi\sin\theta + mr\dot{\phi}\cos\phi\sin\theta + r\dot{\theta}\sin\phi\cos\theta) \\ &\qquad - mr\sin\phi\sin\theta(\dot{r}\cos\phi\sin\theta - mr\dot{\phi}\sin\phi\sin\theta + r\dot{\theta}\cos\phi\cos\theta) \\ &= mr^2(\cos^2\phi + \sin^2\phi)(\sin^2\theta)\dot{\phi} = mr^2(\sin^2\theta)\dot{\phi} = p_\phi\,. \end{aligned}$$

Thus, $L_z = x\dot{y} - y\dot{x}$ is equivalent to the canonical momentum p_ϕ, the angular momentum about the z-axis. It is a conserved property, because $L_z = p_\phi = 0$. More generally, L_z will be conserved not just for a free particle, but for any motion subject to a potential energy that does not depend on the rotation angle; that is, $p_\phi = 0$ if $\partial V/\partial\phi = 0$.

The square of the magnitude of the angular momentum vector,

$$L^2 = L_x^2 + L_y^2 + L_z^2\,,$$

takes a simple form in polar coordinates. With a little bit of algebra one can show that

$$L^2 = m^2r^4(\dot{\theta}^2 + \dot{\phi}^2\sin^2\theta). \tag{13.45}$$

Comparing Eq. (13.45) with Eq. (13.30) shows that

$$T = \frac{1}{2m} p_r^2 + \frac{1}{2I} L^2, \tag{13.46}$$

where

$$I = mr^2 \tag{13.47}$$

is called the ***moment of inertia***[12] of the particle. Just as the mass of a particle is a measure of its resistance to motion in a straight line, the moment of inertia is a measure of its resistance to rotation.

Eqs. (13.44) are usually written in vector notation, as the ***cross product*** of the position vector and the angular momentum vector. The cross product of two vectors is defined as

$$(a_x, a_y, a_z) \times (b_x, b_y, b_z) = (a_y b_z - a_z b_y,\ -a_x b_z + a_z b_x,\ a_x b_y - a_y b_x)\,. \tag{13.48}$$

It is a vector that points in the direction perpendicular to both of the original vectors. We can write

$$\mathbf{L} = \mathbf{r} \times \mathbf{p}\,, \tag{13.49}$$

where $\mathbf{r} = (x, y, z)$ is the position vector and $\mathbf{p} = (p_x, p_y, p_z)$ is the linear momentum vector. The magnitude of $\mathbf{L}$ can be expressed as a *dot* product,

$$L = \sqrt{\mathbf{L} \cdot \mathbf{L}} = \sqrt{L_x^2 + L_y^2 + L_z^2}. \tag{13.50}$$

Example 13.6. *Rigid-body rotation.** Consider a collection of particles, such as a rigid molecule, rotating together about an axis through the center of mass. The velocity vector $\mathbf{v}_j$ of particle j is perpendicular to the axis and to the radial vector $\mathbf{r}_j$ that points from the center of mass to the particle. The ***angular velocity*** is defined as the vector $\boldsymbol{\omega}$ such that

$$\mathbf{v}_j = \boldsymbol{\omega} \times \mathbf{r}_j\,. \tag{13.51}$$

It is the same for all particles in a rigid body and it points along the axis. The total angular momentum vector is the sum over the angular momenta of the particles

$$\mathbf{L} = \sum_j \mathbf{L}_j = \sum_j \mathbf{r}_j \times \mathbf{p}_j = \sum_j \mathbf{r}_j \times (m_j \mathbf{v}_j) = \sum_j m_j \mathbf{r}_j \times (\boldsymbol{\omega} \times \mathbf{r}_j)\,. \tag{13.52}$$

Thanks to a vector identity

$$\mathbf{a} \cdot (\mathbf{b} \times \mathbf{c}) = \mathbf{b} \cdot (\mathbf{c} \times \mathbf{a}) \tag{13.53}$$

(Exercise 13.10) the kinetic energy can be written in terms of angular momentum as

$$T = \tfrac{1}{2} \sum_j m_j \mathbf{v}_j \cdot \mathbf{v}_j = \tfrac{1}{2} \sum_j m_j \mathbf{v}_j \cdot (\boldsymbol{\omega} \times \mathbf{r}_j) = \tfrac{1}{2} \sum_j m_j \boldsymbol{\omega} \cdot (\mathbf{r}_j \times \mathbf{v}_j) = \tfrac{1}{2} \boldsymbol{\omega} \cdot \mathbf{L}\,. \tag{13.54}$$

13.3 Differentials in Thermodynamics

In thermodynamics the common practice is to develop the theory in terms of differentials. The first law of thermodynamics, for example, is usually expressed as the differential equation $dU = dq + dw$, where U is a state function called the *internal energy* of the system, q is the energy transfer between the system and the surroundings through random thermal motion (as measured by temperature changes), and w is the energy transfer due to orderly motion, such as changes in volume against an external force or

[12] "Moment" is used here in the sense of "importance" or "weight." It should not be confused with the word *momentum*.

passage of electric current through a wire. We could state the first law as $\Delta U = q + w$, in terms of quantities with numerical values, but the statement in terms of differentials is usually more convenient for the purpose of deriving other theoretical relationships.

For example, enthalpy is defined as $H = U + pV$. The change in enthalpy for a process is $\Delta H = \Delta U + \Delta(pV)$, but $\Delta(pV) = p_\mathrm{f}V_\mathrm{f} - p_\mathrm{i}V_\mathrm{i}$ is complicated, because it involves simultaneous changes in both p and V. Using differentials,

$$H + dH = U + dU + (p + dp)(V + dV) = (U + pV) + (dU + pdV + Vdp) + dV dp.$$

We can now collect terms according to the power of infinitesimals (in the same manner as we manipulated Taylor series). The singly infinitesimal terms give us the equation

$$dH = dU + pdV + Vdp, \tag{13.55}$$

which involves changes in V and p individually. Simultaneous changes, $dV dp$, are ignored. From a theoretical standpoint, this is a significant simplification.

The most important differential in chemical thermodynamics is that for the change in Gibbs free energy in response to changes in p and T. For a single-component system,

$$dG = V dp - SdT, \tag{13.56}$$

where S is the entropy of the system. The criterion for a process to be spontaneous is $dG < 0$. This differential expresses dG in terms of the easily measurable variables p and T. Note that

$$\left(\frac{\partial G}{\partial p}\right)_T = V, \qquad \left(\frac{\partial G}{\partial T}\right)_p = -S. \tag{13.57}$$

These follow immediately from the definition of the partial derivative.

Keep in mind that a differential is inherently infinitesimal. It has no numerical value. At the end of the analysis one typically integrates to convert the differential into something that can be compared with experiment. For example,

$$\Delta U = \int_{U_\mathrm{i}}^{U_\mathrm{f}} dU$$

has a numerical value that can in principle be measured experimentally. dU is just a theoretical construct. A statement such as $dU = 583$ J is inherently absurd, as it claims that an infinitesimal has a nonzero numerical value.

13.4 Transport Equations

Here we consider phenomena in which some material or property is transported through a continuous medium. The medium is some substance (gas, liquid, or solid) that is treated as if it were continuous, rather than composed of discrete particles. A wide variety of such processes can be described with similar differential equations. Instead of calculating a trajectory for an individual particle, one calculates the bulk flow of the property of interest.

One way to simplify the problem is to use a ***compartmental analysis***, in which we designate some region of the system as a "compartment" and then focus on the rate of transport into and out of the compartment, ignoring the details within. This is often used in chemical engineering, where the compartment is a reaction vessel into which solutions are fed through input pipes and removed through output pipes. It can also be used to model the ventillation of a toxic gas from an enclosed volume (such as carbon monoxide or radon from a house) and the transport of pollutants in the environment. The basic idea of compartmental analysis is simply that the rate of change in the amount of a substance is equal to its inflow minus its outflow.

Example 13.7. *Water pollution.* Consider a pond into which polluted water is being fed by a stream. Let Φ be the flow rate of the stream (e.g., in m^3/h), let γ be the molar concentration of the pollutant in the stream, V, the volume of the pond, and $n(t)$, the number of moles of pollutant in the pond at time t. We want to know the pollutant concentration $c(t)$ in the pond. Assume the outflow from the pond equals the inflow, so that V is constant. The inflow of pollutant in time period dt is $\Phi\gamma dt$ while the outflow of pollutant is $\Phi c(t)dt$, which means that $dn = \Phi\gamma dt - \Phi c(t)dt$. But $dn = Vdc$. Dividing by dt gives the differential equation

$$\dot{c} = (\Phi/V)[\gamma - c(t)] . \qquad (13.58)$$

If compartmentalization is not appropriate, then the flow can be analyzed through an infinitesimal region. This results in differential equations involving spatial coordinates as well as time.

Example 13.8. *Groundwater flow.** The property being transported is mass of water and the medium is soil. Let x be the coordinate in the flow direction and let z be the vertical direction. Our measure of flow will be the ***specific discharge***, with symbol q, which is defined as the volume of water passing through a unit of area perpendicular to the x direction per unit time. It has SI units of $m^3/(m^2 s) = m/s$.

The flow of water is caused by differences in water pressure as a function of x. Water pressure is difficult to measure directly. Consider, however, a closely related property: the difference in height of the water in a well dug at one x value compared with the height of the water in a well at some other point. The greater the water pressure, the higher the water level. Typically, the "well" is actually a pipe, called a "standpipe," placed in a narrow shaft drilled into the ground. The level of water in a standpipe is called the ***hydraulic head*** at that point. It is given the symbol ϕ.

The specific discharge due to a difference in hydraulic head at two different points is found to be approximately

$$q = k(\phi_1 - \phi_2)/\ell , \qquad (13.59)$$

where ℓ is the distance between two points x_1 and x_2, ϕ_j is the head at x_j, and k is an empirical constant. Eq. (13.59) is ***Darcy's law***. It was discovered by the French engineer Henry Darcy in 1856 and remains an important principle of groundwater mechanics. k ranges from 10^{-2} m/s for gravel to 10^{-9} m/s for clay.

Let x_1 be an arbitrary point x and let $x_2 = x + dx$ be an infinitesimal distance away. Then $\phi_2 - \phi_1 = \phi(x + dx) - \phi(x) = d\phi$. Darcy's law becomes a differential equation $q = -k\, d\phi/dx$. One can determine q from this equation by measuring ϕ at various points x and then interpolating a function $\phi(x)$ between the values. For flow in any arbitrary direction, we replace $d\phi/dx$ with $\vec{\nabla}\phi$, called the ***hydraulic gradient***. Then,

$$\mathbf{q} = -k\vec{\nabla}\phi\,. \tag{13.60}$$

Darcy's law provides a general equation of motion for the water, but the principle of conservation of mass provides a constraint on the possible solutions. If the soil is already saturated, then no additional mass of water can be absorbed by the soil. Consider water arriving at a point x with specific discharge $q(x)$. The flow into the point x must equal the flow out of x into $x + dx$. Therefore, $q(x + dx) = q(x)$, which implies that $\frac{dq}{dx} = 0$ or, in general, $\vec{\nabla}q = 0$. Substituting Darcy's law into this gives

$$\nabla^2\phi = 0\,, \tag{13.61}$$

which in principle could give a theoretical result for ϕ. This differential equation is called ***Laplace's equation***. Its appearance is deceptively simple; in practice it can be rather difficult to solve.

Now suppose that it is raining. Let $Q(\mathbf{x})$ be the rate at which rainwater infiltrates into the groundwater at the point $\mathbf{x} = (x, y, z)$. The specific discharge at $\mathbf{x}$ will be increased by Q, so that $\mathbf{q}(\mathbf{x} + d\mathbf{x}) - \mathbf{q}(\mathbf{x}) = Q(\mathbf{x})$. It follows that

$$\nabla^2\phi = -Q(\mathbf{x})/k\,. \tag{13.62}$$

This is an example of a class of differential equations called ***Poisson equations***.

Example 13.9. *Solute transport.** Our goal here is to derive a differential equation for solute concentration, c, as a function of position and time. Let us assume for now that the solute is moving in the x-direction at the same velocity as the solvent. Such transport of solute, due to flow of the solvent, is called ***advection***. Let $J(x,t)$ be the solute flux; that is, the rate at which solute passes through an arbitrary cross-sectional area perpendicular to the x-axis, at position x and time t, per unit time and per unit area. The flux of solution into position x from the left must equal the flux out of position x to the right. Because the solute moves with the solvent, this implies that J at any given x is constant, and its differential is zero. Therefore,

$$0 = dJ = \left(\frac{\partial J}{\partial t}\right)_x dt + \left(\frac{\partial J}{\partial x}\right)_t dx.$$

Dividing through by dt,

$$\left(\frac{\partial J}{\partial t}\right)_x = -v\left(\frac{\partial J}{\partial x}\right)_t. \tag{13.63}$$

$v = dx/dt$ is the solvent velocity. The flux of solute must be proportional to the concentration of solute—if we double the number of solute particles per unit volume, we double the number that pass through the cross-sectional area. Thus we obtain an equation for $c(x,t)$,

$$\partial c/\partial t = -v\,\partial c/\partial x. \tag{13.64}$$

This analysis has been for flow in the x direction. For arbitrary direction, with solvent velocity vector $\mathbf{v}$, we simply replace the derivative with respect to x with the gradient,

$$\partial c/\partial t = -\mathbf{v}\cdot\vec{\nabla}c\,. \tag{13.65}$$

Another cause of solute transport is ***diffusion***, the tendency, due to entropy, of solute to have spontaneous net flow out of a region of high concentration into a neighboring region of low concentration. This can be driven by thermal motion of solvent molecules or by macroscopic turbulence of bulk solvent. In either case, the solute flux is found to be

$$J = -D\partial c/\partial x\,, \tag{13.66}$$

where D is an empirical constant characteristic of the system. It is possible to prove in general that the proportionality factor between J and c is the solute velocity, dx/dt. It then follows from Eq. (13.64) that

$$\partial c/\partial t = D\,\partial^2 c/\partial x^2. \tag{13.67}$$

In three dimensions, the equation is

$$\mathbf{J} = -D\vec{\nabla}c, \qquad \partial c/\partial t = D\nabla^2 c. \tag{13.68}$$

These are ***Fick's laws.***[13] For a full description, the contributions from all causes must be added together. Let v_{rxn} be the sum of chemical reaction rate laws for all reactions affecting the solute concentration. Then

$$\partial c/\partial t = -\mathbf{v}\cdot\vec{\nabla}c + D\nabla^2 c + v_{\text{rxn}}\,. \tag{13.69}$$

Similar equations can be derived for "transport" of nonmaterial phenomena such as heat or sound.

Exercises

13.1 Derive Eq. (13.3) by integrating the extent-of-reaction differential, Eq. (13.1).

13.2 To increase the rate of the reaction $H_2 + Br_2 \to 2HBr$ would it be more effective to increase $[H_2]$ or to increase $[Br_2]$, or would it be equally effective either way?

13.3 Write the rate law for $H_2 + Br_2 \to 2HBr$ for a situation in which a 5.0 L container with hydrogen gas at partial pressure 4.5 bar and temperature 296 K has bromine gas fed into it at a steady rate of 0.10 mol per hour.

13.4 Derive Lagrange's equations for a free particle in cylindrical coordinates.

13.5 Give Lagrange's and Hamilton's equations for a particle constrained to the surface of a sphere of radius R but otherwise free.

13.6 Give Lagrange's equations for a two-dimensional damped oscillator of mass m, with $V(r,\phi) = \frac{1}{2}k_r r^2 + \frac{1}{2}k_\phi \phi^2$ and friction $\mathcal{F}_r = -\gamma\dot{r}$.

13.7 Prove that p_q is a constant of motion if and only if there is no explicit appearance of q in the Hamiltonian.

13.8 Prove that the angular momentum vector is conserved if $\mathbf{r}\times\mathbf{F}$ (the *torque*) is zero.

13.9 Derive Eq. (13.45), expressing L^2 in spherical polar coordinates.

13.10* Derive the identity $\mathbf{a}\cdot(\mathbf{b}\times\mathbf{c}) = \mathbf{b}\cdot(\mathbf{c}\times\mathbf{a})$.

13.11* Prove that $J = c\frac{dx}{dt}$. *(Hint: Consider a thin slab of solution of thickness Δx during a short time interval Δt.)*

---

[13]Discovered by the German physicist and medical researcher A. E. Fick (1829-1901), who is also known as the inventor of the contact lens.

Chapter 14

Solving Differential Equations, I

Solving a differential equation is referred to as *integrating* the equation. There is no systematic procedure that will give the exact solution of an arbitrary differential equation. The situation for integrating equations is analogous to what we saw with integration of functions: We have an assortment of "tricks" that work in certain special cases but it is very common to encounter differential equations for which it is impossible to express the solution in terms of elementary functions. Some such equations occur often enough in applications that the solutions have been studied in detail, given names, and are considered standard special functions. Otherwise, the solution must be computed using numerical methods or estimated with approximation methods such as power series. This chapter focuses on analytical solution methods. Numerical methods and approximation methods are treated in Chapter 15.

14.1 Basic Concepts

Consider, as a simple example, a function $y(x)$ that satisfies the differential equation

$$\frac{dy}{dx} = \alpha y\,, \tag{14.1}$$

in which the parameter α is a constant. The function y is called the ***dependent variable*** while x is called the ***independent variable***. The solution is

$$y = ce^{\alpha x}, \tag{14.2}$$

as is easily verified by substituting it into the differential equation. c is any arbitrary constant. The first thing to notice is that the nature of the solution depends very strongly on the value of α. Similar-looking equations can lead to qualitatively different solutions.

Example 14.1. *Solutions to the differential equation of the exponential.* For negative real α, $y(x) = ce^{\alpha x}$ is a decaying exponential, which rapidly drops toward zero while for positive real α it increases very rapidly toward infinity. If α is pure imaginary, $\alpha = ib$, then $y(x) = c[\cos(bx) + i\sin(bx)]$, which oscillates with constant amplitude, but if $\alpha = a + ib$ has nonzero real part a, then $y(x) = ce^{a}[\cos(bx) + i\sin(bx)]$, which oscillates with exponentially increasing or decreasing amplitude, depending on the sign of $a = \mathrm{Re}\,\alpha$. The figure below compares $y(x) = e^{\alpha x}$ for complex and pure real α with negative real part (solid curves) and positive real part (dashed curves).

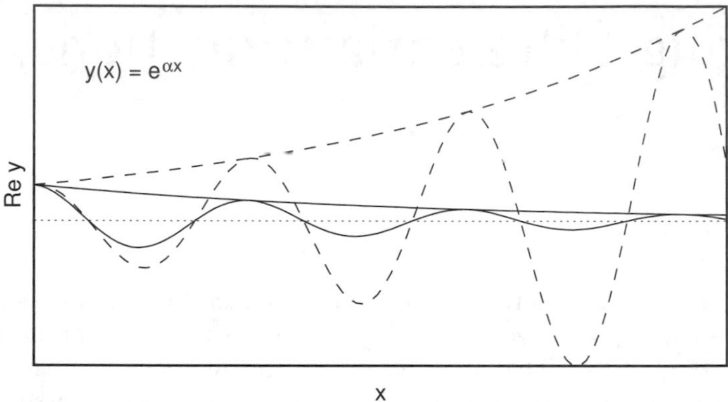

The solution to this equation illustrates another basic concept: A differential equation does not, by itself, provide enough information to determine a physical solution to a problem; the solution will contain one or more arbitrary constants, called ***constants of integration***.[1] In the solution $y(x) = ce^{\alpha x}$, the prefactor c is a constant of integration. Its value is not determined by the differential equation. The constant α, however, is not a constant of integration; it is a parameter whose value *is* set by the differential equation.

Example 14.2. *Constant of integration for a reaction rate law.* Consider a chemical reaction in which the molar concentration of substance A happens to follow the rate law

$$\frac{d[A]}{dt} = -k[A]\,.$$

The solution has the form $[A]_t = ce^{-kt}$. At $t = 0$ we have $[A]_0 = ce^0 = c$. If we measure $[A]$ at time $t = 0$, then that measured value $[A]_0$ determines the value of the constant of integration $c = [A]_0$.

Example 14.3. *Constants of integration for a trajectory.* Consider Newton's equation for a one-dimensional trajectory $x(t)$ subject to a force $F(t)$,

$$\frac{d^2x}{dt^2} = \frac{1}{m}F(t)\,.$$

The independent variable is the time t while the dependent variable is the position x. Suppose that some particular function $x_{\mathrm{a}}(t)$ is found to satisfy this equation. Then $x(t) = x_{\mathrm{a}}(t) + c_1 t + c_0$, where c_1 and c_0 are arbitrary constants of integration, will also be solutions, because the last two terms will be destroyed by the derivative. For a trajectory, it is typical to specify a value for the initial position, $x(0) = x_0$, and a value for the initial velocity, $\dot{x}(0) = v_0$, in order to determine the two constants:

$$c_0 = x_0 - x_{\mathrm{a}}(0), \qquad c_1 = v_0 - \dot{x}_{\mathrm{a}}(0)\,.$$

In general, values of constants of integration are determined by constraints that must be specified in addition to the differential equation itself. These constraints are called ***boundary conditions***. The most common way to

[1] This is analogous to the undetermined constants in indefinite integrals of functions.

impose them is to specify the value of the function and, if necessary, its derivatives at a single point. This is called an ***initial-value problem***, as typically the "point" in question corresponds to the initial conditions of the problem, as in Example 14.3. One can also set the constants of integration by specifying values for the dependent variable or its derivative at *different* values of the independent variable. This is called a ***boundary-value problem***. In Example 14.3 we could have specified x at two different times, or $\dot{x}$ at two different times, or $\dot{x}$ at one time and x at another.

There is an elaborate terminology for classifying differential equations. Classification is important because different classes of equations require different solution strategies. Table 14.1 summarizes the most common terms. A differential equation with only one independent variable is called an ***ordinary differential equation***, abbreviated ***ODE***. If the number of independent variables is greater than one, then we have a ***partial differential equation***, abbreviated ***PDE***. In general, one can expect that the difficulty of solving a differential equation to increase very significantly as the number of independent variables increases.

It is often convenient to write differential equations using operator notation. For example,

$$y'' + (1 - x^2)y' - 3x^4 y = \cos(\pi x) \tag{14.3}$$

can be written $\mathcal{M}\,y = \cos(\pi x)$, where $\mathcal{M} = \frac{d^2}{dx^2} + (1 - x^2)\frac{d}{dx} - 3x^4$. Any $\mathcal{M}$

Table 14.1: Terminology for classifying differential equations. The comments reference relevant sections and chapters.

Term	Definition	Comment
ordinary	Only one independent variable.	Relatively straightforward to solve. (14.3, 14.4, 15)
partial	Contains partial derivatives.	Can be very difficult. (14.5, 17.3, 20)
order	Order of highest-order derivative.	Tells number of boundary conditions needed for a linear ODE. (14.4)
linear operator	$\mathcal{M}\,(c_1 y_1 + c_2 y_2) = c_1\mathcal{M}\,y_1 + c_2\mathcal{M}\,y_2$	Can use superposition principle. (14.2)
homogeneous	$\mathcal{M}\,y = 0$	Superposition of solutions is also a solution. (14.2, 19, 20)
eigenvalue equation	$(\mathcal{M} - \alpha)\,y = 0$	If linear, can be treated with linear algebra. (19, 20)
inhomogeneous	$\mathcal{M}\,y = g(x)$	$y = y_{\text{part}} + \sum_j c_j\, y_{\text{comp},j}$ (14.2, 14.3, 14.5, 15.3.2)

such that

$$\mathcal{M}(y_1 + y_2) = \mathcal{M}y_1 + \mathcal{M}y_2 , \qquad \mathcal{M}(cy) = c\mathcal{M}y , \tag{14.4}$$

where c is a constant, is called a ***linear operator***. A differential equation that can be written $\mathcal{M}y = g(x)$ in terms of a linear operator $\mathcal{M}$ and a function g is called a ***linear differential equation***. For an equation to be linear, y and its derivatives can only appear to the first power. The following equations are each *non*linear:

$$y' - y^2 = x , \tag{14.5a}$$

$$(y')^2 - y = x , \tag{14.5b}$$

$$y' - \cos y = x . \tag{14.5c}$$

On the other hand, Eq. (14.3), which looks more complicated, is in fact linear. The equation is called "linear" as long as it is linear in the *dependent* variable, regardless of whether or not it is linear in the *independent* variable.

The ***order*** of a differential equation is the order of its highest-order derivative. Eq. (14.3) is of second order while Eqs. (14.5) are of first order.

14.2 The Superposition Principle

It is useful to distinguish between equations that are ***homogeneous***,

$$\mathcal{M}y = 0 , \tag{14.6}$$

and ***inhomogeneous***,

$$\mathcal{M}y = g(x) , \tag{14.7}$$

where g is not identically zero. An equation of the form

$$\mathcal{M}y = \alpha y , \tag{14.8}$$

where α is a constant, is called an ***eigenvalue equation***. These are encountered quite often in science and engineering applications. An eigenvalue equation is a type of homogeneous equation, because it can always be written as $(\mathcal{M} - \alpha)y = 0$, in terms of an operator $(\mathcal{M} - \alpha)$. Linear eigenvalue equations can be treated systematically using linear algebra. They will be discussed in detail in Chapters 19 and 20.

For linear homogeneous equations we have an important theorem called the ***superposition principle***:

Theorem 14.2.1. *Suppose that $y_1(x)$ and $y_2(x)$ are different particular solutions of an equation $\mathcal{M}y = 0$ where $\mathcal{M}$ is a linear operator. Then the sum $c_1y_1 + c_2y_2$, for arbitrary constants c_1 and c_2, is also a solution.*

Proof. Given that $\mathcal{M}y_1 = 0$ and $\mathcal{M}y_2 = 0$, and using the definition of a linear operator,

$$\mathcal{M}(c_1y_1 + c_2y_2) = c_1\mathcal{M}y_1 + c_2\mathcal{M}y_2 = 0 .$$

□

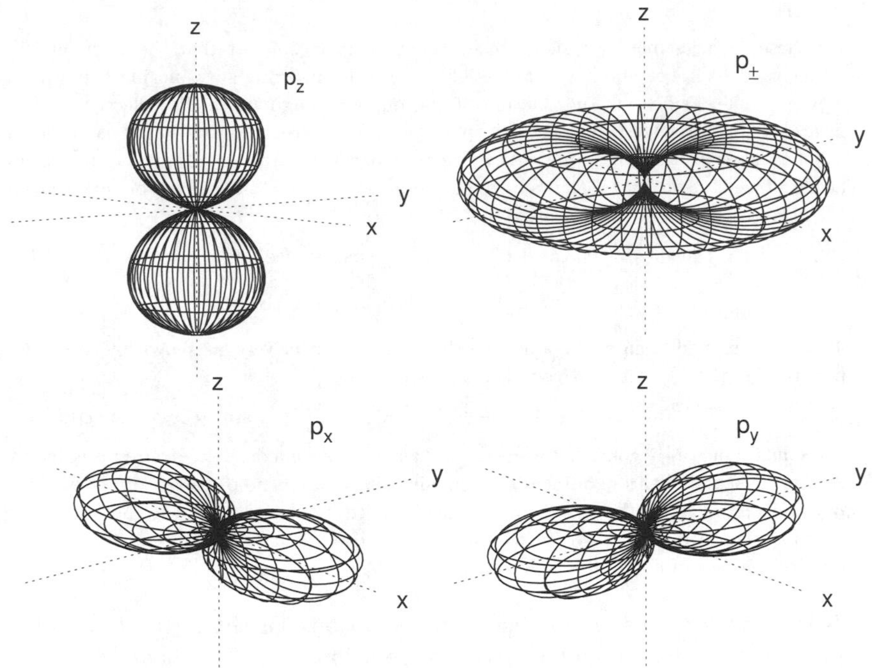

Figure 14.1: Angular dependence of various p orbitals.

An expression of the form $c_1 y_1 + c_2 y_2$ is called a ***linear superposition***, or a ***linear combination***,[2] of the functions y_1 and y_2.

Example 14.4. *Linear superpositions of p orbitals.* The "2p orbitals" are three functions that describe possible solutions for the spatial distribution of the electron for the first excited state of a hydrogen atom. They are functions $\psi(r, \theta, \phi)$ that solve the differential equation

$$\left(-\frac{1}{2}\nabla^2 - \frac{1}{r} + \frac{1}{8}\right)\psi = 0\,. \tag{14.9}$$

∇^2 is the Laplacian operator, Eq. (3.39). This equation is linear and homogeneous. Using a systematic procedure (to be described in Section 20.3), one can derive the following particular solutions:

$$\psi_0 = r\cos\theta\, e^{-r/2}, \tag{14.10}$$

$$\psi_+ = 2^{-1/2} r e^{i\phi}\sin\theta\, e^{-r/2}, \qquad \psi_- = 2^{-1/2} r e^{-i\phi}\sin\theta\, e^{-r/2}. \tag{14.11}$$

The fact that they are solutions can be verified by plugging them into Eq. (14.9). The physical interpretation is that $|\psi(r, \theta, \phi)|^2$ is an acceptable solution for the probability distribution of finding the electron at any given location.

These functions have distinctive angular dependences. To plot $|\psi|^2$ would require a four-dimensional space $(x, y, z, |\psi|^2)$. A common technique for displaying angular dependence (in three dimensions) is to set r at a fixed value and then plot a surface such that the distance from the origin to any point on the surface, for any given values of θ and ϕ, is equal to the value of $|\psi(\theta, \phi)|^2$. The upper panels of Fig. 14.1 show the surfaces that result from Eqs. (14.10) and (14.11). ($|\psi_+|^2 = |\psi_-|^2$ because $\psi_+ = \psi_-^*$.)

[2] These phrases are synonymous, but "superposition" is more commonly used when speaking specifically of solutions to differential equations while "combination" is more commonly used in the more general context of linear algebra.

Do these pictures make physical sense? They seem to suggest that the electron will either tend to linger about the z-axis or will reside in a ring surrounding the z-axis. However, the z-axis is a construction of the human imagination. The electron is not aware of any preferred direction in space! This apparent contradiction is resolved by the superposition principle. Any linear superposition of ψ_0, ψ_+, and ψ_- is an acceptable solution, and we can construct superpositions to have any arbitrary spatial orientation. For example, consider the following:

$$\psi_x = 2^{-1/2}\psi_+ + 2^{-1/2}\psi_- = r\cos\phi\sin\theta\, e^{-r/2}, \tag{14.12}$$

$$\psi_y = \frac{1}{i}\,2^{-1/2}\psi_+ - \frac{1}{i}\,2^{-1/2}\psi_- = r\sin\phi\sin\theta\, e^{-r/2}. \tag{14.13}$$

These are aligned along the x-axis and the y-axis, respectively, as shown by the lower panels of Fig. 14.1. Alternatively, the superposition

$$2^{-1/2}\psi_x + 2^{-1/2}\psi_y = \tfrac{1}{2}\,(1-i)\psi_+ + \tfrac{1}{2}\,(1+i)\psi_- = 2^{-1/2}\, r\sin\theta(\cos\phi + \sin\phi)e^{-r/2}$$

is aligned along a diagonal between the x-axis and the y-axis. Any desired alignment can be constructed through linear superpositions. There is no preferred direction. ψ_x, ψ_y, and ψ_0 are called the p_x, p_y, and p_z orbitals, respectively. They are the particular solutions traditionally shown in chemistry textbooks, but they are only three of an infinite number of acceptable solutions.

Now consider linear *in*homogeneous equations, in the form of Eq. (14.7). Suppose that we have found a particular solution $y_{\text{part}}(x)$, so that $\mathcal{M}\, y_{\text{part}} = g$. Then we can construct other solutions by adding to this particular solution any arbitrary linear superposition of solutions to the corresponding homogeneous equation. Let $y_{\text{comp},j}$ be various solutions to $\mathcal{M}\, y = 0$. These are called ***complementary solutions*** of the inhomogeneous equation. Then

$$y = y_{\text{part}} + \sum_j c_j\, y_{\text{comp},j} \tag{14.14}$$

is also a solution to the inhomogeneous equation:

$$\mathcal{M}\, y = \mathcal{M}\, y_{\text{part}} + \mathcal{M} \sum_j c_j\, y_{\text{comp},j} = g + \sum_j c_j \mathcal{M}\, y_{\text{comp},j} = g\,,$$

because $\mathcal{M}\, y_{\text{comp},j} = 0$.

The superposition principle yields a collection of *possible* solutions. Boundary conditions are what determine which of the possible solutions correspond to actual solutions of the physical problem.

14.3 First-Order ODE's

Differential equations are usually solved using computers, but there exist special cases in which analytic expressions for the solutions can be systematically derived by hand. In particular, there is a category of first-order ODE's for which the problem of solving the differential equation can be converted to a problem of evaluating integrals of functions. If the integrals can be evaluated analytically then an analytic solution to the ODE is obtained.

Consider a first-order homogeneous ODE

$$y' = A(x)f(y) \tag{14.15a}$$

or, equivalently,

$$A(x)dx = B(y)dy\,, \tag{14.15b}$$

with $B(y) = 1/f(y)$. The second form is obtained by multiplying through by $dx/f(y)$ and noting that $(dx)y' = (dx)(dy/dx) = dy$. In Eq. (14.15b), the left-hand side depends only on x while the right-hand side depends only on y. Thus, the two variables have been separated from each other. An equation that can be put into such a form is called a ***separable*** ODE. It is solved by separately integrating each side:

$$\int_{x_0}^{x} A(u)du = \int_{y_0}^{y} B(v)dv. \tag{14.16}$$

This gives a solution $y(x)$ with the boundary condition $y(x_0) = y_0$.

Example 14.5. *Integrated rate laws.* For a rate law of the form kc_A^a, involving only a single substance A, we can solve directly for $c_A(t)$ by separating the variables. For example, if A + B → *products* has empirical rate law $\dot{c}_A = -kc_A^3$, then $-c_A^{-3}dc_A = kdt$. Integrating the right-hand side gives $k\int_0^t ds = kt$. (The dummy variable has been renamed s here in order to avoid confusion with the upper range of integration, which is an arbitrary specified value of t.) Integrating the left-hand side gives $\frac{1}{2}c_A(t)^{-2} - \frac{1}{2}c_A(0)^{-2}$. It follows that $c_A^{-2} = c_{A,0}^{-2} + 2kt$. Solving this for $c_A(t)$ gives The solution is $c_A(t) = c_{A,0}/(1 + 2c_{A,0}^2 kt)^{1/2}$, where $c_{A,0} = c_A(0)$.

If the rate law involves more than one substance, we can formulate the problem in terms of the extent of reaction ξ as the dependent variable. In general, rate laws have the form

$$\frac{d\xi}{dt} = f(\xi), \qquad \frac{1}{f(\xi)}\,d\xi = dt, \tag{14.17}$$

where ξ is an extent-of-reaction variable. This is always separable if $f(\xi)$ has no explicit dependence on time. Consider the differential equation from Example 13.1 for a reaction 2A + 3B → *products* with an empirically determined rate law $\dot{\xi} = kc_A c_B^2$. The equation can be separated as

$$dt = \frac{d\xi}{k(c_{A,0} - 2\xi)(c_{B,0} - 3\xi)^2}\,. \tag{14.18}$$

The boundary condition is $\xi = 0$ at $t = 0$, which establishes the ranges of the integrations. The integral of the right-hand side can be evaluated analytically using partial fractions:[3]

$$\frac{1}{(c_{A,0} - 2x)(c_{B,0} - 3x)^2} = \frac{\alpha}{c_{A,0} - 2x} + \frac{\beta + \gamma x}{(c_{B,0} - 3x)^2}\,,$$

$$\alpha = 4\delta^{-2}, \quad \beta = \tfrac{3}{4}(3c_{A,0} - 4c_{B,0})\alpha\,, \quad \gamma = \tfrac{9}{2}\alpha\,, \quad \delta = 3c_{A,0} - 2c_{B,0}\,,$$

$$\begin{aligned} kt &= \alpha\int_0^{\xi}\frac{dx}{c_{A,0} - 2x} + \beta\int_0^{\xi}\frac{dx}{(c_{B,0} - 3x)^2} + \gamma\int_0^{\xi}\frac{xdx}{(c_{B,0} - 3x)^2} \\ &= -\frac{1}{2}\alpha\int_{c_{A,0}}^{c_{A,0}-2\xi} u^{-1}du - \frac{1}{3}\beta\int_{c_{B,0}}^{c_{B,0}-3\xi} v^{-2}dv - \frac{1}{9}\gamma\int_{c_{B,0}}^{c_{B,0}-3\xi}\frac{c_{B,0} - v}{v^2}\,dv \\ &= -\frac{1}{2}\alpha\ln(1 - 2\xi/c_{A,0}) + \frac{3\beta + c_{B,0}\gamma}{9c_{B,0}}\left(\frac{1}{1 - 3\xi/c_{B,0}} - 1\right) + \frac{1}{9}\gamma\ln(1 - 3\xi/c_{B,0}). \end{aligned}$$

---

[3]See Example 4.2 and Exercise 4.3, or, if you prefer, use computer algebra.

Substituting in the expressions for α, β, and γ, we find that

$$kt = 2\delta^{-2}[\ln(1 - 3\xi/c_{\mathrm{B},0}) - \ln(1 - 2\xi/c_{\mathrm{A},0})] + \frac{1}{c_{\mathrm{B},0}\,\delta}\left(\frac{3\xi/c_{\mathrm{B},0}}{1 - 3\xi/c_{\mathrm{B},0}}\right). \tag{14.19}$$

After a little more algebra, we can write this as

$$kt = \frac{2}{\delta^2}\ln\frac{c_{\mathrm{B},t}/c_{\mathrm{B},0}}{c_{\mathrm{A},t}/c_{\mathrm{A},0}} + \frac{1}{\delta}\left(\frac{1}{c_{\mathrm{B},t}} - \frac{1}{c_{\mathrm{B},0}}\right), \tag{14.20}$$

in terms of the concentrations at time t, $c_{\mathrm{A},t} = c_{\mathrm{A},0} - 2\xi$ and $c_{\mathrm{B},t} = c_{\mathrm{B},0} - 3\xi$.

This is an analytic solution, but what it gives us is an explicit solution for $t(\xi)$. We seek $\xi(t)$. At least we have converted the *differential* equation into an *algebraic* equation, which is significant progress, as we have a variety of very effective techniques for solving algebraic equations. One approach is to use a numerical root-finding algorithm such as described in Section 5.5, which would give a numerical value for ξ at any desired numerical value of t. Another approach is to derive a Taylor series for ξ in powers of $(t - t_1)$. This would give an explicit analytic expression valid over a range of time values in the vicinity of a specified t_1. For example, let us expand about $t_1 = 0$. By definition, the extent of reaction $\xi(t)$ is zero at $t = 0$. We will expand Eq. (14.19) in powers of ξ and then invert the series to obtain a power series in t. The first-order series is

$$kt \sim \frac{1}{c_{\mathrm{A},0}c_{\mathrm{B},0}^2}\xi + \mathcal{O}(\xi^2).$$

Thus, to first order, ξ is proportional to t. This implies that $\mathcal{O}(\xi)$ is proportional to $\mathcal{O}(t)$, $\mathcal{O}(\xi^2)$ is proportional to $\mathcal{O}(t^2)$, and so on. Therefore,

$$\xi \sim c_{\mathrm{A},0}c_{\mathrm{B},0}^2 kt + \mathcal{O}(t^2). \tag{14.21}$$

Let us calculate the series through second order:

$$kt \sim \frac{1}{c_{\mathrm{A},0}c_{\mathrm{B},0}^2}\xi + \frac{3c_{\mathrm{A},0} + 2c_{\mathrm{B},0}}{c_{\mathrm{A},0}^2 c_{\mathrm{B},0}^3}\xi^2 + \mathcal{O}(\xi^3), \qquad \xi \sim c_{\mathrm{A},0}c_{\mathrm{B},0}^2 kt - \frac{3c_{\mathrm{A},0} + c_{\mathrm{B},0}}{c_{\mathrm{A},0}c_{\mathrm{B},0}}\xi^2 + \mathcal{O}(\xi^3).$$

Substituting Eq. (14.21) for ξ on the right-hand side and replacing $\mathcal{O}(\xi^3)$ with $\mathcal{O}(t^3)$ gives

$$\xi \sim c_{\mathrm{A},0}c_{\mathrm{B},0}^2 kt - (3c_{\mathrm{A},0} + c_{\mathrm{B},0})c_{\mathrm{A},0}c_{\mathrm{B},0}^3 k^2t^2 + \mathcal{O}(t^3). \tag{14.22}$$

We could continue in this way to arbitrary order in t. (In fact, there is an easier way to obtain this Taylor series, directly from the differential equation without first solving for the analytic solution. This is described in Section 15.3.1.)

Now consider a first-order linear *in*homogeneous ODE,

$$\frac{dy}{dx} = f(x)y + h(x), \tag{14.23}$$

which would be separable if it were not for the presence of the added function $h(x)$. This can be converted into a separable equation using a technique called ***variation of parameters***. First, solve the homogeneous equation

$$\frac{dy_{\mathrm{comp}}}{dx} = f(x)y_{\mathrm{comp}}$$

for the complementary solution $c\,y_{\mathrm{comp}}$, where c is a constant of integration. Then replace the constant c with a function $u(x)$ and attempt to solve the differential equation with a particular solution in the form

$$y_{\mathrm{part}}(x) = u(x)y_{\mathrm{comp}}(x). \tag{14.24}$$

Substituting this into the differential equation gives

$$u' y_{\rm comp} + u y'_{\rm comp} = f u y_{\rm comp} + h\,.$$

But $y'_{\rm comp} = f y_{\rm comp}$. It follows that $u' y_{\rm comp} = h$, which implies that

$$du = \frac{h(x)}{y_{\rm comp}(x)}\, dx\,. \tag{14.25}$$

u is separated from x. Integrate both sides to obtain $u(x)$ and substitute the result into Eq. (14.24) to obtain the particular solution.

14.4 Higher-Order ODE's

The order can tell us how many integration constants there will be:

Theorem 14.4.1. *The general solution of a linear ordinary differential equation of order n contains n constants of integration. Any particular solution corresponds to a particular choice for the values of these constants.*

This tells us that if we find a general solution with n arbitrary constants, then we know that this is the only solution—we have solved the problem once and for all and do not need to look for other solutions. A *non*linear ordinary differential equation of order n will also have a solution that can be specified by n constants of integration, but this solution is not unique; there may also be other solutions with different functional forms.

Second-order equations are very commonly encountered in practical applications. In particular, Newton's equation for the motion of a particle subject to a potential energy function $V(x)$,

$$m\frac{d^2x}{dt^2} = -\frac{dV}{dx}\,, \tag{14.26}$$

is a second-order ODE for the trajectory $x(t)$. There is no general analytical solution method for second-order equations, but there are well-known techniques that work in special cases, and which are discussed at length in textbooks and reference works devoted to differential equations.[4]

Perhaps the most commonly used technique is to guess a functional form for the solution and include undetermined parameters in it. Substituting the function into the differential equation gives an algebraic equation for the parameters. If the functional form is incorrect, then the algebraic equation will have no solution. However, if the algebraic equation can be solved, then those parameter values yield a solution to the differential equation.

---

[4]See, for example, D. Zwillinger, *Handbook of Differential Equations*, 2nd ed. (Academic Press, San Diego, 1992).

Example 14.6. *Classical mechanical harmonic oscillator.* The harmonic oscillator in one dimension is defined by the potential energy function $V(x) = \frac{1}{2}kx^2$, where k is some positive constant. Newton's equation for the trajectory of a particle of mass m subject to this potential is

$$\frac{d^2x}{dt^2} = -\frac{k}{m}x\,. \tag{14.27}$$

We seek a function $x(t)$ that remains unchanged except for a constant prefactor under the action of the second derivative. What functional form has this property? The sine and cosine functions come to mind. Let us guess the solution

$$x(t) = \cos(\omega t + \gamma)$$

with as yet undetermined parameters ω and γ. Substituting this into the differential equation, we obtain the algebraic equation

$$-\omega^2\cos(\omega t + \gamma) = -\frac{k}{m}\cos(\omega t + \gamma)\,,$$

or $\omega^2 = k/m$. This gives a solution for any value of γ. Because the differential equation is homogeneous, we can multiply our solution by an arbitrary constant and it will still be valid. The general solution is

$$x(t) = A\cos(\omega t + \gamma)\,, \qquad \omega = \sqrt{k/m}\,, \tag{14.28}$$

with the parameters A and γ determined by the boundary conditions. For example, if we require $x(t_0) = x_0$ and $x'(t_0) = v_0$, for specified values x_0 and v_0 of position and speed at time t_0, then we have the two simultaneous equations

$$x_0 = A\cos(\omega t_0 + \gamma)\,, \qquad v_0 = -\omega A\sin(\omega t_0 + \gamma)\,,$$

which can be solved for A and γ:

$$\gamma = \arctan\left(-\frac{v_0}{x_0\omega}\right) - \omega t_0\,, \qquad A = \frac{x_0}{\cos(\omega t_0 + \gamma)}\,.$$

The differential equation in this case was linear. According to Theorem 14.4.1, a linear second-order ODE has its solution uniquely specified by two constants of integration. Our constants of integration were A and γ, and we have set values for both of them. The theorem tells us that this is the unique solution for the trajectory. (ω is *not* a constant of integration, because its value is fixed for all possible solutions of the differential equation. Its value is determined not by boundary conditions but by the values of k and m in the differential equation.)

Suppose that we were not so clever as to realize that we could include the parameter γ in the argument of the cosine, and that we chose simply $x(t) = A\cos(\omega t)$. As before, we find $\omega = \pm(k/m)^{1/2}$, but now we have only the one undetermined parameter A to serve as a constant of integration. According to Theorem 14.4.1, this is not sufficient to fully solve the problem. Note, however, that $x(t) = A\sin(\omega t)$ also satisfies the differential equation. Thus we have two particular solutions. Our differential equation is linear and homogeneous. According to the superposition principle, the linear superposition

$$x(t) = A_1\cos(\omega t) + A_2\sin(\omega t) \tag{14.29}$$

is also a solution, and it contains two parameters, A_1 and A_2, that can serve as the two constants of integration.

But Theorem 14.4.1 claims that the solution is unique! Is there a contradiction here? In fact, our two solutions are equivalent. This can be demonstrated using Euler's formula. We know that

$$\cos(\omega t) = \mathrm{Re}\, e^{i\omega}, \qquad \sin(\omega t) = \cos(\omega t - \pi/2) = \mathrm{Re}\, e^{i(\omega t - \pi/2)},$$

and

$$\cos(\omega t + \gamma) = \mathrm{Re}\, e^{i(\omega t + \gamma)}.$$

Is it possible to require that

$$Ae^{i(\omega t+\gamma)} = A_1 e^{i\omega t} + A_2 e^{i(\omega t - \pi/2)}\ ?$$

This would make Eqs. (14.28) and (14.29) equivalent. Multiplying both sides by $e^{-i\omega t}$, we obtain $Ae^{i\gamma} = A_1 - iA_2$, and therefore,

$$A = |Ae^{i\gamma}| = \sqrt{A_1^2 + A_2^2}, \qquad \gamma = \arg\left(Ae^{i\gamma}\right) = \arg(A_1 - iA_2) = \arctan(-A_2/A_1).$$

With these formulas, the two functional forms are equivalent.

Of course, it will not be possible in general to successfully guess the functional form of the solution of an arbitrary differential equation. One might instead try to look up the answer in a table or try to solve the equation with computer algebra. However, it is very easy to construct a differential equation for which it is impossible to express the solution analytically in terms of elementary functions. Even the simple equation $y'' = xy$ is in this category. For some such equations the solutions have been characterized in detail; their properties can be looked up in reference works[5] and their numerical values computed using computer algebra. Some examples are given in Table 14.2.

Table 14.2: Examples of second-order ordinary differential equations whose solutions are well-known special functions.[5] n is an arbitrary non-negative integer while a and b are arbitrary real numbers. Because these are second-order linear ODE's, their general solutions are arbitrary linear superpositions of two different kinds of standard solutions.

Equation	Name of solutions
$y'' = xy$	Airy functions: $\mathrm{Ai}(x)$, $\mathrm{Bi}(x)$
$x^2y'' + xy' + (x^2 - n^2)y = 0$	Bessel functions: $J_n(x)$, $Y_n(x)$
$xy'' + (b - x)y' - ay = 0$	Kummer functions: $M(a,b;x)$, $U(a,b;x)$
$y'' + 2xy' - 2ny = 0$	Integrals of erfc: $\mathrm{i}^n\ \mathrm{erfc}(\pm x)$, $\mathrm{i}^0\ \mathrm{erfc}(x) = \mathrm{erfc}(x)$, $\mathrm{i}^n\ \mathrm{erfc}(x) = \int_x^\infty \mathrm{i}^{n-1}\ \mathrm{erfc}(\xi)d\xi$

[5]The standard reference is *Handbook of Mathematical Functions*, ed. M. Abramowitz and I. A. Stegun (Dover, New York, 1965).

[5]This list is not at all exhaustive. A few more will be given later, in Table 17.3. See A. D. Polyanin and V. F. Zaitsev, *Handbook of Exact Solutions for Ordinary Differential Equations*, 2nd ed. (Chapman & Hall/CRC Press, Boca Raton, FL, 2003), for thousands more.

14.5 Partial Differential Equations

The first step in analyzing a PDE is usually to try to convert it into a system of simultaneous ODE's. The most common technique for doing this is called ***separation of variables***.[6] This consists of guessing that the solution $y(x_1, x_2, x_3, \dots)$ can be expressed as a separable product,

$$y(x_1, x_2, x_3, \dots) = Y_1(x_1)Y_2(x_2)Y_3(x_3)\cdots . \tag{14.30}$$

Example 14.7. *Separation of variables in Fick's second law.* Fick's laws were given by Eqs. (13.68). They describe the transport of solute from a region of higher concentration to a neighboring region of lower concentration. Consider Fick's second law for the case of solute diffusion in one dimension,

$$\frac{\partial c}{\partial t} = D\frac{\partial^2 c}{\partial x^2} . \tag{14.31}$$

$c(x,t)$ is the concentration of a solute at position x and time t, and D is an empirical constant. Let us propose a solution of the form

$$c(x,t) = X(x)T(t) . \tag{14.32}$$

Substituting this into the differential equation gives

$$X(x)\frac{dT(t)}{dt} = DT(t)\frac{d^2X(x)}{dx^2} ,$$

which we can rearrange as

$$\frac{1}{T(t)}\frac{dT(t)}{dt} = \frac{D}{X(x)}\frac{d^2X(x)}{dx^2} .$$

This is a curious equation. The left side is a function only of t while the right-hand side is a function only of x. However, t and x are both independent variables—we could hold t at some fixed value while varying x arbitrarily, and for any value of x this equation would still have to be valid. The only way this can happen is if both sides of the equation are independently equal to a constant,

$$\frac{1}{T(t)}\frac{dT(t)}{dt} = \kappa , \qquad \frac{D}{X(x)}\frac{d^2X(x)}{dx^2} = \kappa . \tag{14.33}$$

Thus, the problem has been reduced to solving a set of two ODE's.

We have already seen the first of these equations, in Example 14.1. The solution is $T(t) = T_0 e^{\kappa t}$. The nature of this solution depends on whether the real part of κ is positive or negative. If $\operatorname{Re}\kappa$ is positive, then at infinite time the concentration will become infinite, which is not physically reasonable. We will only consider here solutions with negative and pure real κ, and write the solution as

$$T(t) = T_0 e^{-\lambda t} , \qquad \lambda = -\kappa , \tag{14.34}$$

where λ is positive. Let us write the second ODE as

$$\frac{d^2X}{dx^2} = -aX, \qquad a = \lambda/D .$$

[6]Despite the similar name, the technique described here, for PDE's, is different from the separation of first-order ODE's described in Section 14.3.

This is also also an equation we have already seen, in Example 14.6. The solution for X is

$$X(x) = A\cos(a^{1/2}x + \gamma)\,, \tag{14.35}$$

where A is an arbitrary prefactor. Thus we find a particular solution for the solute concentration

$$c(x,t) = A\cos\left(\sqrt{\lambda/D}\,x + \gamma\right)e^{-\lambda t} + c_\infty + b_\infty x\,. \tag{14.36}$$

The prefactor T_0 in $T(t)$ has been omitted because it is redundant (it would simply multiply the prefactor A), and a linear polynomial with arbitrary constants c_∞ and b_∞ has been added because it will be destroyed by the derivatives. $c_\infty + b_\infty x$ is the equilibrium concentration, at infinite time. We have constants of integration b_∞, and c_∞ and, for each choice of the separation constant λ, we have additional constants of integration A, γ. However, λ is an arbitrary positive number. The general solution is a superposition over an arbitrary set of λ_j values,

$$c(x,t) = c_\infty + b_\infty x + \sum_j A_j \cos\left(\sqrt{\lambda_j/D}\,x + \gamma_j\right)e^{-\lambda_j t}. \tag{14.37}$$

To use this solution with realistic boundary conditions requires additional analysis. We will discuss this further in Example 17.4 as an illustration of Fourier series.

Separation of variables is a very powerful technique when it works, but usually it does not work. As with ODE's, there are various tricks that can give analytical solutions of PDE's in special cases, but usually one must resort to numerical methods or approximation methods.

Exercises

14.1 Prove that Eq. (14.3) is linear, by showing that it satisfies Eqs. (14.4).

14.2 Is $yy' = g(x)$ a linear equation?

14.3 Verify that Eqs. (14.10) and (14.11) are solutions to Eq. (14.9).

14.4 Which of the following are eigenvalue equations?

(a) $y'' + 7y - x = 0$ (b) $(\frac{d}{dx} - 7)y = 0$ (c) $(\frac{d}{dx} - 3x - 7)y = 0$

(d) $\frac{dy}{dx} = 7xy$ (e) $y^{(4)} - 3y^{(3)} + 2y'' + x^2y' - 7y = 0$

14.5 Consider an object of mass 2.3 kg undergoing harmonic oscillations driven by the potential energy function $\frac{1}{2}kx^2$ with $k = 4.5$ J/m^2. Determine the trajectory $x(t)$ subject to the boundary conditions that $x = 67$ cm at $t = 0$ and that this be the farthest the object ever gets from the origin.

14.6 Consider a reaction with stoichiometry A + 2B $\to$ C and with an empirically determined rate law $-\frac{d}{dt}[\mathrm{A}] = k[\mathrm{A}]$. Suppose that the initial reactant concentrations were $[\mathrm{A}]_0$=0.1000 mol/L and $[\mathrm{B}]_0$=0.0500 mol/L, and that after 60 s [B] had dropped to 0.0207 mol/L. Determine k.

14.7 Consider a reaction with stoichiometry A+2B $\to$ C and empirical rate law $\dot{\xi} = kc_\mathrm{A}c_\mathrm{B}$, with $k = 0.12$ L/mol·s. Obtain an exact analytic expression for the solution for [A] as a function of time t in terms of the initial concentrations $c_{\mathrm{A},0}$ and $c_{\mathrm{B},0}$. [This is doable by hand, but you might rather use computer algebra to invert the expression for $t(\xi)$.]

14.8 For each of the following state whether or not the equation is separable, and if it is, derive an analytic solution for $y(x)$ with boundary condition $y(1) = 2$.

(a) $y' + xy + x = 0$ (b) $y' + 6 = 0$ (c) $xy' + (1-x)^2y^3 = 0$ (d) $y' + (3-y)^2 = 0$

14.9 Solve the equation $\alpha y'' = xy$, where α is a constant. Give a general solution for $y(x)$ containing arbitrary constants of integration.

14.10 Consider the differential equation for the $2p$ orbitals, Eq. (14.9). (a) What is the order of the equation? (b) How many constants of integration are there in the general solution? (c) Is your answer consistent with Theorem 14.4.1? Explain.

14.11 Consider the equation $\frac{\partial^2 z}{\partial x^2} - \frac{\partial^2 z}{\partial y^2} = 4z$.

(a) Convert this into a pair of ODE's.

(b) Obtain an explicit solution for $z(x, y)$.

14.12 The following is Fick's second law in two-dimensional space:

$$\frac{\partial c}{\partial t} = D\left(\frac{\partial^2 c}{\partial x^2} + \frac{\partial^2 c}{\partial y^2}\right)$$

Transform this into a set of three ODE's, with two arbitrary separation constants.

Chapter 15

Solving Differential Equations, II

15.1 Numerical Solution

15.1.1 Basic Algorithms

Consider a first-order ODE

$$dy = g(x, y)dx$$

with initial condition $y(x_0) = y_0$. Replace the infinitesimals dy and dx with finite steps Δy and Δx and number the steps with an index j,

$$\Delta x_{j+1} = x_{j+1} - x_j \,, \qquad \Delta y_{j+1} = y_{j+1} - y_j \,,$$

where y_j is an abbreviated notation for $y(x_j)$. Then the differential equation can be approximated with a ***finite difference equation***

$$\Delta y_j = g(x_j, y_j)\Delta x_j \,. \tag{15.1}$$

To simplify the analysis, let us assume for now that the step sizes Δx_j are all the same, and equal to some specified value s. Then the difference equation can be written

$$y_{j+1} = y_j + sg(x_j, y_j). \tag{15.2}$$

It becomes equivalent to the differential equation in the limit $s \to 0$.

This is the theoretical justification of a simple (but rather inefficient) computational algorithm known as ***Euler's method***:

1. Specify the initial condition $y_0 = y(x_0)$.
2. Extrapolate to y_1 using $y_1 \approx y_0 + sg(x_0, y_0)$.
3. Extrapolate from y_1 to y_2 using $y_2 \approx y_1 + sg(x_1, y_1)$.

Continue, according to Eq. (15.2), until the desired final value of x is reached.

Example 15.1. *Euler's method.* Consider the equation

$$y' = \tfrac{1}{11}(x-5)(1155 + 97x - 1093x^2 + 499x^3 - 78x^4 + 4x^5)e^{-(2x-9)^2/22} = 0\,,$$

which can be solved analytically. The approximate solutions from Euler's method using two different step sizes are shown in Fig. 15.1 along with the exact solution.

$g(x, y) = dy/dx$ is the derivative of $y(x)$. Between x_j and x_{j+1}, Euler's method approximates $y(x)$ as a straight line with slope $g(x_j, y_j)$. If $g(x, y)$ changes significantly within the step interval, then the linear extrapolation

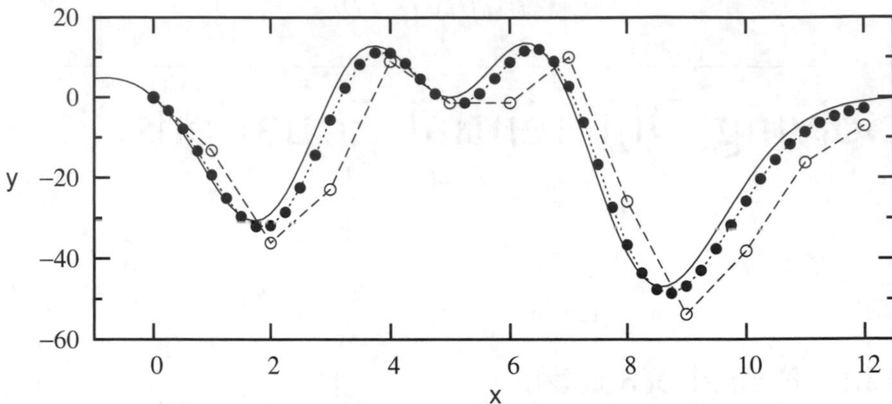

Figure 15.1: Integrating a differential equation using Euler's method. The exact solution (solid curve) is compared with the numerical approximation from a step size of 1.0 and a step size of 0.25.

can send the approximate solution off some distance from the exact solution. We can prevent this by using a small step size, as illustrated in Fig. 15.1, but that requires more evaluations of the function, which slows the computation.

A strategy for improving the efficiency is illustrated in Fig. 15.2, which shows the first Euler step from Fig. 15.1 for step size $s = 1$ (the line from the solid circle to the open circle). Euler's method uses the slope at x_0 to extrapolate from x_0 to x_1. Suppose g depends only on x. If we use instead the slope at $x_0 + \frac{1}{2}s$, halfway between x_0 and x_1, with

$$y(x_j + s) \approx y(x_j) + sg\left(x_j + \tfrac{1}{2}s\right) , \tag{15.3}$$

then the accuracy at x_1 is much higher. This is intuitively reasonable, and we can justify it theoretically by noting that the Taylor series

$$y(x_j + s) \approx y(x_j) + sg(x_j + \tfrac{1}{2}s) \sim y(x_j) + s[g(x_j) + g'(x_j)\tfrac{1}{2}s + \mathcal{O}(s^2)]$$

agrees through second order with the exact Taylor series,

$$y(x_j + s) \sim y(x_j) + y'(x_j)s + \tfrac{1}{2}y''(x_j)s^2 + \mathcal{O}(s^3),$$

because $y' = g$, and $s\mathcal{O}(s^2) = \mathcal{O}(s^3)$. If g is a function of both x and y, then the following expression gives agreement through second order:

$$y_{j+1} = y_j + sg\left(x_j + \tfrac{1}{2}s,\, y_j + \tfrac{1}{2}\kappa_j\right), \qquad \kappa_j = sg(x_j, y_j). \tag{15.4}$$

This is the ***second-order Runge-Kutta method.***[1] Eq. (15.4) requires evaluation of g at two different points for each step, but this is usually more than compensated for by the higher accuracy, which allows a larger step size.

[1]It was developed by German mathematicians Carl Runge (1856-1927) and Martin Kutta (1867-1944).

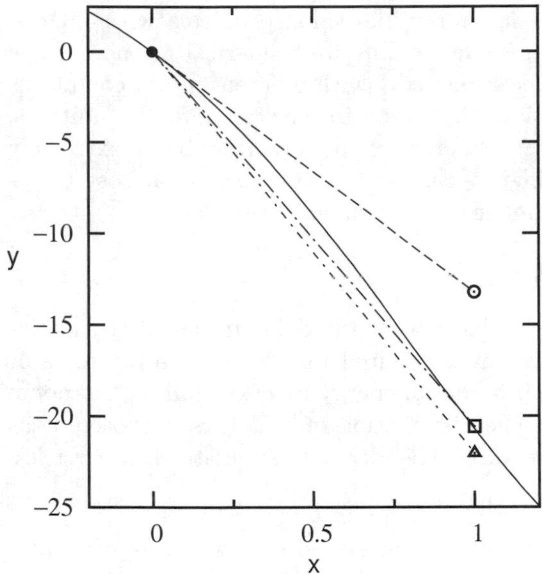

Figure 15.2: Comparison of a single step using Euler's method (open circle), the second-order Runge-Kutta method (triangle), and fourth-order Runge-Kutta (square).

This analysis can be generalized by forcing agreement with the exact Taylor series to higher order. The expression

$$y_{j+1} \approx y_j + s\left[\tfrac{1}{6}g(x_j) + \tfrac{2}{3}g(x_j + \tfrac{1}{2}s) + \tfrac{1}{6}g(x_j + s)\right] \tag{15.5}$$

gives agreement through fourth order. The slope here is a weighted average of the slopes at the points x_j, $x_j + \frac{1}{2}s$, and $x_j + s$. In general, for g a function of both x and y, the ***fourth-order Runge-Kutta method*** is given by

$$y_{j+1} = y_j + \tfrac{1}{6}\kappa_{j,1} + \tfrac{1}{3}\kappa_{j,2} + \tfrac{1}{3}\kappa_{j,3} + \tfrac{1}{6}\kappa_{j,4}\,, \tag{15.6}$$

$$\kappa_{j,1} = sg(x_j, y_j), \qquad \kappa_{j,2} = sg\big(x_j + \tfrac{1}{2}s,\; y_j + \tfrac{1}{2}\kappa_{j,1}\big),$$

$$\kappa_{j,3} = sg\big(x_j + \tfrac{1}{2}s,\; y_j + \tfrac{1}{2}\kappa_{j,2}\big), \qquad \kappa_{j,4} = sg(x_j + s,\; y_j + \kappa_{j,3}).$$

If g depends on x and y, then the fourth-order method is the one most often used as it tends to give a good balance of accuracy against computation speed. Runge-Kutta methods are relatively dependable and reasonably efficient. In commercial software they are usually implemented with adaptive step size, so that a smaller s is automatically used in regions where the solution is changing rapidly. There are other methods that can be more computationally efficient, but they are more prone to failure if the solution is complicated.

Computer algebra systems typically have some built-in capability to automatically choose an algorithm that is suited for your equation. Nevertheless, it is sometimes worthwhile to write your own algorithm using a method appropriate for your particular application. A method that is commonly used for molecular dynamics computations is described in the following section.

The methods described here are for initial-value problems with ODE's. Boundary-value problems are more demanding for numerical methods. The integration needs to be performed repeatedly with different initial conditions in order to find the initial conditions that result in the desired final conditions.

Numerical integration of *partial* differential equations can be an extremely demanding computational problem. Expect the computational cost to increase dramatically with the number of independent variables.

15.1.2 The Leapfrog Method*

The ***leapfrog method***[2] is particularly well suited for molecular dynamics simulations, in which trajectories are computed for a collection of atoms or molecules subject to a theoretical potential energy function and with random initial positions and velocities. The distribution of velocities is chosen so as to be consistent with the temperature. The idea is to simulate each particle's trajectory with

$$x(t+s) \approx x(t) + v\big(t + \tfrac{1}{2}s\big)\, s, \tag{15.7}$$

as in second-order Runge-Kutta, but to determine the velocities v using a separate simultaneous numerical integration of $v(t)$.

This is done in such a way that third-order terms in the Taylor series of x and v are zero. We begin with the third-order Taylor series

$$x(t+s) \sim x(t) + \dot{x}(t)s + \tfrac{1}{2}a(t)s^2 + \tfrac{1}{6}\,\dddot{x}(t)s^3 + \mathcal{O}(s^4),$$

$$v(t+s) \sim v(t) + a(t)s + \tfrac{1}{2}\ddot{v}(t)s^2 + \tfrac{1}{6}\,\dddot{v}(t)s^3 + \mathcal{O}(s^4),$$

where $a(t) = \ddot{x}(t) = -m^{-1}\partial V/\partial x$ is the acceleration, calculated from the potential energy function. We can eliminate the odd-order terms by adding the Taylor series for $x(t-s)$ to $x(t+s)$ and for $v(t-s)$ to $v(t+s)$, with the results

$$x(t+s) \sim 2x(t) - x(t-s) + a(t)s^2 + \mathcal{O}(s^4), \tag{15.8}$$

$$v(t+s) \sim 2v(t) - v(t-s) + \ddot{v}(t)s^2 + \mathcal{O}(s^4). \tag{15.9}$$

Let us continue in similar fashion to express $\ddot{v}$ in terms of accelerations, this time subtracting to eliminate even-order terms and using a step size of $s/2$:

$$a(t+s/2) = \dot{v}(t+s/2) \sim \dot{v}(t) + \ddot{v}(t)(s/2) + \tfrac{1}{2}\,\dddot{v}(t)(s/2)^2 + \mathcal{O}(s^3),$$

$$a(t+s/2) - a(t-s/2) \sim \ddot{v}(t)\, s + \mathcal{O}(s^3). \tag{15.10}$$

Our strategy, according to Eq. (15.7), is to express the trajectory in terms of $v(t+s/2)$. Eq. (15.9) is valid for any t. Let us replace t with $t - s/2$,

$$v(t+s/2) \sim 2v(t-s/2) - v(t-3s/2) + [a(t) - a(t-s)]s + \mathcal{O}(s^4), \tag{15.11}$$

using Eq. (15.10) to replace $\ddot{v}$. Note that the error remains $\mathcal{O}(s^4)$.

[2] One of several popular algorithms that are refinements of a method developed by Norwegian mathematician and physicist Carl Størmer (1874-1957). For a detailed discussion of the advantages of the leapfrog, see A. K. Mazur, *J. Comp. Phys.* **136**, 354-365 (1997).

Finally, we write our algorithm in terms of discrete values, using the notation

$$x_j = x(js), \qquad v_j = v(js), \qquad a_j = a(js)\,.$$

With $t = js$ in Eqs. (15.8) and (15.11), we obtain the coupled equations

$$x_{j+1} = 2x_j - x_{j-1} + a_j s^2, \qquad v_{j+\frac{1}{2}} = 2v_{j-\frac{1}{2}} - v_{j-\frac{3}{2}} + (a_j - a_{j-1})s. \quad (15.12)$$

These can be expressed in a much simpler form:

$$x_{j+1} = x_j + v_{j+\frac{1}{2}}s\,, \qquad v_{j+\frac{1}{2}} = v_{j-\frac{1}{2}} + a_j s, \quad (15.13)$$

which is equivalent to Eqs. (15.12). This is verified as follows:

$$v_{j+\frac{1}{2}} = v_{j-\frac{1}{2}} + a_j s = 2v_{j-\frac{1}{2}} - v_{j-\frac{1}{2}} + a_j s = 2v_{j-\frac{1}{2}} - \left(v_{j-1-\frac{1}{2}} + a_{j-1}s\right) + a_j s$$

and

$$\begin{aligned} x_{j+1} &= 2x_j - x_j + v_{j+\frac{1}{2}}s = 2x_j - \left(x_{j-1} + v_{j-1+\frac{1}{2}}s\right) + v_{j+\frac{1}{2}}s \\ &= 2x_j - x_{j-1} - v_{j-\frac{1}{2}}s + \left(v_{j-\frac{1}{2}} + a_j s\right)s. \end{aligned}$$

Thus, we have proved that the algorithm given by Eqs. (15.13) is equivalent to Eqs. (15.8) and (15.11), which have error of only $\mathcal{O}(s^4)$. Eqs. (15.13) are the leapfrog algorithm. $v(t)$ and $x(t)$ are propagated in alternating steps

$$x_0 \rightarrow x_1, \quad v_{0+\frac{1}{2}} \rightarrow v_{1+\frac{1}{2}}, \quad x_1 \rightarrow x_2, \quad v_{1+\frac{1}{2}} \rightarrow v_{2+\frac{1}{2}}, \quad \ldots,$$

with the indices playing a game of leapfrog. To obtain a three-dimensional trajectory, the y and z coordinates are propagated similarly.

The only catch here is that if the initial condition for x is specified at $t = 0$, then the initial condition for v must be specified at $t = s/2$. For many problems, such as, for example, computing the trajectory of a canon ball, this might be awkward—the initial condition would specify $v(0)$, not $v(s/2)$. However, in a constant-temperature molecular dynamics simulation the velocities are random with the same probability distribution at $t = s/2$ as at $t = 0$, so this is not a problem.

15.1.3 Systems of Differential Equations

Numerical integration can also be applied to *systems* of differential equations. Consider the simultaneous equations

$$\frac{du}{dx} = g(x, u, w), \qquad \frac{dw}{dx} = h(x, u, w)\,. \quad (15.14)$$

x is the independent variable while u and w are both dependent variables. We make alternating steps, extrapolating from u_j to u_{j+1} and from w_j to w_{j+1}. For example, Euler's method could be implemented as follows:

$$\begin{aligned} u_1 &= u_0 + s\,g(x_0, u_0, w_0)\,, \quad w_1 = w_0 + s\,h(x_0, u_0, w_0)\,, \\ u_2 &= u_1 + s\,g(x_1, u_1, w_1)\,, \quad w_2 = w_1 + s\,h(x_1, u_1, w_1)\,, \\ &\vdots \end{aligned} \quad (15.15)$$

A higher-order ODE can be integrated by transforming it into a set of first-order ODE's. For example, given an equation

$$\frac{d^2y}{dx^2} = g(y, y', x),$$

the usual practice is to define a new dependent variable $u = y'$, to get a pair of first-order equations

$$u' = g(y, u, x), \quad y' = u. \tag{15.16}$$

A difficulty sometimes encountered with a system of ODE's is that the different equations might merit different step sizes. This commonly occurs, for example, when integrating a set of equations describing rate laws for coupled chemical reactions. Reaction rate constants can vary by many orders of magnitude. A very fast reaction, with a large rate constant, can be expected to need a much smaller step size than a slow reaction. A set of differential equations in which a single step size is not appropriate for all the equations is said to be ***stiff***. Sophisticated numerical methods have been developed for dealing with stiff sets of equations. They add to the computational cost but are much more efficient than the alternative of using in all the equations the step size of the equation that needs the smallest steps.

15.2 Chemical Reaction Mechanisms

In Sections 13.1 and 14.3 the topic of chemical reaction rates was considered from a macroscopic perspective. The rate of reaction was expressed in terms of an empirical rate law for the overall process, in terms of molar concentrations of initial reactants and final products, with no attempt to describe the details of the atomic interactions and rearrangements. Here we take a microscopic perspective. We will begin with a set of ***elementary reactions***, that is, reaction equations that purport to describe an actual atomic process, and then derive from them an approximate theoretical rate law.

Consider an elementary bimolecular collision,

$$\mathrm{A} + \mathrm{B} \to \mathrm{X} + \mathrm{Y},$$

in which two molecules, A and B, collide, immediately undergo a rearrangement of atoms, and then separate as two new molecules, X and Y. The rate at which products will be generated will be proportional to the number of collisions between molecules of type A and molecules of type B per unit time. This in turn will be proportional to the molar concentration [A] and the molar concentration [B]; in other words,

$$\dot{\xi} = \frac{d[\mathrm{Y}]}{dt} = k[\mathrm{A}][\mathrm{B}], \tag{15.17}$$

where k is a proportionality constant. By similar reasoning, we conclude for an elementary reaction $\mathrm{A} + \mathrm{A} \to \mathrm{X} + \mathrm{Y}$ that $\dot{\xi} = k[\mathrm{A}]^2$.

If a molecule spontaneously decomposes in a unimolecular process,

$$\mathrm{A} \to \mathrm{X} + \mathrm{Y}\,,$$

the reaction rate is proportional to just the single reactant concentration,

$$\dot{\xi} = k[\mathrm{A}]\,. \tag{15.18}$$

Another possibility is that the rate is completely independent of any reactant concentrations. This could occur, for example, in a surface-catalyzed reaction in which the catalytic sites are saturated,

$$\mathrm{A}(\text{adsorbed on surface}) \to \mathrm{X} + \mathrm{Y}\,.$$

This is not affected by the concentration of the free A molecules. It does depend on the concentration (amount per unit area) of A on the surface, but for a saturated surface that concentration is constant. The process is described by a zeroth-order expression, $\dot{\xi} = k$.

If a single elementary reaction is all that is occurring, then the analysis is the same as in Section 14.3; we integrate the rate law analytically. However, an overall chemical reaction very often consists of two or more elementary reactions taking place simultaneously. The description of the overall process in terms of elementary reactions is called a ***reaction mechanism***.

Example 15.2. *Coupled differential equations for a reaction mechanism.* Consider the following mechanism:

$$\mathrm{A} \underset{k_{-1}}{\overset{k_1}{\rightleftharpoons}} \mathrm{C} + \mathrm{X}$$

$$\mathrm{C} + \mathrm{C} \xrightarrow{k_2} \mathrm{Y}$$

This can be described in terms of a set of coupled first-order ODE's, one equation for each chemical species:

$$\frac{d[\mathrm{A}]}{dt} = -k_1[\mathrm{A}] + k_{-1}[\mathrm{C}][\mathrm{X}], \qquad \frac{d[\mathrm{C}]}{dt} = k_1[\mathrm{A}] - k_{-1}[\mathrm{C}][\mathrm{X}] - 2k_2[\mathrm{C}]^2,$$

$$\frac{d[\mathrm{X}]}{dt} = k_1[\mathrm{A}] - k_{-1}[\mathrm{C}][\mathrm{X}], \qquad \frac{d[\mathrm{Y}]}{dt} = k_2[\mathrm{C}]^2. \tag{15.19}$$

Note that a minus sign is needed if the elementary rate law destroys the species in question. Thus, the "k_1" reaction contributes the term $-k_1[\mathrm{A}]$ to $d[\mathrm{A}]/dt$ but $+k_1[\mathrm{A}]$ to $d[\mathrm{C}]/dt$ and $d[\mathrm{X}]/dt$. In the $d[\mathrm{C}]/dt$ equation, k_2 is multiplied by -2 because the rate law for the second step is $-\frac{1}{2}d[\mathrm{C}]/dt = k_2[\mathrm{C}]^2$, with the stoichiometric factor $-1/2$.

It is reasonably straightforward to integrate these equations numerically (although if the values of the rate constants are not all comparable in magnitude, then a numerical algorithm that can handle stiff equations should be used). Numerical integration of such coupled equations is routine for chemical processes involving many dozens of elementary reactions.

Numerical integration yields a numerical value for each species' concentration at any given time. For many kinds of applications this is sufficient. However, it is often useful to have a closed-form analytical expression for an overall rate law, such as the empirical rate laws we considered in Section 13.1. Considering that the elementary rate constants are difficult to calculate from fundamental theory and are usually treated as empirical parameters, the approximations needed to derive an approximate analytical expression might not significantly increase the error beyond that already present in the values of the rate constants. More importantly, having a single differential equation for the extent of reaction variable, even if only approximate, can give qualitative insight that might not be obvious from the set of coupled equations.

The most commonly used approximation strategy for analyzing reaction mechanisms is the ***steady-state approximation***. It is based on the assumption that the rates of destruction of intermediate species will approximately equal their rates of creation.

Example 15.3. *Steady-state analysis of a reaction mechanism.* In the mechanism given in Example 15.2, C is an intermediate. Let us assume that at the start of the process substance A is all that is present. C will be created by the "k_1" reaction but it will eventually be completely destroyed by the "k_2" reaction. Given enough time, the overall stoichiometry of the process will be $2A \to 2X + Y$. Very early in the process, the rate of creation of C will be greater than the rate of destruction. (Initially there is no C present, which means the rate of destruction, $k_{-1}[C][X] + 2k_2[C]^2$ will start at zero.) Very late in the process the destruction rate will be larger than the creation rate, as A dwindles. However, in between usually the rates of creation and destruction for an intermediate are about in balance. This means [C] is approximately constant,

$$\frac{d[C]}{dt} = k_1[A] - k_{-1}[C][X] - 2k_2[C]^2 \approx 0\,, \tag{15.20}$$

which implies

$$[C] \approx \frac{1}{4k_2}\left(-k_{-1}[X] + \sqrt{k_{-1}^2[X]^2 + 8k_2k_1[A]}\right). \tag{15.21}$$

The quadratic formula in principle puts "±" in front of the square root. However, the rate constants and concentrations are by definition positive quantities. The choice of the negative sign would give a negative concentration for [C], which is impossible.

The overall reaction rate is

$$\dot{\xi} = \frac{d[Y]}{dt} = k_2[C]^2. \tag{15.22}$$

This is not, strictly speaking, a stoichiometric rate law, because C, being an intermediate, does not appear in the stoichiometric reaction equation $2A \to 2X + Y$. Substituting the steady-state approximation, Eq. (15.21), for [C] gives an approximate rate law in terms of the concentrations of A and X.

To understand the implications of this result we will simplify it with Taylor series. Let us write Eq. (15.21) as

$$[C] \approx \frac{k_{-1}[X]}{4k_2}\left[\left(1 + \frac{8k_2k_1}{k_{-1}^2}\,\frac{[A]}{[X]^2}\right)^{1/2} - 1\right] \tag{15.23}$$

and let $\alpha = (8k_2k_1/k_{-1}^2)[A]/[X]^2$ be the expansion variable. If $\alpha < 1$, then the term in brackets has the convergent expansion

$$(1+\alpha)^{1/2} - 1 \sim \left(1 + \tfrac{1}{2}\alpha - \tfrac{1}{8}\alpha^2 + \cdots\right) - 1 = \tfrac{1}{2}\alpha + \mathcal{O}(\alpha^2)$$

and

$$[\mathrm{C}]^2 \approx \frac{k_{-1}^2[\mathrm{X}]^2}{16k_2^2}\left[\tfrac{1}{2}\alpha + \mathcal{O}(\alpha^2)\right]^2 \sim \frac{k_{-1}^2[\mathrm{X}]^2}{16k_2^2}\left[\tfrac{1}{4}\alpha^2 + \mathcal{O}(\alpha^3)\right] = \frac{k_{-1}^2[\mathrm{X}]^2}{16k_2^2}\,\tfrac{1}{4}\alpha^2[1+\mathcal{O}(\alpha)].$$

It follows that

$$\dot{\xi} = k_2[\mathrm{C}]^2 \approx \frac{k_1^2 k_2}{k_{-1}^2}\left(\frac{[\mathrm{A}]}{[\mathrm{X}]}\right)^2 [1 + \mathcal{O}(\alpha)]\,. \tag{15.24}$$

Alternatively, we could write Eq. (15.21) as

$$[\mathrm{C}] \approx \frac{k_1^{1/2}[\mathrm{A}]^{1/2}}{(2k_2)^{1/2}}\left[\left(1 + \frac{k_{-1}^2[\mathrm{X}]^2}{8k_2k_1[\mathrm{A}]}\right)^{1/2} - \frac{k_{-1}[\mathrm{X}]}{(8k_2k_1[\mathrm{A}])^{1/2}}\right] \tag{15.25}$$

and define an expansion variable $\beta = k_{-1}[\mathrm{X}]/(8k_2k_1[\mathrm{A}])^{1/2}$. Then if $\beta < 1$ we have the expansion

$$(1+\beta^2)^{1/2} - \beta \sim 1 - \beta + \tfrac{1}{2}\beta^2 + \cdots \sim 1 - \mathcal{O}(\beta)\,,$$

which implies

$$[\mathrm{C}]^2 \sim \frac{k_1[\mathrm{A}]}{2k_2}[1 - \mathcal{O}(\beta)]^2 \sim \frac{k_1[\mathrm{A}]}{2k_2}[1 - \mathcal{O}(\beta)]$$

and

$$\dot{\xi} = k_2[\mathrm{C}]^2 \approx \tfrac{1}{2}k_1[\mathrm{A}][1 - \mathcal{O}(\beta)]\,. \tag{15.26}$$

Suppose we wish to experimentally determine the values of the constants k_1, k_{-1}, and k_2. These limiting steady-state results can be helpful. Prepare a series of mixtures with different values of [A] but all with $[\mathrm{X}] \gg [\mathrm{A}]$, so that $\beta \ll 1$. A statistical fit of $\dot{\xi}$ as a linear function of [A] gives a value for k_1. Next, prepare mixtures with $[\mathrm{A}] \gg [\mathrm{X}]$, so that $\alpha \ll 1$. A fit of $\dot{\xi}$ vs. $([\mathrm{A}]/[\mathrm{X}])^2$ will give us a value for the slope $\kappa = k_1^2k_2/k_{-1}^2$. Substitute $k_2 = \kappa k_{-1}^2/k_1^2$ into the coupled set of differential equations, Eqs. (15.19) and replace k_1 and κ with the numerical values just determined. We now have only a single undetermined parameter, k_{-1}. Prepare a mixture with [A] and [X] of the same magnitude and measure $\dot{\xi}$. Then guess a value for k_{-1}, numerically integrate Eqs. (15.19) to obtain a theoretical value for $\dot{\xi}$, and compare with the measured $\dot{\xi}$. Repeat the numerical integration with a different guess for k_{-1}. Do this repeatedly until the deviation between theory and experiment is minimized.

15.3 Approximation Methods

A common strategy for dealing with difficult differential equations is to introduce an approximation to simplify the analysis. A large variety of approximation methods have been developed. The steady-state approximation, described in the previous section, is one example. Methods based on linear algebra will be discussed in subsequent chapters. Here we discuss two systematic procedures that yield solutions in the form of power series.

15.3.1 Taylor Series*

It is generally possible to derive a Taylor series for the solution directly from the differential equation. Considering that an exact analytical expression for the solution very often is unobtainable, and an accurate numerical solution while obtainable in principle can often be impractical due to extremely high computational cost, the fact that Taylor series will give a solution even for very difficult (for example, nonlinear and partial) differential equations makes this method a valuable tool. A disadvantage is that the solution is approximate, most accurate in the vicinity of a given point and progressively less accurate away from that point. Furthermore, the series will often be divergent and the location of the singularity responsible for the divergence can be harder to indentify than in the case of a Taylor series of a function calculated directly from an analytical expression.

Consider a first-order ODE,

$$y'(x) = f(x, y), \tag{15.27}$$

perhaps nonlinear, with initial-value boundary condition $y(x_0) = y_0$. Let us expand the solution about the point x_0. We seek a solution in the form

$$y(x) \sim y(x_0) + y'(x_0)(x - x_0) + \tfrac{1}{2}y''(x_0)(x - x_0)^2 + \cdots + \tfrac{1}{n!}y^{(n)}(x_0)(x - x_0)^n.$$

The zeroth-order term is given by the boundary condition and the first-order coefficient is given directly by the differential equation,

$$y'(x_0) = f\big(x_0, y(x_0)\big) = f(x_0, y_0)\,.$$

For the second-order coefficient, we need the second derivative of the solution, $y''(x) = dy'/dx$. Let us express this in terms of differentials. Taking the differential of both sides of Eq. (15.27), we obtain

$$dy' = df = \frac{\partial f}{\partial x}dx + \frac{\partial f}{\partial y}dy\,, \tag{15.28}$$

which implies that

$$y''(x) = \frac{dy'}{dx} = \frac{\partial f}{\partial x} + \frac{\partial f}{\partial y}\frac{dy}{dx} = \frac{\partial f}{\partial x} + \frac{\partial f}{\partial y}f\,. \tag{15.29}$$

Evaluating this at the point $(x, y) = (x_0, y_0)$ gives

$$y''(x_0) = \left.\frac{\partial f}{\partial x}\right|_{(x_0,y_0)} + \left.\frac{\partial f}{\partial y}\right|_{(x_0,y_0)} f(x_0, y_0)\,. \tag{15.30}$$

For third order, we write

$$dy'' = \frac{\partial y''}{\partial x}dx + \frac{\partial y''}{\partial y}dy\,, \tag{15.31}$$

and, from Eq. (15.29),

$$\frac{\partial y''}{\partial x} = \frac{\partial^2 f}{\partial x^2} + \frac{\partial^2 f}{\partial x \partial y}f + \frac{\partial f}{\partial y}\frac{\partial f}{\partial x}\,, \qquad \frac{\partial y''}{\partial y} = \frac{\partial^2 f}{\partial x \partial y} + \frac{\partial^2 f}{\partial y^2}f + \frac{\partial f}{\partial y}\frac{\partial f}{\partial y}\,. \tag{15.32}$$

Substituting these into Eq. (15.31), dividing by dx, and evaluating the result at (x_0, y_0) gives the value of $y'''(x_0)$, and so on for higher orders.

Example 15.4. *Taylor-series integration of rate laws.* Let us again consider the rate law from Example 13.1, which was integrated analytically in Example 14.5. The equation for $\xi(t)$ is

$$\frac{d\xi}{dt} = k(c_{\mathrm{A},0} - 2\xi)(c_{\mathrm{B},0} - 3\xi)^2.$$

Evaluating the right-hand side at the initial condition $(t, \xi) = (0, 0)$ gives

$$\dot{\xi}(0) = kc_{\mathrm{A},0}c_{\mathrm{B},0}^2.$$

Using Eq. (15.30) with x replaced by t and y replaced by ξ, we find that the second derivative at the initial condition is

$$\begin{aligned}
\ddot{\xi}(0) &= 0 + k\frac{\partial}{\partial\xi}(c_{\mathrm{A},0} - 2\xi)(c_{\mathrm{B},0} - 3\xi)^2\Big|_{\xi=0} kc_{\mathrm{A},0}c_{\mathrm{B},0}^2 \\
&= [-2k(c_{\mathrm{B},0} - 3\xi)^2 - 6k(c_{\mathrm{A},0} - 2\xi)(c_{\mathrm{B},0} - 3\xi)]\big|_{\xi=0} kc_{\mathrm{A},0}c_{\mathrm{B},0}^2 \\
&= -2k^2 c_{\mathrm{A},0}c_{\mathrm{B},0}^3(c_{\mathrm{B},0} + 3c_{\mathrm{A},0}).
\end{aligned}$$

Therefore,

$$\xi(t) \sim kc_{\mathrm{A},0}c_{\mathrm{B},0}^2 t - k^2 c_{\mathrm{A},0}c_{\mathrm{B},0}^3(c_{\mathrm{B},0} + 3c_{\mathrm{A},0})t^2 + \mathcal{O}(t^3).$$

This agrees with the result from expanding the exact solution in Example 14.5 but was derived with much less work.

This method is applicable to coupled equations. Consider

$$u' = g(x, u, v), \qquad v' = h(x, u, v),$$

with initial conditions $u(x_0) = u_0$, $v(x_0) = v_0$. We derive a series for each function,

$$\begin{aligned}
u(x) &\sim u(x_0) + u'(x_0)(x - x_0) + \tfrac{1}{2}u''(x_0)(x - x_0)^2 + \cdots, \\
v(x) &\sim v(x_0) + v'(x_0)(x - x_0) + \tfrac{1}{2}v''(x_0)(x - x_0)^2 + \cdots,
\end{aligned}$$

with $u'(x_0) = g(x_0, u_0, v_0)$ and $v'(x_0) = h(x_0, u_0, v_0)$. The differential for u' is

$$du' = dg = \frac{\partial g}{\partial x}dx + \frac{\partial g}{\partial u}du + \frac{\partial g}{\partial v}dv.$$

Dividing by dx gives

$$u'' = \frac{\partial g}{\partial x} + \frac{\partial g}{\partial u}g + \frac{\partial g}{\partial v}h.$$

Similarly, for v,

$$v'' = \frac{\partial h}{\partial x} + \frac{\partial h}{\partial u}g + \frac{\partial h}{\partial v}h.$$

These are evaluated at x_0 and substituted into the Taylor series.

Higher-order equations can be solved by converting them to a system of first-order equations. Taylor series can also be systematically derived for PDE's and for boundary-value problems, although the analysis is more complicated. For linear equations the Taylor series can usually be derived more simply by substituting the series into the equation with unknown coefficients and then collecting terms according to the power of $(x - x_0)$, which yields a recursion relation for the coefficients.

15.3.2 Perturbation Theory*

Suppose that your differential equation could be solved analytically if it were not for the presence of one particular term, and furthermore, suppose that you have reason to believe that the offending term has only a small effect on the solution. Let us call the extra term a "perturbation." The "unperturbed" solution, obtained by solving the equation with the perturbation omitted, is a reasonable initial approximation. A ***perturbation theory*** is a procedure that allows one to recursively add correction terms to the unperturbed solution in order to systematically obtain higher accuracy.

Consider a harmonic oscillator subject to friction, with the boundary condition that at time $t = 0$ the particle is at $x = A$ and is not yet moving. The force is $F = -kx + \gamma\dot{x}$, where k is the harmonic-oscillator force constant and γ is a constant that indicates the strength of the friction. Newton's equation is $m\ddot{x} = F$, which can be written

$$\ddot{x} + \omega^2 x - \epsilon\dot{x} = 0\,, \qquad \omega = (k/m)^{1/2}, \quad \epsilon = \gamma/m\,. \tag{15.33}$$

Our strategy is to assume that ϵ is a small, such that we can expand $x(t)$ in a power series in ϵ,

$$x(t) \sim \sum_{j=0}^{n} \epsilon^j u_j(t)\,. \tag{15.34}$$

This differs from a Taylor series in that the expansion parameter, ϵ, is independent of the variable, t. The variable appears only in the series coefficients, $u_j(t)$. Eq. (15.34) is called a ***perturbation series*** and n is called the *order* of the pertubation theory. The series is substituted into the differential equation and then terms are collected according to powers of ϵ. This usually (but not always) yields a recursive solution for each u_j in terms of the u_i with $i < j$.

Example 15.5. *Perturbation theory of harmonic oscillator with friction.* Let us develop a first-order perturbation theory for Eq. (15.33). If $\epsilon = 0$, then we have the unperturbed solution $u_0(t) = c\,\cos(\omega t + \gamma)$. Let us impose the boundary conditions on u_0, with $\dot{u}(0) = 0$ and $u_0(0) = A$. These imply $\gamma = 0$, $c = A$, and $u_0(t) = A\cos(\omega t)$. This is the oscillator with friction turned off. Now substitute the perturbation series into the differential equation:

$$\left(\frac{d^2}{dt^2} + \omega^2\right)(u_0 + \epsilon u_1) - \epsilon\frac{d}{dt}u_0 \sim \mathcal{O}(\epsilon^2)\,.$$

Collect terms according to power of ϵ:

$$\epsilon^0: \qquad \ddot{u}_0 + \omega^2 u_0 = 0, \tag{15.35}$$

$$\epsilon^1: \qquad \ddot{u}_1 + \omega^2 u_1 + A\omega\sin(\omega t) = 0. \tag{15.36}$$

The zeroth-order equation is just the unperturbed harmonic oscillator. The first-order equation, if we can solve it, gives u_1.

The equation for u_1 is inhomogeneous. Using the method of variation of parameters, we guess a solution in the form

$$u_1(t) = \eta(t)\cos(\omega t),$$

in terms of some as yet unknown function η. Substituting this into Eq. (15.36) gives us a differential equation for η,

$$\ddot{\eta} = 2\omega\left(\dot{\eta} - \tfrac{1}{2}A\right)\tan(\omega t)\,.$$

It is easy enough to guess a solution $\eta = At/2$, which makes both sides of the equation become zero. Thus we find a particular solution

$$u_{1,\text{part}}(t) = \tfrac{1}{2}At\cos(\omega t)\,.$$

We have to construct a general solution with two new constants of integration to make it possible for $u_0(t) + \epsilon u_1(t)$ to satisfy the two boundary conditions. To provide these constants, we add to the particular solution two complementary solutions,

$$u_1(t) = \tfrac{1}{2}At\cos(\omega t) + B\cos(\omega t) + C\sin(\omega t)\,.$$

In order to have $A = x(0) \sim u_0(0) + \epsilon u_1(0) = A + \epsilon B$, we set B to zero. The condition $\dot{x}(0) = 0$ requires $C = -A/(2\omega)$. In conclusion, the first-order perturbation series is

$$x(t) \sim A\cos(\omega t) + \tfrac{1}{2}A\left[t\cos(\omega t) - \omega^{-1}\sin(\omega t)\right]\epsilon\,. \qquad (15.37)$$

In the figure below this approximate solution for the case $\epsilon = 0.1$ (the dashed curve) is compared with the exact solution (the solid curve) and the solution with no friction (the dotted curve).

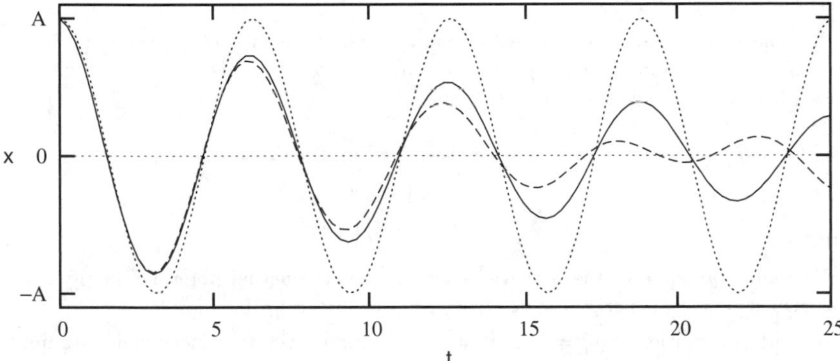

Exercises

15.1 Consider the differential equation of Example 15.1. Prepare a table showing the $y(x_j)$ values from Euler's method, from second-order Runge-Kutta, and from fourth-order Runge-Kutta for an integration from $x = 0$ to $x = 3$, using step sizes $s = 1$, $s = 1$, and $s = 3$, respectively. Judging from the exact solution shown in Fig. 15.1, which method gives the best result?

15.2 Consider a reaction with stoichiometry 3A + B + C → D + 2E and an empirically determined rate law $\dot{\xi} = kc_\text{A}^2/c_\text{E}$.

(a) Write the differential equation for the extent of reaction variable.

(b) Solve the equation analytically for $t(\xi)$.

(c) Solve the differential equation numerically using Euler's method and using the second-order Runge-Kutta method. (Choose an arbitrary value for k.) Write the computer algorithm yourself. Compare with the exact solution using various different step sizes.

15.3 Consider two particles in one dimension. They are each subject to the harmonic potential $V_{\rm a}(x) = (x-5)^2$ and, in addition, they repel each other according to the potential $V_{\rm b}(x_1, x_2) = 1/|x_2 - x_1|$. (a) Write a set of four coupled first-order differential equations for this system. (b) Use a numerical algorithm to compute the trajectories of each of the particles. Repeat the computation several times using different initial conditions and different masses. Plot your trajectories $x_1(t)$ and $x_2(t)$.

15.4* Using the programing language in your computer algebra system (or a traditional programing language, if you know one), write a program to implement the leapfrog algorithm for the motion of a single particle subject to some arbitrary potential energy function and arbitrary intial conditions for x_0 and $v_{1/2}$. Plot the trajectory as a function of time. Do this for several different potentials.

15.5* Consider the harmonic oscillator with friction coefficient $\epsilon = 0.1$, frequency $\omega = 1$, and initial conditions $x(0) = 1$, $v(0) = 0$, which was treated in Example 15.5.

(a) Compute the exact trajectory using a numerical algorithm, and plot the result.

(b) Calculate a first-order Taylor series for the trajectory, expanding about $t = 5\pi$. Plot the exact trajectory, the result from perturbation theory, and your result from the Taylor series in a single graph.

(c) Plot the kinetic energy, the potential energy, and the total energy (their sum) for the exact (numerically computed) trajectory as functions of t, all together in a single graph. Give a physical explanation of what you see.

15.6* Consider an anharmonic oscillator described by the differential equation

$$\ddot{x} + x - \epsilon x^2 = 0,$$

with $\epsilon = 0.1$.

(a) Give an expression for the potential energy. Plot the actual potential energy and the unperturbed potential energy in a single graph for $-6 \leq x \leq 16$.

(b) For initial conditions $x(0) = 0$, $\dot{x}(0) = -2.0$, compute the trajectory using a numerical algorithm and derive an approximate trajectory using perturbation theory. Plot the two results for $x(t)$ together in a single graph for $0 \leq t \leq 40$ and plot the potential energy, the kinetic energy, and the total energy together in a separate graph.

(c) Repeat for $x(0) = 0$, $\dot{x}(0) = -5.7735$. Compare and discuss your results.

Part IV

Linear Algebra

Chapter 16

Vector Spaces

This chapter presents an introduction to the field of mathematics known as ***linear algebra***. Learning linear algebra is akin to learning a foreign language—the basic concepts are familiar but the words and symbols used to describe them are not. Furthermore, words with familiar meaning from elementary mathematics, such as "product," "vector," and "space," are given broader definitions that include the elementary meanings but go well beyond them in unexpected ways. We begin with a review of the familiar—coordinate vectors and their arithmetic—and then develop a more general notion of vectors that in subsequent chapters will have a variety of applications.

16.1 Cartesian Coordinate Vectors

A familiar example of a vector is a three-dimensional vector in Cartesian coordinate space,[1]

$$\vec{x} = (x, y, z), \tag{16.1}$$

where the ***components***, x, y, and z, are real numbers. $\vec{x}$ can be thought of as an arrow from the point $(0,0,0)$ (the *origin* of the coordinate system) reaching to the point (x, y, z). The length of the arrow is

$$|\vec{x}| = \sqrt{x^2 + y^2 + z^2}\,, \tag{16.2}$$

called the ***norm*** of the vector.[2] This can be generalized to N dimensions with $\vec{u} = (a_1, a_2, \ldots, a_N)$ and

$$|\vec{u}| = \sqrt{a_1^2 + a_2^2 + \cdots + a_N^2}\,. \tag{16.3}$$

Various arithmetic operations can be performed on Cartesian coordinate vectors. Consider $\vec{u} = (a_1, a_2, a_3)$ and $\vec{v} = (b_1, b_2, b_3)$.[3] Addition of two vectors yields another vector according to

$$(a_1, a_2, a_3) + (b_1, b_2, b_3) = (a_1 + b_1,\ a_2 + b_2,\ a_3 + b_3). \tag{16.4}$$

Multiplication of a vector by a number c yields a new vector according to

$$c\,(a_1, a_2, a_3) = (ca_1,\ ca_2,\ ca_3). \tag{16.5}$$

An expression $a\vec{u} + b\vec{v}$, combining vector addition with multiplication by the arbitrary numbers a and b, is called a *linear combination* of $\vec{u}$ and $\vec{v}$.

---

[1] Note that "x" on the right-hand side represents only one of the three numbers (the "x-coordinate") while the symbol $\vec{x}$ represents the vector as a whole. The notation $\mathbf{x}$, in boldface, is often used instead of $\vec{x}$.

[2] This is a generalization of the concept of absolute value, which for a number c can be expressed as $|c| = \sqrt{c^2}$. Only the positive branch of the square root is being used here.

[3] The subscripts $_1$, $_2$, $_3$ here indicate the x, y, z coordinates. We are switching now to this notation to make the generalization to high dimension more obvious.

Multiplication of a vector by a vector is more complicated. There are two different kinds of vector products for Cartesian coordinate vectors. The ***dot product*** maps a pair of vectors into a number,

$$(a_1, a_2, a_3) \cdot (b_1, b_2, b_3) = a_1 b_1 + a_2 b_2 + a_3 b_3 \; ; \tag{16.6}$$

the ***cross product*** maps a pair of vectors into another vector, according to

$$(a_1, a_2, a_3) \times (b_1, b_2, b_3) = (a_2 b_3 - a_3 b_2,\; -a_1 b_3 + a_3 b_1,\; a_1 b_2 - a_2 b_1)\,. \tag{16.7}$$

Note that

$$\vec{u} \cdot \vec{u} = a_1^2 + a_2^2 + a_3^2 = |\vec{u}|^2. \tag{16.8}$$

More generally, one can show using trigonometry that Eq. (16.6) implies

$$\vec{u} \cdot \vec{v} = |\vec{u}||\vec{v}| \cos \alpha \,, \tag{16.9}$$

where α is the angle between the two vectors.

From Eq. (16.7) it is clear that $\vec{u} \times \vec{u} = (0, 0, 0)$. In general, if $\vec{v} = c\vec{u}$ (which means that $\vec{u}$ and $\vec{v}$ are collinear), then $\vec{u} \times \vec{v} = c\vec{u} \times \vec{u} = (0, 0, 0)$. If u and v are not collinear, then $\vec{u} \times \vec{v}$ points in a direction perpendicular to the plane containing $\vec{u}$ and $\vec{v}$. The cross product is of interest to us mainly because the angular momentum vector $\vec{L}$ can be conveniently expressed as

$$\vec{L} = \vec{x} \times \vec{p}, \tag{16.10}$$

in terms of particle position $\vec{x}$ and linear momentum $\vec{p}$. (See Section 13.2.3.)

16.2 Sets

This section introduces basic terminology that will be used in the rest of the chapter. A ***set*** is an unordered collection—of numbers, or of functions, or of anything else. To list the elements of a set, we will enclose them in braces, $\{a_1, a_2, a_3, \ldots, a_N\}$. The integer subscripts here label the elements a_j, but the order in which they are listed is of no significance; $\{a_1, a_2\}$ is the same set as $\{a_2, a_1\}$. Sometimes the abbreviated notation $\{a_j\}$ will be used, with the understanding that j will range over all appropriate values.

A set can have a finite number of elements or an infinite number of elements. For example, the set of all integers, $\{0, -1, 1, -2, 2, -3, 3, \ldots\}$, is infinite. We give special symbols to the more important infinite sets: The symbol $\mathbb{Z}$ is used for the set of all integers, $\mathbb{C}$ for the set of all complex numbers, $\mathbb{R}$ for the set of all real numbers, and $\mathbb{Q}$ for the set of all rational numbers. (A *rational number* is any number that can be exactly expressed as a ratio of integers.) Note that $\mathbb{Z}$ is a subset of $\mathbb{Q}$, while $\mathbb{Q}$ is a subset of $\mathbb{R}$, and $\mathbb{R}$ is a subset of $\mathbb{C}$. The symbol $\subset$ means "is a subset of" while $\in$ means "is an element of." Thus, $\mathbb{R} \subset \mathbb{C}$ while $3 + 2i \in \mathbb{C}$. Any of the sets $\mathbb{C}$, $\mathbb{R}$, or $\mathbb{Q}$ (but not $\mathbb{Z}$) will be referred to as ***fields***. (The technical definition of a field is somewhat broader than this, but for our purposes it is sufficient to consider only these three fields.) An element of a field is called a ***scalar***.

Parentheses will be used to indicate ***ordered sets***. For example, a point with Cartesian coordinates $x = 1.2$, $y = -3.4$, $z = 5.6$ could be written as $(1.2, -3.4, 5.6)$. This set is "ordered" in the sense that interchanging the order of the numbers would change the location of the point, in contrast to the unordered set $\{1.2, -3.4, 5.6\}$. The infinite set of ordered pairs (x, y), where x and y are in $\mathbb{R}$, is given the symbol $\mathbb{R}^2$. The set of all ordered triples (x, y, z) of real numbers is designated $\mathbb{R}^3$, and in general, the set of all ordered sets of N real numbers is designated $\mathbb{R}^N$. Similarly, ordered sets of complex numbers are designated $\mathbb{C}^N$.

16.3 Groups

Our goal is to develop a formal definition of "vector" that applies to Cartesian vectors but is more general. We begin by defining basic characteristics of $\mathbb{R}^N$. Specifically, $\mathbb{R}^N$ has the following properties: It is an "abelian group;" it is a kind of abelian group called a "vector space;" it is a kind of vector space called a "Hilbert space." In this section we discuss the concept of a *group*.

Definition. A ***group*** is a set V together with an operation called ***addition*** (+) such that:

1. For any pair of elements u and v in V, there is an element $u + v$ that is also in V. (In other words, V is ***closed*** to addition.)

2. In applying the addition operation to any three elements u, v, and w in V, the order in which they are added has no effect on the result, so that $(u + v) + w = u + (v + w)$. (In other words, addition is ***associative***.)

3. There exists an element $\mathbf{0}$ in V such that $v + \mathbf{0} = v$ for any v in V. (Existence of zero.)

4. For any v in V, there is an element t in V such that $v+t = \mathbf{0}$. (Existence of an ***additive inverse***.)

With addition defined by Eq. (16.4) and the usual rules for adding real numbers, it is easy to show that $\mathbb{R}^N$ is a group. For example, for $\mathbb{R}^3$ the zero element is $(0, 0, 0)$. The additive inverse of, for example, $(1, -7, 3)$ is $(-1, 7, -3)$.

While the definition of a group requires that an operation "+" exist, it does not say specifically what that operation does, beyond that it must satisfy the four axioms. This leaves considerable freedom to apply the concept of a group to sets in which the "+" bears little resemblance to arithmetic addition.

Example 16.1. *A molecular symmetry group.* The concept of a group is particularly useful in the analysis and classification of symmetry. It provides a systematic way to describe the symmetry of a molecule (or a crystal). In this case the elements of the group are not numbers but ***symmetry operations***. These are rotations, reflections, and various other transformations that leave the molecule unchanged.

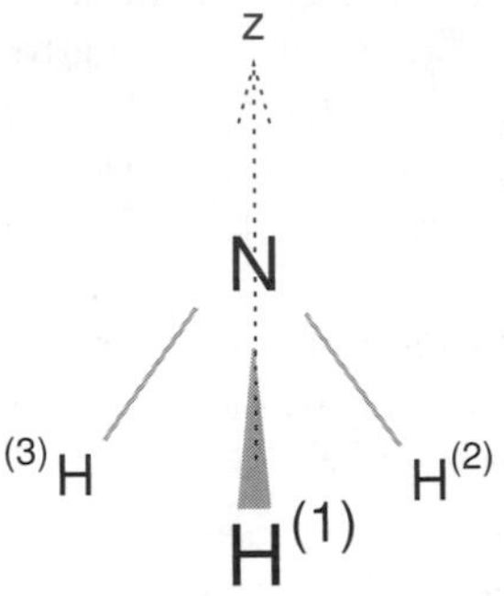

Consider an ammonia molecule. Define the z-axis of a Cartesian coordinate system as passing through the nitrogen atom and the center of the triangle of the three hydrogen atoms. If we rotate the molecule about the z-axis by $2\pi/3 = 120°$ it will seem identical to before the rotation. There will be no way to tell that the molecule has been rotated, because the H atoms are indistinguishable. (We label them here only for the purpose of describing the symmetry operations.) Let us call this operation C_3, as it is one-third of a full rotation. Similarly, a rotation by 240° leaves the molecule indistinguishable from its previous configuration. This operation is called C_3^2. Note that C_3^2 is equivalent to rotating by 120° and then again by another 120°. In other words,

$$C_3^2 = C_3 + C_3 \ .$$

This will be our definition of "addition" for a molecular symmetry group: perform one operation and then perform another.[4] Operating with C_3^2 and then by C_3 returns all three H atoms to their original positions, which is the same as performing no operation at all. Let us call the act of performing no operation at all E. Thus,

$$C_3 + C_3^2 = E \ .$$

E will be the zero element of the group, because "adding" it to another operation has no effect,

$$E + C_3 = C_3 \ , \qquad E + C_3^2 = C_3^2 \ .$$

Another symmetry operation we can perform on this molecule is reflection through a plane. Consider the plane that includes the z-axis and the initial position of $\mathrm{H}^{(1)}$. $\sigma^{(1)}$ is the operation that has the effect of interchanging the other two H atoms.

The molecular symmetry group for the ammonia molecule, called $\mathbf{C}_{3v}$, consists of six operations:

$$\mathbf{C}_{3v} = \{E,\, C_3,\, C_3^2,\, \sigma^{(1)},\, \sigma^{(2)},\, \sigma^{(3)}\}.$$

The following table shows the effect of "adding" any two of the group elements:[5]

	E	C_3	C_3^2	$\sigma^{(1)}$	$\sigma^{(2)}$	$\sigma^{(3)}$
E	E	C_3	C_3^2	$\sigma^{(1)}$	$\sigma^{(2)}$	$\sigma^{(3)}$
C_3	C_3	C_3^2	E	$\sigma^{(3)}$	$\sigma^{(1)}$	$\sigma^{(2)}$
C_3^2	C_3^2	E	C_3	$\sigma^{(2)}$	$\sigma^{(3)}$	$\sigma^{(1)}$
$\sigma^{(1)}$	$\sigma^{(1)}$	$\sigma^{(3)}$	$\sigma^{(2)}$	E	C_3	C_3^2
$\sigma^{(2)}$	$\sigma^{(2)}$	$\sigma^{(1)}$	$\sigma^{(3)}$	C_3^2	E	C_3
$\sigma^{(3)}$	$\sigma^{(3)}$	$\sigma^{(2)}$	$\sigma^{(1)}$	C_3	C_3^2	E

The row label is applied first and then the column label. For example, $\sigma^{(3)} + \sigma^{(1)}$ equals C_3^2, at the intersection of the $\sigma^{(1)}$ row and the $\sigma^{(3)}$ column. (The notation

[4]In the case of molecular symmetry groups it is common practice to refer to this "+" operation as "multiplication." The reason will become apparent in Chapter 18, when we see that the addition of symmetry operations can be expressed in terms of multiplication of matrices. Because a group has only a single defined operation, we are free to call that operation by whatever name we wish.

[5]This is usually called a *group multiplication table.*

$\sigma^{(3)} + \sigma^{(1)}$ means that the operation on the right, $\sigma^{(1)}$, is performed first.)

There are approximately 50 different chemically important symmetry groups.

Definition. An ***abelian group***[6] is a group such that for any u and v in V, addition is ***commutative***; that is, $u + v = v + u$.

For the group $\mathbf{C}_{3v}$ in Example 16.1, note that

$$\sigma^{(1)} + \sigma^{(2)} = C_3, \qquad \sigma^{(2)} + \sigma^{(1)} = C_3^2 \,. \tag{16.11}$$

The order of the operations affects the result. Addition is not commutative for this group. The group is not abelian. Many of the molecular symmetry groups are abelian but $\mathbf{C}_{3v}$ is one that is not. $\mathbb{R}^N$ is an abelian group, because addition of real numbers is always commutative, for example, $1 + 2 = 2 + 1$.

16.4 Vector Spaces

Definition. A ***vector space*** over a field F is an abelian group V with an operation ***scalar multiplication***, defined such that:

1. For any element v in V and any scalar c in F, there exists a corresponding element cv in V. (V is closed to scalar multiplication.)
2. $a(u + v) = au + av$ and $(a + b)u = au + bu$ for any u and v in V and any scalars a and b in F. (Scalar multiplication is ***distributive***.)
3. $(ab)v = a(bv)$ for any v in V and any scalars a and b in F. (Scalar multiplication is associative.)
4. For any v in V, the element $(-1)v$ is its additive inverse, $v + (-1)v = \mathbf{0}$.
5. $1\,v = v$ for any v in V.

Any element of a vector space is called a ***vector***. We define ***subtraction*** of vectors as the operation $u - v = u + (-1)v$.

Having now defined addition and scalar multiplication, we can introduce a more general alegebraic operation that combines these two operations:

Definition. Consider vectors $v_1, v_2, \ldots, v_n$ in a vector space, and scalars $c_1, c_2, \ldots, c_N$. The vector

$$v = \sum_{i=1}^{N} c_i \, v_i \tag{16.12}$$

is called a ***linear combination*** of the vectors v_i.

It follows from the closure properties that any linear combination is also an element of the vector space.

[6]Named after the Norwegian mathematician Niels Henrik Abel (1802-1829).

16.5 Functions as Vectors

Let $\mathbb{P}^N$ be the set of all polynomials of degree less than or equal to positive integer N. In other words, $\mathbb{P}^N$ contains all possible functions of the form

$$p(x) = \sum_{j=0}^{N} c_j x^j. \tag{16.13}$$

We will only consider the case of $x \in \mathbb{R}$ but we will allow the coefficients c_j to be any arbitrary complex numbers. Let us prove that $\mathbb{P}^N$ is a vector space over the field $\mathbb{C}$. First we show that this set constitutes an abelian group. Consider, for example, the sum of two quadratic polynomials,

$$(a_0 + a_1x + a_2x^2) + (b_0 + b_1x + b_2x^2) = c_0 + c_1x + c_2x^2,$$

with $c_0 = a_0 + b_0$, $c_1 = a_1 + b_1$, and $c_2 = a_2 + b_2$. The sum is another polynomial and its degree is no greater than that of the original polynomials. Adding polynomials that are each of degree 2 or less cannot produce a polynomial of higher degree. Thus we have closure with respect to addition. Associativity and commutativity follow from the usual rules for arithmetic. The zero element of the set is the polynomial with all coefficients equal to zero. The additive inverse is obtained by changing the sign of each coefficient, because

$$(a_0 + a_1x + a_2x^2) + (-a_0 - a_1x - a_2x^2) = 0.$$

Therefore, $\mathbb{P}^N$ is an abelian group.

Next we consider scalar multiplication. Multiplication of a polynomial by a constant yields another polynomial and does not increase the degree; for example,

$$c\,(a_0 + a_1x + a_2x^2) = ca_0 + (ca_1)x + (ca_2)x^2,$$

which is still a polynomial of degree 2. Therefore, the set is closed with respect to scalar multiplication. Distributivity and associativity follow from the rules for arithmetic; for example,

$$(a_0 + a_1x + a_2x^2) + (b_0 + b_1x + b_2x^2) = (b_0 + b_1x + b_2x^2) + (a_0 + a_1x + a_2x^2),$$

and so on. Multiplication of a polynomial by -1 changes the sign of each coefficient, which gives the additive inverse. Finally, multiplying by the number 1 leaves a polynomial unchanged. Therefore, $\mathbb{P}^N$ is a vector space. Thus, we learn that polynomials are vectors!

The following example shows that other kinds of functions can also be treated as vectors.

Example 16.2. *Some other function spaces that qualify as vector spaces:*
The set of all polynomials from $\mathbb{P}^N$ *each multiplied by* e^{-x^2}. Note that

$$ap_1(x)e^{-x^2} + bp_2(x)e^{-x^2} = \left[ap_1(x) + bp_2(x)\right]e^{-x^2}$$

for any p_1 and p_2 in $\mathbb{P}^N$. The term in brackets is a linear combination of polynomials. It is therefore straightforward to show that the behavior with respect to addition and scalar multiplication is completely analogous to that of $\mathbb{P}^N$.

The set of all functions $f_i(x)$, $x \in \mathbb{R}$, such that $\int_{-\infty}^{\infty} f_i(x)dx$ is finite, with scalar field $\mathbb{R}$. Note that

$$\int_{-\infty}^{\infty} [f_j(x) + f_k(x)]dx = \int_{-\infty}^{\infty} f_j(x)dx + \int_{-\infty}^{\infty} f_k(x)dx,$$

which is also finite. The set is closed to addition. The scalar product $\int_{-\infty}^{\infty} cf_i(x)dx = c\int_{-\infty}^{\infty} f_i(x)dx$ is finite if c is finite, and all real numbers c are finite. (Although the real numbers have no upper limit, ∞ is *not* technically an element of $\mathbb{R}$.) The set is closed to scalar multiplication. The other conditions follow from the rules of arithmetic.

Example 16.3. *A function space that is not a vector space:*
The set of real-valued functions such that $f(x) < 5$ for all $x \in \mathbb{R}$, with scalar field $\mathbb{R}$. Consider $f(x) = 4 - x^2$ and $g(x) = 4 - x^4$. Both have maximum value of 4, at $x = 0$, and therefore are in the set, but their sum, $f + g = 8 - x^2 - x^4$, has maximum value 8. The set is not closed to addition.

With function spaces, the conditions to focus on are the two closure properties. The other eight conditions are usually not a problem.

16.6 Hilbert Spaces

Hilbert spaces[7] are vector spaces that allow not just addition of a vector to a vector and multiplication of a vector by a scalar, but also multiplication of a vector by a *vector*. First we must define how to multiply a vector by another vector. Neither the dot product nor the cross product will do; they are particular to Cartesian coordinate vectors. Our strategy is to construct a generalization of the dot product that retains key features of the Cartesian vector definition but is abstract enough to apply to other kinds of vector spaces. The generalization is called an "inner" product, with the notation $\langle u, v\rangle$ instead of the dot.

Definition. A mapping $\langle u, v\rangle$ that maps a pair of vectors u, v of a vector space V into a scalar field F is an ***inner product*** if

1. $\langle u, v\rangle = \langle v, u\rangle^*$. (The inner product is ***conjugate-symmetric***.)
2. $\langle u, v + w\rangle = \langle u, v\rangle + \langle u, w\rangle$ for any u, v, and w in V.
3. $\langle u, av\rangle = a\langle u, v\rangle$ and $\langle au, v\rangle = a^*\langle u, v\rangle$ for any a in F.
4. $\langle \mathbf{0}, v\rangle = 0$.
5. If $v \neq \mathbf{0}$ then $\langle v, v\rangle$ yields a positive real number.[8] (The inner product is ***positive definite***.)

[7] Named after the German mathematician David Hilbert (1862-1943) who developed a rigorous logical framework for the linear algebra of functions. When quantum mechanics was discovered in the late 1920's, it soon became clear that Hilbert's highly abstract approach to functions (which had seemed to some of his colleagues to be more theological than mathematical) had important practical applications.

[8] This condition applies to the inner product of a vector with *itself*. For $u \neq v$, the inner product $\langle u, v\rangle$ could in principle yield complex numbers.

A vector space for which an inner product has been defined is called an ***inner product space***. Note in the fourth axiom the distinction between the vector **0** and the number 0.

We can combine axioms 2 and 3 to obtain

$$\langle au+bv,w\rangle = a^*\langle u,w\rangle + b^*\langle v,w\rangle, \quad \langle w,au+bv\rangle = a\langle w,u\rangle + b\langle w,v\rangle. \quad (16.14)$$

A mapping is said to be ***conjugate-linear*** if it satisfies Eqs. (16.14). If the complex conjugation is omitted, it is just said to be ***linear***.[9]

Example 16.4. *The dot product qualifies as an inner product.* Consider, for example, the vector space $\mathbb{R}^2$. Let $\vec{u} = (a_1, a_2)$, $\vec{v} = (b_1, b_2)$, and $\vec{w} = (c_1, c_2)$ be arbitrary vectors, and let γ be an arbitrary real number.

1. $\vec{u} \cdot \vec{v} = (a_1, a_2) \cdot (b_1, b_2) = a_1 b_1 + a_2 b_2 = b_1 a_1 + b_2 a_2 = \vec{v} \cdot \vec{u}$.

2. $\vec{u} \cdot (\vec{v} + \vec{w}) = a_1(b_1 + c_1) + a_2(b_2 + c_2) = (a_1 b_1 + a_2 b_2) + (a_1 c_1 + a_2 c_2) = \vec{u} \cdot \vec{v} + \vec{u} \cdot \vec{w}$.

3. $\vec{u} \cdot (\gamma \vec{v}) = a_1 \gamma b_1 + a_2 \gamma b_2 = \gamma(a_1 b_1 + a_2 b_2) = \gamma(\vec{u} \cdot \vec{v})$.

4. $(0, 0) \cdot (b_1, b_2) = 0\, b_1 + 0\, b_2 = 0$.

5. $\vec{v} \cdot \vec{v} = b_1^2 + b_2^2$, which is greater than zero as long as b_1 and b_2 are not both zero.

$\mathbb{R}^N$ (over the field $\mathbb{R}$) for any N qualifies as an inner product space. So does $\mathbb{C}^N$ (over the field $\mathbb{C}$), thanks to the complex conjugation in axioms 1 and 3. The complex conjugation ensures that $\langle v, v\rangle$ will be positive definite. For example, for any $v \in \mathbb{C}$ we have $v = a+ib$ and $\langle a+ib, a+ib\rangle = (a-ib)(a+ib) = a^2 + b^2$. If the elements of the space V and the field F have no imaginary parts, then the complex conjugation is irrelevant.

We are free to specify the mapping from vector space to scalar field however we wish, as long as it satisfies the five axioms. That is a strength of this kind of analysis. We can define the mapping differently according to what is appropriate for a given kind of application. Consider the vector space of polynomials, $\mathbb{P}^N$. The inner product will have to be some sort of operation that maps functions into numbers. An obvious choice for such an operation is integration, because the definite integral of a function is a number. But there are many different kinds of definite integrals that would be acceptable.

Example 16.5. *An inner product for functions.* Consider

$$\langle p_1, p_2\rangle = \int_{-1}^{1} p_1(x)^*\, p_2(x)dx \quad (16.15)$$

as a possible choice. This satisfies all the requirements of the inner product. For example, let $p_1(x) = 4 + 3x$ and $p_2(x) = 2 + x$, so that

$$\langle p_1, p_2\rangle = \int_{-1}^{1} (8 + 10x + 3x^2)dx = 18\,.$$

[9]The "linearity" refers to the scalars. The right-hand sides of the equations are linear in a and b (or a^* and b^*).

The polynomials p_1 and p_2 are mapped to the scalar 18. The order in which the polynomials are multiplied does not matter, because $p_1 p_2 = (4 + 3x)(2 + x)$ is equal to $p_2 p_1 = (2 + x)(4 + 3x)$, which proves that this is a symmetric mapping. Linearity and positive definiteness are also easy to demonstrate, and clearly, $\int_{-1}^{1} 0\,dx = 0$.

A ***Hilbert space*** is an inner product space that is ***complete***, which is to say that if a sequence of vectors in the space approaches a limit, then that limit is also an element of the space. Almost all the inner product spaces we will consider here will also be Hilbert spaces.[10]

Definition. The ***norm*** of a vector in a Hilbert space is

$$|v| = \sqrt{\langle v, v \rangle}\,. \tag{16.16}$$

The fact that the inner product is positive definite ensures that the norm is a non-negative real number. For Cartesian vectors, the norm is the length of the vector. Eq. (16.16) allows us to apply the concept of "length" to functions.

Definition. A pair of nonzero vectors is ***orthogonal*** if their inner product is equal to zero.

According to Eq. (16.9), if the dot product of two Cartesian vectors with nonzero norm is equal to zero, then the cosine of the angle between them must be zero. In other words, the two vectors must be perpendicular, with $\alpha = \pm 90°$. Thus, for Cartesian coordinate vectors, "orthogonal" means the same thing as "perpendicular." This gives us a geometric analogy for thinking about functions. Two functions in a Hilbert space are "perpendicular" to each other if their inner product is zero. We can go further and use Eq. (16.9) to specify the ***angle*** between nonzero vectors v_1, v_2 of a Hilbert space of functions. Replacing the dot product in Eq. (16.9) with the inner product, we have

$$\langle v_1, v_2 \rangle = |v_1||v_2| \cos\alpha\,. \tag{16.17}$$

We can define the angle between the vectors as

$$\alpha = \arccos\left(\frac{\langle v_1, v_2 \rangle}{|v_1||v_2|}\right). \tag{16.18}$$

This is valid for any Hilbert space. The ***Cauchy-Schwarz inequality***,

$$|\langle v_1, v_2 \rangle| \le |v_1||v_2|\,, \tag{16.19}$$

can be proved to hold for any elements of any Hilbert space. This ensures that $|\cos\alpha| \le 1$, which is necessary for Eq. (16.17) to make sense.

[10] As an example of an inner product space that is not a Hilbert space, consider the set $\mathbb{Q}^3$ of vectors (c_1, c_2, c_3) in which the c_j are rational numbers, over the scalar field $\mathbb{Q}$ with the dot product as the inner product. One can construct a sequence of vectors that comes infinitely close to, for example, $(\sqrt{2}, e, \pi)$, but this limit is not an element of the vector space, because its components are irrational.

The following properties of geometric length also hold for any Hilbert space V:

1. $|v_1 + v_2| \le |v_1| + |v_2|$ for all v_1, v_2 in V. (The triangle inequality.)
2. $|cv| = |c||v|$ for vector v and scalar c.
3. $|v| > 0$ if $v \neq \mathbf{0}$.
4. $|\mathbf{0}| = 0$.

16.7 Basis Sets

The goal of this section is to generalize the concept of "coordinate." We begin with a reexamination of Cartesian coordinates. The standard unit vectors of $\mathbb{R}^3$ are

$$\hat{\mathbf{x}} = (1,0,0), \quad \hat{\mathbf{y}} = (0,1,0), \quad \hat{\mathbf{z}} = (0,0,1). \tag{16.20}$$

Any vector (a,b,c) can be written as a linear combination of the unit vectors,

$$(a,b,c) = a\,\hat{\mathbf{x}} + b\,\hat{\mathbf{y}} + c\,\hat{\mathbf{z}}. \tag{16.21}$$

Thus, the coordinates of a vector are the scalars that multiply the unit vectors. This is how we will define coordinates in arbitrary Hilbert spaces.

Because all vectors in $\mathbb{R}^3$ can be expressed in this manner, the set $\{\hat{\mathbf{x}}, \hat{\mathbf{y}}, \hat{\mathbf{z}}\}$ is said to ***span*** the set $\mathbb{R}^3$. However, $\{(1,0,0),(0,1,0),(0,0,1)\}$ is not the only set that spans $\mathbb{R}^3$. We could have chosen $\hat{\mathbf{x}} = (1,0,0)$, $\hat{\mathbf{y}} = (0,1,0)$, $\hat{\mathbf{z}} = (0,0,2)$. Then instead of Eq. (16.21),

$$(a,b,c) = a\hat{\mathbf{x}} + b\hat{\mathbf{y}} + \tfrac{1}{2}c\hat{\mathbf{z}}. \tag{16.22}$$

Note that the set $\{(1,0,0),(0,1,0)\}$, with only two vectors, does not span the vector space; nor does the set $\{(1,0,0),(0,1,0),(1,1,0)\}$, because any linear combination of these vectors will have 0 as the third element; nor does the set $\{(1,1,1),(1,0,1),(2,2,2)\}$.[11] We could, however, have chosen a larger set, such as $\{(1,0,0),(0,1,0),(0,0,1),(0,0,2)\}$. The inclusion of both $(0,0,1)$ and $(0,0,2)$ is not necessary, but this set does span the vector space.

These examples suggest the need for terminology to distinguish between such situations.

Definition. Consider a set of vectors $S = \{v_1, v_2, \ldots, v_k\}$. A vector u is said to be ***linearly dependent*** on S if it can be expressed as a linear combination of the elements of S,

$$u = \sum_{j=1}^{k} c_j\, v_j, \tag{16.23}$$

where the c_j are scalars. If u cannot be expressed as such a linear combination, then u is ***linearly independent*** of S.

[11] Try to express $(0,0,1)$ as a linear combination of $(1,1,1)$, $(1,0,1)$, and $(2,2,2)$. It is not possible, because any linear combination of the set will have the same value for the first and third components.

S is said to be a ***linearly independent set*** if each element of the set is linearly independent of all other elements.

From the definition of linear dependence, one can prove the following:

Theorem 16.7.1. *The set of vectors $S = \{v_1, v_2, \ldots, v_k\}$ is linearly dependent if and only if there exists a set of scalars $c_1, c_2, \ldots, c_k$ such that*

$$\sum_{j=1}^{k} c_j \, v_j = \mathbf{0} \tag{16.24}$$

and the c_j are not all equal to zero.

Proof. Choose some arbitrary value m of the summation index and rearrange Eq. (16.24) as

$$v_m = \sum_{j=1,\ j\neq m}^{k} b_j \, v_j \,, \qquad b_j = -\frac{c_j}{c_m} \,. \tag{16.25}$$

According to Eq. (16.23), this implies that the set is linearly dependent. If there is no set $\{c_j\}$ such that Eq. (16.24) is true, then v_m cannot be expressed as a linear combination of the $v_{j\neq m}$. The set is then linearly independent. □

Example 16.6. *Linear dependence.* Theorem 16.7.1 can be used to test a set for linear dependence.[12] Consider the set $\{(2,-1,6), (-19,2,17), (-62,1,26)\}$. It is not obvious by inspection whether one of the vectors can be expressed in terms of the other two. Consider the equation

$$c_1\,(2,-1,6) + c_2\,(-19,2,17) + c_3\,(-62,1,26) = (0,0,0)\,,$$

which would imply

$$2c_1 - 19c_2 - 62c_3 = 0, \quad -c_1 + 2c_2 + c_3 = 0, \quad 6c_1 + 17c_2 + 26c_3 = 0\,.$$

These three equations in three unknowns have the solution $c_1 = 7,\ c_2 = 4,\ c_3 = -1$. Because these values are not all zero, we conclude the set is linearly dependent, with $(-62,1,26) = 7\,(2,-1,6) + 4\,(-19,2,17)$. In contrast, consider the set $\{(2,0,0), (0,4,0), (0,0,3)\}$. We obtain the simultaneous equations $\{2c_1 = 0,\ 4c_2 = 0,\ 3c_3 = 0\}$, which has the solution $c_1 = c_2 = c_3 = 0$. The set is linearly independent.

Definition. A set of linearly independent vectors that spans a vector space is said to be a ***basis set***, or a ***basis***, for the space.[13]

To say that a set "spans" a space is to say that any vector in the space can be expressed as a linear combination of vectors in the set.

Definition. The ***dimension*** of a vector space is the minimum number of vectors required to span the space.

The number of elements in a basis set is equal to the dimension of the vector space, because any additional elements beyond the minimum number could be expressed as linear combinations of the others.

[12] However, a more convenient linear dependence test will be given in Section 18.5.

[13] The plural of "basis" is "bases," pronounced *BAY-seez.*

Example 16.7. *Bases for* $\mathbb{R}^3$. The dimension of this space is 3. The sets $\{(1,0,0), (0,1,0),(0,0,1)\}$ and $\{(1,0,0),(0,1,0),(0,0,2)\}$ both qualify as bases, because their elements are linearly independent and they span the space, but $\{(1,0,0),(0,1,0)\}$ is not a basis, because a vector (a,b,c) with nonzero c cannot be expressed as a linear combination of the set. The set $\{(1,0,0), (0,1,0),(0,0,1),(0,0,2)\}$ spans the vector space but is not a basis, because its last two elements are linearly dependent.

The dimension of a vector space need not be finite. For example, the vector space of all possible polynomials of any degree, which we can designate $\mathbb{P}^\infty$, has infinite dimension.

Example 16.8. *A basis for* $\mathbb{P}^\infty$. The infinite set $B = \{1,x,x^2,x^3,\ldots\}$ spans $\mathbb{P}^\infty$, because any polynomial can be expressed as a linear combination of monomials x^j. Furthermore, B is a linearly independent set. For example, it is impossible to express x^3 as a linear combination of 1, x, and x^2. Linear combination only involves scalar multiplication and addition. It does not allow for x to be multiplied by x^2.

In actual computations we usually will have to truncate an infinite-dimension basis at some finite dimension. The set $\{1,x,x^2,x^3,x^4,x^5\}$, for example, could be used to analyze functions in an infinite-dimension Hilbert space. This is an ***incomplete basis*** for the space. Strictly speaking it is a basis only for the six-dimensional subspace $\mathbb{P}^5$. It is not complete for $\mathbb{P}^\infty$ because any polynomial of degree higher than 5 cannot be *exactly* expressed as a linear combination of the elements of the incomplete basis. The linear combination could, however, give a useful approximation. The accuracy would improve as the dimension of the incomplete basis is increased.

Vectors v_1, v_2 of a Hilbert space are orthogonal if the inner product satisfies $\langle v_1, v_2\rangle = 0$. The Cartesian coordinate unit vectors, which have the dot product as their inner product, clearly are orthogonal, for example,

$$(1,0,0)\cdot(0,0,1) = 1\cdot 0 + 0\cdot 0 + 0\cdot 1 = 0.$$

Note also that they each have a norm of unity, $|\hat{\mathbf{x}}| = 1$, $|\hat{\mathbf{y}}| = 1$, $|\hat{\mathbf{z}}| = 1$.

Definition. A vector v is ***normalized*** if $|v| = 1$.

Any nonzero vector in a Hilbert space can be normalized (that is, transformed into a normalized vector) by dividing it by the square root of its norm. The vector

$$\hat{v} = v/\langle v,v\rangle^{1/2} \tag{16.26}$$

has a norm of 1 by construction, because

$$\langle \hat{v},\hat{v}\rangle = \left\langle \frac{v}{\langle v,v\rangle^{1/2}}, \frac{v}{\langle v,v\rangle^{1/2}} \right\rangle = \frac{1}{\langle v,v\rangle^{1/2}\langle v,v\rangle^{1/2}}\langle v,v\rangle = \frac{\langle v,v\rangle}{\langle v,v\rangle} = 1.$$

A set of vectors in which all pairs of elements are orthogonal and each element is normalized is called an ***orthonormal*** set. Thus the inner product of orthonormal vectors v_j and v_k is equal to zero if the vectors are different and equal to one if they are the same. This can be expressed with the notation

$$\langle v_j, v_k \rangle = \delta_{j,k} \;, \tag{16.27}$$

where $\delta_{j,k}$, called the ***Kronecker delta***,[14] is defined as

$$\delta_{j,k} = 1 \text{ for } j = k \,, \quad \delta_{j,k} = 0 \text{ for } j \neq k \,, \tag{16.28}$$

for integers j and k.

Orthonormal bases are the generalization of the concept of unit vectors. They are particularly convenient due to the following theorem:

Theorem 16.7.2. *Let $\{\hat{u}_1, \hat{u}_2, \hat{u}_3, \dots\}$ be an orthonormal basis for a Hilbert space V of dimension N. Then any vector v in V can be expressed as*

$$v = \sum_{j=1}^{N} a_j \hat{u}_j \;, \tag{16.29a}$$

where

$$a_j = \langle \hat{u}_j, v \rangle \,. \tag{16.29b}$$

This theorem applies whether the dimension N is finite or infinite.

Example 16.9. *Using inner products to determine coordinates.* The vector $(3, 7, 2)$ in $\mathbb{R}^3$, expressed as a linear combination of unit vectors, is

$$(3, 7, 2) = 3(1, 0, 0) + 7(0, 1, 0) + 2(0, 0, 1) \,.$$

Its y-coordinate can be calculated by taking an inner product with the unit vector $\hat{y}$,

$$\langle \hat{y}, (3, 7, 2) \rangle = (0, 1, 0) \cdot (3, 7, 2) = 7.$$

In this example the answer was quite obvious in advance, but the procedure is especially useful with Hilbert spaces of functions, in which case the answer is often not at all obvious. It only works if the basis set is orthonormal.

This analysis suggests a more general definition of "coordinate":

Definition. Let $B = \{\hat{u}_1, \hat{u}_2, \hat{u}_3, \dots\}$ be an orthonormal basis for a Hilbert space V. The ***coordinates*** of a vector v in V with respect to B are the ordered set of scalars $(a_1, a_2, a_3, \dots)$ such that $a_j = \langle \hat{u}_j, v \rangle$.

[14] Named in honor of Prussian logician and mathematician Leopold Kronecker (1823-1891). He argued that mathematics should be formulated only in terms of countable quantities, without such enigmatic concepts as irrational numbers and infinity. He is famous for his statement, "God made the integers, all else is the work of man." Note that the Kronecker delta, $\delta_{j,k}$, with integer subscripts, is rather different from the Dirac delta, $\delta(x)$, introduced in Section 4.4.

To ***resolve*** a vector into components is to express the vector as a linear combination of basis elements.[15]

Example 16.10. *Vector resolution in* $\mathbb{R}^3$. The vector $(2,3,4)$ has the resolution

$$(2,3,4) = 2(1,0,0) + 3(0,1,0) + 4(0,0,1)$$

in terms of the usual choice of unit vectors. However, the resolution of a given vector is not unique; it depends on the choice of basis set. In the orthonormal basis

$$\{\hat{u}_1, \hat{u}_2, \hat{u}_3\} = \left\{ \left(\frac{1}{\sqrt{2}}, \frac{1}{\sqrt{2}}, 0\right), \left(\frac{1}{\sqrt{6}}, -\frac{1}{\sqrt{6}}, \sqrt{\frac{2}{3}}\right), \left(-\frac{1}{\sqrt{3}}, \frac{1}{\sqrt{3}}, \frac{1}{\sqrt{3}}\right) \right\}$$

the resolution is $(2,3,4) = 3.53553\,\hat{u}_1 + 2.85774\,\hat{u}_2 + 2.88675\,\hat{u}_3$. The coefficients are calculated from dot products with a normalized basis element; for example,

$$\hat{u}_1 \cdot (2,3,4) = \frac{2}{\sqrt{2}} + \frac{3}{\sqrt{2}} + 0 = 5/\sqrt{2} = 3.53553\,.$$

Exercises

16.1 Which of the following sets are closed to addition?

(a) The set of all functions that can be expressed as $1/p$ where p is a polynomial.

(b) The set of all functions that can be expressed as a ratio of polynomials.

(c) The set of all functions with a first-order pole at $x = 3$ but otherwise nonsingular.

16.2 Demonstrate that the set $\{\hat{u}_1, \hat{u}_2, \hat{u}_3\}$ of Example 16.10 is orthonormal.

16.3 Consider the Hilbert space $\mathbb{R}^2$ (over field $\mathbb{R}$) but with the following unorthodox definition for the inner product between arbitrary vectors $\vec{x} = (x_1, x_2)$ and $\vec{y} = (y_1, y_2)$:

$$\langle \vec{x}, \vec{y} \rangle = p x_1 y_1 - q x_2 y_1 - r x_1 y_2 + s x_2 y_2 + t\,,$$

where the parameters p, q, r, s, and t are non-negative real numbers. What conditions on the five parameters are necessary for this to qualify as an inner product?

16.4 In terms of the basis $\{\hat{u}_1, \hat{u}_2, \hat{u}_3\}$ of Example 16.10, give the coordinates of $(1,0,2)$.

16.5 Prove that $(0,0,0)$ will always have all coordinates equal to zero in any basis for $\mathbb{R}^3$.

16.6 Derive Eq. (16.9). *(Hint: Let the z-axis point in the direction of one of the vectors.)*

16.7 A methane molecule can be modeled as cube with H atoms at four of the eight corners. Treat each C—H bond as a vector and use Eq. (16.17) to find the bond angle.

16.8 Show that $\mathbf{C}_{3v}$ satisfies the fourth axiom of the definition of a group (existence of additive inverse) by determining the inverse of each of the six elements of the group.

16.9 Is $\mathbb{Q}^3$ a Hilbert space over the field $\mathbb{R}$? Explain.

16.10 Prove Theorem 16.7.2.

[15] *Resolve* comes from the Latin verb *solvere*, which means to untie or set free. The basis elements are bound together in the vector. The linear combination sets them free.

Chapter 17

Spaces of Functions

The branch of mathematics that deals with vector spaces of functions is called *functional analysis.* This chapter introduces some of its basic ideas and techniques. The goal is to develop various Hilbert spaces of functions that can be used to systematically approximate unknown functions.

17.1 Orthogonal Polynomials

In this section we will construct orthonormal bases for $\mathbb{P}^\infty$, the vector space of all polynomials over scalar field $\mathbb{C}$. We will need to specify a choice for the inner product, to make $\mathbb{P}^\infty$ a Hilbert space, and each different choice will result in a different basis. We begin in each case with the infinite-dimension basis of monomials,

$$B = \{1, x, x^2, x^3, \dots\}, \tag{17.1}$$

and then systematically transform B into an orthonormal set.

The first inner product we will consider is

$$\langle p_1, p_2 \rangle = \int_{-1}^{1} p_1(x)^* \, p_2(x) dx. \tag{17.2}$$

The set B is not orthogonal under this inner product. For example,

$$\langle 1, x^2 \rangle = \int_{-1}^{1} x^2 \, dx = \tfrac{2}{3}, \tag{17.3}$$

which is not zero. Neither are the elements of B normalized, for example,

$$\langle 1, 1 \rangle = \int_{-1}^{1} dx = 2. \tag{17.4}$$

To construct an orthogonal and normalized set of vectors, to serve as the unit vectors for a Hilbert space of polynomials, we will use a procedure called ***Gram-Schmidt orthogonalization.***[1] It starts with an arbitrary set $\{v_0, v_1, v_2, \dots\}$ of linearly independent vectors in $\mathbb{P}^\infty$. For the sake of simplicity we will make the obvious choice

$$v_0 = 1, \quad v_1 = x, \quad v_2 = x^2, \quad v_3 = x^3, \ \dots, \tag{17.5}$$

but any basis set would do. The strategy is to transform the linearly indepen-

[1] Named for the German mathematician Erhard Schmidt (1876-1959) and the Danish actuary Jørgen Pedersen Gram (1850-1916). Schmidt completed his Ph.D. under the supervision of Hilbert and spent much of his subsequent career systematizing Hilbert's ideas into the modern theory of Hilbert spaces. The approach presented in this present chapter originates largely from a series of papers he published in 1907 and 1908. Gram had earlier developed a related orthogonalization procedure in the course of his work in the insurance industry.

dent (but nonorthogonal and unnormalized) set $\{v_j\}$ into a set $\{u_j\}$ that is orthogonal (but unnormalized) and then into a set $\{\phi_j\}$ that is orthonormal.

Starting with $u_0 = v_0$, we normalize it to obtain ϕ_0,

$$\phi_0 = v_0/\langle v_0, v_0\rangle^{1/2} = 1/\sqrt{2}\,, \tag{17.6}$$

so that $\langle \phi_0, \phi_0\rangle = \int_{-1}^{1}\left(\frac{1}{\sqrt{2}}\right)^2 dx = 1$. We then let

$$u_1 = v_1 + c_{1,0}\,\phi_0\,, \tag{17.7}$$

where $c_{1,0}$ is some as yet unspecified scalar. Requiring u_1 to be orthogonal to ϕ_0 gives an equation that determines the value of $c_{1,0}$,

$$0 = \langle \phi_0, u_1\rangle = \langle \phi_0, v_1\rangle + c_{1,0}\langle \phi_0, \phi_0\rangle = \frac{1}{\sqrt{2}}\int_{-1}^{1} x\,dx + c_{1,0} = c_{1,0}\,.$$

Therefore, $u_1 = v_1 = x$, and normalizing it gives

$$\phi_1 = u_1/\langle u_1, u_1\rangle^{1/2} = \sqrt{\tfrac{3}{2}}\,x\,. \tag{17.8}$$

Continuing in this fashion, we let

$$u_2 = v_2 + c_{2,1}\phi_1 + c_{2,0}\,\phi_0\,. \tag{17.9}$$

Orthogonalizing with respect to ϕ_0 and ϕ_1 gives *two* equations,

$$0 = \langle \phi_0, u_2\rangle = \langle \phi_0, v_2\rangle + c_{2,1}\langle \phi_0, \phi_1\rangle + c_{2,0}\langle \phi_0, \phi_0\rangle,$$

$$0 = \langle \phi_1, u_2\rangle = \langle \phi_1, v_2\rangle + c_{2,1}\langle \phi_1, \phi_1\rangle + c_{2,0}\langle \phi_1, \phi_0\rangle.$$

Because the set $\{\phi_0, \phi_1\}$ is orthonormal, we have $\langle \phi_0, \phi_1\rangle = 0$, $\langle \phi_1, \phi_0\rangle = 0$, and $\langle \phi_1, \phi_1\rangle = 1$. Therefore,

$$0 = \left\langle \frac{1}{\sqrt{2}}\,,\ x^2\right\rangle + c_{2,0} = \frac{1}{\sqrt{2}}\int_{-1}^{1} x^2\,dx + c_{2,0} = \frac{2}{3\sqrt{2}} + c_{2,0}$$

and

$$0 = \left\langle \sqrt{\frac{3}{2}}\,x,\ x^2\right\rangle + c_{2,1} = \sqrt{\frac{3}{2}}\int_{-1}^{1} x^3\,dx + c_{2,1} = 0 + c_{2,1}\,,$$

which determines the values of $c_{2,0}$ and $c_{2,1}$, with the result

$$u_2 = v_2 + 0\,\phi_1 - \frac{2}{3\sqrt{2}}\,\phi_0 = x^2 - \tfrac{1}{3}\,. \tag{17.10}$$

Normalizing u_2 gives

$$\phi_2 = \frac{x^2 - \frac{1}{3}}{\left[\int_{-1}^{1} \left(x^2 - \frac{1}{3}\right)^2 dx\right]^{1/2}} = \sqrt{\frac{5}{2}} \left(\frac{3}{2}x^2 - \frac{1}{2}\right). \tag{17.11}$$

Next we write

$$u_3 = v_3 + c_{3,2}\phi_2 + c_{3,1}\phi_1 + c_{3,0}\phi_0 \,. \tag{17.12}$$

Orthogonalizing with respect to $\{\phi_0, \phi_1, \phi_2\}$ gives a set of *three* equations, which determines the values of $c_{3,2}$, $c_{3,1}$, and $c_{3,0}$. Normalizing u_3 gives

$$\phi_3 = \sqrt{\frac{7}{2}} \left(\frac{5}{2}x^3 - \frac{3}{2}x\right), \tag{17.13}$$

and so on. It is customary to express these as

$$P_m(x) = \left(m + \tfrac{1}{2}\right)^{-1/2} \phi_m(x), \tag{17.14}$$

because $P_0(x) = 1$, $P_1(x) = x$, $P_2(x) = \frac{3}{2}x^2 - \frac{1}{2}$, $P_3(x) = \frac{5}{2}x^3 - \frac{3}{2}x$, etc., turns out to be a well-known set of functions called the ***Legendre polynomials***.[2] The P_m are orthogonal but they are not normalized. The explicit expression for them is

$$P_m(x) = \frac{1}{2^m} \sum_{j=0}^{[m/2]} (-1)^j \binom{m}{j} \binom{2m-2j}{m} x^{m-2j}, \tag{17.15}$$

in terms of binomial coefficients. The notation $[m/2]$ represents the largest integer less than or equal to $m/2$. The normalized Legendre polynomials are

$$\mathcal{P}_m = \left(m + \tfrac{1}{2}\right)^{1/2} P_m \,, \tag{17.16}$$

replacing the generic notation ϕ with the specific symbol $\mathcal{P}$.

A more general choice for an inner product is

$$\langle p_1, p_2 \rangle = \int_a^b p_1(x)^* p_2(x) w(x) dx, \tag{17.17}$$

with an arbitrary factor $w(x)$, called a ***weight function***, and an arbitrary integration range. The inclusion of $w(x)$ can allow for an infinite integration range. Although polynomials become infinite in the limits $x \to \pm\infty$, the choice of a $w(x)$ that goes to zero sufficiently quickly in these limits can ensure a convergent integral.

[2] Developed by the French mathematician Adrien-Marie Legendre (1752-1833) to describe planetary orbits. The Legendre polynomials occur frequently in physics problems due to the fact that they are the coefficients of an important Taylor series,

$$\frac{1}{\sqrt{1 - 2az + z^2}} \sim \sum_{j=0}^{\infty} P_j(a)\, z^j.$$

Each choice for the inner product leads to a different set of orthogonal polynomials. Various sets commonly seen in physical applications are described in Table 17.1. The inner product

$$\langle p_1, p_2 \rangle = \int_0^\infty p_1(x)^* p_2(x) e^{-x} dx \tag{17.18}$$

leads to orthonormal functions $L_m(x)$ called ***Laguerre polynomials***.[3] It is customary to include a factor of $(-1)^m$ in the definition,

$$L_m(x) = (-1)^m \phi_m(x), \tag{17.19}$$

where the ϕ_m here are the orthonormal polynomials obtained from Gram-Schmidt orthogonalization. This makes $L_m(0) = 1$ for all m. Multiplying a function by (-1) has no effect on its orthonormality. The Laguerre polynomials are orthonormal.[4]

Using $w(x) = x^k e^{-x}$, with positive integer k, gives orthonormal polynomials ϕ_m that generate the ***associated Laguerre polynomials***,

$$L_m^{(k)}(x) = (-1)^m \sqrt{\frac{(m+k)!}{m!}}\, \phi_m(x). \tag{17.20}$$

Note that $L_m^{(0)}$ is equivalent to L_m. In fact, k can be any real number greater than -1, but the those with integer indices are ones most commonly encountered. The standard notation is L_m^k, in which the superscript "k" is the index that specifies the weight function of the inner product; it does not indicate a power. The parentheses are included here to avoid confusion.

The inner product

$$\langle p_1, p_2 \rangle = \int_{-\infty}^\infty p_1(x)^* p_2(x) e^{-x^2} dx \tag{17.21}$$

leads to orthonormal ϕ_m that generate the ***Hermite polynomials***,[5]

$$H_m(x) = (-1)^m 2^{m/2} \pi^{1/4} (m!)^{1/2} \phi_m(x). \tag{17.22}$$

The complicated prefactor makes all the coefficients integers, with $H_0 = 1$.

---

[3]Named for the French mathematician Edmond Laguerre (1834-1886), who encountered them in the course of a study of divergent series and then appreciated that their orthogonality made them useful for practical applications.

[4]Sometimes the L_m are defined with prefactor $(-1)^m m!$, in order to make all of the coefficients of the polynomials have integer values. In that case, the L_m would not be normalized. For the same reason, the associated Laguerre polynomials (see below) are sometimes defined with $(-1)^m m!$ instead of the prefactor in Eq. (17.20).

[5]Named for the French mathematician Charles Hermite (1822-1901), who published studies of their properties in 1864. His biography is compelling. After one year as a university student, he was expelled on account of a physical disability that made it difficult for him to walk. Only after years of independent study was he able to pass the examinations for a bachelor's degree, but then he went on to become a professor of international renown, famous for his teaching ability as well as for many important contributions to number theory and algebra. The name is pronounced *err-meet*.

Table 17.1: Some common sets of orthogonal polynomials that result from inner products in the form of Eq. (17.17) for given choices of a, b, and $w(x)$. The normalization prefactor $\mathcal{N}_m$ is also shown.

Legendre polynomials:

$a = -1 \quad b = 1 \quad w(x) = 1 \quad \mathcal{N}_m = \left(m + \tfrac{1}{2}\right)^{1/2}$

$P_0 = 1 \quad P_1 = x \quad P_2 = (3x^2 - 1)/2$

$P_3 = (5x^3 - 3x)/2 \quad P_4 = (35x^4 - 30x^2 + 3)/8$

Laguerre polynomials:

$a = 0 \quad b = \infty \quad w(x) = e^{-x} \quad \mathcal{N}_m = 1$

$L_0 = 1 \quad L_1 = -x + 1 \quad L_2 = (x^2 - 4x + 2)/2$

$L_3 = (-x^3 + 9x^2 - 18x + 6)/6 \quad L_4 = (x^4 - 16x^3 + 72x^2 - 96x + 24)/24$

Associated Laguerre polynomials:

$a = 0 \quad b = \infty \quad w(x) = x^k e^{-x} \quad \mathcal{N}_m^{(k)} = \sqrt{m!/(m+k)!}$

$L_0^{(k)} = 1 \quad L_1^{(k)} = -x + k + 1 \quad L_2^{(k)} = [x^2 - 2(k+2)x + (k+1)(k+2)]/2$

$L_3^{(k)} = [-x^3 + 3(k+3)x^2 - 3(k+2)(k+3)x + (k+1)(k+2)(k+3)]/6$

$$L_4^{(k)} = [x^4 - 4(k+4)x^3 + 6(k+3)(k+4)x^2 \\ -4(k+2)(k+3)(k+4)x + (k+1)(k+2)(k+3)(k+4)]/24$$

Hermite polynomials:

$a = -\infty \quad b = \infty \quad w(x) = e^{-x^2} \quad \mathcal{N}_m = (-1)^m 2^{-m/2} \pi^{-1/4} (m!)^{-1/2}$

$H_0 = 1 \quad H_1 = 2x \quad H_2 = 4x^2 - 2$

$H_3 = 8x^3 - 12x \quad H_4 = 16x^4 - 48x^2 + 12$

Chebyschev polynomials of the first kind:

$a = -1 \quad b = 1 \quad w(x) = (1 - x^2)^{-1/2} \quad \mathcal{N}_0 = \sqrt{1/\pi} \quad \mathcal{N}_{m>0} = \sqrt{2/\pi}$

$T_0 = 1 \quad T_1 = x \quad T_2 = 2x^2 - 1 \quad T_3 = 4x^3 - 3x \quad T_4 = 8x^4 - 8x^2 + 1$

Chebyschev polynomials of the second kind:

$a = -1 \quad b = 1 \quad w(x) = (1 - x^2)^{1/2} \quad \mathcal{N}_m = \sqrt{2/\pi}$

$U_0 = 1 \quad U_1 = 2x \quad U_2 = 4x^2 - 1 \quad U_3 = 8x^3 - 4x \quad U_4 = 16x^4 - 12x^2 + 1$

The ***Chebyshev polynomials***[6] result from $w(x) = (1-x^2)^{\mp 1/2}$ and integration from -1 to 1, with $T_0 = \sqrt{\pi}\,\phi_0$ and $T_{m>0} = \sqrt{\pi/2}\,\phi_m$ for the first kind and $U_m = \sqrt{\pi/2}\,\phi_m$ for the second kind.

Explicit expressions in their customary forms and their normalized forms, respectively, are as follows:

$$L_m^{(k)}(x) = \sum_{j=0}^{m}(-1)^j \binom{m+k}{m-j}\frac{1}{j!}x^j, \quad \mathcal{L}_m^{(k)} = \sqrt{m!/(m+k)!}\,L_m^{(k)}, \tag{17.23}$$

$$H_m(x) = \sum_{j=0}^{[m/2]} \frac{(-1)^j m!}{j!(m-2j)!}(2x)^{m-2j}, \quad \mathcal{H}_m = (-1)^m 2^{-m/2}(m!)^{-1/2}\pi^{-1/4}H_m, \tag{17.24}$$

$$T_m(x) = \frac{m}{2}\sum_{j=0}^{[m/2]}(-1)^j \frac{(m-j-1)!}{j!(m-2j)!)}(2x)^{m-2j},$$

$$\mathcal{T}_0 = \sqrt{1/\pi}, \quad \mathcal{T}_{m>0} = \sqrt{2/\pi}\,T_m, \tag{17.25}$$

$$U_m(x) = \sum_{j=0}^{[m/2]}(-1)^j \frac{(m-j)!}{j!(m-2j)!}(2x)^{m-2j}, \quad \mathcal{U}_m = \sqrt{2/\pi}\,U_m. \tag{17.26}$$

Because they are constructed from the basis set $\{1, x, x^2, x^3, \dots\}$, any orthogonal polynomial of index m will have degree m. Suppose that we take an orthogonal polynomial of index m and multiply it by x. The degree of the resulting polynomial will be $m+1$, which means that it will be linearly independent of the set of orthogonal polynomials of degree m and lower. Therefore, the set consisting of the polynomial of degree m multiplied by x along with the polynomials of degree less than or equal to m is a basis for $\mathbb{P}^{m+1}$. Any polynomial of degree $m+1$ can be expressed as a linear combination of these basis elements. In particular, we can express the orthogonal polynomial of index $m+1$ in terms of this basis. For example, using Table 17.1 we see that

$$H_4(x) = 2xH_3(x) - 6H_2(x).$$

This is a ***recursion relation***, in which we express the function with some given index in terms of functions with lower indices. The standard orthogonal polynomials all have simple recursion relations, which can be convenient in practical applications. They are listed in Table 17.2. Note that the associated Laguerre polynomials have recursions for the upper and for the lower index.

[6]Developed by Russian mathematician Pafnuty Chebyshev (1821-1894). They are often used in numerical analysis. The Chebyshev polynomials of the first kind have the interesting property that $T_n(\cos\phi) = \cos(n\phi)$. Chebyshev made important contributions in both pure and applied mathematics and in probability theory. He was apparently the first person to recognize the usefulness of a general theory of orthogonal polynomials and, in fact, he independently developed the "Hermite" polynomials five years before Hermite did. He too had a disability that made it very difficult to walk. The name is pronounced (approximately) *chebi-SHOFE*.

Table 17.2: Recursion relations for orthogonal polynomials.

$$(m+1)P_{m+1}(x) = (2m+1)xP_m(x) - mP_{m-1}(x)$$

$$(m+1)L_{m+1}^{(k)}(x) = (1+2m+k-x)L_m^{(k)} - (m+k)L_{m-1}^{(k)}$$

$$L_m^{(k+1)} = L_m^{(k)} + L_{m-1}^{(k-1)}$$

$$H_{m+1}(x) = 2xH_m(x) - 2mH_{m-1}(x)$$

$$T_{m+1}(x) = 2xT_m(x) - T_{m-1}(x)$$

$$U_{m+1}(x) = 2xU_m(x) - U_{m-1}(x)$$

Table 17.3: Differential equations satisfied by orthogonal polynomials.

P_m:	$(1-x^2)y'' - 2xy' + m(m+1)y = 0$
$L_m^{(k)}$:	$xy'' + (k+1-x)y' + my = 0$
H_m:	$y'' - 2xy' + 2my = 0$
T_m:	$(1-x^2)y'' - xy' + m^2y = 0$
U_m:	$(1-x^2)y'' - 3xy' + m(m+2)y = 0$

Now consider derivatives of orthogonal polynomials. The derivative of a polynomial is a polynomial of lower degree. It is always possible to express a polynomial in terms of its derivatives multiplied by powers of x and constants. For example,

$$H_4 = \frac{1}{4}x\frac{d}{dx}H_4 - \frac{1}{8}\frac{d^2}{dx^2}H_4 .$$

If one were to encounter a differential equation $y'' - 2xy' + 8y = 0$ for an unknown function $y(x)$, one could immediately write $y(x) = H_4(x)$ as a solution. Table 17.3 lists differential equations satisfied by the various orthogonal polynomials. They can be derived by taking the derivatives of the explicit expressions given by Eqs. (17.15) and (17.23) through (17.26).

17.2 Function Resolution

We can resolve a function into a linear combination of orthonormal basis vectors in a manner analogous to the way we resolved a Cartesian coordinate vector in $\mathbb{R}^3$ into a linear combination of unit vectors. The only differences are that for a space of functions the unit vectors are themselves functions and the inner product is no longer the dot product but some specified mapping of a pair of functions into a scalar. This can be a powerful technique for obtaining approximate solutions to difficult problems in applied mathematics.

Example 17.1. *Resolution of a polynomial.* Any polynomial can be resolved into a finite linear combination of orthogonal polynomials. Consider the monomial x^3. In a basis of Legendre polynomials, the coefficients of the linear combination are $a_k = \langle \mathcal{P}_k, x^3 \rangle$,

$$a_0 = \left(\tfrac{1}{2}\right)^{1/2} \int_{-1}^{1} x^3 dx = 0, \qquad a_1 = \left(\tfrac{3}{2}\right)^{1/2} \int_{-1}^{1} x^4 dx = \left(\tfrac{3}{2}\right)^{1/2} \tfrac{2}{5},$$

$$a_2 = \left(\tfrac{5}{2}\right)^{1/2} \int_{-1}^{1} \tfrac{(3x^2-1)}{2} x^3 dx = 0, \quad a_3 = \left(\tfrac{7}{2}\right)^{1/2} \int_{-1}^{1} \tfrac{(5x^3-3x)}{2} x^3 dx = \left(\tfrac{7}{2}\right)^{1/2} \tfrac{4}{35}.$$

($\mathcal{P}_k$ must be used in the inner product, not the unnormalized P_k.) The resolution of x^3 is

$$x^3 = \sum_{k=0}^{3} a_k \mathcal{P}_k = \left(\tfrac{3}{2}\right)^{1/2} \tfrac{2}{5} \left(\tfrac{3}{2}\right)^{1/2} P_1 + \left(\tfrac{7}{2}\right)^{1/2} \tfrac{4}{35} \left(\tfrac{7}{2}\right)^{1/2} P_3 = \tfrac{3}{5} P_1 + \tfrac{2}{5} P_3 .$$

Each different basis set gives a different resolution. The calculation is the same except that a different inner product is used. For example, in terms of Hermite polynomials,

$$a_0 = \pi^{-1/4} \int_{-\infty}^{\infty} x^3 e^{-x^2} dx = 0, \quad a_1 = \tfrac{-1}{2^{1/2}\pi^{1/4}} \int_{-\infty}^{\infty} 2x^4 e^{-x^2} dx = \tfrac{-1}{2^{1/2}\pi^{1/4}} \tfrac{3\pi^{1/2}}{2},$$

etc., giving

$$x^3 = -2^{1/2}\pi^{1/4} \tfrac{3}{4} \mathcal{H}_1 - 3^{1/2}\pi^{1/4} \tfrac{1}{8} \mathcal{H}_3 = \tfrac{3}{4} H_1 + \tfrac{1}{8} H_3 .$$

Definition. $C[a,b]$ is the Hilbert space that contains all functions that are continuous for all x in a finite interval $a \le x \le b$, over the field $\mathbb{C}$ and with inner product $\langle f, g \rangle = \int_a^b f^* g \, dx$.

Theorem 17.2.1. *(Weierstrass[7] approximation theorem) Any function in $C[a,b]$ can be approximated to arbitrarily high accuracy by a polynomial.*

It follows that $\{1, x, x^2, x^3, \dots\}$ is a basis for $C[a,b]$. Any $f \in C[a,b]$ can be expressed as

$$f(x) \approx \sum_{k=0}^{N} a_k x^k \tag{17.27}$$

for some set of coefficients a_k. Approximate equality "$\approx$" becomes exact equality "$=$" in the limit $N \to \infty$. For a polynomial of degree n, the exact result is obtained with the finite truncation $N = n$ but for other functions in $C[a,b]$ the exact result requires an infinite number of terms. Because each of the x^k can be expressed as a linear combination of orthogonal polynomials, it follows that the orthogonal polynomials are also bases for $C[a,b]$.

Definition. L^2 is the Hilbert space that contains all piecewise-continuous[8] functions f such that $\langle f, f \rangle$ is finite, with inner product $\langle f, g \rangle = \int_{-\infty}^{\infty} f^* g \, dx$ and field $\mathbb{C}$. The functions in L^2 are said to be ***square integrable***. If the inner product contains a weight function, the space is called a *weighted* L^2 space, indicated with the notation $L^2[w(x)]$. If the range of integration in the inner product is $a \le x \le b$, then the space is designated $L^2[a, b; w(x)]$.

[7] Proved by German mathematician Karl Weierstrass (1815-1897).

[8] *Piecewise*-continuous means that the function has at most a finite number of discontinuities in the range of integration.

The restriction to $C[a,b]$ used by Weierstrass in Theorem 17.2.1 was later shown to be unnecessarily severe. In fact, a similar theorem also applies to L^2 spaces.[9] The approximating polynomial can be expressed as a linear combination of orthogonal polynomials generated from the inner product with the range and weight of the L^2 space. Functions that are not polynomials are represented by a linear combination of an inifite number of basis elements. In practice, a finite truncation can often yield a reasonable approximation.

Example 17.2. *Chebyshev approximation of a discontinuous function.* Consider the function

$$f(x) = \begin{cases} 2+2x, & -1 \le x < -0.5, \\ \frac{2}{3}(x-0.25), & -0.5 \le x \le 0.75, \\ \frac{1}{3}, & 0.75 < x \le 1, \end{cases}$$

which is only piecewise continuous. Let us resolve this into a linear combination of Chebyshev polynomials of the first kind. The coordinate with respect to the normalized basis element $\mathcal{T}_m$ is

$$a_m = \langle \mathcal{T}_m, f\rangle = \int_{-1}^{1} \mathcal{T}_m(x) f(x)(1-x^2)^{-1/2} dx\,.$$

For example,

$$\begin{aligned} a_3 = &\int_{-1}^{-0.5} \sqrt{\frac{2}{\pi}}(4x^3-x)(2+2x)(1-x^2)^{-1/2}dx \\ &+ \int_{-0.5}^{0.75} \sqrt{\frac{2}{\pi}}(4x^3-x)\tfrac{2}{3}(x-0.25)(1-x^2)^{-1/2}dx \\ &+ \int_{0.75}^{1} \sqrt{\frac{2}{\pi}}(4x^3-x)\tfrac{1}{3}(1-x^2)^{-1/2}dx \\ = &\ 0.076683\,. \end{aligned}$$

Truncating the linear combination at $\mathcal{T}_4$ gives

$$f(x) \approx 0.288322\,\mathcal{T}_0 + 0.065478\,\mathcal{T}_1 + 0.236603\,\mathcal{T}_2 + 0.076683\,\mathcal{T}_3 - 0.306525\,\mathcal{T}_4\,.$$

The following plot compares fourth-order (dotted curve) and 20th-order (dashed curve) truncations:

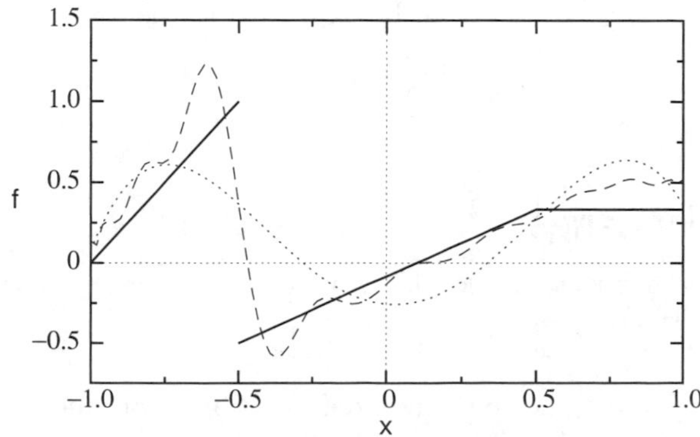

[9]For a more careful statement of these theorems see, for example, J. T. Cushing, *Applied Analytical Mathematics for Physical Scientists* (Wiley, New York, 1975), Chap. 4.

Resolution in a space of orthogonal polynomials gives us yet another technique for approximating a function. It is easily computerized and can be applied even when the function is only available in terms of numerical values, without knowing the function's analytical form.

Chebyshev polynomials are often used for function approximation because they give very nearly a ***minimax*** polynomial, that is, a polynomial that minimizes the maximum deviation from the true function over the range of x. Although Chebyshev polynomials are defined only for $-1 \leq x \leq 1$, with a change of variable they can be used to approximate functions over any arbitrary range of x. The other sets of orthogonal polynomials are not so often used for direct approximation of functions, but they are very useful for approximating the solutions of differential equations.

It is interesting to compare with Taylor series, which also express functions as polynomials. Advantages of orthogonal polynomial resolution are:

- It can be applied to functions for which singularities cause the Taylor series to diverge, as long as the norm of the function, $\langle f, f \rangle$, is finite.
- It spreads the approximation error more evenly throughout the range. The Taylor series is most accurate in the neighborhood of $x = 0$ and less accurate as the distance from the origin increases.

However, polynomial resolution requires information about the true function throughout the approximation range while the Taylor series is constructed from information about the function in the vicinity of a single special point. Also, Taylor series give deeper insight into the properties of the true function. The coefficients are proportional to derivatives, and large-order behavior of the series can give information about the function's singularity structure.

It is common to see weight factors included in the basis set. Consider the Hermite polynomials, a basis for $L^2[0, \infty; e^{-x^2}]$. If a function $f(x)$ is an element of this space, then $e^{-x^2/2} f(x)$ is an element of $L^2[0, \infty]$, with no weight in the inner product. Therefore, the set $\{\mathcal{H}_k e^{-x^2/2}\}$ is a basis for $L^2[0, \infty]$, and any $g \in L^2[0, \infty]$ can be expanded in the form

$$g(x) = \sum_{k=0}^{\infty} a_k \mathcal{H}_k e^{-x^2/2}. \tag{17.28}$$

17.3 Fourier Series

Polynomials are not the only kinds of important function spaces. Consider the set

$$\{1,\ e^{i\phi},\ e^{-i\phi},\ e^{2i\phi},\ e^{-2i\phi},\ e^{3i\phi},\ e^{-3i\phi}, \dots\} \tag{17.29}$$

as a basis for a vector space over a field $\mathbb{C}$ with inner product

$$\langle u, v \rangle = \int_0^{2\pi} u^*(\phi) v(\phi) d\phi\,. \tag{17.30}$$

Note that for integers $j \neq k$,

$$\langle e^{ij\phi}, e^{ik\phi} \rangle = \int_0^{2\pi} e^{i(k-j)\phi} d\phi = \frac{1}{i(k-j)} \left[e^{2\pi i(k-j)} - 1 \right] = 0, \tag{17.31}$$

because $e^{2\pi in} = \cos(2n\pi) + i \sin(2n\pi) = 1 + 0i$ for integer n. If $j = k$, then

$$\langle e^{ik\phi}, e^{ik\phi} \rangle = \int_0^{2\pi} d\phi = 2\pi. \tag{17.32}$$

Therefore, the set $\{e^{ik\phi}/\sqrt{2\pi}\}$ for $k = 0, \pm 1, \pm 2, \ldots$ is orthonormal. Furthermore, this set turns out to be a basis for all periodic functions such that $f(\phi) = f(\phi + 2\pi)$ in the Hilbert space $L^2[0, 2\pi]$. Any such function can be expressed as

$$f(\phi) = c_0 + \sum_{k=1}^{\infty} c_k \, e^{ik\phi} + \sum_{k=1}^{\infty} c_{-k} \, e^{-ik\phi}. \tag{17.33}$$

This is called a ***Fourier series.***[10] The coefficients are given by

$$c_k = \left\langle \frac{1}{\sqrt{2\pi}} e^{ik\phi}, f \right\rangle = \frac{1}{2\pi} \int_0^{2\pi} e^{-ik\phi} f(\phi) d\phi. \tag{17.34}$$

The factor of $1/2\pi$ comes from the factor $(2\pi)^{-1/2}$ in the normalized basis function $e^{ik\phi}/\sqrt{2\pi}$ in Eq. (17.33), which we have absorbed into c_k, and another factor $(2\pi)^{-1/2}$ from the normalized basis function in the inner product.

Alternatively, we could use the basis set

$$\{1, \; \cos\phi, \; \sin\phi, \; \cos 2\phi, \; \sin 2\phi, \; \cos 3\phi, \; \sin 3\phi, \; \ldots\}.$$

The inner product $\langle \sin j\phi, \cos k\phi \rangle$ is zero for any integers j and k, and if $j \neq k$ then we have $\langle \cos j\phi, \cos k\phi \rangle = \langle \sin j\phi, \sin k\phi \rangle = 0$. For $j = k$,

$$\langle 1, 1 \rangle = 2\pi, \qquad \langle \cos k\phi, \cos k\phi \rangle = \langle \sin k\phi, \sin k\phi \rangle = \pi\,.$$

To normalize the basis, the first element is multiplied by $(2\pi)^{-1/2}$ while all the others are multiplied by $\pi^{-1/2}$. We write the series as

$$f(\phi) = \tfrac{1}{2} a_0 + \sum_{k=1}^{\infty} a_k \, \cos k\phi + \sum_{k=1}^{\infty} b_k \, \sin k\phi\,, \tag{17.35}$$

with

$$a_k = \frac{1}{\pi} \int_0^{2\pi} \cos(k\phi) f(\phi) d\phi, \quad b_k = \frac{1}{\pi} \int_0^{2\pi} \sin(k\phi) f(\phi) d\phi\,. \tag{17.36}$$

This looks rather different from Eq. (17.33), but in fact they are equivalent—

[10]Pronounced *foo-RYĀY*. Developed by the French mathematician and theoretical physicist Joseph Fourier (1768-1830) to describe the transport of heat in a solid.

substituting $\cos k\theta = \frac{1}{2}(e^{ik\phi}+e^{-ik\phi})$, $\sin k\theta = \frac{i}{2}(e^{-ik\phi}-e^{ik\phi})$, into Eq. (17.35) gives

$$c_0 = \tfrac{1}{2}a_0; \quad c_{\pm k} = \tfrac{1}{2}(a_k \mp ib_k), \quad k > 0\,.$$

Which form of Fourier series one uses is a matter of convenience. If $f(\phi)$ is pure real, then one usually uses the series with the sines and cosines.

The main use for Fourier series is, as one might expect, to approximate functions that are periodic. They are widely used in analyzing wave phenomena, such as electromagnetic or acoustic waves. The period of the function does not have to be from 0 to 2π. With an appropriate change of variable, the range of the analysis can be transformed to any arbitrary value.

Example 17.3. *Fourier analysis over an arbitrary range.* Suppose that we want to analyze a function $f(x)$ that we know to be periodic with a period R; that is, $f(x+R) = f(x)$. Consider the basis set

$$\{\hat{u}_k\} = \left\{\frac{1}{R^{1/2}}\, e^{2\pi ikx/R}\right\}, \qquad k = 0, \pm1, \pm2, \pm3, \ldots, \tag{17.37}$$

and the inner product

$$\langle u, v\rangle = \int_0^R u^*(x)v(x)dx\,.$$

The inner product between two basis elements with the change of variable $\phi = 2\pi x/R$ gives the same integrals as we had for the standard Fourier series,

$$\langle \hat{u}_j, \hat{u}_k\rangle = \frac{1}{R}\int_0^R e^{i2\pi x(k-j)/R}\, dx = \frac{1}{2\pi}\int_0^{2\pi} e^{i(k-j)\phi}\, d\phi = \delta_{j,k}\,, \tag{17.38}$$

where $\delta_{j,k}$ is the Kronecker delta. Therefore, $\{\hat{u}_k\}$ is an orthonormal basis. The Fourier series can be written

$$f(x) = c_0 + \sum_{k=\pm1}^{\pm\infty} c_k e^{i2\pi kx/R}, \qquad c_k = \frac{1}{R}\int_0^R e^{-i2\pi kx/R} f(x)dx, \tag{17.39}$$

with the factor $R^{-1/2}$ from the basis function absorbed into the c_k. Alternatively, it can be expressed as

$$f(x) = \tfrac{1}{2}a_0 + \sum_{k=1}^{\infty} a_k \cos(2\pi kx/R) + \sum_{k=1}^{\infty} b_k \sin(2\pi kx/R), \tag{17.40}$$

$$a_k = \frac{2}{R}\int_0^R \cos(2\pi kx/R)f(x)dx, \quad b_k = \frac{2}{R}\int_0^R \sin(2\pi kx/R)f(x)dx. \tag{17.41}$$

Suppose that you want to approximate f over some range R_1 to R_2 but f is *not* periodic. One way to deal with this is to choose some larger range $\tilde{R}_1 \le x \le \tilde{R}_2$ and then define the function outside the original range such that f is periodic. The Fourier series will have no physical meaning outside the original range but within the physical range the Fourier series can give a reasonable approximation. Alternatively, one might apply the Fourier analysis to a new function $g(x) = f(x) - h(x)$ with an arbitrary specified function h chosen to ensure that g is periodic.

Another application of Fourier series is to solve certain common differential equations.

Example 17.4. *Solute transport boundary condition.** Let us continue the analysis begun in Example 14.7. For the differential equation $\partial c/\partial t = D\partial^2 c/\partial x^2$ we derived a solution

$$c(x,t) = c_\infty + b_\infty x + \sum_{j=1}^{N} A_j \cos\left(\sqrt{\lambda_j/D}\, x + \gamma_j\right) e^{-\lambda_j t} \qquad (17.42)$$

for the solute concentration, but we did not specify values for the parameters c_∞, b_∞, A_j, γ_j, and λ_j, which depend on the boundary conditions. Consider an approximately rectangular lake in which c depends only on the distance from the shore perpendicular to the x-axis, with no dependence on the z-axis (the depth) or the y-axis. (This is so that we can treat it as a one-dimensional problem. However, this solution method can be generalized to higher dimensionality.) Suppose that some toxic substance has been released at the shore, $x = 0$, evenly distributed along the shore. At some time $t = 0$ we examine the distribution of the substance and determine that it is constant in a narrow strip extending to x_0 and then zero beyond that. Thus, the observed initial condition is

$$c_{\text{obs}}(x,0) = \begin{cases} c_0\,, & 0 \le x \le x_0\,, \\ 0\,, & x_0 < x \le R\,, \end{cases} \qquad (17.43)$$

with c_0 in some appropriate units such as g/L. R is the length of the lake.

We also know what $c(x,t)$ should be in the limit of infinite t—given an infinite amount of time, the toxic substance will become evenly mixed throughout the lake. In the $t \to \infty$ limit the terms in the summation in Eq. (17.42) all disappear, on account of the factors of $e^{-\lambda_j t}$, leaving just $c_\infty + b_\infty x$. (This is why the constants were given the ∞ subscripts.) For this to be constant with respect to x, we need $b_\infty = 0$.

The formal solution for $c(x,t)$ at $t = 0$ is then

$$c(x,0) = c_\infty + \sum_{j=1}^{N} A_j \cos\left(\sqrt{\lambda_j/D}\, x + \gamma_j\right).$$

which looks very much like a Fourier series. Our strategy will be to choose the parameter values so that this becomes a Fourier series that we can fit to the initial conditions.

The initial conditions are not periodic. Suppose, however, that we define the function $c(x,t)$ for negative x as the mirror image of the function for positive x; that is, $c(-x,t) = c(x,t)$. Then $c(-R,0) = c(R,0)$. $c(x,0)$ can be treated as a periodic function of period $2R$, with the range $-R \le x \le R$ being one period. Let us set all the γ_j to zero and choose the λ_j such that

$$\sqrt{\lambda_j/D} = 2\pi j/(2R). \qquad (17.44)$$

This gives

$$c(x,0) = \frac{1}{2}A_0 + \sum_{j=1}^{N} A_j \cos(j\pi x/R), \qquad (17.45)$$

with $A_0/2 = c_\infty$. This expression is periodic with period $2R$ and is symmetric about $x = 0$. It is exactly a Fourier series, in the form of Eq. (17.40), but with R replaced with $2R$. The sine terms all have zero coefficients, to ensure symmetry about $x = 0$.

The coefficients are given by

$$A_j = \frac{2}{2R}\int_{-R}^{R} \cos(\pi j x/R)\, c_{\text{obs}}(x)\, dx\,. \qquad (17.46)$$

Note that

$$A_0 = \frac{1}{R}\int_{-x_0}^{x_0} \cos(0)\, c_0\, dx = \frac{1}{R} c_0 \int_{-x_0}^{x_0} dx = \frac{2x_0}{R}\, c_0 \,,$$

which implies that $c_\infty = (x_0/R)c_0$. This is just what we expect in the $t \to \infty$ limit; the solute is spread evenly throughout the lake and the final concentration is the original value reduced by the ratio of the initial size x_0 of the polluted region to the full length of the lake.

The value for N is arbitrary, but it must be large enough to give an acceptably accurate description of the initial condition $c_{\rm obs}(x)$. The figure below compares $c_{\rm obs}$ with $c(x, 0)$ for $N = 5$ (the dotted curve) and $N = 50$ (the dashed curve):

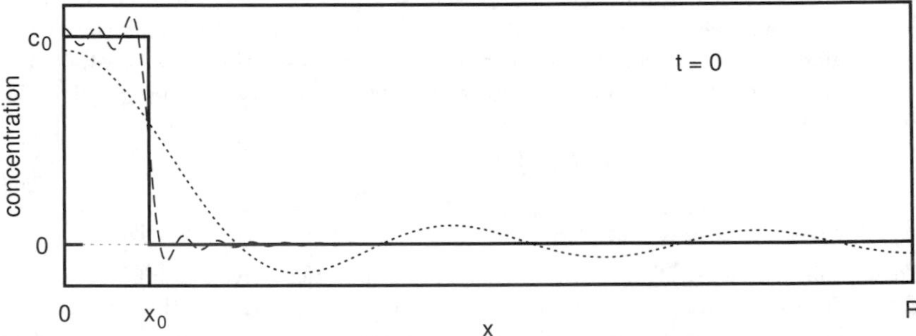

For large N the Fourier series gives an accurate approximation over almost the full range, although just before x_0, the point of discontinuity in $c_{\rm obs}$, it has trouble, overshooting just before x_0 and undershooting just after it. This behavior is characteristic of Fourier series (and of other kinds of orthogonal function series approximations—see the Chebyshev series in Example 17.2) in the neighborhood of a discontinuity. It is known as the Gibbs phenomenon.[11]

For the full solution for the concentration at any x and t, we plug in the A_j's obtained from Eq. (17.46) into the general expression,

$$c(x,t) = \frac{x_0}{R}c_0 + \sum_{j=1}^{N} A_j \cos(j\pi x/R)\, e^{-(j\pi/R)^2 Dt}. \qquad (17.47)$$

The following plot shows the result from using $N = 250$. $c(x, t)$ is shown at $t = 0$, at an intermediate time t_1, and at a much later time t_2.

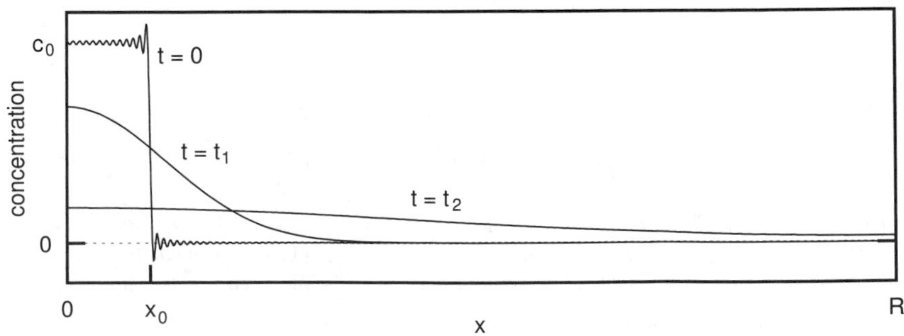

[11]It was studied in detail by Harvard professor Emile Bôcher in 1906. He called it the Gibbs phenomenon because it was brought to his attention by a paper of Gibbs.

With t even just slightly greater than zero, the discontinuity is smoothed out, the Gibbs phenomenon becomes insignificant, and a smaller basis size can give adequate accuracy. In any case, the discontinuity in $c_{\rm obs}$ is just an artifact. A more accurate measurement of the initial concentration distribution would give a $c_{\rm obs}(x)$ that drops to zero as a continuous function in the neighborhood of x_0.

17.4 Spherical Harmonics

We saw in Section 4.1.2 that the differential volume for integrating in spherical polar coordinates is $r^2 \sin\theta \, d\theta \, d\phi \, dr$. If r is constant, then the differential in the integral is just $\sin\theta \, d\theta \, d\phi$. For problems in which motion is constrained to the surface of a sphere we will often encounter integrals with this differential. For this reason it useful to consider an inner product

$$\langle f, g \rangle = \int_0^{2\pi} \int_0^{\pi} f^*(\theta,\phi) g(\theta,\phi) \sin\theta \, d\theta d\phi. \tag{17.48}$$

The *spherical harmonics* are a particular set of functions of the variables θ and ϕ that span an L^2 Hilbert space with this inner product.

Let us first determine the ϕ dependence of the basis functions. Note that the integral over ϕ in Eq. (17.48) is identical to the inner product used to construct the Fourier series. This means that for any specified value of θ, any functions f and g in the Hilbert space can be expressed as Fourier series,

$$f(\theta,\phi) = \sum_{m=-\infty}^{\infty} a_m(\theta) \frac{1}{\sqrt{2\pi}} e^{im\phi}, \quad g(\theta,\phi) = \sum_{n=-\infty}^{\infty} b_n(\theta) \frac{1}{\sqrt{2\pi}} e^{in\phi}. \tag{17.49}$$

The θ dependence is located only in the Fourier coefficients.

Consider the simple case in which there is no dependence on ϕ. Then all of the a_m are zero except a_0, because only $e^{i0\phi} = 1$ is independent of ϕ. We have $f = a_0(\theta)/\sqrt{2\pi}$, $g = b_0(\theta)/\sqrt{2\pi}$, and

$$\langle f, g \rangle = \int_0^{\pi} a_0^*(\theta) \, b_0(\theta) \sin\theta d\theta.$$

Make the change of variable

$$x = \cos\theta \,, \quad dx = -\sin\theta d\theta \,, \quad a_0(\theta) = \alpha_0(\cos\theta) \,, \quad b_0(\theta) = \beta_0(\cos\theta) \,,$$

with limits of integration from $\cos 0 = 1$ to $\cos\pi = -1$. The inner product becomes

$$\langle f, g \rangle = \int_{-1}^{1} \alpha_0(x)^* \, \beta_0(x) dx,$$

which is the inner product that generates the Legendre polynomials. Any function $f(\theta)$ (with no ϕ dependence) in this Hilbert space can be expressed in a Legendre polynomial basis with $x = \cos\theta$,

$$f(\theta) = \sum_{k=0}^{\infty} \alpha_{0,k} P_k(\cos\theta),$$

where the $\alpha_{0,k}$ are constants.

We can proceed in similar fashion for functions with ϕ dependence, now including the higher-order terms of the Fourier series. The inner product becomes

$$\begin{aligned}\langle f, g\rangle &= \sum_{m,n} \langle a_m, b_n\rangle \left\langle \frac{1}{\sqrt{2\pi}} e^{-im\phi}, \frac{1}{\sqrt{2\pi}} e^{in\phi} \right\rangle \\ &= \sum_m \langle a_m, b_m\rangle = \sum_m \int_0^{\pi} a_m^*(\theta)\, b_m(\theta) \sin\theta d\theta. \qquad (17.50)\end{aligned}$$

Thus, the inner product has the same integral as before. However, it is *not* valid to also express the Fourier coefficients with $m \neq 0$ in terms of the P_k. The derivation of appropriate basis functions for the higher-order coefficients is rather complicated. It turns out that the use of the P_k will not result in a complete set to span the space of functions of the two coordinates θ and ϕ. However, this can be remedied with a minor modification.

Let us define the ***associated Legendre functions*** as[12]

$$P_k^{(m)}(x) = (1 - x^2)^{m/2} \frac{d^m}{dx^m} P_k(x). \qquad (17.51)$$

for $m \geq 0$. These turn out to be orthogonal with the same inner product as the P_k. Note that we must have $k \geq m$, so that P_k, a polynomial of degree k, is not made zero by the derivative. The $P_k^{(m)}$ are also defined for negative values of the upper index; in that case the function is related to the one with corresponding positive index according to the definition

$$P_k^{(-m)}(x) = (-1)^m \frac{(k-m)!}{(k+m)!} P_k^{(m)}(x). \qquad (17.52)$$

The normalized version of these functions is

$$\mathcal{P}_k^{(m)}(x) = \sqrt{\frac{(k+\frac{1}{2})(k-m)!}{(k+m)!}}\, P_k^{(m)}(x). \qquad (17.53)$$

Substituting into the Fourier series, we obtain

$$f(\theta, \phi) = \sum_{m=-\infty}^{\infty} \frac{1}{\sqrt{2\pi}} e^{im\phi} (-1)^m \sum_{k=|m|}^{\infty} a_{k,m} \mathcal{P}_k^{(m)}(\cos\theta), \qquad (17.54)$$

and it turns out that any function in the Hilbert space can be expressed in this way.[13] The factor $(-1)^m$ in this equation is arbitrary—an orthogonal

[12] The standard notation is P_k^m.

[13] See Section 19.4 and Exercise 19.7.

Table 17.4: Spherical harmonics.

$$Y_0^{(0)} = \frac{1}{\sqrt{4\pi}}$$

$$Y_1^{(0)} = \sqrt{\tfrac{3}{4\pi}}\cos\theta \qquad Y_1^{(1)} = -\sqrt{\tfrac{3}{8\pi}}\sin\theta\, e^{i\phi} \qquad Y_1^{(-1)} = \sqrt{\tfrac{3}{8\pi}}\sin\theta\, e^{-i\phi}$$

$$Y_2^{(0)} = \sqrt{\tfrac{5}{16\pi}}(\cos^2\theta - 1)$$

$$Y_2^{(1)} = -\sqrt{\tfrac{15}{8\pi}}\sin\theta\cos\theta\, e^{i\phi} \qquad Y_2^{(-1)} = \sqrt{\tfrac{15}{8\pi}}\sin\theta\cos\theta\, e^{-i\phi}$$

$$Y_2^{(2)} = -\sqrt{\tfrac{15}{32\pi}}\sin^2\theta\, e^{2i\phi} \qquad Y_2^{(-2)} = \sqrt{\tfrac{15}{32\pi}}\sin^2\theta\, e^{-2i\phi}$$

$$Y_3^{(0)} = \sqrt{\tfrac{7}{16\pi}}(5\cos^3\theta - 3\cos\theta)$$

$$Y_3^{(1)} = -\sqrt{\tfrac{21}{64\pi}}(5\cos^2\theta - 1)\sin\theta\, e^{i\phi} \qquad Y_3^{(-1)} = \sqrt{\tfrac{21}{64\pi}}(5\cos^2\theta - 1)\sin\theta\, e^{-i\phi}$$

$$Y_3^{(2)} = -\sqrt{\tfrac{105}{32\pi}}\sin^2\theta\cos\theta\, e^{2i\phi} \qquad Y_3^{(-2)} = \sqrt{\tfrac{105}{32\pi}}\sin^2\theta\cos\theta\, e^{-2i\phi}$$

$$Y_3^{(3)} = -\sqrt{\tfrac{35}{64\pi}}\sin^3\theta\, e^{3i\phi} \qquad Y_3^{(-3)} = \sqrt{\tfrac{35}{64\pi}}\sin^3\theta\, e^{-3i\phi}$$

vector can be multiplied by -1 and it will remain orthogonal—but is usually included as a matter of convention. (Of course, it could simply be absorbed into the unspecified constants $a_{k,m}$.)

Eq. (17.54) is called Laplace's series.[14] It is usually written in the form

$$f(\theta,\phi) = \sum_{m=-\infty}^{\infty}\sum_{k=|m|}^{\infty} a_{k,m}Y_k^{(m)}(\theta,\phi), \tag{17.55}$$

where

$$Y_k^{(m)}(\theta,\phi) = (-1)^m\mathcal{P}_k^{(m)}(\cos\theta)\,\frac{1}{\sqrt{2\pi}}\,e^{im\phi}. \tag{17.56}$$

The functions $Y_k^{(m)}(\theta,\phi)$ are called the ***spherical harmonics***.[15] Explicit expressions for some of them are given in Table 17.4. They form a complete orthonormal basis for piecewise continuous functions of θ and ϕ with the inner product

$$\langle u,v\rangle = \int_0^{2\pi}\int_0^{\pi} u^*(\theta,\phi)v(\theta,\phi)\sin\theta d\theta d\phi. \tag{17.57}$$

[14]It was developed by Laplace in the 1780's to solve the differential equation $\nabla^2 f = 0$ in polar coordinates.

[15]This reason for this evocative name, coined by Kelvin in the mid-19th century, is somewhat obscure. Solutions to Laplace's equation were known as "harmonic" functions, because when expressed in Cartesian coordinates they satisfy equations that look like Newton's equation for the harmonic oscillator, which is a model for the vibrating string of a musical instrument. In German the more sensible name *Kugelfunktion* (sphere function) is used. The standard notation is Y_k^m.

The differential equation satisfied by the $Y_k^{(m)}(\theta,\phi)$ is

$$\frac{1}{\sin\theta}\frac{\partial}{\partial\theta}\left(\sin\theta\,\frac{\partial Y_k^{(m)}}{\partial\theta}\right)+\frac{1}{\sin^2\theta}\frac{\partial^2 Y_k^{(m)}}{\partial\phi^2}+k(k+1)\,Y_k^{(m)}=0. \qquad (17.58)$$

The differential operator here just happens to be the angular part of the Laplacian, ∇^2, as given by Eq. (3.39). (This fact will prove useful!)

Exercises

17.1 (a) Prove that any element of an orthonormal set can be multiplied by (-1) without affecting its orthonormality. (b) Prove also that multiplication by i does not affect orthonormality.

17.2 Consider the linearly independent set $\{(1,1,1),(3,2,1),(2,0,4)\}$.

(a) Use Gram-Schmidt orthogonalization to transform it into an orthonormal basis for $\mathbb{R}^3$.

(b) What are the coordinates of the point $(-5,1,-7)$ in the basis from part (a)? Obtain your answer using an inner product, and then confirm the result by showing the linear combination of unit vectors that yields the point in question.

(c) Express the conventional unit vectors (1,0,0), (0,1,0), and (0,0,1) as linear combinations of your basis vectors.

17.3 Using the inner product as defined for the Laguerre polynomials, apply Gram-Schmidt orthogonalization to the basis $\{1,x,x^2,x^3\}$ but start with x^3 as the first (unnormalized) vector. Feel free to use computer algebra to evaluate the integrals.

17.4 What is the "angle" between the Hermite polynomials H_3 and H_4?

17.5 Express x^4 as a linear combination of Legendre polynomials.

17.6 Approximate the square well $V(x)=\begin{cases}-1, & -1\le x\le 0.5\\ 1, & |x|>0.5\end{cases}$

over the range $-1\le x\le 1$ using the following methods. In each case carry out the analysis for various basis dimensions and compare the approximate and the exact results for $V(x)$ by plotting them.

(a) Resolution in a Legendre polynomial basis.

(b) Resolution in a basis of Chebyshev polynomials of the first kind.

(c) A Fourier series.

17.7* Repeat the transport analysis of Example 17.4 with the pollutant initially in a narrow band somewhere *within* the lake.

17.8* Repeat the transport analysis of Example 17.4 in two dimensions, for a square lake with the pollutant initially in a square region of dimension $x_0\times x_0$ at one of the shores.

17.9 (a) Demonstrate that $Y_3^{(1)}$ and $Y_3^{(2)}$ are orthogonal. (b) Demonstrate that $Y_3^{(1)}$ and $Y_2^{(1)}$ are orthogonal.

Chapter 18

Matrices

18.1 Matrix Representation of Operators

Matrices are a convenient tool for linear algebra calculations. They are usually the method of choice for representing operators in finite-dimension vector spaces. As an example, let us construct an operator $\mathbf{R}(\phi)$ that rotates any Cartesian coordinate vector by angle ϕ in the xy-plane. An easy way to do this is to specify the effect of the rotation on the unit vectors $(1,0)$ and $(0,1)$. The effect of the operator, as shown in Fig. 18.1, is to carry out the mappings

$$(1,0) \stackrel{\mathbf{R}(\phi)}{\longrightarrow} (\cos\phi, \sin\phi) = \cos\phi\ (1,0) + \sin\phi\ (0,1)\,, \tag{18.1}$$

$$(0,1) \stackrel{\mathbf{R}(\phi)}{\longrightarrow} (-\sin\phi, \cos\phi) = -\sin\phi\ (1,0) + \cos\phi\ (0,1)\,. \tag{18.2}$$

Having determined the effect of the operator on the unit vectors, we can now determine the effect on an arbitrary vector (u_1, u_2). First we express the arbitrary vector as a linear combination of the unit vectors,

$$(u_1, u_2) = u_1\ (1,0) + u_2\ (0,1). \tag{18.3}$$

Then we operate with $\mathbf{R}(\phi)$:

$$\begin{aligned}\mathbf{R}(\phi)\ (u_1,u_2) &= u_1\ \mathbf{R}(\phi)\ (1,0) + u_2\ \mathbf{R}(\phi)\ (0,1)\\ &= u_1\ [\cos\phi\ (1,0) + \sin\phi\ (0,1)] + u_2\ [-\sin\phi\ (1,0) + \cos\phi\ (0,1)]\,.\end{aligned}$$

Collecting the terms according to the unit vectors, we obtain

$$\begin{aligned}\mathbf{R}(\phi)\ (u_1,u_2) &= (u_1\cos\phi - u_2\sin\phi)(1,0) + (u_1\sin\phi + u_2\cos\phi)(0,1)\\ &= (u_1\cos\phi - u_2\sin\phi,\ u_1\sin\phi + u_2\cos\phi)\,.\end{aligned} \tag{18.4}$$

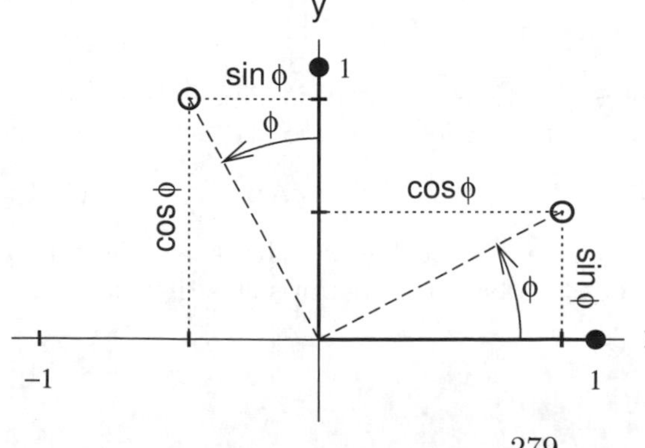

Figure 18.1: The effect of rotation by angle ϕ in the xy-plane on the unit vectors $(1,0)$ and $(0,1)$.

The analysis as carried out above was cumbersome. It can be expressed more simply using a ***matrix***. A matrix is a array of scalars enclosed in parentheses. Operators in a two-dimensional vector space can be represented by four numbers in two rows,

$$\mathbf{A} = \begin{pmatrix} a_{1,1} & a_{1,2} \\ a_{2,1} & a_{2,2} \end{pmatrix}, \tag{18.5}$$

such that $\mathbf{A}\,\mathbf{u} = (a_{1,1}u_1 + a_{1,2}u_2,\ a_{2,1}u_1 + a_{2,2}u_2)$. This can be conveniently expressed as

$$\begin{pmatrix} a_{1,1} & a_{1,2} \\ a_{2,1} & a_{2,2} \end{pmatrix} \begin{pmatrix} u_1 \\ u_2 \end{pmatrix} = \begin{pmatrix} a_{1,1}u_1 + a_{1,2}u_2 \\ a_{2,1}u_1 + a_{2,2}u_2 \end{pmatrix}. \tag{18.6}$$

The representation of $\mathbf{u}$ in Eq. (18.6) as a column of its coordinates,

$$\mathbf{u} = \begin{pmatrix} u_1 \\ u_2 \end{pmatrix}, \tag{18.7}$$

is called a ***column vector***. The dot product of the first row of the matrix with the column vector $\mathbf{u}$ yields the first element of the column vector of the result, and the dot product of the second row with $\mathbf{u}$ yields the second element of the result. Eq. (18.5) is a 2×2 matrix, since it has two rows and two columns. An operator in vector spaces of dimension N would be represented by an $N \times N$ matrix. Comparing with Eq. (18.4), we see that the ***rotation matrix*** is

$$\mathbf{R}(\phi) = \begin{pmatrix} \cos\phi & -\sin\phi \\ \sin\phi & \cos\phi \end{pmatrix}. \tag{18.8}$$

Example 18.1. *Rotation in xy-plane.* Let us rotate the vector $(2, 3)$ by $153°$. The rotation matrix is

$$\mathbf{R}(153°) = \begin{pmatrix} -0.891 & -0.454 \\ 0.454 & -0.891 \end{pmatrix}$$

and the rotated vector is

$$\begin{pmatrix} -0.891 & -0.454 \\ 0.454 & -0.891 \end{pmatrix} \begin{pmatrix} 2 \\ 3 \end{pmatrix}$$

$$= \begin{pmatrix} -1.78 - 1.36 \\ 0.908 - 2.67 \end{pmatrix} = \begin{pmatrix} -3.14 \\ -1.77 \end{pmatrix}.$$

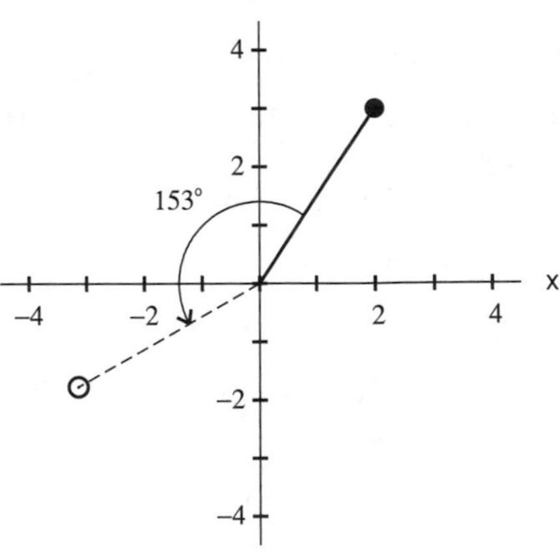

The reason that we can express $\mathbf{R}(\phi)$ as a matrix is that it is a ***linear operator***. Any operator $\mathbf{A}$ with the properties

$$\mathbf{A}\,(\mathbf{u} + \mathbf{v}) = \mathbf{A}\,\mathbf{u} + \mathbf{A}\,\mathbf{v}, \qquad \mathbf{A}\,(c\mathbf{u}) = c\mathbf{A}\mathbf{u}, \tag{18.9}$$

for vectors $\mathbf{u}$ and $\mathbf{v}$ and scalar c, is said to be *linear*. Thanks to these two properties, if we know what $\mathbf{A}$ does to a basis of a vector space then we know what it does to any vector in the space. This is because any vector can be expressed as a linear combination of the basis vectors.

The reason the rotation matrix has dimensions 2×2 is that it operates on vectors of dimension 2 and gives back vectors also of dimension 2. In other words, it maps vectors from $\mathbb{R}^2$ into other vectors in $\mathbb{R}^2$. Similarly, any linear operator that maps vectors from $\mathbb{R}^N$ to other vectors in $\mathbb{R}^N$ (or, more generally, $\mathbb{C}^N$ to $\mathbb{C}^N$) can be expressed as a matrix of dimension $N \times N$, with N rows and N columns.

Example 18.2. *Moment of inertia tensor.** In Example 13.6 we introduced the angular velocity vector $\boldsymbol{\omega} = (\omega_x, \omega_y, \omega_z)$ to describe the rotation of a rigid body. For the angular momentum vector of a rigid body we obtained the rather messy expression

$$\mathbf{L} = \sum_j m_j \mathbf{r}_j \times (\boldsymbol{\omega} \times \mathbf{r}_j),$$

where $\mathbf{r}_j = (x_j, y_j, z_j)$ is the coordinate vector of the jth of the particles that comprise the body. It is straightforward (albeit tedious) to show that $\mathbf{r}_j \times (\boldsymbol{\omega} \times \mathbf{r}_j)$ is the column vector

$$\begin{pmatrix} (y_j^2 + z_j^2)\omega_x - x_j y_j \omega_y - x_j z_j \omega_z \\ -x_j y_j \omega_x + (x_j^2 + z_j^2)\omega_y - y_j z_j \omega_z \\ -x_j z_j \omega_x - y_j z_j \omega_y + (x_j^2 + y_j^2)\omega_z \end{pmatrix}.$$

It follows that

$$\mathbf{L} = \boldsymbol{\mathcal{I}}\boldsymbol{\omega}, \qquad \boldsymbol{\mathcal{I}} = \begin{pmatrix} \sum_j m_j (y_j^2 + z_j^2) & -\sum_j m_j x_j y_j & -\sum_j m_j x_j z_j \\ -\sum_j m_j x_j y_j & \sum_j m_j (x_j^2 + z_j^2) & -\sum_j m_j y_j z_j \\ -\sum_j m_j x_j z_j & -\sum_j m_j y_j z_j & \sum_j m_j (x_j^2 + y_j^2) \end{pmatrix}. \tag{18.10}$$

The 3×3 matrix $\boldsymbol{\mathcal{I}}$ is called the ***moment of inertia tensor***. $\boldsymbol{\mathcal{I}}$ is a linear operator that maps the three-dimensional angular velocity vector $\boldsymbol{\omega}$ to the three-dimensional angular momentum vector $\mathbf{L}$.

If $\mathbf{A}$ maps vectors of $\mathbb{R}^N$ to vectors of $\mathbb{R}^M$ (or $\mathbb{C}^N$ to $\mathbb{C}^M$), the matrix representation of $\mathbf{A}$ will have N columns and M rows. Let $\{\mathbf{e}_n\}$, $n = 1, \ldots, N$, be a basis for $\mathbb{R}^N$ (or $\mathbb{C}^N$) such that each $\mathbf{e}_n$ is a column vector with 1 in the nth position and 0 in all other positions. Similarly, let $\{\mathbf{f}_m\}$, $m = 1, \ldots, M$, be a basis for $\mathbb{R}^M$ (or $\mathbb{C}^M$). If the effect of the operator $\mathbf{A}$ on $\mathbf{e}_n$ is

$$\mathbf{e}_n \xrightarrow{\mathbf{A}} a_{1,n}\mathbf{f}_1 + a_{2,n}\mathbf{f}_2 + \cdots + a_{M,n}\mathbf{f}_M, \tag{18.11}$$

then the nth column of the matrix $\mathbf{A}$ consists of $(a_{1,n}, a_{2,n}, \ldots, a_{M,n})$. For example, for a mapping of $\mathbb{R}^3$ to $\mathbb{R}^3$,

$$\mathbf{A}\,\mathbf{e}_2 = \begin{pmatrix} a_{1,1} & a_{1,2} & a_{1,3} \\ a_{2,1} & a_{2,2} & a_{2,3} \\ a_{3,1} & a_{3,2} & a_{3,3} \end{pmatrix} \begin{pmatrix} 0 \\ 1 \\ 0 \end{pmatrix} = \begin{pmatrix} a_{1,2} \\ a_{2,2} \\ a_{3,2} \end{pmatrix} = a_{1,2}\mathbf{f}_1 + a_{2,2}\mathbf{f}_2 + a_{3,2}\mathbf{f}_3 .$$

An arbitrary vector $b_1\mathbf{e}_1 + b_2\mathbf{e}_2 + \cdots + b_N\mathbf{e}_N$ in $\mathbb{R}^N$ (or $\mathbb{C}^N$) is mapped to a vector in $\mathbb{R}^M$ (or $\mathbb{C}^M$) according to

$$\begin{pmatrix} a_{1,1} & \cdots & a_{1,N} \\ \vdots & & \vdots \\ a_{M,1} & \cdots & a_{M,N} \end{pmatrix} \begin{pmatrix} b_1 \\ \vdots \\ b_N \end{pmatrix} = \begin{pmatrix} \sum_{j=1}^{N} a_{1,j} b_j \\ \vdots \\ \sum_{j=1}^{N} a_{M,j} b_j \end{pmatrix}. \tag{18.12}$$

A matrix of dimensions $M \times N$, with M rows and N columns, is called a ***square matrix*** if $M = N$ and a ***rectangular matrix*** if $M \neq N$.

18.2 Matrix Algebra

Let us operate twice with $\mathbf{R}(\phi)$. The result is

$$\begin{aligned} \mathbf{R}(\phi)\,\mathbf{R}(\phi)\,u = \mathbf{R}(\phi) &\begin{pmatrix} u_1 \cos\phi - u_2 \sin\phi \\ u_1 \sin\phi + u_2 \cos\phi \end{pmatrix} \\ &= \begin{pmatrix} \cos\phi\,(u_1 \cos\phi - u_2 \sin\phi) - \sin\phi\,(u_1 \sin\phi + u_2 \cos\phi) \\ \sin\phi\,(u_1 \cos\phi - u_2 \sin\phi) + \cos\phi\,(u_1 \sin\phi + u_2 \cos\phi) \end{pmatrix} \\ &= \begin{pmatrix} u_1(\cos^2\phi - \sin^2\phi) - 2u_2 \sin\phi\cos\phi \\ 2u_1 \sin\phi\cos\phi + u_2(\cos^2\phi - \sin^2\phi) \end{pmatrix}. \end{aligned} \tag{18.13}$$

Comparing with the trigonometric identities for angle addition, Eqs. (2.39) and (2.40), we see that

$$\mathbf{R}(\phi)\,\mathbf{R}(\phi)\,u = \begin{pmatrix} u_1 \cos 2\phi - u_2 \sin 2\phi \\ u_1 \sin 2\phi + u_2 \cos 2\phi \end{pmatrix} = \mathbf{R}(2\phi)\,u\,. \tag{18.14}$$

Thus, two rotations by ϕ are equivalent to a single rotation by 2ϕ, as we should expect!

Eq. (18.13) implies that the matrix representation of the double rotation is

$$\mathbf{R}(\phi)\,\mathbf{R}(\phi) = \begin{pmatrix} \cos^2\phi - \sin^2\phi & -2\sin\phi\cos\phi \\ 2\sin\phi\cos\phi & \cos^2\phi - \sin^2\phi \end{pmatrix}. \tag{18.15}$$

We can obtain this directly from the matrix representation of $\mathbf{R}(\phi)$ as follows:

$$\begin{pmatrix} \cos\phi & -\sin\phi \\ \sin\phi & \cos\phi \end{pmatrix} \begin{pmatrix} \cos\phi & -\sin\phi \\ \sin\phi & \cos\phi \end{pmatrix} = \begin{pmatrix} \cos^2\phi - \sin^2\phi & -2\sin\phi\cos\phi \\ 2\sin\phi\cos\phi & \cos^2\phi - \sin^2\phi \end{pmatrix}.$$

The "1,1" element of the result is the dot product of row 1 of the first matrix and column 1 of the second matrix,

$$(\cos\phi \quad -\sin\phi) \begin{pmatrix} \cos\phi \\ \sin\phi \end{pmatrix} = (\cos\phi,\ -\sin\phi) \cdot (\cos\phi,\ \sin\phi) = \cos^2\phi - \sin^2\phi\,.$$

The "1,2" element is the product of row 1 and column 2,

$$(\cos\phi \quad -\sin\phi)\begin{pmatrix}-\sin\phi\\ \cos\phi\end{pmatrix} = (\cos\phi, -\sin\phi)\cdot(-\sin\phi, \cos\phi) = -2\cos\phi\,\sin\phi\,.$$

The "2,1" element is the product of row 2 and column 1, and the "2,2" element is the product of row 2 and column 2.

An alternative notation for the dot product of arbitrary vectors (b_1, b_2) and (a_1, a_2) is to write one vector as a ***row vector*** and the other as a column vector:

$$(b_1 \quad b_2)\begin{pmatrix}a_1\\ a_2\end{pmatrix} = b_1 a_1 + b_2 a_2\,. \tag{18.16}$$

The product of any pair of 2×2 matrices $\mathbf{A}$ and $\mathbf{B}$ can be written

$$\begin{aligned}\mathbf{B\,A} &= \begin{pmatrix}b_{1,1} & b_{1,2}\\ b_{2,1} & b_{2,2}\end{pmatrix}\begin{pmatrix}a_{1,1} & a_{1,2}\\ a_{2,1} & a_{2,2}\end{pmatrix}\\ &= \begin{pmatrix}(b_{1,1} \quad b_{1,2})\begin{pmatrix}a_{1,1}\\ a_{2,1}\end{pmatrix} & (b_{1,1} \quad b_{1,2})\begin{pmatrix}a_{1,2}\\ a_{2,2}\end{pmatrix}\\ (b_{2,1} \quad b_{2,2})\begin{pmatrix}a_{1,1}\\ a_{2,1}\end{pmatrix} & (b_{2,1} \quad b_{2,2})\begin{pmatrix}a_{1,2}\\ a_{2,2}\end{pmatrix}\end{pmatrix}\\ &= \begin{pmatrix}b_{1,1}a_{1,1} + b_{1,2}a_{2,1} & b_{1,1}a_{1,2} + b_{1,2}a_{2,2}\\ b_{2,1}a_{1,1} + b_{2,2}a_{2,1} & b_{2,1}a_{1,2} + b_{2,2}a_{2,2}\end{pmatrix}.\end{aligned} \tag{18.17}$$

$\mathbf{A}$ and $\mathbf{B}$ need not have the same numbers of rows and columns, as long as the number of columns in the matrix on the left equals the number of rows in the matrix on the right. Multiplication of an $M\times K$ matrix on the left by a $K\times N$ matrix on the right yields an $M\times N$ matrix.

Example 18.3. *The product of a 4×2 matrix and a 2×3 matrix is 4×3.*

$$\begin{pmatrix}4 & 3\\ 0 & 2\\ 1 & 5\\ -2 & 1\end{pmatrix}\begin{pmatrix}3 & 2 & 1\\ 2 & 0 & 5\end{pmatrix} = \begin{pmatrix}12+6 & 8+0 & 4+15\\ 0+4 & 0+0 & 0+10\\ 3+10 & 2+0 & 1+25\\ -6+2 & -4+0 & -2+5\end{pmatrix} = \begin{pmatrix}18 & 8 & 19\\ 4 & 0 & 10\\ 13 & 2 & 26\\ -4 & -4 & 3\end{pmatrix}.$$

Matrix multiplication can be expressed in summation notation. Let $\mathbf{A}$ and $\mathbf{B}$ be arbitrary matrices of dimensions $M\times K$ and $K\times N$, respectively, with elements $a_{i,j}$ and $b_{i,j}$. Then $\mathbf{C} = \mathbf{BA}$ is the $M\times N$ matrix with elements

$$c_{p,q} = \sum_{k=1}^{K} b_{p,k}\,a_{k,q}\,. \tag{18.18}$$

We can develop an algebra of matrices. Multiplication of a pair matrices is defined as above. Multiplication of a matrix by a scalar gives a matrix in which each element is multiplied by the scalar,

$$c \begin{pmatrix} a_{1,1} & a_{1,2} \\ a_{2,1} & a_{2,2} \end{pmatrix} = \begin{pmatrix} c\,a_{1,1} & c\,a_{1,2} \\ c\,a_{2,1} & c\,a_{2,2} \end{pmatrix}. \tag{18.19}$$

Adding matrices of the same dimensions adds elements in the same positions,

$$\begin{pmatrix} a_{1,1} & a_{1,2} \\ a_{2,1} & a_{2,2} \end{pmatrix} + \begin{pmatrix} b_{1,1} & b_{1,2} \\ b_{2,1} & b_{2,2} \end{pmatrix} = \begin{pmatrix} a_{1,1}+b_{1,1} & a_{1,2}+b_{1,2} \\ a_{2,1}+b_{2,1} & a_{2,2}+b_{2,2} \end{pmatrix}. \tag{18.20}$$

The examples being used here are 2×2, but the generalization to arbitrary dimensions $M \times N$ is straightforward. The $M \times N$ ***zero matrix*** is the matrix $\mathbf{0}$ with all elements equal to zero. Adding it to another matrix leaves that matrix unchanged,

$$\mathbf{A} + \mathbf{0} = \mathbf{A}\,. \tag{18.21}$$

Multiplying a matrix by a zero matrix (of appropriate dimension), from either side, yields a zero matrix. The additive inverse of a matrix is obtained by multiplying it by the scalar -1.

We define the ***identity matrix***, with symbol $\mathbf{I}$, as a square matrix that is 1 along the diagonal and zero everywhere else. The 2×2 identity matrix, for example, is

$$\mathbf{I} = \begin{pmatrix} 1 & 0 \\ 0 & 1 \end{pmatrix}. \tag{18.22}$$

Multiplying a matrix by $\mathbf{I}$ leaves it unchanged,

$$\mathbf{IA} = \mathbf{A}\,, \qquad \mathbf{AI} = \mathbf{A}\,. \tag{18.23}$$

Matrix multiplication is associative but generally not commutative; that is, $\mathbf{A}(\mathbf{BC})$ equals $(\mathbf{AB})\mathbf{C}$ but $\mathbf{BA}$ does not in general equal $\mathbf{AB}$. [An exception is multiplication by $\mathbf{I}$, which according to Eq. (18.23) is commutative.] In contrast to arithmetic with scalars, the order in which matrices are multiplied matters.

18.3 Matrix Operations

Transpose and *inversion* are operations that map a matrix into another matrix while the *trace* and the *determinant* are operations that map a matrix into a scalar. Here the first three of these operations are described. The determinant is the subject of Section 18.5.

The ***transpose*** of a matrix, designated $\mathbf{A}^{\mathbf{T}}$, is the matrix obtained by interchanging the rows and columns:

$$\begin{pmatrix} a_{1,1} & a_{1,2} \\ a_{2,1} & a_{2,2} \end{pmatrix}^{\mathbf{T}} = \begin{pmatrix} a_{1,1} & a_{2,1} \\ a_{1,2} & a_{2,2} \end{pmatrix}, \qquad \begin{pmatrix} a_{1,1} & a_{1,2} \\ a_{2,1} & a_{2,2} \\ a_{3,1} & a_{3,2} \end{pmatrix}^{\mathbf{T}} = \begin{pmatrix} a_{1,1} & a_{2,1} & a_{3,1} \\ a_{1,2} & a_{2,2} & a_{3,2} \end{pmatrix}.$$

For square matrices this is the same as reflecting through the diagonal. In general, if the elements of $\mathbf{A}$ are $a_{i,j}$, then the elements of $\mathbf{A}^\mathbf{T}$ are $a_{j,i}$. (Note that the transpose of an $M \times N$ matrix is an $N \times M$ matrix.) It is easy to demonstrate that

$$(\mathbf{A}+\mathbf{B})^\mathbf{T} = \mathbf{A}^\mathbf{T} + \mathbf{B}^\mathbf{T}, \tag{18.24}$$

$$(\mathbf{BA})^\mathbf{T} = \mathbf{A}^\mathbf{T}\mathbf{B}^\mathbf{T}. \tag{18.25}$$

Note that in Eq. (18.25) the matrices reverse their positions.

Example 18.4. *Transpose of a sum.* Let us derive Eq. (18.24) for a 2×2 matrix:

$$\left[\begin{pmatrix} a_{1,1} & a_{1,2} \\ a_{2,1} & a_{2,2} \end{pmatrix} + \begin{pmatrix} b_{1,1} & b_{1,2} \\ b_{2,1} & b_{2,2} \end{pmatrix}\right]^\mathbf{T}$$

$$= \begin{pmatrix} a_{1,1}+b_{1,1} & a_{1,2}+b_{1,2} \\ a_{2,1}+b_{2,1} & a_{2,2}+b_{2,2} \end{pmatrix}^\mathbf{T} = \begin{pmatrix} a_{1,1}+b_{1,1} & a_{2,1}+b_{2,1} \\ a_{1,2}+b_{1,2} & a_{2,2}+b_{2,2} \end{pmatrix},$$

$$\begin{pmatrix} a_{1,1} & a_{1,2} \\ a_{2,1} & a_{2,2} \end{pmatrix}^\mathbf{T} + \begin{pmatrix} b_{1,1} & b_{1,2} \\ b_{2,1} & b_{2,2} \end{pmatrix}^\mathbf{T}$$

$$= \begin{pmatrix} a_{1,1} & a_{2,1} \\ a_{1,2} & a_{2,2} \end{pmatrix} + \begin{pmatrix} b_{1,1} & b_{2,1} \\ b_{1,2} & b_{2,2} \end{pmatrix} = \begin{pmatrix} a_{1,1}+b_{1,1} & a_{2,1}+b_{2,1} \\ a_{1,2}+b_{1,2} & a_{2,2}+b_{2,2} \end{pmatrix}.$$

A ***symmetric matrix*** is a square matrix $\mathbf{A}$ such that $\mathbf{A} = \mathbf{A}^\mathbf{T}$. In other words, $a_{i,j} = a_{j,i}$. A ***diagonal matrix*** is a square matrix in which all elements not on the diagonal are zero.

The transpose of a column vector is a row vector, and vice versa,

$$\begin{pmatrix} a_1 \\ a_2 \\ a_3 \end{pmatrix}^\mathbf{T} = \begin{pmatrix} a_1 & a_2 & a_3 \end{pmatrix}, \qquad \begin{pmatrix} a_1 & a_2 & a_3 \end{pmatrix}^\mathbf{T} = \begin{pmatrix} a_1 \\ a_2 \\ a_3 \end{pmatrix}. \tag{18.26}$$

Row vectors can multiply a matrix from the left-hand side, for example,

$$\begin{pmatrix} c_1 & c_2 \end{pmatrix} \begin{pmatrix} a_{1,1} & a_{2,1} \\ a_{1,2} & a_{2,2} \end{pmatrix} = \begin{pmatrix} c_1 a_{1,1} + c_2 a_{2,1} & c_1 a_{1,2} + c_2 a_{2,2} \end{pmatrix}. \tag{18.27}$$

The result is another row vector.

The multiplicative inverse,[1] designated $\mathbf{A}^{-1}$, is a matrix that transforms a square matrix $\mathbf{A}$ into $\mathbf{I}$ by multiplication,

$$\mathbf{A}^{-1}\mathbf{A} = \mathbf{I}, \qquad \mathbf{A}\mathbf{A}^{-1} = \mathbf{I}. \tag{18.28}$$

While for a nonzero scalar x the multiplicative inverse is simply the reciprocal $1/x$, for a matrix the inverse is not so obvious. There are systematic algorithms for computing a matrix inverse that are straightforward to implement on a computer but are tedious to perform by hand. Matrix inversion is readily performed by computer algebra systems.

[1] The phrase "inverse of a matrix" is usually understood to mean the *multiplicative* inverse. For the additive inverse one should always use the qualifier "additive."

Example 18.5. *Inverse of a square matrix.* The inverse of the matrix $\begin{pmatrix} 1 & 2 & 3 \\ 4 & 5 & 4 \\ 3 & 2 & 1 \end{pmatrix}$ is $\begin{pmatrix} 3/8 & -1/2 & 7/8 \\ -1 & 1 & -1 \\ 7/8 & -1/2 & 3/8 \end{pmatrix}$, as is confirmed by multiplying them together:

$$\begin{pmatrix} 3/8 & -1/2 & 7/8 \\ -1 & 1 & -1 \\ 7/8 & -1/2 & 3/8 \end{pmatrix} \begin{pmatrix} 1 & 2 & 3 \\ 4 & 5 & 4 \\ 3 & 2 & 1 \end{pmatrix} = \begin{pmatrix} 1 & 0 & 0 \\ 0 & 1 & 0 \\ 0 & 0 & 1 \end{pmatrix}.$$

However, just as the scalar 0 has no reciprocal, there exist matrices that have no inverse. The following existence theorem is therefore quite important:

Theorem 18.3.1. *Let* $\mathbf{A}$ *be an* $N \times N$ *matrix. Then there exists a unique matrix* $\mathbf{A}^{-1}$ *such that* $\mathbf{A}^{-1}\mathbf{A} = \mathbf{I}$ *if and only if the columns of* $\mathbf{A}$*, considered as vectors in* $\mathbb{R}^N$ *or* $\mathbb{C}^N$*, are a linearly independent set.*

This theorem also holds for the rows of the matrix. A simple test for linear independence of columns or rows will be given in Section 18.5.

A matrix for which no inverse exists is said to be ***singular***. Do not confuse this with the concept of a singular function. These are two different meanings of "singular." Sometimes the terms *invertible* and *noninvertible* are used instead of *nonsingular* and *singular*, respectively, to avoid this confusion.

It can be useful to also have a procedure for mapping a matrix to a *scalar*. One way to do this is the ***trace***, designated $\operatorname{tr}\mathbf{A}$, which is simply the sum of the diagonal elements, of a square matrix. Off-diagonal elements have no effect on the value. For example,

$$\operatorname{tr}\begin{pmatrix} 4 & 3 & 2 \\ 2 & 7 & 5 \\ 1 & 1 & -1 \end{pmatrix} = 4 + 7 - 1 = 10\,.$$

This simple operation has important applications in the theory of molecular symmetry. The trace is often called the **character** of the matrix.[2]

18.4 Pseudoinverse*

$\mathbf{A}^{-1}$ is defined only for a square matrix. This is because if $\mathbf{A}$ is rectangular, then the inverse is not unique—it turns out that Eqs. (18.28) can have more than one solution for $\mathbf{A}^{-1}$. For rectangular matrices, it is useful to define something called the *pseudo*inverse:

Definition: For an $M \times N$ matrix $\mathbf{A}$, the **pseudoinverse** is the $N \times M$ matrix $\mathbf{A}^+$ that satisfies the four conditions[3]

$$\mathbf{A}\mathbf{A}^+\mathbf{A} = \mathbf{A}, \quad \mathbf{A}^+\mathbf{A}\mathbf{A}^+ = \mathbf{A}^+, \quad (\mathbf{A}\mathbf{A}^+)^\mathbf{T} = \mathbf{A}\mathbf{A}^+, \quad (\mathbf{A}^+\mathbf{A})^\mathbf{T} = \mathbf{A}^+\mathbf{A}. \tag{18.29}$$

[2]The motivation for this nomenclature will become clear in Section 19.2.

[3]Strictly speaking, this is called the *Moore-Penrose pseudoinverse*, to distinguish it from other, less commonly used, generalizations of the inverse.

By substituting Eqs. (18.28) for $\mathbf{A}^+\mathbf{A}$ or $\mathbf{A}\mathbf{A}^+$, it is easy to show that the true inverse $\mathbf{A}^{-1}$, of a *square* matrix $\mathbf{A}$, satisfies all of these conditions. If $\mathbf{A}$ is rectangular, the following theorem tells us how to calculate $\mathbf{A}^+$:

Theorem 18.4.1. *For any rectangular matrix* $\mathbf{A}$ *there exists at most one solution* $\mathbf{A}^+$ *to Eqs. (18.29). If* $\mathbf{A}$ *is broad (i.e., more columns than rows), then*

$$\mathbf{A}^+ = \mathbf{A}^\mathbf{T}(\mathbf{A}\mathbf{A}^\mathbf{T})^{-1}, \tag{18.30a}$$

which exists if the square matrix $\mathbf{A}\mathbf{A}^\mathbf{T}$ *has an inverse. If* $\mathbf{A}$ *is narrow (fewer columns than rows), then*

$$\mathbf{A}^+ = (\mathbf{A}^\mathbf{T}\mathbf{A})^{-1}\mathbf{A}^\mathbf{T}, \tag{18.30b}$$

which exists if the square matrix $\mathbf{A}^\mathbf{T}\mathbf{A}$ *has an inverse.*

Example 18.6. *Inverse of a narrow matrix.* Consider a 2×1 matrix (i.e., a two-dimensional column vector),

$$\mathbf{A} = \begin{pmatrix} 2 \\ 3 \end{pmatrix}, \qquad \mathbf{A}^+ = \left[(2\ \ 3) \begin{pmatrix} 2 \\ 3 \end{pmatrix} \right]^{-1} (2\ \ 3) = 13^{-1}\,(2\ \ 3) = \left(\tfrac{2}{13}\ \ \tfrac{3}{13} \right),$$

which is easily shown to satisfy Eqs. (18.29). For example,

$$\mathbf{A}^+\mathbf{A} = \left(\tfrac{2}{13}\ \ \tfrac{3}{13} \right) \begin{pmatrix} 2 \\ 3 \end{pmatrix} = 1, \qquad \mathbf{A}\mathbf{A}^+ = \begin{pmatrix} 2 \\ 3 \end{pmatrix} \left(\tfrac{2}{13}\ \ \tfrac{3}{13} \right) = \begin{pmatrix} 4/13 & 6/13 \\ 6/13 & 9/13 \end{pmatrix}.$$

This last matrix is symmetric; obviously, $\mathbf{A}\mathbf{A}^+ = (\mathbf{A}\mathbf{A}^+)^\mathbf{T}$. Compare $\mathbf{A}^+$ with $\mathbf{B} = \left(\tfrac{1}{2}\ \ 0 \right)$. Note that $\mathbf{B}\mathbf{A} = 1$. Therefore, $\mathbf{B}$ can also be thought of as an inverse of $\mathbf{A}$, but it is not the unique pseudoinverse as defined by Eqs. (18.29), because $\mathbf{A}\mathbf{B}$ gives an unsymmetric matrix.

The pseudoinverse has important applications in statistical analysis.

Example 18.7. *The method of least squares as a matrix computation.* Consider a data set $\{(x_1, f_1), (x_2, f_2), \ldots, (x_N, f_N)\}$ with standard deviations $\{\hat\sigma_1, \hat\sigma_2, \ldots, \hat\sigma_N\}$, and basis set $\{u_0(x), u_1(x), \ldots, u_M(x)\}$. The basis functions could be monomials x^j or any other set of linearly independent functions. Using the method of weighted least squares (Section 10.1.2), the fitting function and quality-of-fit functional are, respectively,

$$f(x) = \sum_{j=1}^{M+1} \hat c_{j-1} u_{j-1}(x), \qquad \chi^2 = \sum_{k=1}^{N} \frac{1}{\hat\sigma_k^2} \left[f_k - \sum_{j=1}^{M+1} \hat c_{j-1} u_{j-1}(x_k) \right]^2.$$

Our goal is to find the best set of values $\hat c_i$, as determined by the normal equations,

$$0 = \sum_{k=1}^{N} \frac{1}{\hat\sigma_k^2} \left[f_k - \sum_{j=1}^{M+1} \hat c_{j-1} u_{j-1}(x_k) \right] u_{m-1}(x_k), \quad m = 1, 2, \ldots, M+1.$$

Define an $N \times (M+1)$ matrix $\mathbf{C}$ and a column vector $\mathbf{b}$ of length N according to

$$C_{k,j} = u_{j-1}(x_k)/\hat\sigma_k\,, \qquad b_k = f_k/\hat\sigma_k\,. \tag{18.31}$$

$\mathbf{C}$ is called the **design matrix** of the fit. For a fit (as opposed to an interpolation), we need more data points than coefficients. This requires that $N > M + 1$. $\mathbf{C}$ is therefore a narrow rectangular matrix. Let $\hat{\mathbf{c}}$ be the column vector of length $M+1$ of the unknown coefficients $\hat c_i$. We can then write the normal equations in matrix notation as $\mathbf{C}^\mathbf{T}\mathbf{b} = (\mathbf{C}^\mathbf{T}\mathbf{C})\hat{\mathbf{c}}$, or more simply, in terms of a pseudoinverse, as

$$\hat{\mathbf{c}} = \mathbf{C}^+\,\mathbf{b}. \tag{18.32}$$

This can be solved using computer algorithms. Singular value decomposition (see Section 18.7) is usually the method of choice, to avoid propagation of roundoff error.

18.5 Determinants

The ***determinant*** of a 2×2 matrix is

$$\begin{vmatrix} a & b \\ c & d \end{vmatrix} = ac - bd\,. \tag{18.33}$$

The determinant of a matrix of dimensions $N \times N$ is defined in terms of a set of N determinants of matrices of dimensions $(N-1) \times (N-1)$. Consider an arbitrary $N \times N$ matrix $\mathbf{A}$. Let $|\mathbf{A}|_{i,j}$ denote the determinant of the matrix obtained by removing row i and column j from $\mathbf{A}$. The determinant of $\mathbf{A}$ can be defined as

$$|\mathbf{A}| = \sum_{j=1} (-1)^{j-1} a_{1,j} |\mathbf{A}|_{1,j}\,, \tag{18.34}$$

where the $a_{1,j}$ are the elements of the first row of $\mathbf{A}$.

Example 18.8. *Determinants.*

$$\begin{vmatrix} 4 & 3 & 2 \\ 2 & 7 & 5 \\ 2 & 3 & -2 \end{vmatrix} = 4\begin{vmatrix} 7 & 5 \\ 3 & -2 \end{vmatrix} - 3\begin{vmatrix} 2 & 5 \\ 2 & -2 \end{vmatrix} + 2\begin{vmatrix} 2 & 7 \\ 2 & 3 \end{vmatrix}$$
$$= 4(-14-15) - 3(-4-10) + 2(6-14) = -90,$$

$$\begin{vmatrix} 4 & 3 & 2 & 6 \\ 2 & 7 & 5 & 0 \\ 1 & 1 & -1 & 3 \\ 3 & 8 & -4 & 7 \end{vmatrix} = 4\begin{vmatrix} 7 & 5 & 0 \\ 1 & -1 & 3 \\ 8 & -4 & 7 \end{vmatrix} - 3\begin{vmatrix} 2 & 5 & 0 \\ 1 & -1 & 3 \\ 3 & -4 & 7 \end{vmatrix} + 2\begin{vmatrix} 2 & 7 & 0 \\ 1 & 1 & 3 \\ 3 & 8 & 7 \end{vmatrix} - 6\begin{vmatrix} 2 & 7 & 5 \\ 1 & 1 & -1 \\ 3 & 8 & -4 \end{vmatrix}.$$

Notice the alternating signs of the prefactors.

Determinants have a number of special properties:

Theorem 18.5.1. *If a row or column consists only of zeros, then the value of the determinant is zero.*

Theorem 18.5.2. *Interchanging any two rows or any two columns changes the value of the determinant by a multiplicative factor of* -1.

Theorem 18.5.3. *Multiplying all the elements of a row or all the elements of a column by a scalar c changes the value of the determinant by a multiplicative factor of c.*

Theorem 18.5.4. *Adding to a row any linear combination of the other rows or adding to a column any linear combination of other columns leaves the value of a determinant unchanged.*

Theorem 18.5.5. *The determinant of a matrix equals that of its transpose,* $|\mathbf{A}| = |\mathbf{A}^{\mathrm{T}}|$.

Theorem 18.5.6. *The determinant of the product of square matrices equals the product of their determinants,* $|\mathbf{AB}| = |\mathbf{A}||\mathbf{B}|$.

Determinants have a variety of uses, due to these theorems. A particularly fundamental use is to determine whether or not the inverse of a matrix exists. Recall that the inverse exists if the columns of the matrix are linearly independent. How can we tell if the vectors are linearly independent? Consider the matrices

$$\begin{pmatrix} 1 & 0 & 0 \\ 0 & 0 & 1 \\ 0 & 1 & 0 \end{pmatrix}, \qquad \begin{pmatrix} 1 & 2 & -1 \\ 3 & 6 & -3 \\ 5 & -4 & 16 \end{pmatrix}.$$

In the first matrix the columns are obviously linearly independent. In the second matrix the columns are linearly dependent, because the last column can be expressed as a linear combination of the first two, according to

$$\begin{pmatrix} -1 \\ -3 \\ 16 \end{pmatrix} = 2 \begin{pmatrix} 1 \\ 3 \\ 5 \end{pmatrix} - \frac{3}{2} \begin{pmatrix} 2 \\ 6 \\ -4 \end{pmatrix}.$$

However, this is not at all obvious from looking at the matrix. For this reason, the following theorem is quite useful:

Theorem 18.5.7. *The columns of a square matrix comprise a linearly independent set, and the rows comprise a linearly independent set, if and only if the determinant of the matrix is nonzero.*

For example,

$$\begin{vmatrix} 1 & 2 & -1 \\ 3 & 6 & -3 \\ 5 & -4 & 16 \end{vmatrix} = 1 \begin{vmatrix} 6 & -3 \\ -4 & 16 \end{vmatrix} - 2 \begin{vmatrix} 3 & -3 \\ 5 & 16 \end{vmatrix} - 1 \begin{vmatrix} 3 & 6 \\ 5 & -4 \end{vmatrix}$$

$$= 1 \cdot 84 - 2 \cdot 63 - 1 \cdot (-42) = 0.$$

Proof of Theorem 18.5.7. Let $\mathbf{a}_j$, $j = 1, 2, \ldots N$, be the columns (or rows) of an $N \times N$ matrix. Suppose they comprise a linearly dependent set. Then, according to Theorem 16.7.1, there exists a set of scalars c_j, not all zero, such that

$$\sum_{j=1}^{N} c_j \, \mathbf{a}_j = \mathbf{0}.$$

Let $b_j = c_j / c_k$, where c_k is one of the nonzero scalars. Dividing the equation by c_k gives

$$\mathbf{a}_k + \sum_{j=1}^{k-1} b_j \, \mathbf{a}_j + \sum_{j=k+1}^{N} b_j \, \mathbf{a}_j = \mathbf{0}.$$

Thus, an entire column (or row) $\mathbf{a}_k$ can have all its elements made to be zero by adding to it some particular linear combination of the other columns (or rows). According to Theorem 18.5.4, this leaves the determinant's value unchanged, and according to Theorem 18.5.1, the determinant's value is zero. □

Combining Theorem 18.5.7 with the existence theorem for the matrix inverse, Theorem 18.3.1, we have the following:

Theorem 18.5.8. *A square matrix is singular if and only if its determinant is zero.*

18.6 Orthogonal and Unitary Matrices

Consider a real inner product space; that is, one defined such that the field (the set of scalars) is $\mathbb{R}$. The space of Cartesian coordinate vectors is one example. The space of all polynomials with real coefficients could be another. An ***orthogonal transformation*** is an operation that preserves the inner product on a real inner product space; in other words, if Q is the operator and v_1 and v_2 are any elements of the space, then

$$\langle Q\,v_1, Q\,v_2\rangle = \langle v_1, v_2\rangle. \tag{18.35}$$

Let us express this in matrix notation. Let $\mathbf{Q}$ be the matrix representation of Q. Then, for real vectors, we have

$$\langle v_1, v_2\rangle = \mathbf{v}_1^{\mathbf{T}}\,\mathbf{v}_2\,, \tag{18.36}$$

and

$$\langle Q\,v_1, Q\,v_2\rangle = (\mathbf{Q}\,\mathbf{v}_1)^{\mathbf{T}}\,\mathbf{Q}\,\mathbf{v}_2 = \mathbf{v}_1^{\mathbf{T}}\,\mathbf{Q}^{\mathbf{T}}\,\mathbf{Q}\,\mathbf{v}_2\,, \tag{18.37}$$

thanks to Eq. (18.25). A matrix that represents an orthogonal transformation is called an ***orthogonal matrix***. (Note that this use of "orthogonal" is different from that in the phrase "orthogonal vectors.") Comparing Eqs. (18.35), (18.36), and (18.37), we see that if $\mathbf{Q}$ is orthogonal, then

$$\mathbf{Q}^{\mathbf{T}}\,\mathbf{Q} = \mathbf{I}\,. \tag{18.38}$$

Comparing with the definition of an inverse matrix, Eq. (18.28), we obtain the following:

Theorem 18.6.1. $\mathbf{Q}$ *is an orthogonal matrix if and only if all its elements are real and*

$$\mathbf{Q}^{-1} = \mathbf{Q}^{\mathbf{T}}. \tag{18.39}$$

Orthogonal matrices are a special class of square matrices with many applications. The rotation matrix, for example, is orthogonal, which one should expect, because a rotation has no effect on the length $\langle v, v\rangle^{1/2}$ of a vector or on the angle

$$\phi = \arccos\left(\langle v_1, v_2\rangle/\sqrt{\langle v_1, v_1\rangle\langle v_2, v_2\rangle}\,\right) \tag{18.40}$$

between two vectors. Using Theorems 18.5.5 and 18.5.6, one can prove the following:

Theorem 18.6.2. *The determinant of an orthogonal matrix is either* $+1$ *or* -1.

Example 18.9. *The determinant of the two-dimensional rotation matrix.*

$$\begin{vmatrix} \cos\phi & -\sin\phi \\ \sin\phi & \cos\phi \end{vmatrix} = \cos\phi\,\cos\phi - (-\sin\phi)\sin\phi = \cos^2\phi + \sin^2\phi = 1.$$

Another class of square matrices, called *Hermitian matrices*,[4] are of importance in quantum mechanics.

Definition. The ***Hermitian conjugate***[5] of a matrix **M**, designated by[6] $\mathbf{M}^\dagger$, is the complex conjugate of the matrix transpose.

In other words, if the elements of **M** are the complex numbers M_{ij}, then the elements of $\mathbf{M}^\dagger$ are M_{ji}^*. The Hermitian conjugate of a column vector **v** is the row vector $\mathbf{v}^\dagger$ of the complex conjugates of the elements. Because the inner product is defined as *conjugate*-symmetric, the inner product of v with itself can be written

$$\langle v, v \rangle = \mathbf{v}^\dagger \mathbf{v}. \tag{18.41}$$

Definition. An ***Hermitian matrix***[7] is a matrix that is equal to its Hermitian conjugate.

In other words, **M** is Hermitian if $M_{ij} = M_{ji}^*$. (If the elements of the matrix are real numbers, then "Hermitian" just means the same thing as "symmetric.") Hermitian matrices will be discussed in detail in Chapter 19.

A ***unitary transformation*** is defined as a transformation that preserves the inner product in a complex inner product space. It is the complex counterpart of the concept of an orthogonal transformation. A matrix representation of a unitary transformation is called a ***unitary matrix***. Let Q now be a unitary transformation and let **Q** be its matrix representation. Then

$$\langle Q\, v, Q\, v \rangle = (\mathbf{Q}\, \mathbf{v})^\dagger\, \mathbf{Q}\, \mathbf{v} = \mathbf{v}^\dagger\, \mathbf{Q}^\dagger\, \mathbf{Q}\, \mathbf{v}. \tag{18.42}$$

For a unitary matrix,

$$\mathbf{Q}^\dagger \mathbf{Q} = \mathbf{I}\,. \tag{18.43}$$

Theorem 18.6.3. **Q** *is a unitary matrix if and only if*

$$\mathbf{Q}^{-1} = \mathbf{Q}^\dagger. \tag{18.44}$$

Theorem 18.6.4. *The determinant of a unitary matrix has absolute value of 1.*

---

[4]Named for Charles Hermite, who studied their properties in the 1850's, as an exercise in pure abstract mathematics. Hermite espoused a mystical philosophy of mathematics in which numbers have an existence of their own, independent of physical phenomena or human imaginations. He had no idea that 70 years later these matrices would become an important tool of physicists and physical chemists.

[5]Also called the *adjoint*.

[6]Sometimes the notation $\mathbf{M}^\mathbf{H}$ is used instead, in honor of Hermite.

[7]Also called a *self-adjoint* matrix.

18.7 Simultaneous Linear Equations

Consider the simultaneous equations

$$a_{1,1}x_1 = b_1\,, \qquad a_{2,1}x_1 + a_{2,2}x_2 = b_2\,, \tag{18.45}$$

where the a's and b's are constants of given values while the x_k are unknowns. We can solve the first equation, $x_1 = b_1/a_{1,1}$, and then substitute the solution into the second, to obtain

$$x_2 = (b_2 - a_{2,1}b_1/a_{1,1})/a_{2,2}\,. \tag{18.46}$$

If both equations contain both variables,

$$a_{1,1}x_1 + a_{1,2}x_2 = b_1\,, \qquad a_{2,1}x_1 + a_{2,2}x_2 = b_2\,, \tag{18.47}$$

then we could eliminate x_2 from the first equation by multiplying the second by $a_{1,2}/a_{2,2}$ and subtracting it from the first, to obtain equations in the form of Eqs. (18.45). This is a viable approach with only a small number of variables but it is very tedious when the number of variables is large.

The analysis can be considerably simplified by expressing the problem as a matrix equation. Consider a system of N equations in N unknowns x_k,

$$\sum_{k=1}^{N} a_{j,k}x_k = b_j\,, \qquad j = 1, 2, 3, \ldots, N. \tag{18.48}$$

This can be written as

$$\mathbf{A}\,\mathbf{x} = \mathbf{b}\,, \tag{18.49}$$

where $\mathbf{A}$ is an $N \times N$ matrix and $\mathbf{x}$ and $\mathbf{b}$ are column vectors,

$$\mathbf{A} = \begin{pmatrix} a_{1,1} & a_{1,2} & a_{1,3} & \cdots & a_{1,N} \\ a_{2,1} & a_{2,2} & a_{2,3} & \cdots & a_{2,N} \\ \vdots & \vdots & \vdots & \ddots & \vdots \\ a_{N,1} & a_{N,2} & a_{N,3} & \cdots & a_{N,N} \end{pmatrix}, \qquad \mathbf{x} = \begin{pmatrix} x_1 \\ x_2 \\ \vdots \\ x_N \end{pmatrix}, \qquad \mathbf{b} = \begin{pmatrix} b_1 \\ b_2 \\ \vdots \\ b_N \end{pmatrix}.$$

In principle, we can solve for $\mathbf{x}$ by multiplying both sides of Eq. (18.49) from the left by $\mathbf{A}^{-1}$, so that

$$\mathbf{x} = \mathbf{A}^{-1}\mathbf{b}\,. \tag{18.50}$$

Thus, the problem of solving simultaneous linear equations is equivalent to the problem of solving for the inverse of a matrix.

An immediate implication of Eq. (18.50) is the following:

Theorem 18.7.1. *A set of N simultaneous linear algebraic equations in N unknowns has a unique solution if and only if the matrix of the coefficients multiplying the unknowns is nonsingular.*

We can tell if the system of equations has a solution simply by computing the determinant of the coefficient matrix. If the determinant is zero, then there is no unique solution. The theorem also gives insight into *why* a set of equations might fail to have a solution. The row-column dot products,

$$(a_{k,1}, a_{k,2}, \dots, a_{k,N}) \cdot (x_1, x_2, \dots, x_N), \qquad k = 1, 2, \dots, N,$$

can each be thought of as a linear combination of the elements of a basis set $\{x_1, x_2, \dots, x_N\}$. If the matrix $\mathbf{A}$ is singular, then these linear combinations are not all linearly independent. This gives two possibilities: (1) two of the equations are inconsistent, because the same linear combination is set equal to two different constants, or (2) two of the equations are equivalent, except for a constant multiplying both sides of the equation, in which case the problem is underdetermined—there are really just $N-1$ equations to determine N unknowns.

Example 18.10. *Linear equations with no unique solution.* Consider the simultaneous equations

$$2x_1 + 3x_2 = 5, \qquad 4x_1 + 6x_2 = 8.$$

If we divide the second equation by 2, we obtain $2x_1 + 3x_2 = 4$, which is inconsistent with the first equation. No solution is possible. Consider the equations

$$2x_1 + 3x_2 = 5, \qquad 4x_1 + 6x_2 = 10.$$

Dividing the second equation by 2 gives $2x_1 + 3x_2 = 5$, which is equivalent to the first equation. This means any x_1 and x_2 such that $x_2 = (5 - 2x_1)/3$ will be acceptable. There are infinitely many such solutions. The determinant of the coefficient matrix in either case is

$$\begin{vmatrix} 2 & 3 \\ 4 & 6 \end{vmatrix} = 2 \cdot 6 - 3 \cdot 4 = 0.$$

The coefficient matrix is singular.

Given a nonsingular coefficient matrix, we could solve the set of simultaneous equations using Eq. (18.50), by computing the matrix inverse, but a more efficient computational strategy is to use an **LU decomposition**. "LU" stands for "lower-triangular, upper-triangular." A matrix $\mathbf{A}$ is said to be ***lower triangular*** if all its elements above the diagonal are zero ($a_{i,j} = 0$ if $j > i$); it is said to be ***upper triangular*** if all the elements below the diagonal are zero ($a_{i,j} = 0$ if $j < i$). The idea is to express $\mathbf{A}$ as the product of a lower-triangular matrix and an upper triangular matrix.

Theorem 18.7.2. *A nonsingular $N \times N$ matrix $\mathbf{A}$ can be expressed as a product $\mathbf{A} = \mathbf{LU}$ where $\mathbf{L}$ is lower triangular and $\mathbf{U}$ is upper triangular, if and only if each of the following determinants are nonzero:*

$$\begin{vmatrix} a_{1,1} & a_{1,2} \\ a_{2,1} & a_{2,2} \end{vmatrix}, \quad \begin{vmatrix} a_{1,1} & a_{1,2} & a_{1,3} \\ a_{2,1} & a_{2,2} & a_{2,3} \\ a_{3,1} & a_{3,2} & a_{3,3} \end{vmatrix}, \quad \dots, \quad \begin{vmatrix} a_{1,1} & a_{1,2} & \cdots & a_{1,N} \\ a_{2,1} & a_{2,2} & \cdots & a_{2,N} \\ \vdots & \vdots & \ddots & \vdots \\ a_{N,1} & a_{N,2} & \cdots & a_{N,N} \end{vmatrix}.$$

Given a decomposition of $\mathbf{A}$ into $\mathbf{L}$ and $\mathbf{U}$, the solution for $\mathbf{x}$ is quite simple. The simultaneous equations can be expressed as

$$\mathbf{LUx} = \mathbf{b}\,. \tag{18.51}$$

First we solve the matrix equation $\mathbf{Ly} = \mathbf{b}$ for $\mathbf{y}$. Because $\mathbf{L}$ is triangular, this is trivial to do. It is analogous to the situation in Eqs. (18.45). Start with the dot product with the first row of $\mathbf{L}$, which contains only one nonzero element, $\ell_{1,1}$. The solution for y_1 is simply $b_1/\ell_{1,1}$. The dot product with each successive row of $\mathbf{L}$ involves just one additional unknown. Once all the elements of $\mathbf{y}$ are computed, we solve the equation $\mathbf{Ux} = \mathbf{y}$ for $\mathbf{x}$. The method is the same, except the solution begins with the *last* row of $\mathbf{U}$, with $x_N = y_N/u_{N,N}$. The advantage of this approach is that a very efficient computer algorithm (*Crout's algorithm*) exists for computing matrices $\mathbf{L}$ and $\mathbf{U}$.

The drawback of LU decomposition is the fact that the accuracy of the solution can be seriously compromised by roundoff error. Computer arithmetic normally uses only about 16 decimal digits of precision. Because the solution is obtained iteratively, with x_k computed from all the previous $x_{j>k}$, roundoff error can be propagated from one step to the next at rapid rate. It is generally a good idea to polish the solution $\mathbf{x}_{\text{approx}}$ initially obtained from LU decomposition by plugging it back into the original matrix equation to see how well it works. Let

$$\mathbf{x}_{\text{approx}} = \mathbf{x} + \boldsymbol{\delta}\,, \tag{18.52}$$

where $\mathbf{x}$ is the desired exact result. Substituting $\mathbf{x} = \mathbf{x}_{\text{approx}} - \boldsymbol{\delta}$ into $\mathbf{Ax} = \mathbf{b}$ gives

$$\mathbf{A}\boldsymbol{\delta} = \mathbf{A}\mathbf{x}_{\text{approx}} - \mathbf{b}\,. \tag{18.53}$$

The right-hand side is a known column vector. Thus we have an equation that can be solved for the correction $\boldsymbol{\delta}$ with another round of LU decomposition.

There is an alternative method, called ***singular value decomposition***, that is less efficient than LU decomposition but allows for more careful control over the effects of roundoff error.[8] It is based on the following theorem:

Theorem 18.7.3. *For any square matrix* $\mathbf{A}$ *there exists an orthogonal matrix* $\mathbf{V}$ *that transforms* $\mathbf{A}$ *into the product of an orthogonal matrix* $\mathbf{Q}$ *and a diagonal matrix* $\mathbf{W}$ *such that*

$$\mathbf{AV} = \mathbf{QW}\,. \tag{18.54}$$

Computer algorithms are available to obtain $\mathbf{V}$, $\mathbf{Q}$, and $\mathbf{W}$. Now suppose that we multiply both sides of $\mathbf{AV} = \mathbf{QW}$ from the left by $\mathbf{A}^{-1}$, from the right by $\mathbf{W}^{-1}$, and then from the right by $\mathbf{Q}^{-1} = \mathbf{Q}^{\mathrm{T}}$. We find that

$$\mathbf{A}^{-1} = \mathbf{VW}^{-1}\mathbf{Q}^{\mathrm{T}} \tag{18.55}$$

and thereby obtain the solution $\mathbf{x} = \mathbf{A}^{-1}\mathbf{b}$ for the simultaneous equations.

[8] Solving linear equations is just one of the various applications of singular value decomposition. It is often used in statistical analysis and it is a better method than Gram-Schmidt orthogonalization for computing orthonormal basis sets. (Gram-Schmidt should only be used for analytic derivations.) See W. H. Press *et al.*, *Numerical Recipes* (Cambridge University Press, Cambridge, 2007), Chaps. 2 and 15; and L. N. Trefethen and D. Bau III, *Numerical Linear Algebra* (SIAM, Philadelphia, 1997).

This works as long as $\mathbf{A} = \mathbf{QWV^T}$ is nonsingular. If $\mathbf{A}$ is singular then its inverse $\mathbf{A}^{-1}$ does not exist.

$\mathbf{A}$ is singular if and only if its determinant is zero. According to Theorem 18.5.6,

$$|\mathbf{A}| = |\mathbf{QWV^T}| = |\mathbf{Q}||\mathbf{W}||\mathbf{V^T}|. \tag{18.56}$$

Being orthogonal matrices, $\mathbf{Q}$ and $\mathbf{V^T}$ each have determinants of ± 1. Therefore, $|\mathbf{A}| = 0$ if and only if $|\mathbf{W}| = 0$. The determinant of a diagonal matrix is simply the product of its diagonal elements,

$$|\mathbf{W}| = \prod_{j=1}^{N} w_j \,. \tag{18.57}$$

This is zero if and only if at least one of the w_j is zero.

The **condition number** of a square matrix $\mathbf{A}$ is defined as the absolute value of the ratio of the w_k with the largest magnitude to the w_k with the smallest magnitude in its singular value decomposition. For a singular matrix the condition number is infinite. Suppose, however, that the matrix is nonsingular but its condition number is greater than $1/\epsilon$ where ϵ is the magnitude of roundoff error. In that case $\mathbf{A}$ is said to be **ill-conditioned**. It is impossible to distinguish between the value $1/\epsilon$ and a value of infinity within the precision of the computer. (Computers typically have $\epsilon \approx 10^{-12}$.) LU decomposition will give a solution even if the matrix is ill conditioned, but one or more of the x_k will probably be inaccurate even after polishing.

It is easy to verify that the inverse of a diagonal matrix is the diagonal matrix of the reciprocals:

$$\mathbf{W} = \begin{pmatrix} w_1 & & & \\ & w_2 & \mathbf{0} & \\ & \mathbf{0} & \ddots & \\ & & & w_N \end{pmatrix}, \quad \mathbf{W}^{-1} = \begin{pmatrix} 1/w_1 & & & \\ & 1/w_2 & \mathbf{0} & \\ & \mathbf{0} & \ddots & \\ & & & 1/w_N \end{pmatrix}. \tag{18.58}$$

Let $\hat{w}_k$ be the normalized w_k, given by $\hat{w}_k = w_k/\max_j |w_j|$. Then for any $|\hat{w}_k| < \epsilon$, the corresponding x_k will meaningless on account of roundoff error. A simple way to avoid propagating this error into the rest of the vector $\mathbf{x}$ is to modify the matrix $\mathbf{W}^{-1}$ as follows:

> If $\hat{w}_k < \epsilon$, then replace the $1/w_k$ in $\mathbf{W}^{-1}$ with a value of exactly zero.

This forces x_k to be exactly zero and thereby prevents it from contaminating the other x_j's. It may seem counterintuitive to replace a value of $1/w_k$ that is almost infinite with a value of zero, but this is a reliable procedure for obtaining solutions for ill-conditioned systems of equations. The resulting $\mathbf{A}^{-1}\mathbf{b}$ gives no solution for those x_j whose value of $1/w_j$ has been set to zero but it yields accurate solutions for the other x_j's.

Exercises

18.1 Prove that the set of all $N \times N$ orthogonal matrices forms a non-abelian group, with the "addition" operation defined as matrix multiplication.

18.2 Consider the row vectors $\mathbf{u} = (7 \;\; 1 \;\; 0 \;\; 3)$ and $\mathbf{v} = (2 \;\; 2 \;\; 3 \;\; 4)$. Calculate $\mathbf{u}\mathbf{v}^{\mathbf{T}}$ and $\mathbf{u}^{\mathbf{T}}\mathbf{v}$. *(Hint: The former yields a scalar, the latter, a matrix.)*

18.3 For the row vector $\mathbf{v} = (3 \;\; 1 \;\; 2 \;\; 1)$ and the matrix $\mathbf{M} = \begin{pmatrix} 1 & 0 & 5 & 2 \\ 2 & 1 & 1 & 3 \\ 1 & 0 & 4 & 0 \\ 0 & 1 & 3 & 1 \end{pmatrix}$ calculate:

(a) $\mathbf{M}\mathbf{v}^{\mathbf{T}}$ (b) $\mathbf{v}\mathbf{v}^{\mathbf{T}}$ (c) $\mathbf{v}\mathbf{M}$ (d) $\mathbf{M}^{\mathbf{T}}\mathbf{M}$ (e) tr $\mathbf{M}^{\mathbf{T}}$ (f) $|\mathbf{M}|$ (g) $|\mathbf{M}^{\mathbf{T}}|$

18.4 Prove that tr $\mathbf{AB}$ = tr $\mathbf{BA}$. (This holds true even if $\mathbf{AB}$ is not equal to $\mathbf{BA}$. Demonstrate with an example.)

18.5 Demonstrate that $(\mathbf{BA})^{\mathbf{T}} = \mathbf{A}^{\mathbf{T}}\mathbf{B}^{\mathbf{T}}$ for an arbitrary pair of 2×2 matrices.

18.6 Prove that $(\mathbf{AB})^{-1} = \mathbf{B}^{-1}\mathbf{A}^{-1}$, assuming $\mathbf{A}$ and $\mathbf{B}$ are nonsingular.

18.7 Derive matrix representations for operators that perform the following rotations of a three-dimensional Cartesian coordinate vector: (a) Rotation by an angle ϕ about the z-axis. (b) Rotation by an angle ϕ about the x-axis. (c) Rotation by an angle ϕ_z about the z-axis followed by a rotation by an angle ϕ_x about the x-axis.

18.8 Demonstrate Theorems 18.5.1 through 18.5.6 using arbitrary (but nonsingular) 3×3 matrices of your choosing.

18.9 Prove that a unitary transformation, in order to preserve the norm of a complex vector, must be represented by a unitary matrix.

18.10 Suppose that $\mathbf{A}$ is a *non*-Hermitian matrix. Prove that $\mathbf{B} = \mathbf{A} + \mathbf{A}^{\dagger}$ and $\mathbf{C} = i(\mathbf{A} - \mathbf{A}^{\dagger})$ are Hermitian.

18.11 (a) Derive Eq. (18.58), i.e., the inverse of a diagonal matrix is the diagonal matrix of reciprocals. (b) Prove that the determinant of a diagonal matrix is equal to the product of its diagonal elements.

18.12 Set up a system of five linear equations in five unknowns. Give arbitrary values to all of the 25 coefficients except one. Solve the equations (using computer algebra) using a range of different values for the unspecified coefficient. Plot the base-10 logarithm of the five normalized diagonal elements, $\hat{w}_k$, as a function of the unspecified coefficient and find a range of values where one of the $\hat{w}_k$'s drops below $\epsilon = 10^{-6}$.

Chapter 19

Eigenvalue Equations

An ***eigenvalue equation*** is an equation of the form

$$Gv = cv\,. \tag{19.1}$$

G is an operator on a vector space, v is a nonzero vector, and c is a nonzero scalar. Note that the same vector v appears on each side of the equation. v is called an ***eigenvector*** and c is called an ***eigenvalue***.[1] If this equation is true then that means that G transforms the vector v into a scalar multiple of itself. Usually, such an equation can be solved only for particular values of c, and otherwise has no solution.

We will only consider here linear operators G. Then for any nonzero scalar a, we have

$$G\,(av) = aGv = acv = c\,(av). \tag{19.2}$$

Thus, if v is an eigenvector, so is av. The equation determines the eigenvalues but the eigenvector is not completely determined; it has an arbitrary prefactor.

Eigenvalue equations appear frequently in science and engineering. They are of special interest to chemists due to their central role in quantum mechanics. The particular nature of the eigenvalue is the mathematical expression of the concept of quantization of physical properties. We will begin with matrix eigenvalue equations, in which a matrix $\mathbf{G}$ operates on a finite-dimension vector space, and then consider differential operators operating on infinite-dimension vector spaces of functions. Finally, we will see how the infinite-dimension problem can be approximated with a matrix equation.

19.1 Matrix Eigenvalue Equations

Consider the matrix equation

$$\mathbf{Mv} = c\mathbf{v}, \tag{19.3}$$

where $\mathbf{M}$ is a matrix of dimension $n \times n$, $\mathbf{v}$ is a column vector of dimension n, and c is a scalar.

Example 19.1. *A 2×2 matrix eigenvalue equation.* Suppose that

$$\mathbf{M} = \begin{pmatrix} 5 & -6 \\ 2 & -2 \end{pmatrix}, \qquad \mathbf{v} = \begin{pmatrix} 2 \\ 1 \end{pmatrix}. \tag{19.4}$$

---

[1] In German the prefix *Eigen* means "particular" or "peculiar to." An eigenvalue equation is an equation that typically has a solution only for particular values of the scalar. The English word "eigenvalue" is a peculiar partial translation of the German word *Eigenwert* coined by David Hilbert. (*Wert* means "value.") However, equations of this type had already been studied in the 19th century by French mathematicians, who used the terms *racine caractéristique* ("characteristic root") and *equation caractéristique.*

Then,
$$\begin{pmatrix}5 & -6\\ 2 & -2\end{pmatrix}\begin{pmatrix}2\\1\end{pmatrix} = \begin{pmatrix}4\\2\end{pmatrix} = 2\begin{pmatrix}2\\1\end{pmatrix}.$$

$\mathbf{v}$ is clearly an eigenvector of $\mathbf{M}$ with eigenvalue of 2.

Given only the matrix, one way to systematically solve for the eigenvectors and their corresponding eigenvalues is to write the components of the vector as variables and solve the resulting set of linear algebraic equations. In this case,

$$\begin{pmatrix}5 & -6\\ 2 & -2\end{pmatrix}\begin{pmatrix}v_1\\v_2\end{pmatrix} = \begin{pmatrix}5v_1 - 6v_2\\ 2v_1 - 2v_2\end{pmatrix} = \begin{pmatrix}cv_1\\cv_2\end{pmatrix}, \tag{19.5}$$

which implies that

$$5v_1 - 6v_2 = cv_1, \qquad 2v_1 - 2v_2 = cv_2\,. \tag{19.6}$$

The second equation implies

$$v_1 = (\tfrac{1}{2}c + 1)v_2\,, \tag{19.7}$$

and substituting this into the first equation gives $[(5-c)(\frac{1}{2}c+1) - 6]v_2 = 0$, which, for nonzero v_2, implies a quadratic equation

$$c^2 - 3c + 2 = 0\,. \tag{19.8}$$

The quadratic formula gives $c = \frac{1}{2}(3 \pm \sqrt{9-8})$, implying that the eigenvalue can be either 2 or 1. Thus, $\mathbf{M}$ has two possible eigenvalues. Substituting $c = 2$ into Eq. (19.7) gives $v_1 = 2v_2$. Therefore, any vector $a\,(2,1)$, where a is a nonzero scalar, is an eigenvector of $\mathbf{M}$ with eigenvalue $c = 2$. For $c = 1$ we find $2v_1 = 3v_2$; the corresponding eigenvectors are $b\,(3,2)$, where b is an arbitrary nonzero scalar. We ignore the trivial solution $v_1 = 0$, $v_2 = 0$.

The method of solution in Example 19.1 is rather laborious, especially for dimension higher than 2. The important point to note is that the key to the solution was reducing the problem to finding the roots of a polynomial, Eq. (19.8). In general, for any $n \times n$ matrix, there exists a polynomial of degree n such that the roots of the polynomial are all the possible eigenvalues of the matrix. This is called the ***characteristic polynomial*** of the matrix.

We now construct a general method for finding this polynomial. Let us write Eq. (19.3) as

$$\mathbf{N}\mathbf{v} = \mathbf{0}\mathbf{v}, \qquad \mathbf{N} = \mathbf{M} - c\,\mathbf{I}, \tag{19.9}$$

where $\mathbf{I}$ is the identity matrix and $\mathbf{0}$ is the zero matrix. The zero matrix has no inverse. If it did, then we would have $\mathbf{0}\,\mathbf{0}^{-1} = \mathbf{I}$, but by definition, multiplying any matrix by $\mathbf{0}$ gives $\mathbf{0}$. The fact that $\mathbf{0}$ is singular implies that $\mathbf{N}$ must also be singular. Theorem 18.5.7 stated that a singular matrix has a determinant of zero. We therefore conclude the following:

Theorem 19.1.1. *If c is an eigenvalue of a matrix $\mathbf{M}$, then $|\mathbf{M} - c\,\mathbf{I}| = 0$.*

Example 19.2. *Characteristic polynomial from a determinant.* Let

$$\mathbf{N} = \begin{pmatrix}5 & -6\\ 2 & -2\end{pmatrix} - c\begin{pmatrix}1 & 0\\ 0 & 1\end{pmatrix} = \begin{pmatrix}5-c & -6\\ 2 & -2-c\end{pmatrix},$$

$$|\mathbf{N}| = (5-c)(-2-c) - (-6)(2) = 2 - 3c + c^2 = 0\,,$$

which is equivalent to Eq. (19.8).

It is easy to prove in general that for an $n \times n$ matrix $\mathbf{M}$, the determinant of $\mathbf{M} - c\mathbf{I}$ is a polynomial in c of degree n. The fundamental theorem of algebra then implies the following:

Theorem 19.1.2. *An $n \times n$ matrix $\mathbf{M}$ can have at most n different eigenvalues and, if the elements of $\mathbf{M}$ are all real numbers, then the eigenvalues are either real numbers or complex-conjugate pairs.*

The qualification "at most" is needed because it is possible that some of the roots of the characteristic polynomial are multiple roots or are zero. The possibility of multiple roots means that it is possible for more than one eigenvector to have the same eigenvalue.

Theorem 19.1.3. *Eigenvectors with different eigenvalues are linearly independent.*

Proof. Consider a matrix $\mathbf{M}$ such that each eigenvector $\mathbf{v}_i$ has an eigenvalue μ_i such that no two of of the μ_i are equal. Suppose that the set $\{\mathbf{v}_i\}$ of eigenvectors is linearly dependent. Then, according to the definition of linear dependence, Eq. (16.23), there exists a subset of linearly independent eigenvectors in terms of which $\mathbf{v}_1$ can be expressed as a linear combination. Label the eigenvectors in this subset with subscripts $2, 3, \ldots, k$. Then

$$\mathbf{v}_1 = \sum_{i=2}^{k} a_i \mathbf{v}_i \ , \tag{19.10}$$

where the a_i are nonzero scalars. By assumption, $\mathbf{M}\mathbf{v}_1 = \mu_1 \mathbf{v}_1$. Substituting Eq. (19.10) for $\mathbf{v}_1$ gives

$$\mathbf{M}\mathbf{v}_1 = \mu_1 \sum_{i=2}^{k} a_i \mathbf{v}_i \ . \tag{19.11}$$

Operating on the right-hand side of Eq. (19.10) gives

$$\mathbf{M} \sum_{i=2}^{k} a_i \mathbf{v}_i = \sum_{i=2}^{k} a_i \mathbf{M}\mathbf{v}_i = \sum_{i=2}^{k} a_i \mu_i \mathbf{v}_i \ . \tag{19.12}$$

Eqs. (19.11) and (19.12) must be equal. Therefore,

$$\sum_{i=2}^{k} a_i (\mu_i - \mu_1) \mathbf{v}_i = \mathbf{0} \, ,$$

where $\mathbf{0}$ is the zero vector. But according to Theorem 16.7.1, this implies the subset is linearly dependent, which is a contradiction. Therefore, the full set $\{\mathbf{v}_1, \mathbf{v}_2, \ldots \mathbf{v}_n\}$ must be linearly independent. □

If $\mathbf{M}$ happens to be Hermitian (i.e., if $M_{ij} = M_{ji}^*$), then we can make several additional statements about the eigenvalue equation, given here without proof:

Theorem 19.1.4. *All eigenvalues of an Hermitian matrix are real numbers.*

Theorem 19.1.5. *All eigenvectors of an Hermitian matrix, except those that differ only by a scalar prefactor, are linearly independent.*

Theorem 19.1.6. *Eigenvectors of an Hermitian matrix that have different eigenvalues are orthogonal to each other.*

Distinct eigenvectors with the same eigenvalue are said to be ***degenerate***. ("Distinct" here means that they differ by more than just a scalar prefactor). Distinct degenerate eigenvectors are not necessarily orthogonal (although they are linearly independent if the matrix is Hermitian). Consider two eigenvectors, $\mathbf{v}_1$ and $\mathbf{v}_2$, of an Hermitian matrix $\mathbf{M}$ with the same eigenvalue μ. Then, $\mathbf{M}\mathbf{v}_1 = \mu\,\mathbf{v}_1$ and $\mathbf{M}\mathbf{v}_2 = \mu\,\mathbf{v}_2$. But this means that any linear combination of $\mathbf{v}_1$ and $\mathbf{v}_2$ is also an eigenvector with eigenvalue μ, because

$$\mathbf{M}(a\,\mathbf{v}_1 + b\,\mathbf{v}_2) = a\mathbf{M}\mathbf{v}_1 + b\mathbf{M}\mathbf{v}_2 = \mu\,(a\,\mathbf{v}_1 + b\,\mathbf{v}_2) \tag{19.13}$$

for any scalars a and b. Therefore, it will always be possible to construct orthogonal eigenvectors, all with eigenvalue μ, from degenerate eigenvectors using the Gram-Schmidt orthogonalization procedure, of Section 17.1.

Theorem 19.1.5 states that all n eigenvectors of an $n \times n$ Hermitian matrix are linearly independent, whether or not they have different eigenvalues. n is also the dimension of the vector space on which the matrix operates. If we have a set of n linearly independent vectors in a vector space of dimension n, then, according to the discussion in Section 16.7, this set must be a basis for the space. This proves the following theorem, which is useful in quantum mechanical applications:

Theorem 19.1.7. *The distinct eigenvectors of an Hermitian matrix form a basis for the vector space.*

This theorem can be used to calculate the effect of an operator on an arbitrary vector, as long as one knows the eigenvectors and corresponding eigenvalues of the operator. Any vector $\mathbf{v}$ in the vector space can be expressed as a linear combination of the basis vectors,

$$\mathbf{v} = \sum_{j=1}^{n} a_j \mathbf{v}_j\,. \tag{19.14}$$

Let $\{\mathbf{v}_j\}$ be the eigenvectors of an $n \times n$ Hermitian matrix $\mathbf{M}$, with corresponding eigenvalues $\{\mu_j\}$. Then,

$$\mathbf{M}\mathbf{v} = \sum_{j=1}^{n} a_j \mu_j \mathbf{v}_j\,. \tag{19.15}$$

If n is large, this can be more efficient than explicitly multiplying $\mathbf{v}$ by $\mathbf{M}$.

19.2 Matrix Diagonalization

Theorem 19.1.1 tells how to obtain the characteristic polynomial of a matrix, but solving for the roots of this polynomial is usually not a good way to find the eigenvalues. With only a little bit of additional analysis we can efficiently obtain a solution for the eigenvalues and, as well, for the eigenvectors.

Consider a vector $\mathbf{v}$ and some nonsingular square matrix $\mathbf{S}$, and define a new vector $\mathbf{w} = \mathbf{S}\mathbf{v}$. Suppose that $\mathbf{w}$ is an eigenvector of some matrix $\mathbf{M}$ with eigenvalue μ. Then $\mathbf{M}\mathbf{w} = \mu\mathbf{w}$ and

$$\mathbf{MSv} = \mu\,\mathbf{Sv}. \tag{19.16}$$

Because $\mathbf{S}$ is nonsingular, its inverse exists. Multiplying both sides of the equation by $\mathbf{S}^{-1}$ gives

$$\mathbf{S}^{-1}\mathbf{MSv} = \mu\,\mathbf{v}\,. \tag{19.17}$$

Examination of this equation leads to an interesting conclusion:

Theorem 19.2.1. *Consider two square matrices* $\mathbf{M}$ *and* $\mathbf{S}$ *of the same dimensions, with* $\mathbf{S}$ *nonsingular. The matrix* $\mathbf{S}^{-1}\mathbf{MS}$ *has the same eigenvalues as the matrix* $\mathbf{M}$.

A transformation of the form $\mathbf{S}^{-1}\mathbf{MS}$ is called a ***similarity transformation***. Matrices that are related to each other by a similarity transformation are said to be ***similar***.

Example 19.3. *Eigenvalues of similar matrices.* Let

$$\mathbf{M} = \begin{pmatrix} 5 & -6 \\ 2 & -2 \end{pmatrix}, \qquad \mathbf{S} = \begin{pmatrix} 3 & 1 \\ 2 & 5 \end{pmatrix}.$$

$\mathbf{M}$ is the same matrix we studied in the previous two examples.

$$\mathbf{S}^{-1} = \begin{pmatrix} 5/13 & -1/13 \\ -2/13 & 3/13 \end{pmatrix}, \qquad \mathbf{S}^{-1}\mathbf{MS} = \begin{pmatrix} 1 & -9 \\ 0 & 2 \end{pmatrix}.$$

The characteristic polynomial of $\mathbf{S}^{-1}\mathbf{MS}$ is $(1-c)(2-c)$, with roots $c = 2$ and $c = 1$, the same as we found for $\mathbf{M}$ in Examples 19.1 and 19.2. However, the eigevectors are now $a\,(-9,1)$ for $c = 2$ and $b\,(0,1)$ for $c = 1$ (for arbitrary scalar prefactors a and b). These are not the same as before. Similarity transformations preserve the eigenvalues but not the eigenvectors.

Now consider the following theorem:

Theorem 19.2.2. *Let* $\mathbf{M}$ *be an Hermitian matrix of dimension* $n \times n$. *Then there exists an* $n \times n$ *unitary matrix* $\mathbf{Q}$ *such that* $\mathbf{Q}^{-1}\mathbf{MQ}$ *is a diagonal matrix. The columns of* $\mathbf{Q}$ *are orthonormal eigenvectors of* $\mathbf{M}$ *and the diagonal elements of* $\mathbf{Q}^{-1}\mathbf{MQ}$ *are the eigenvalues of* $\mathbf{M}$.

It follows from this that the problem of finding eigenvalues and eigenvectors is equivalent to the problem of ***diagonalizing*** a matrix; i.e., finding a similarity transformation that converts the matrix representation of the operator into a diagonal matrix. Matrix diagonalization is a relatively straightforward computation. Efficient computer algorithms are available for finding $\mathbf{Q}$.

This important theorem may seem surprising, but its proof is not difficult:

Proof. (Theorem 19.2.2.) Let us work backward from the conclusion. Suppose the columns of $\mathbf{Q}$ comprise an orthonormal set of eigenvectors $\mathbf{q}_i$ of $\mathbf{M}$, with each $\mathbf{q}_i$ represented as a column vector with elements $q_{1,i}$, $q_{2,i}, \ldots, q_{n,i}$. Let μ_i be the eigenvalue of $\mathbf{q}_i$. Then $\mathbf{M}\mathbf{q}_i = \mu_i \mathbf{q}_i$, and, therefore,

$$\mathbf{MQ} = \begin{pmatrix} \mu_1 q_{1,1} & \mu_2 q_{1,2} & \cdots & \mu_n q_{1,n} \\ \mu_1 q_{2,1} & \mu_2 q_{2,2} & \cdots & \mu_n q_{2,n} \\ \vdots & \vdots & \ddots & \vdots \\ \mu_1 q_{n,1} & \mu_2 q_{n,2} & \cdots & \mu_n q_{n,n} \end{pmatrix}.$$

Note that the right-hand side of this equation is identical to the result of multiplying a diagonal matrix of the eigenvalues by the matrix $\mathbf{Q}$,

$$\begin{pmatrix} q_{1,1} & q_{1,2} & \cdots & q_{1,n} \\ q_{2,1} & q_{2,2} & \cdots & q_{2,n} \\ \vdots & \vdots & \ddots & \vdots \\ q_{n,1} & q_{n,2} & \cdots & q_{n,n} \end{pmatrix} \begin{pmatrix} \mu_1 & 0 & \cdots & 0 \\ 0 & \mu_2 & \cdots & 0 \\ \vdots & \vdots & \ddots & \vdots \\ 0 & 0 & \ldots & \mu_n \end{pmatrix}$$

$$= \begin{pmatrix} \mu_1 q_{1,1} & \mu_2 q_{1,2} & \cdots & \mu_n q_{1,n} \\ \mu_1 q_{2,1} & \mu_2 q_{2,2} & \cdots & \mu_n q_{2,n} \\ \vdots & \vdots & \ddots & \vdots \\ \mu_1 q_{n,1} & \mu_2 q_{n,2} & \cdots & \mu_n q_{n,n} \end{pmatrix}.$$

Let $\widetilde{\mathbf{M}}$ be the diagonal matrix of the eigenvalues. We have now shown that $\mathbf{MQ} = \mathbf{Q}\widetilde{\mathbf{M}}$. The fact that the columns of $\mathbf{Q}$ are, by assumption, orthogonal implies that they are linearly independent, and therefore that $\mathbf{Q}$ is nonsingular. The inverse of $\mathbf{Q}$ exists and $\mathbf{Q}^{-1}\mathbf{MQ} = \widetilde{\mathbf{M}}$. To prove that $\mathbf{Q}$ is unitary, note that $\mathbf{Q}^\dagger\mathbf{Q}$ yields a matrix in which the i, j element is $\mathbf{q}_i^\dagger \mathbf{q}_j = \langle \mathbf{q}_i, \mathbf{q}_j \rangle$. By assumption the columns of $\mathbf{Q}$ are orthonormal. Therefore, $\langle \mathbf{q}_i, \mathbf{q}_j \rangle = \delta_{i,j}$, which are the elements that make up the identity matrix. Hence, $\mathbf{Q}^\dagger\mathbf{Q} = \mathbf{I}$. This proves that $\mathbf{Q}$ is unitary. □

The most common use of matrix diagonalization is to calculate eigenvalues and eigenvectors, but it has various other applications as well, as illustrated by the following examples.

Example 19.4.* *Rigid-body moments of inertia.* In Example 18.2 it was found that the angular momentum vector of a rigid body is given by $\mathbf{L} = \boldsymbol{\mathcal{I}}\boldsymbol{\omega}$, where $\boldsymbol{\mathcal{I}}$ is the moment of inertia tensor and $\boldsymbol{\omega}$ is the angular frequency. This result can be greatly simplified by diagonalizing $\boldsymbol{\mathcal{I}}$. Let

$$\widetilde{\boldsymbol{\mathcal{I}}} = \mathbf{Q}^{-1}\boldsymbol{\mathcal{I}}\mathbf{Q} = \begin{pmatrix} I_{xx} & 0 & 0 \\ 0 & I_{yy} & 0 \\ 0 & 0 & I_{zz} \end{pmatrix}. \tag{19.18}$$

The diagonal elements, I_{xx}, I_{yy}, I_{zz} are called the ***moments of inertia*** of the rigid body. The kinetic energy of the body, as explained in Example 13.6, is $T = \frac{1}{2}\boldsymbol{\omega} \cdot \mathbf{L}$.

It follows that

$$T = \tfrac{1}{2}\boldsymbol{\omega}^{\mathrm{T}}\boldsymbol{\mathcal{I}}\boldsymbol{\omega} = \tfrac{1}{2}\boldsymbol{\omega}^{\mathrm{T}}\mathbf{Q}\mathbf{Q}^{-1}\boldsymbol{\mathcal{I}}\mathbf{Q}\mathbf{Q}^{-1}\boldsymbol{\omega} = \tfrac{1}{2}\widetilde{\boldsymbol{\omega}}^{\mathrm{T}}\widetilde{\boldsymbol{\mathcal{I}}}\widetilde{\boldsymbol{\omega}}, \tag{19.19}$$

where $\widetilde{\boldsymbol{\omega}}$ is the vector

$$\widetilde{\boldsymbol{\omega}} = \mathbf{Q}^{-1}\boldsymbol{\omega} = \mathbf{Q}^{\mathrm{T}}\boldsymbol{\omega}. \tag{19.20}$$

($\mathbf{Q}^{-1} = \mathbf{Q}^{\mathrm{T}}$ because $\mathbf{Q}$ is unitary and real.) $\boldsymbol{\omega}$ is the angular frequency for rotation about some arbitrarily chosen z-axis. $\widetilde{\boldsymbol{\omega}}$ is the angular frequency in a new coordinate system in which the original Cartesian axes are rotated by the transformation matrix $\mathbf{Q}$. The unit vectors for the new axes in terms of the original coordinate system are

$$\tilde{\mathbf{x}} = \mathbf{Q}^{-1}\begin{pmatrix}1\\0\\0\end{pmatrix}, \quad \tilde{\mathbf{y}} = \mathbf{Q}^{-1}\begin{pmatrix}0\\1\\0\end{pmatrix}, \quad \tilde{\mathbf{z}} = \mathbf{Q}^{-1}\begin{pmatrix}0\\0\\1\end{pmatrix}. \tag{19.21}$$

These new axes are called the ***principal axes of rotation*** for the rigid body. It follows from Eq. (19.19) that in this coordinate system rotational kinetic energy has a very simple expression:

$$T = \tfrac{1}{2}\left(I_{xx}\widetilde{\omega}_x^2 + I_{yy}\widetilde{\omega}_y^2 + I_{zz}\widetilde{\omega}_z^2\right), \tag{19.22}$$

where $\widetilde{\omega}_x$, $\widetilde{\omega}_y$, and $\widetilde{\omega}_z$ are the components of $\widetilde{\boldsymbol{\omega}}$.

Example 19.5.* *Principal axes of rotation for formyl chloride.* In Example 3.3 we constructed an internal Cartesian coordinate system for the formyl chloride molecule with origin at the center of mass. The positions of the atoms were given by the vectors

$$\vec{x}_{\mathrm{O}} = (-0.48, 1.8, 0), \qquad \vec{x}_{\mathrm{C}} = (-0.48, 0.58, 0),$$
$$\vec{x}_{\mathrm{H}} = (-1.03, -0.30, 0), \qquad \vec{x}_{\mathrm{Cl}} = (0.42, -0.99, 0).$$

The moment of inertia tensor, according to Eq. (18.10), is

$$\boldsymbol{\mathcal{I}} = \begin{pmatrix} 90.270 & 31.409 & 0 \\ 31.409 & 13.686 & 0 \\ 0 & 0 & 103.956 \end{pmatrix},$$

in units of u Å^2. This is diagonalized according to

$$\widetilde{\boldsymbol{\mathcal{I}}} = \mathbf{Q}^{-1}\boldsymbol{\mathcal{I}}\mathbf{Q} = \begin{pmatrix} 104.0 & 0 & 0 \\ 0 & 101.5 & 0 \\ 0 & 0 & 2.45 \end{pmatrix}, \quad \mathbf{Q} = \begin{pmatrix} 0 & 0.94159 & 0.33677 \\ 0 & 0.33677 & -0.94159 \\ 1 & 0 & 0 \end{pmatrix}.$$

The principal axes, from Eq. (19.21), are shown below:

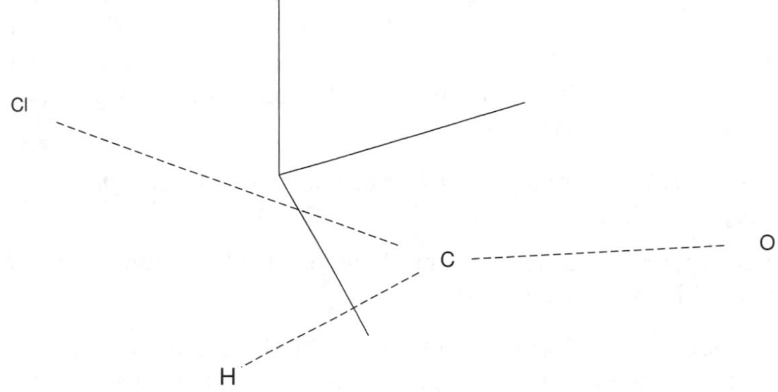

We can describe the rotational motion in terms of simultaneous rotations of the molecule about these three axes, according to Eq. (19.22).

Example 19.6.* *Functions of Hermitian matrices.* In Section 7.3 we defined the "exponential of anything" in terms of a Taylor series. In particular, the exponential of a matrix is given by

$$e^{\mathbf{M}} = \mathbf{I} + \sum_{j=1}^{\infty} \frac{1}{j!} \mathbf{M}^j, \tag{19.23}$$

where $\mathbf{M}^j$ represents multiplication of the matrix by itself j times. This definition is not useful for practical computation on account of the infinite number of terms to evaluate. However, for a *diagonal* matrix $\widetilde{\mathbf{M}}$, computing a power of the matrix is easy:

$$\widetilde{\mathbf{M}}^j = \begin{pmatrix} \mu_1 & 0 & \cdots & 0 \\ 0 & \mu_2 & \cdots & 0 \\ \vdots & \vdots & \ddots & \vdots \\ 0 & 0 & \cdots & \mu_n \end{pmatrix}^j = \begin{pmatrix} \mu_1^j & 0 & \cdots & 0 \\ 0 & \mu_2^j & \cdots & 0 \\ \vdots & \vdots & \ddots & \vdots \\ 0 & 0 & \cdots & \mu_n^j \end{pmatrix}. \tag{19.24}$$

It then follows from Eq. (19.23) that $e^{\widetilde{\mathbf{M}}}$ is the diagonal matrix with

$$1 + \sum_{j=1}^{\infty} \frac{1}{j!} \mu_i^j = e^{\mu_i}$$

as the ith diagonal element. We can compute $e^{\mathbf{M}}$ for an Hermitian $\mathbf{M}$ by diagonalizing $\mathbf{M}$, computing the exponentials of the diagonal elements, and then undoing the similarity transformation:

$$e^{\mathbf{M}} = \mathbf{Q}\, e^{\mathbf{Q}^{-1}\mathbf{M}\mathbf{Q}}\, \mathbf{Q}^{-1}, \tag{19.25}$$

which follows from the fact that

$$\mathbf{Q}\, e^{\mathbf{Q}^{-1}\mathbf{M}\mathbf{Q}}\, \mathbf{Q}^{-1} = \mathbf{Q}\Big(\sum_{j=0}^{\infty} \frac{1}{j!} \overbrace{\mathbf{Q}^{-1}\mathbf{M}\mathbf{Q}\mathbf{Q}^{-1}\mathbf{M}\mathbf{Q}\cdots}^{j\text{ times}}\Big)\mathbf{Q}^{-1} = \sum_{j=0}^{\infty} \frac{1}{j!}\mathbf{M}^j, \tag{19.26}$$

because $\mathbf{Q}\mathbf{Q}^{-1} = \mathbf{I}$. In the same way, other functions of matrices can be defined in terms of their Taylor series, $f(\mathbf{M}) \sim \mathbf{Q}\, f(\widetilde{\mathbf{M}})\, \mathbf{Q}^{-1}$.

The following two theorems are consequences of Theorem 19.2.2.

Theorem 19.2.3. *The trace of an Hermitian matrix is equal to the sum of its eigenvalues.*

Proof. The eigenvalues of a diagonal matrix are simply the elements on the diagonal. In this case the trace is obviously the sum of the eigenvalues. According to Theorem 19.2.1, the eigenvalues are invariant to a similarity transformation. According to Theorem 19.2.2, any Hermitian matrix can be diagonalized by a similarity transormation. □

Theorem 19.2.4. *The trace of an Hermitian matrix* $\mathbf{M}$ *is equal to the trace of* $\mathbf{Q}^{-1}\mathbf{M}\mathbf{Q}$ *for any unitary matrix* $\mathbf{Q}$.

Proof. The trace is equal to the sum of the eigenvalues, and the eigenvalues are unaffected by a similarity transform. □

Theorem 19.2.4 has important applications in the group theory of molecular symmetry. The elements of a group can be divided into subsets, called *classes*, such that within each class all the elements can be transformed one into the other by similarity transformations. In the group $\mathbf{C}_{3v}$, in Example 16.1, the classes are the subsets $\{E\}$, $\{C_3, C_3^2\}$, and $\{\sigma^{(1)}, \sigma^{(2)}, \sigma^{(3)}\}$. The traces of matrices representing the transformations are, in this context, usually referred to as *characters* of the matrices. While the characters are the same within a class, the values of the characters for each class depend on the symmetry properties of the basis vectors on which the matrices operate. This fact makes it possible to use the set of characters to characterize the symmetry of the basis vectors and hence of the physical states being described by them, such as molecular electronic wavefunctions corresponding to different quantum states.[2]

19.3 Differential Eigenvalue Equations

Consider the equation

$$\frac{d}{dx}\, e^{cx} = c\, e^{cx}. \tag{19.27}$$

Clearly this qualifies as an eigenvalue equation according to the definition of Eq. (19.1). In fact, any function of the form ae^{cx} where a and c are constants is an eigenvector of the first derivative operator, with eigenvalue c. In a vector space of functions, the term ***eigenfunction*** is often used instead of eigenvector. Thus, we say that ae^{cx} is an eigenfunction of the operator $\frac{d}{dx}$.

Consider the second derivative operator. The second derivative of ae^{cx} is

$$\frac{d^2}{dx^2}\, ae^{cx} = c^2\, ae^{cx}. \tag{19.28}$$

This is an eigenvalue equation with eigenvalue c^2. Consider the function

$$f(x) = a\cos(cx) + b\sin(cx), \tag{19.29}$$

where a, b, and c are arbitrary numbers. This is not an eigenfunction of $\frac{d}{dx}$, because

$$\frac{d}{dx}\, f = c\,[-a\sin(cx) + b\cos(cx)]\,. \tag{19.30}$$

$-a\sin(cx) + b\cos(cx)$ is not proportional to f. However, f is an eigenfunction of the *second* derivative,

$$\frac{d^2}{dx^2}\, f = -c^2\, f\,, \tag{19.31}$$

with eigenvalue $-c^2$, because $f'' = c\,[-ca\cos(cx) - cb\sin(cx)]$. Another solution of Eq. (19.31) is

$$f(x) = ae^{icx} + be^{-icx}. \tag{19.32}$$

For these examples the eigenvalues are not very "particular"—a solution for the eigenfunction exists for any choice of eigenvalue. But suppose that the eigenfunction is required to satisfy *boundary conditions*; that is, the function or its first derivative is required to have certain specified values at particular points x. This can impose constraints on the possible solutions.

[2] Among the practical applications of characters are the construction of wavefunctions with a given symmetry and the prediction of selection rules for absorption and emission spectra. See references in the Bibliography for details.

Example 19.7. *Quantum mechanical particle on a ring.* The Schrödinger equation for a particle of mass m confined to a circle of radius r is

$$-\frac{\hbar^2}{2mr^2}\frac{d^2}{d\phi^2}\psi = E\psi\,. \tag{19.33}$$

$\hbar$ is Planck's constant divided by 2π and E is the energy of the particle. We use the Greek letter ψ ("psī") for the eigenfunction, which is the common practice in quantum mechanics. ψ is a function of ϕ, the angular coordinate that specifies the position on the circle. Multiplying both sides of Eq. (19.33) by $-2mr^2/\hbar^2$, we obtain Eq. (19.31) with $c^2 = 2mr^2E/\hbar^2$. The physical nature of this problem requires that a full rotation, through an angle of 2π, must bring the particle back to its original condition. Expressing this mathematically, we have

$$\psi(\phi) = \psi(\phi + 2\pi). \tag{19.34}$$

Using Eq. (19.29) with $f(x)$ replaced by $\psi(\phi)$, this boundary condition implies

$$a\cos(c\phi) + b\sin(c\phi) = a\cos(c\phi + 2\pi c) + b\sin(c\phi + 2\pi c),$$

which will be true only if c is an integer. Thus, the operator $\frac{d^2}{d\phi^2}$ subject to the boundary condition Eq. (19.34) has an eigenvalue equation

$$\frac{d^2}{d\phi^2}\,\psi(\phi) = -j^2\psi(\phi), \qquad j = \pm 1, \pm 2, \pm 3, \ldots, \tag{19.35}$$

with eigenfunctions

$$\psi_j(\phi) = a\cos(j\phi) + b\sin(j\phi)\,. \tag{19.36}$$

If j is not an integer, then there is no solution to the eigenvalue equation. This is a mathematical expression of the physical concept of quantization of energy. The only possible energy levels are $E_j = -\hbar^2 j^2/2mr^2$, where j is a nonzero integer.

19.4 Hermitian Operators

The theory of differential eigenvalue equations closely parallels that of matrix eigenvalue equations. To see this, we must specify an inner product, to make the vector space a Hilbert space. The choice of inner product will depend on the boundary conditions of the problem under consideration. For the particle on a circle the choice

$$\langle f, g\rangle = \int_0^{2\pi} f^*(\phi)g(\phi)d\phi \tag{19.37}$$

would be appropriate.[3] The complex conjugate of f is needed to ensure that the norm, $\langle f, f\rangle$, is always a non-negative real number. For a particle constrained to a line, we use

$$\langle f, g\rangle = \int_{-\infty}^{\infty} f^*(x)g(x)dx\,, \tag{19.38}$$

and for a particle in three-dimensional space, we use

$$\langle f, g\rangle = \int_{-\infty}^{\infty}\int_{-\infty}^{\infty}\int_{-\infty}^{\infty} f^*(x,y,z)g(x,y,z)\,dx\,dy\,dz\,, \tag{19.39}$$

[3]In quantum mechanics the inner product is more commonly denoted $\langle f|g\rangle$ rather than $\langle f, g\rangle$. Here, for consistency with previous chapters, the comma notation is retained.

with the boundary conditions that f and g go to zero at infinity. Because the integral in Eq. (19.39) appears so frequently, the common practice is to use the shorthand notation

$$\langle f, g \rangle = \int f^* g \, d\tau \,, \tag{19.40}$$

where integration over "$d\tau$" is understood to mean that this is a multidimensional definite integral over all space. (τ is the Greek letter *tau*.)

Let us define an ***Hermitian operator*** as an operator G such that

$$\langle f, Gg \rangle = \langle Gf, g \rangle. \tag{19.41}$$

For example, with the inner product in Eq. (19.40), an Hermitian operator satisfies the condition

$$\int f^* G g \, d\tau = \int (Gf)^* g \, d\tau \,. \tag{19.42}$$

We can state the following theorems:

Theorem 19.4.1. *All eigenvalues of an Hermitian operator are real numbers.*

Theorem 19.4.2. *Distinct eigenvectors of an Hermitian operator are linearly independent.*

Theorem 19.4.3. *Eigenvectors of an Hermitian operator that have different eigenvalues are orthogonal to each other.*

Theorem 19.4.4. *The distinct eigenvectors of an Hermitian operator form a basis for the Hilbert space.*

These are the generalizations for Hermitian operators of the theorems 19.1.4, 19.1.5, 19.1.6, and 19.1.7 for Hermitian matrices. The only significant difference between the matrix and operator versions is that for matrices the Hilbert space has some finite dimension n while differential operators can operate on Hilbert spaces of infinite dimension. For second-order differential eigenvalue equations we have an additional theorem, which can be useful for establishing that an operator is Hermitian:

Theorem 19.4.5. *Consider an operator G in a Hilbert space of functions of a coordinate x in some range $[x_{\text{min}}, x_{\text{max}}]$. Suppose that the operator has the form*

$$G = q_0(x) \frac{d^2}{dx^2} + q_1(x) \frac{d}{dx} + q_2(x), \tag{19.43}$$

where the q_j are real functions of x and q_0 is nonzero for $x_{\text{min}} < x < x_{\text{max}}$. Then G is Hermitian if and only if

$$q_0'(x) = q_1(x) \,. \tag{19.44}$$

An operator that satisfies these conditions is said to be ***self-adjoint***.

Example 19.8. *Quantum mechanical harmonic oscillator.* The Schrödinger equation for a one-dimensional harmonic oscillator is

$$\left(-\frac{\hbar^2}{2m}\frac{d^2}{dx^2} + \tfrac{1}{2}kx^2\right)\psi = E\psi\,. \tag{19.45}$$

Before applying linear algebra, let us simplify this equation by consolidating the various constants. This can be done with a change of variable, $x = \alpha\xi$, where α is as yet unspecified. Then

$$\left(-\frac{\hbar^2}{2m\alpha^2}\frac{d^2}{d\xi^2} + \tfrac{1}{2}k\alpha^2\xi^2\right)\psi = E\psi\,.$$

Multiply through by $2m\alpha^2/\hbar^2$ to eliminate the constant in the first term,

$$\left(-\frac{d^2}{d\xi^2} + \frac{m\alpha^4 k}{\hbar^2}\xi^2\right)\psi = \frac{2m\alpha^2}{\hbar^2}E\psi\,.$$

We can choose α so that the constant in the second term disappears. Let $\alpha^4 = \hbar^2/mk$. Then

$$\left(\frac{d^2}{d\xi^2} - \xi^2\right)\psi = -\epsilon\psi\,, \qquad \epsilon = \frac{2}{\hbar}\sqrt{\frac{m}{k}}\,E\,. \tag{19.46}$$

Scaling the equation in this way is not absolutely necessary, but it is well worth the effort. The more miscellaneous constants in the equation, the greater the likelihood of accidently omitting or miscopying one of them.

Let us then consider the operator

$$G = \frac{d^2}{d\xi^2} - \xi^2. \tag{19.47}$$

$ae^{-c\xi^2}$ is an eigenfunction, but only for a particular value of c, because

$$G\,ae^{-c\xi^2} = (-2c + 4c^2\xi^2 - \xi^2)ae^{-c\xi^2}, \tag{19.48}$$

and the terms proportional to ξ^2 in the parentheses disappear only if $c = \frac{1}{2}$, giving

$$G\,ae^{-\xi^2/2} = -ae^{-\xi^2/2}. \tag{19.49}$$

The eigenvalue is -1. Although c is constrained, the prefactor a is arbitrary.

This is not the only possible solution for an eigenfunction of G. Consider $a\xi e^{-\xi^2/2}$:

$$G\,a\xi e^{-\xi^2/2} = -3ae^{-\xi^2/2}, \tag{19.50}$$

and the eigenvalue is -3. How about a general solution $f_j(\xi) = p_j(\xi)e^{-\xi^2/2}$, where $p_j(\xi)$ is a polynomial of arbitrary degree j? To see which polynomials would be acceptable, consider the inner product,

$$\langle f_j, f_k\rangle = \int_{-\infty}^{\infty} f_j^*(\xi)f_k(\xi)d\xi = \int_{-\infty}^{\infty} p_j^*(\xi)p_k(\xi)e^{-\xi^2}d\xi\,. \tag{19.51}$$

Suppose that none of the eigenvectors are degenerate. Then Theorem 19.4.3 implies that $\langle f_j, f_k\rangle = \delta_{j,k}$, i.e.,

$$\int_{-\infty}^{\infty} p_j^*(\xi)p_k(\xi)e^{-\xi^2}d\xi = \delta_{j,k}\,. \tag{19.52}$$

This is equivalent to Eq. (17.21), the weighted inner product that defined the Hermite polynomials. Therefore, any p_j that are proportional to Hermite polynomials H_j are acceptable solutions, with

$$f_j(\xi) = aH_j(\xi)e^{-\xi^2/2}. \tag{19.53}$$

Substituting $(mk/\hbar^2)^{1/2}x$ for ξ gives the eigenfunctions in terms of the original coordinate x. The eigenvalue corresponding to f_j, which can be found by operating with G, is $-(2j+1)$. The spectrum of possible energy levels is

$$E = \hbar\sqrt{\frac{k}{m}}\,(j+\tfrac{1}{2}), \quad j = 1, 2, 3, \ldots . \tag{19.54}$$

Now compare Eq. (19.47) with Eq. (19.43). We have $q_0 = 1$ and $q_1 = 0$. These satisfy Eq. (19.44), the criterion for a self-adjoint operator. Therefore G is Hermitian and its eigenfunctions are a basis for the Hilbert space. This proves the claim made in Section 17.1 that the functions $H_j(\xi)e^{-\xi^2/2}$ constitute a basis for $L^2[-\infty, \infty]$.

19.5 The Variational Principle*

Most eigenvalue equations that arise in science and engineering cannot be solved exactly in terms of standard mathematical functions. A common strategy is to guess a functional form for the eigenvector containing arbitrary parameters and then choose the parameter values that give the best approximation for that functional form. One can argue that the "best" choice of the parameter set is that which minimizes the lowest eigenvalue. This is the ***variational principle***. Its theoretical foundation is the following theorem:

Theorem 19.5.1. *Consider an eigenvalue equation $G\psi = c\psi$ with a spectrum of real eigenvalues $c = c_0,\ c_1,\ c_2, \ldots$ and corresponding eigenvectors $\psi = \psi_0,\ \psi_1,\ \psi_2, \ldots$. Assume that G is an Hermitian operator and that $c_0 \le c_1 \le c_2$, etc. Let V be the Hilbert space spanned by the eigenvectors. For any vector Φ in V,*

$$c_0 \le \frac{\langle \Phi, G\Phi \rangle}{\langle \Phi, \Phi \rangle} . \tag{19.55}$$

Proof. The set of eigenvectors $\{\psi_j\}$ form a basis for the Hilbert space, according to Theorem 19.4.4. Therefore, any arbitrary vector Φ in V can be expressed as

$$\Phi = \sum_{j=0}^{\infty} b_j \psi_j , \tag{19.56}$$

in terms of scalars b_j. (The summation index starts at $j = 0$ because the lowest-eigenvalue solution is conventionally labeled with subscript 0.) Operating on Φ with G and then taking its inner product with Φ gives

$$\begin{aligned}\langle \Phi, G\Phi \rangle &= \sum_{j=0}^{\infty}\sum_{k=0}^{\infty} \langle b_j\psi_j, G\, b_k\psi_k \rangle = \sum_{j=0}^{\infty}\sum_{k=0}^{\infty} b_j^* b_k \langle \psi_j, G\psi_k \rangle \\ &= \sum_{j=0}^{\infty}\sum_{k=0}^{\infty} b_j^* b_k \langle \psi_j, c_k\psi_k \rangle = \sum_{j=0}^{\infty}\sum_{k=0}^{\infty} c_k b_j^* b_k \langle \psi_j, \psi_k \rangle .\end{aligned} \tag{19.57}$$

Assume $\{\psi_j\}$ is orthonormal. (An orthonormal set can be constructed using

Gram-Schmidt orthogonalization in the case that any eigenvectors are degenerate.) Then $\langle\psi_j,\psi_k\rangle = \delta_{j,k}$, and

$$\langle\Phi, G\Phi\rangle = \sum_{k=0}^{\infty} c_k |b_k|^2. \tag{19.58}$$

The inner product of Φ with itself is

$$\langle\Phi,\Phi\rangle = \sum_{j=0}^{\infty}\sum_{k=0}^{\infty}\langle b_j\psi_j,\, b_k\psi_k\rangle = \sum_{j=0}^{\infty}\sum_{k=0}^{\infty} b_j^* b_k \delta_{j,k} = \sum_{k=0}^{\infty}|b_k|^2. \tag{19.59}$$

Let us add the quantity $(c_0 - c_0)\langle\Phi,\Phi\rangle$, which is of course equal to zero, to Eq. (19.58),

$$\langle\Phi, G\Phi\rangle = (c_0 - c_0)\langle\Phi,\Phi\rangle + \sum_{k=0}^{\infty} c_k|b_k|^2 = c_0\langle\Phi,\Phi\rangle - c_0\sum_{k=0}^{\infty}|b_k|^2 + \sum_{k=0}^{\infty} c_k|b_k|^2$$

$$= c_0\langle\Phi,\Phi\rangle + \sum_{k=0}^{\infty}(c_k - c_0)|b_k|^2.$$

It follows that

$$c_0 = \frac{\langle\Phi, G\Phi\rangle}{\langle\Phi,\Phi\rangle} - \gamma, \qquad \gamma = \frac{\sum_{k=0}^{\infty}(c_k - c_0)|b_k|^2}{\sum_{k=0}^{\infty}|b_k|^2}. \tag{19.60}$$

Because c_0 is the lowest eigenvalue, we know that $c_k - c_0 \geq 0$ for all k. Therefore, $\gamma \geq 0$. This implies Eq. (19.55). □

The variational principle gives a practical strategy for solving the Schrödinger equation,

$$\hat{H}\Psi = E\Psi\,,$$

where the eigenvalue E is the energy of a quantum mechanical system (such as an atom) and $\hat{H}$ is is an Hermitian linear operator called the *Hamiltonian operator.* (See Section 20.1.1.) Let Φ be some guess for Ψ in the form of a linear combination

$$\Phi = \sum_{j=1}^{N} a_j f_j\,, \tag{19.61}$$

where $\{f_1, f_2, f_3, \ldots, f_N\}$ is a finite set of orthonormal functions. If the variational parameters a_j appear only as linear prefactors (i.e., the f_j contain no variational parameters), then the analysis can be formulated in terms of matrix algebra. Let

$$\tilde{E} = \langle\Phi, \hat{H}\Phi\rangle/\langle\Phi,\Phi\rangle\,. \tag{19.62}$$

$\tilde{E}$ can be thought of as the approximation for the ground-state energy that results from a given choice of the trial function. (Note that this is a functional, $\tilde{E}[\Phi]$.) Eq. (19.62) can be written

$$\langle\Phi, \hat{H}\Phi\rangle = \tilde{E}\,\langle\Phi,\Phi\rangle. \tag{19.63}$$

Let us now substitute the finite linear combination Eq. (19.61) for Φ. Then

$$\langle \Phi, \Phi \rangle = \sum_{j=1}^{N} \sum_{k=1}^{N} a_j^* a_k \langle f_j, f_k \rangle = \sum_{k=1}^{N} a_k^* a_k \,. \tag{19.64}$$

Let us express the a_k as a column vector $\mathbf{a}$ and let $\mathbf{H}$ be the matrix with elements

$$H_{jk} = \langle f_j, \hat{H} f_k \rangle / \langle f_j, f_k \rangle. \tag{19.65}$$

Eq. (19.63) can be written

$$\sum_{j=1}^{N} \sum_{k=1}^{N} a_j^* H_{j,k} a_k = \tilde{E} \sum_{k=1}^{N} a_k^* a_k \,. \tag{19.66}$$

which is equivalent to the matrix equation

$$\mathbf{a}^\dagger \mathbf{H} \mathbf{a} = \tilde{E}\, \mathbf{a}^\dagger \mathbf{a} \,, \tag{19.67}$$

where $\mathbf{a}^\dagger$ is the Hermitian conjugate of $\mathbf{a}$.

Example 19.9. *Matrix formulation of the variational principle for a basis of dimension 2.* Eq. (19.66) is

$$a_1^* H_{1,1} a_1 + a_1^* H_{1,2} a_2 + a_2^* H_{2,1} a_1 + a_2^* H_{2,2} a_2 = \tilde{E} a_1^* a_1 + \tilde{E} a_2^* a_2 \,. \tag{19.68}$$

The left-hand side of the matrix equation, Eq. (19.67), is

$$\begin{pmatrix} a_1^* & a_2^* \end{pmatrix} \begin{pmatrix} H_{1,1} & H_{1,2} \\ H_{2,1} & H_{2,2} \end{pmatrix} \begin{pmatrix} a_1 \\ a_2 \end{pmatrix} = \begin{pmatrix} a_1^* & a_2^* \end{pmatrix} \begin{pmatrix} H_{1,1} a_1 + H_{1,2} a_2 \\ H_{2,1} a_1 + H_{2,2} a_2 \end{pmatrix}$$

$$= a_1^* (H_{1,1} a_1 + H_{1,2} a_2) + a_2^* (H_{2,1} a_1 + H_{2,2} a_2) \,,$$

which is equal to the left-hand side of Eq. (19.68), and the right-hand side of the matrix equation is

$$\tilde{E} \begin{pmatrix} a_1^* & a_2^* \end{pmatrix} \begin{pmatrix} a_1 \\ a_2 \end{pmatrix} = \tilde{E} \,(a_1^* a_1 + a_2^* a_2) \,,$$

which is equal to the right-hand side of Eq. (19.68).

We can write $\mathbf{a}^\dagger \tilde{E} \mathbf{a}$ instead of $\tilde{E} \mathbf{a}^\dagger \mathbf{a}$, because $\tilde{E}$ is just a scalar. Eq. (19.67) can then be written as $\mathbf{a}^\dagger (\mathbf{H} - \tilde{E} \mathbf{I}) \mathbf{a} = 0$, where $\mathbf{I}$ is the identity matrix. This implies that $\mathbf{H} - \tilde{E} \mathbf{I}$ is equal to the zero matrix, and therefore

$$\mathbf{H} \mathbf{a} = \tilde{E} \mathbf{a} \,. \tag{19.69}$$

This is a matrix eigenvalue equation. The optimal choice of the $\{a_j\}$ must correspond to an eigenvector of $\mathbf{H}$. The optimal choice according to the variational principle is the set that gives the minimum value of $\tilde{E}$. To obtain the optimal solution, all we need to do is choose the eigenvector corresponding to the lowest eigenvalue.

This suggests the following procedure for solving Schrödinger's equation:

1. Choose an orthonormal basis set $\{f_j\}$ of finite dimension N.
2. Compute the Hamiltonian matrix elements $H_{jk} = \langle f_j, \hat{H} f_k \rangle$.
3. Diagonalize $\mathbf{H}$, to obtain its eigenvalues and eigenvectors.

The lowest eigenvalue $\tilde{E}_0$ is an upper bound to the true ground-state energy E_0. Increasing N lowers the value of $\tilde{E}_0$, making it a more accurate approximation for E_0. The column vector $\mathbf{a}_0$ corresponding to $\tilde{E}_0$ gives an approximate solution for the ground-state eigenfunction,

$$\Psi \approx \sum_{j=1}^{N} a_{0,j} f_j \, , \tag{19.70}$$

where the $a_{0,j}$ are the elements of $\mathbf{a}_0$. Higher eigenvalues and eigenvectors of $\mathbf{H}$ are approximate solutions for excited states. However, in practice the accuracy of the approximations decreases considerably as the energy difference from the ground-state increases.

Exercises

19.1 Use computer algebra to calculate eigenvectors of arbitrary Hermitian matrices of your choosing. Construct diagonalizing matrices $\mathbf{Q}$ and show that satisfy Theorem 19.2.2.

19.2 Show that $4e^{i7\phi}$ is a particle-on-a-ring eigenfunction. Determine the eigenvalue.

19.3 Prove Eq. (19.50).

19.4 Which of the following operators are Hermitian, according to Theorem 19.4.5?

$$(i)\ \frac{d^2}{dx^2} + e^{-x} \qquad (ii)\ \frac{d^2}{dx^2} + e^{-x}\frac{d}{dx} \qquad (iii)\ x\frac{d^2}{dx^2} + \frac{d}{dx} + e^{-x}$$

19.5 Using computer algebra, find the eigenvalues and eigenvectors of arbitrary Hermitian matrices of your choosing. Show that Theorems 19.1.4, 19.1.5, and 19.1.6 are satisfied.

19.6 In deriving the one-dimensional harmonic oscillator eigenfunctions, Eq. (19.53), we assumed they were nondegenerate. Are there any degenerate solutions?

19.7* It follows from Eq. (17.58) that the spherical harmonics are the eigenfunctions of the operator

$$\Omega = \frac{1}{\sin\theta}\frac{\partial}{\partial\theta}\left(\sin\theta\,\frac{\partial}{\partial\theta}\right) + \frac{1}{\sin^2\theta}\frac{\partial^2}{\partial\phi^2} \, .$$

If Ω is Hermitian, then Theorem 19.4.4 implies that the spherical harmonics are a basis set for a Hilbert space. Show that Ω is Hermitian for the Hilbert space of functions of the angles θ and ϕ such that the functions and their first derivatives are invariant to rotation by 360°; i.e., $f(\phi + 2\pi) = f(\phi)$ and $f'(\phi + 2\pi) = f'(\phi)$. Use the inner product $\langle f, g \rangle = \int_0^{2\pi} \int_0^{\pi} f^* g \sin\theta \, d\theta d\phi$.

19.8* Use a variational calculation to estimate the ground-state eigenvalue of the anharmonic oscillator Schrödinger equation $\left(\frac{d^2}{d\xi^2} - \xi^2 + 0.2\,\xi^4\right)\psi = -\epsilon\psi$.

Chapter 20

Schrödinger's Equation

20.1 Quantum Mechanics

Quantum mechanics is a mathematical system, usually expressed in the language of linear algebra, that is found in practice to describe the physical world more accurately than classical mechanics. The mathematical formalism is developed from several postulates. The justification for these postulates is that the resulting theory gives predictions that have so far agreed with all experiments that have been carried out to test it.

20.1.1 Quantum Mechanical Operators

In *classical* mechanics (see Section 13.2.2) the energy E of a particle at position $\vec{x}(t)$ and momentum $\vec{p}(t)$ is equal to the numerical value of the Hamiltonian function,

$$E = H(\vec{x}, \vec{p}) = T(\vec{x}, \vec{p}) + V(\vec{x}, \vec{p}), \tag{20.1}$$

where T is the kinetic energy function and V is the potential energy function. For a system consisting of a single particle of mass m, we have, in Cartesian coordinates,

$$\vec{x} = \big(x(t), y(t), z(t)\big), \qquad \vec{p} = \big(p_x(t), p_y(t), p_z(t)\big), \tag{20.2}$$

where t is time. The kinetic energy in classical mechanics is

$$T = \frac{1}{2m}\left(p_x^2 + p_y^2 + p_z^2\right). \tag{20.3}$$

In classical mechanics the goal is to solve for the trajectory, $\vec{x}(t)$, using the equations of Newton, Lagrange, or Hamilton.

In quantum mechanics, *energy* remains a meaningful concept but *trajectory* does not. Schrödinger[1] conjectured that Eq. (20.1) should be replaced by an eigenvalue equation,

$$\hat{H}\Psi = E\Psi, \tag{20.4}$$

where $\hat{H}$ is the ***Hamiltonian operator***, obtained from the classical Hamiltonian function by replacing the momentum components with differential

[1] Austrian physicist Erwin Schrödinger (1887-1961). He developed this theory in a remarkable series of four papers published in the first half of 1926. A French graduate student, Louis de Broglie, had recently proposed that particles had intrinsic wave-like properties associated with them. Schrödinger figured that where there is a wave there must be a differential equation, and then set about constructing a differential equation that would yield de Broglie's particle wave as a solution. For a fascinating and readable account of Schrödinger's reasoning and the context of his work, see W. Moore, *Schrödinger: Life and Thought* (Cambridge University Press, Cambridge, 1989), Chap. 6.

operators

$$p_x \to \hat{p}_x = \frac{\hbar}{i}\frac{\partial}{\partial x}, \qquad p_y \to \hat{p}_y = \frac{\hbar}{i}\frac{\partial}{\partial y}, \qquad p_z \to \hat{p}_z = \frac{\hbar}{i}\frac{\partial}{\partial z}. \tag{20.5}$$

i is $\sqrt{-1}$, and $\hbar$ (pronounced "h-bar") is $1.05457163 \times 10^{-34}$ J s, which is Planck's constant divided by 2π. Note the use of hats to distinguish an operator from the value of the property to which it corresponds. To derive the operator $\hat{H}$, we write down the classical function $H = T + V$, in terms of momenta, and then make the substitutions of Eqs. (20.5). The kinetic energy *function* becomes the ***kinetic energy operator***

$$\hat{T} = \frac{1}{2m}\left(\hat{p}_x\hat{p}_x + \hat{p}_y\hat{p}_y + \hat{p}_z\hat{p}_z\right) = -\frac{\hbar^2}{2m}\left(\frac{\partial^2}{\partial x^2} + \frac{\partial^2}{\partial y^2} + \frac{\partial^2}{\partial z^2}\right) = -\frac{\hbar^2}{2m}\nabla^2 \tag{20.6}$$

(∇^2 is the Laplacian operator, introduced in Section 3.5) and the Hamiltonian operator for a particle of mass m is

$$\hat{H} = -\frac{\hbar^2}{2m}\nabla^2 + V. \tag{20.7}$$

If, as is typically the case, the potential energy does not depend on momentum, then $V(x, y, z)$ is the same as in classical mechanics.

The eigenvalue in Schrödinger's equation is the physical value of the property (energy) that corresponds to the operator. In general in quantum mechanics, physical properties that are constants of motion will be eigenvalues of an operator.

Example 20.1. *Constants of motion for a free particle.* We saw in Section 13.2 that for a free particle the energy and the three Cartesian components of linear momentum are constants of motion. If these are to be constants of motion in quantum mechanics, then there must exist a function Ψ that is simultaneously an eigenfunction of each of the four corresponding quantum operators: $\hat{H}$, $\hat{p}_x$, $\hat{p}_y$, and $\hat{p}_z$. Consider

$$\Psi = Ae^{ik_x x + ik_y y + ik_z z}, \tag{20.8}$$

where A, k_x, k_y, and k_z are constants. This is an eigenfunction of $\hat{H}$,

$$\begin{aligned}\hat{H}\Psi &= -\frac{\hbar^2}{2m}\left(\frac{\partial^2\Psi}{\partial x^2} + \frac{\partial^2\Psi}{\partial y^2} + \frac{\partial^2\Psi}{\partial z^2}\right) = -\frac{\hbar^2}{2m}\left(-k_x^2\Psi - k_y^2\Psi - k_z^2\Psi\right) \\ &= \frac{\hbar^2}{2m}\left(k_x^2 + k_y^2 + k_z^2\right)\Psi \end{aligned} \tag{20.9}$$

and of the linear momentum operators,

$$\hat{p}_x\Psi = \hbar k_x\Psi, \qquad \hat{p}_y\Psi = \hbar k_y\Psi, \qquad \hat{p}_z\Psi = \hbar k_z\Psi. \tag{20.10}$$

We conclude that the particle's energy is $E = (p_x^2 + p_y^2 + p_z^2)/2m = \vec{p}\cdot\vec{p}/2m$ and its momentum components are $p_x = \hbar k_x$, $p_y = \hbar k_y$, and $p_z = \hbar k_z$. The eigenfunction corresponding to a free particle with momentum vector $\vec{p}$ and energy $\vec{p}\cdot\vec{p}/2m$ is

$$\Psi = Ae^{i(p_x x + p_y y + p_z z)/\hbar} = Ae^{i\vec{p}\cdot\vec{x}/\hbar}. \tag{20.11}$$

The prefactor A is arbitrary.

We saw in Section 13.2.3 that the z-component of angular momentum in classical mechanics is

$$L_z = m(x\dot{y} - y\dot{x}) = xp_y - yp_x\,.$$

The corresponding quantum operator is

$$\hat{L}_z = \frac{\hbar}{i}\left(x\frac{\partial}{\partial y} - y\frac{\partial}{\partial x}\right). \tag{20.12}$$

Similarly, the classical definitions $L_x = yp_z - zp_y$ and $L_y = -xp_z + zp_x$ imply

$$\hat{L}_x = \frac{\hbar}{i}\left(y\frac{\partial}{\partial z} - z\frac{\partial}{\partial y}\right), \qquad \hat{L}_y = \frac{\hbar}{i}\left(z\frac{\partial}{\partial x} - x\frac{\partial}{\partial z}\right). \tag{20.13}$$

Let us transform these to polar coordinates, according to the procedure in Section 3.5. For $\hat{L}_z$ all the terms involving derivatives with respect to r or θ cancel out, leaving only

$$\hat{L}_z = \frac{\hbar}{i}\frac{\partial}{\partial\phi}\,. \tag{20.14}$$

Transforming $\hat{L}_x$ and $\hat{L}_y$ to polar coordinates gives

$$\hat{L}_x = -\frac{\hbar}{i}\left(\sin\phi\frac{\partial}{\partial\theta} + \cot\theta\cos\phi\frac{\partial}{\partial\phi}\right), \quad \hat{L}_y = \frac{\hbar}{i}\left(\cos\phi\frac{\partial}{\partial\theta} - \cot\theta\sin\phi\frac{\partial}{\partial\phi}\right). \tag{20.15}$$

Finally, let us consider the magnitude of the total angular momentum. To obtain the operator corresponding to the property L^2, we write $L^2 = \vec{L}\cdot\vec{L} = L_xL_x + L_yL_y + L_zL_z$ and then substitute in the quantum operators,

$$\hat{L}^2 = \hat{L}_x\hat{L}_x + \hat{L}_y\hat{L}_y + \hat{L}_z\hat{L}_z\,.$$

We can express this in terms of polar coordinates by substituting into it Eqs. (20.14) and (20.15). For example,

$$\hat{L}_z\hat{L}_z\Psi = \frac{\hbar}{i}\frac{\partial}{\partial\phi}\left(\frac{\hbar}{i}\frac{\partial\Psi}{\partial\phi}\right) = -\hbar^2\frac{\partial^2}{\partial\phi^2}\Psi.$$

$\hat{L}_x\hat{L}_x$ and $\hat{L}_y\hat{L}_y$ are individually rather complicated but when $\hat{L}_x\hat{L}_x$, $\hat{L}_y\hat{L}_y$ and $\hat{L}_z\hat{L}_z$ are added together, many terms cancel out. One finds in the end that

$$\hat{L}^2 = -\hbar^2\left(\frac{\partial^2}{\partial\theta^2} + \cot\theta\frac{\partial}{\partial\theta} + \frac{1}{\sin^2\theta}\frac{\partial^2}{\partial\phi^2}\right). \tag{20.16}$$

This is usually written

$$\hat{L}^2 = -\hbar^2\left[\frac{1}{\sin\theta}\frac{\partial}{\partial\theta}\left(\sin\theta\frac{\partial}{\partial\theta}\right) + \frac{1}{\sin^2\theta}\frac{\partial^2}{\partial\phi^2}\right]. \tag{20.17}$$

To demonstrate that these are equivalent, operate on a function:

$$\frac{1}{\sin\theta}\frac{\partial}{\partial\theta}\left(\sin\theta\frac{\partial\Psi}{\partial\theta}\right) = \frac{1}{\sin\theta}\sin\theta\frac{\partial^2\Psi}{\partial\theta^2} + \frac{1}{\sin\theta}\cos\theta\frac{\partial\Psi}{\partial\theta} = \frac{\partial^2\Psi}{\partial\theta^2} + \cot\theta\frac{\partial\Psi}{\partial\theta}\,.$$

We can express $\hat{T}$ in terms of polar coordinates by substituting in the polar-coordinate expression for ∇^2, Eq. (3.39). Comparing that expression with Eq.(20.17), we see that

$$\hat{T} = -(\hbar^2/2m)\nabla^2 = -\frac{\hbar^2}{2m}\left(\frac{\partial^2}{\partial r^2} + \frac{2}{r}\frac{\partial}{\partial r}\right) + \frac{1}{2mr^2}\hat{L}^2. \tag{20.18}$$

This expression is very useful for describing systems with spherical symmetry (such as atoms).

20.1.2 The Wavefunction

$\hat{H}\Psi(\vec{x}) = E\Psi(\vec{x})$ is called the ***time-independent Schrödinger equation.*** Schrödinger conjectured that the time dependence of Ψ is governed by

$$i\hbar\frac{\partial}{\partial t}\Psi(\vec{x},t) = \hat{H}\Psi(\vec{x},t)\,, \tag{20.19}$$

which is called the ***time-dependent Schrödinger equation.*** If Ψ is an eigenfunction of $\hat{H}$, then

$$i\hbar\frac{\partial}{\partial t}\Psi(\vec{x},t) = E\Psi(\vec{x},t)\,, \tag{20.20}$$

which has the solution

$$\Psi(\vec{x},t) = e^{-iEt/\hbar}\,\Psi(\vec{x},0)\,. \tag{20.21}$$

This has the same form as the function used in classical mechanics to describe wave phenomena such as water waves. For this reason, Schrödinger called Ψ the *Wellenfunktion*, which literally translated to English is ***wavefunction.*** For water waves this represents the square root of the wave amplitude, but in the case of the quantum mechanical wave Schrödinger was not able to attach a clear physical significance to the eigenfunction. A few months after Schrödinger's first paper on this subject appeared in print, Born[2] suggested that

$$P(x,y,z) = \Psi(x,y,z)^*\,\Psi(x,y,z) \tag{20.22}$$

is proportional to the probability distribution for the particle's position. In other words, the larger the amplitude of the quantum wave at some point (x,y,z), the higher the probability of finding the particle in the neighborhood of that point. This is the ***Born interpretation*** of the wavefunction:

> The probability of the particle being found in a region of space C is given by the integral of $Pdxdydz$ over the region C, divided by the integral over all space,

$$\text{Probability}(C) = \int_C \Psi^*\Psi d\tau \Big/ \int \Psi^*\Psi d\tau. \tag{20.23}$$

(Recall that $d\tau$ is just an abbreviation for $dxdydz$.)

[2]German-Jewish theoretical physicist Max Born (1882-1970). He directed a research group at the University of Göttingen that was responsible for much of the pioneering development of the new quantum theory in the 1920s and 1930s. He then fled to England, where his research interests focused on the theory of solids.

For the free particle, we have

$$P(x, y, z) = \Psi^* \Psi \propto e^{-ik_x x - ik_y y - ik_z z} e^{ik_x x + ik_y y + ik_z z} = 1, \tag{20.24}$$

which is the same at all $\vec{x}$. This is not unreasonable, because in this case there is no potential energy to localize the particle in any one region. Consider, however, the more interesting case of a particle in a "bound" state, in which it is constrained by a potential energy function. We will define a ***bound state*** as any state described by a wavefunction that is square integrable. In other words, Ψ for a bound state must go to zero at infinity fast enough that its norm in the Hilbert space $L^2[-\infty, \infty]$,

$$\langle \Psi, \Psi \rangle = \int_{-\infty}^{\infty} \int_{-\infty}^{\infty} \int_{-\infty}^{\infty} \Psi^*(x, y, z) \Psi(x, y, z) dx dy dz, \tag{20.25}$$

is finite.

20.1.3 The Basic Postulates*

Postulate 1. Any physically measurable property of the system is represented by an Hermitian operator.

This is motivated by the fact that physical properties are measured in real numbers, and Theorem 19.4.1 tells us that the eigenvalues of Hermitian operators will always be pure real.

Postulate 2. Classical mechanical properties that can be expressed in terms of spatial coordinates and momenta correspond to quantum operators according to the mappings of Eqs. (20.5).

Postulate 3. The most complete possible description of any bound state of a physical system is given by a function Ψ in a Hilbert space appropriate to that system.

According to Theorem 19.4.4, we can construct a basis by determining the eigenfunctions of any Hermitian operator on the Hilbert space. Let $\hat{G}$ be the operator corresponding to some physical property g. Then all possible values of g are given by the various eigenvalues, g_0, g_1, g_2, ..., of $\hat{G}$. Let $\{\psi_k\}$ be an orthonormal set of eigenfunctions such that $\hat{G}\psi_k = g_k \psi_k$. This set is a basis, and the wavefunction for any arbitrary state of the system can be resolved as a linear combination

$$\Psi = \sum_{k=0}^{\infty} a_k \psi_k \tag{20.26}$$

with coefficients

$$a_k = \langle \psi_k, \Psi \rangle . \tag{20.27}$$

In quantum mechanics, eigenvalue equations are commonly written using a convenient, if somewhat cryptic, notational system introduced by Dirac. Eigenfunctions are given the symbol $|k\rangle$, where k is the index that labels the eigenvalue. Thus, we write

$$\hat{G}|k\rangle = g_k|k\rangle. \tag{20.28}$$

The complex conjugate of $|k\rangle$ is indicated by $\langle k|$. An arbitrary wavefunction Ψ is written as $|\Psi\rangle$, and Ψ^*, as $\langle\Psi|$. The inner product of two wavefunctions Ψ_a and Ψ_b is written $\langle\Psi_a|\Psi_b\rangle$, with a vertical bar instead of a comma. Following Dirac, we call $|\Psi\rangle$ a ***ket***, we call $\langle\Psi|$ a ***bra***, and $\langle\Psi_a|\Psi_b\rangle$, a ***bracket***.

For orthonormal eigenfunctions we have $\langle j|k\rangle = \delta_{j,k}$. Then the norm of Ψ is

$$\langle\Psi|\Psi\rangle = \sum_{j=0}^{\infty}\sum_{k=0}^{\infty} a_j^* a_k \langle j|k\rangle = \sum_{k=0}^{\infty} a_k^* a_k\,. \tag{20.29}$$

For convenience we usually use normalized wavefunctions, so that $\langle\Psi|\Psi\rangle = 1$. Any bound-state Ψ can be normalized by dividing it by the scalar $\langle\Psi|\Psi\rangle^{1/2}$.

> *Postulate 4.* Consider an arbitrary state, represented by a normalized wavefunction Ψ, that is subjected to a measurement of the property g corresponding to an operator $\hat{G}$. Let g_k be the eigenvalues of $\hat{G}$ and let $|k\rangle$ be orthonormal eigenfunctions that span the Hilbert space. The result of the measurement will be one of the values of g_k. In a given experiment it is unpredictable which of these possible values will be obtained, but the probability of obtaining a particular g_k is $a_k^* a_k$, where $a_k = \langle k|\Psi\rangle$.

Suppose a large number of identical systems, each described by a given Ψ, are each subjected to a measurement of g. The average of the results, the ***expectation value*** $\langle g\rangle$, is the sum of possible outcomes weighted by their individual probabilities,

$$\langle g\rangle = \sum_k a_k^* a_k\, g_k\,. \tag{20.30}$$

Note that

$$\hat{G}|\Psi\rangle = \sum_k a_k\,\hat{G}|k\rangle = \sum_k a_k g_k|k\rangle. \tag{20.31}$$

Let us evaluate the inner product of this with the bra $\langle\Psi|$,

$$\langle\Psi|\hat{G}|\Psi\rangle = \sum_j\sum_k a_j^* a_k\, g_k\langle j|k\rangle = \sum_k a_k^* a_k\, g_k\,,$$

which is identical to Eq. (20.30). This gives us the result that[3]

$$\langle g\rangle = \langle\Psi|\hat{G}|\Psi\rangle. \tag{20.32}$$

Example 20.2. *Calculating an expectation value.* Let us consider again a particle of mass m confined to move along a circle of radius r_0 but otherwise free ($V = 0$), which was treated in Example 19.7. The Hamiltonian operator is

$$\hat{H} = \hat{T} = -\frac{\hbar^2}{2m}\left(\frac{\partial^2}{\partial r^2} + \frac{2}{r}\frac{\partial}{\partial r}\right) - \frac{\hbar^2}{2mr^2}\left(\frac{\partial^2}{\partial\theta^2} + \cot\theta\frac{\partial}{\partial\theta} + \frac{1}{\sin^2\theta}\frac{\partial^2}{\partial\phi^2}\right),$$

[3]Note that $\langle g\rangle$, $\langle\Psi|\hat{G}|\Psi\rangle$, $\langle\Psi|\hat{G}\Psi\rangle$, $\langle\Psi,\hat{G}\Psi\rangle$, and $\int\Psi^*\hat{G}\Psi d\tau$ all mean the same thing!

but because r is constrained to the constant value r_0 and θ is constrained to the constant value $\pi/2$, Ψ depends only on ϕ, and derivatives with respect to r or θ give zero. We can therefore write

$$\hat{H} = -\frac{\hbar^2}{2mr_0^2}\frac{d^2}{d\phi^2}. \tag{20.33}$$

The eigenfunctions of $\frac{d^2}{d\phi^2}$ can be expressed as $\psi_j = ce^{ij\phi}$, where c is an arbitrary constant, i is $\sqrt{-1}$, and $j = \pm1, \pm2, \pm3, \ldots$. The eigenvalues of $\frac{d^2}{d\phi^2}$ are $-j^2$. It follows that the eigenvalues of $\hat{H}$ are $\epsilon_j = \hbar^2 j^2/2mr_0^2$.

Let us normalize the ψ_j. In circular polar coordinates the differential factor in the integral is $r_0 d\phi$. Therefore,

$$\langle\psi_j|\psi_j\rangle = |c|^2\int_0^{2\pi}\psi_j^*\psi_j r_0 d\phi = |c|^2\int_0^{2\pi} e^{-ij\phi}e^{ij\phi}r_0 d\phi = |c|^2 r_0\int_0^{2\pi} d\phi = 2\pi r_0|c|^2.$$

Setting $\langle\psi_j|\psi_j\rangle = 1$, we find that $|c|^2 = (2\pi r_0)^{-1}$. The normalized eigenfunctions are

$$|j\rangle = (2\pi r_0)^{-1/2}e^{ij\phi}. \tag{20.34}$$

Consider for example a physical state corresponding to the wavefunction

$$\Psi = 3\,|1\rangle - 3\,|5\rangle + 7\,|-2\rangle\,.$$

What is the energy of this state? Ψ is not an eigenfunction of $\hat{H}$; operating with $\hat{H}$ we get

$$\hat{H}\Psi = 3\epsilon_1|1\rangle - 3\epsilon_5|5\rangle + 7\epsilon_{-2}|-2\rangle = \frac{\hbar^2}{2mr_0^2}\left(3\,|1\rangle - 75\,|5\rangle + 28\,|-2\rangle\right),$$

which is not simply Ψ multiplied by a single overall scalar factor. This means the physical state in question does not have a well-defined energy. If we measure its energy the result will be either ϵ_1, ϵ_5, or ϵ_{-2} but it is unpredictable which of these values any particular measurement will yield. The best we can do is calculate the expectation value of the energy. Let us normalize Ψ. We have $\langle\Psi|\Psi\rangle = 9 + 9 + 49 = 67$. The normalized wavefunction is

$$\Psi = \frac{3}{\sqrt{67}}\,|1\rangle - \frac{3}{\sqrt{67}}\,|5\rangle + \frac{7}{\sqrt{67}}\,|-2\rangle$$

and

$$\langle E\rangle = \langle\Psi|\,\hat{H}\,|\Psi\rangle = +\frac{3^2}{67}\epsilon_1\langle1|1\rangle + \frac{3^2}{67}\epsilon_5\langle5|5\rangle + \frac{7^2}{67}\epsilon_{-2}\langle-2|-2\rangle = \frac{430}{67}\frac{\hbar^2}{2mr_0^2}.$$

Mixed terms such as $\langle1|5\rangle$ are zero because the eigenfunctions are orthogonal.

20.2 Atoms and Molecules

The potential energy for the interaction between a pair of electric charges is the ***Coulomb potential***,[4]

$$V = \frac{q_1 q_2}{4\pi\epsilon}\frac{1}{r_{12}}, \tag{20.35}$$

where q_1 and q_2 are the charges, r_{12} is the distance between them, and ϵ is the ***dielectric constant***, a constant that depends on the medium surrounding the charges. We are interested here in the interaction between an atomic nucleus of atomic number Z and an electron. Z is the number of protons in the nucleus. The charge of a proton is $1.60217653 \times 10^{-19}$ C. This quantity is called the ***elementary charge*** and given the symbol e [not to

[4]Discovered by the French engineer and physicist Charles de Coulomb (1736-1806). In his honor the standard unit of charge is called the *coulomb*, with symbol C.

be confused with the base of of the natural logarithm, exp(1), which is also called e]. For the nucleus-electron interaction, we set $q_1 = Ze$ and $q_2 = -e$. Because there is no medium, the dielectric constant is that of the vacuum, $\epsilon_0 = (10^7\ \mathrm{J^{-1}C^2m^{-1}})/(4\pi c^2)$, where c is the speed of light in vacuum. Thus, the Schrödinger equation for an atom with a single electron is[5]

$$\left(-\frac{\hbar^2}{2m_{\mathrm{e}}}\nabla^2 - \frac{Ze^2}{4\pi\epsilon_0}\frac{1}{r}\right)\Psi = E\Psi\,, \tag{20.36}$$

where m_{e} is the electron mass. We are assuming for the sake of simplicity that the nucleus is infinitely more massive than an electron (in fact, neutrons and protons are each about 2000 times more massive than an electron) so that we can consider the nucleus as stationary and use its position as the origin of a polar coordinate system, with radial coordinate r. Eq. (20.36) describes any one-electron atom or ion; for hydrogen, $Z = 1$, for He^+, $Z = 2$, and so on.

Eq. (20.36) looks much more complicated than it needs to. The clusters of fundamental constants are present only to adjust for the fact that the standard units of energy, distance, and charge are J, m, and C. We can make the constants disappear by choosing a nonstandard unit system that is more appropriate for the problem at hand. Let us introduce a new unit of distance, called the ***Bohr radius***,[6] with symbol a_0, defined as

$$a_0 = 4\pi\epsilon_0\hbar^2/m_{\mathrm{e}}e^2 \approx 5.2912 \times 10^{-11}\,\mathrm{m}\,, \tag{20.37}$$

and a new energy unit, called the ***hartree***,[7] with symbol E_{H}, defined as the potential energy between a proton and an electron separated by distance a_0,

$$E_{\mathrm{H}} = \frac{e^2}{4\pi\epsilon_0 a_0} \approx 4.3597 \times 10^{-18}\,\mathrm{J}\,. \tag{20.38}$$

Let us change the distance variable with the substitution $r = a_0\tilde{r}$, to put distance in units of a_0, and express the eigenvalue in hartrees, with $E = E_{\mathrm{H}}\tilde{E}$. The result is

$$\left(-\frac{1}{2}\widetilde{\nabla}^2 - \frac{Z}{\tilde{r}}\right)\Psi = \tilde{E}\Psi\,. \tag{20.39}$$

($\widetilde{\nabla}^2$ is the Laplacian operator in terms of $\tilde{r}$.) The common practice is to omit all the tildes,

$$\hat{H}\Psi = E\Psi, \qquad \hat{H} = -\frac{1}{2}\nabla^2 - \frac{Z}{r}\,, \tag{20.40}$$

but to state that ***atomic units*** are being used, with the understanding that

[5] Beware that the factor of $4\pi\epsilon_0$ is often omitted from this equation. In that case, the symbol "e" is being used for what we are here calling $e/\sqrt{4\pi\epsilon_0}$.

[6] Named for the Danish physicist Niels Bohr (1885-1962), who prepared the way for the development of quantum mechanics with theories of atomic spectra and of the periodic table based on the assumption that energy levels are quantized.

[7] Named after English mathematician and timpanist Douglas Hartree (1897-1958), who developed a computational strategy for obtaining approximate solutions to the many-particle Schrödinger equation that is today fundamental to most quantum chemistry software. He served as professor of applied mathematics and dean of the faculty of music at the University of Manchester, and later as professor of mathematical physics at Cambridge University.

r is in units of a_0 and E is in units of $E_{\rm H}$. It is interesting to note that the transformation to atomic units could also have been accomplished by writing $e^2/4\pi\epsilon_0 = \hbar^2/m_{\rm e}a_0$ in Eq. (20.36) and then setting $\hbar$, $m_{\rm e}$, and a_0 each equal to unity. This simple trick yields the same equation as does the formal analysis with the change of variable.

For an atom or ion with N electrons, the Schrödinger equation in atomic units has the Hamiltonian operator

$$\hat{H} = \sum_j^N \hat{H}_j + \sum_{j=1}^{N-1}\sum_{k=j+1}^{N} r_{jk}^{-1}, \qquad (20.41)$$

$$\hat{H}_j = -\frac{1}{2}\nabla_j^2 - \frac{Z}{r_j}, \quad r_j = \sqrt{x_j^2 + y_j^2 + z_j^2}, \quad \nabla_j^2 = \frac{\partial^2}{\partial x_j^2} + \frac{\partial^2}{\partial y_j^2} + \frac{\partial^2}{\partial z_j^2},$$

where (x_j, y_j, z_j) is the location of the jth electron in a Cartesian coordinate system with origin at the nucleus. r_{jk} is the distance between electron j and electron k,

$$r_{jk} = \sqrt{(x_j - x_k)^2 + (y_j - y_k)^2 + (z_j - z_k)^2}\,. \qquad (20.42)$$

For a molecule, the usual approach is to assume that the nuclei are stationary for purposes of solving a Schrödinger equation for the electrons.[8] The origin of the electronic coordinate system is put at a fixed point in the framework of the nuclei. The electronic Schrödinger equation, for a molecule with $N_{\rm n}$ nuclei and $N_{\rm e}$ electrons, is then

$$\hat{H}_{\rm e}\Psi = E\Psi\,, \qquad \hat{H}_{\rm e} = -\frac{1}{2}\sum_{j=1}^{N_{\rm e}}\nabla_j^2 - \sum_{j=1}^{N_{\rm e}}\sum_{\kappa=1}^{N_{\rm n}}\frac{Z_\kappa}{R_{\kappa j}} + \sum_{j=1}^{N_{\rm e}-1}\sum_{k=j+1}^{N_{\rm e}} r_{jk}^{-1}, \quad (20.43)$$

where Z_κ is the atomic number of nucleus κ and $R_{\kappa j}$ is the distance between nucleus κ and electron j.

20.3 The One-Electron Atom

20.3.1 Orbitals

Schrödinger's equation for a one-electron atom, Eq. (20.40), is a partial differential equation. If we are to have any hope of solving it analytically we will need to use separation of variables (see Section 14.5) to reduce it to a set of ordinary differential equations. The potential energy, $V(r) = -Z/r$, is spherically symmetric, depending only on r. This suggests we use spherical polar coordinates and try to find a separable solution in the form

$$\Psi(r,\theta,\phi) = R(r)Y(\theta,\phi)\,. \qquad (20.44)$$

[8]This is called the Born-Oppenheimer approximation, named for Max Born and American physicist J. Robert Oppenheimer (1904-1967). On average, nuclei move much more slowly than electrons because they are so much more massive.

The kinetic energy operator in polar coordinates and atomic units can be written

$$\hat{T} = -\tfrac{1}{2}\nabla^2 = -\frac{1}{2}\left(\frac{\partial^2}{\partial r^2} + \frac{2}{r}\frac{\partial}{\partial r} - \frac{1}{r^2}\hat{L}^2\right), \tag{20.45}$$

in terms of the angular momentum operator $\hat{L}^2$, which is given by Eq. (20.17) (but with the $\hbar^2$ set to 1). $\hat{L}^2$ depends only on θ and ϕ. Substituting the separated form of Ψ into $(\hat{T} - Z/r)\Psi = E\Psi$ and rearranging, we can indeed segregate all the radial dependence to one side of the equal sign and all the angular dependence to the other,

$$\frac{2r^2}{R}\left[-\frac{1}{2}\left(\frac{d^2R}{dr^2} + \frac{2}{r}\frac{dR}{dr}\right) - \left(\frac{Z}{r} + E\right)R\right] = -\frac{1}{Y}\hat{L}^2 Y = \gamma. \tag{20.46}$$

Because r and (θ, ϕ) are independent variables, both sides of this equation must be equal to a single constant, which we are here calling γ.

The angular side gives

$$\hat{L}^2 Y = -\gamma Y.$$

We have seen this equation before—it is the differential equation for the spherical harmonics, Eq. (17.58). We can write it as

$$\hat{L}^2 Y_\ell^{(m)}(\theta, \phi) = \ell(\ell+1) Y_\ell^{(m)}(\theta, \phi), \tag{20.47}$$

in terms of the spherical harmonics $Y_\ell^{(m)}$, which we developed in Section 17.4. It is traditional to use the letter ℓ as the integer parameter in the eigenvalue, because it is a quantum number associated with the operator $\hat{L}^2$. Setting $\gamma = -\ell(\ell+1)$ and then multiplying Eq. (20.46) by $R/2r^2$, we obtain the ***radial equation*** for the one-electron atom,

$$-\frac{1}{2}R'' - \frac{1}{r}R' + \left[\frac{\ell(\ell+1)}{2r^2} - \frac{Z}{r} - E\right]R = 0. \tag{20.48}$$

This is an ordinary differential eigenvalue equation, which can be analyzed using techniques developed in preceding chapters. A derivation of the solutions is presented in the following section. They are

$$R_{n,\ell}(r) = \left[\frac{(2\alpha)^{2\ell+3}}{2n}\frac{(n-\ell-1)!}{(n+\ell)!}\right]^{1/2} r^\ell e^{-\alpha r} L_{n-\ell-1}^{(2\ell+1)}(2\alpha r), \quad \alpha = Z/n, \tag{20.49}$$

in atomic units. The $L_j^{(k)}$ are associated Laguerre polynomials. The messy term in brackets is just a normalization factor. These functions are characterized by the angular momentum quantum number ℓ and by a new quantum number n called the ***principal quantum number***. The eigenvalues are

$$E = -\tfrac{1}{2}Z^2/n^2, \qquad n - 1, 2, 3, \ldots. \tag{20.50}$$

Note that E is independent of ℓ, depending only on n. The eigenfunctions,

$$\psi_{n,\ell,m}(r,\theta,\phi) = R_{n,\ell}(r)Y_\ell^{(m)}(\theta,\phi)\,, \tag{20.51}$$

are called ***orbitals***.[9] The functions $\psi_{n,0,0}$, with $\ell = 0$, are called s orbitals, the $\psi_{n,1,m}$, with $\ell = 1$, are called p orbitals, the $\psi_{n,2,m}$, d orbitals, the $\psi_{n,3,m}$, f orbitals. In particular, $\psi_{1,0,0}$ is called the 1s orbital, $\psi_{2,0,0}$, the 2s orbital, $\psi_{2,1,m}$ is called a 2p orbital, and so on. It is shown in the next section that for given value of ℓ, the only allowed n values are those such that $n > \ell$.

The probability of finding an electron within a given distance r_0 from the nucleus is given by an integral,

$$P_{n,\ell}^{(r<r_0)}(r_0) = \int_0^{2\pi}\int_0^{\pi}\int_0^{r_0} |R_{n,\ell}(r)Y_\ell^{(m)}(\theta,\phi)|^2\, r^2 \sin^2\theta dr d\theta d\phi\,.$$

The angles are being integrated over their full ranges, and the spherical harmonics are normalized; therefore, the angular integration evaluates to unity, leaving

$$P_{n,\ell}^{(r<r_0)}(r_0) = \int_0^{r_0} |R_{n,\ell}(r)|^2\, r^2 dr\,, \tag{20.52}$$

which can be expressed in terms of incomplete gamma functions.

20.3.2 The Radial Equation*

Let us begin by simplifying Eq. (20.48) so as to more clearly see the qualitative behavior of the solutions. Let $R(r) = f(r)\chi(r)$,

$$-\frac{1}{2}f\chi'' - \left(\frac{1}{r}f + f'\right)\chi' + \left[\frac{\ell(\ell+1)}{2r^2}f - \frac{Z}{r}f - Ef - \frac{1}{r}f' - \frac{1}{2}f''\right]\chi = 0\,,$$

where f is an arbitrary function that will be chosen to make the differential equation for χ simpler than that for R. Let us choose f to eliminate the term with the first derivative. We want f such that $f = -rf'$. Clearly, $f = r^{-1}$ will serve the purpose. Then

$$-\frac{1}{2}\chi'' + \left[\frac{\ell(\ell+1)}{2r^2} - \frac{Z}{r} - E\right]\chi = 0\,, \qquad R(r) = r^{-1}\chi(r)\,. \tag{20.53}$$

Our strategy will be to determine the behavior of $\chi(r)$ in the limits $r \to \infty$ and $r \to 0$, factor out those behaviors, and thereby obtain a new differential equation that we will be able to solve with linear algebra. Consider the term in brackets in Eq. (20.53). In the limit $r \to \infty$ the first two terms are infinitesimal compared with the third term. This tells us that

$$-\tfrac{1}{2}\chi'' - E\chi \sim 0 \qquad (r \to \infty). \tag{20.54}$$

The eigenvalue E is the energy of the quantum state. We are interested in bound states, which means that E will be a negative real number. Let us

[9]For depictions of the shapes of these functions, see any physical chemistry textbook.

write Eq. (20.54) as

$$\chi'' \sim -2E\chi = 2|E|\chi = \alpha^2\chi\,,$$

where $\alpha = \sqrt{2|E|}$. Then the solution for χ, at large r, is

$$\chi(r) \sim Ae^{-\alpha r}, \qquad R(r) \sim Ar^{-1}e^{-\alpha r}. \tag{20.55}$$

A is an arbitrary constant. Similarly, we can determine the behavior at small r. In that case,

$$-\tfrac{1}{2}\chi'' + \tfrac{1}{2}\ell(\ell+1)r^{-2}\chi \sim 0 \qquad (r \to 0). \tag{20.56}$$

This tells us that

$$\chi(r) \sim Ar^{\ell+1}, \qquad R(r) \sim Ar^{\ell}. \tag{20.57}$$

We proceed using the method of variation of parameters (see Section 14.3), replacing the constant A with a function $u(r)$. Let us use the large-r solution and substitute $\chi(r) = u(r)e^{-\alpha r}$ into Eq. (20.53). Replacing E with $-\alpha^2/2$, we find that

$$-\frac{1}{2}u'' + \alpha u' + \left[\frac{\ell(\ell+1)}{2r^2} - \frac{Z}{r}\right]u = 0\,. \tag{20.58}$$

It is not hard to see that this will be satisfied with a polynomial. The effect of taking a first derivative of a polynomial is to reduce its degree by one, which is also the effect of dividing by r. Thus, the term with u' can be made to cancel out the term with r^{-1} and the term with u'' can be made to cancel out the term with r^{-2}. The analysis at $r \to 0$ implies that the lowest-degree term in this polynomial will be $Ar^{\ell+1}$, because all higher powers of r in the polynomial go to zero faster than the lowest-degree term. We can write

$$\chi(r) = v(r)\,r^{\ell+1}e^{-\alpha r}, \qquad R(r) = v(r)\,r^{\ell}e^{-\alpha r}, \tag{20.59}$$

where $v(r)$ is a polynomial in which the lowest-degree term is a constant.

In Section 17.1 we developed a theory of polynomials. Let us apply it here. The form of Eq. (20.59), a polynomial multiplied by a power and an exponential, suggests associated Laguerre polynomials $L_j^{(k)}$, which have weight factor $x^k e^{-x}$. The associated Laguerre differential equation, from Table 17.3, is

$$\left[x\frac{d^2}{dx^2} + (k+1-x)\frac{d}{dx} + j\right]L_j^{(k)}(x) = 0\,. \tag{20.60}$$

Suppose that we absorb the square root of the weight factor into the eigenfunction, using

$$\Phi_j^{(k)}(x) = x^{k/2}e^{-x/2}L_j^{(k)}(x). \tag{20.61}$$

Substituting $x^{-k/2}e^{x/2}\Phi_j^{(k)}$ for $L_j^{(k)}$ in Eq. (20.60) gives

$$-\frac{1}{2}\frac{d^2\Phi_j^{(k)}}{dx^2} - \frac{1}{2x}\frac{d\Phi_j^{(k)}}{dx} + \left(\frac{k^2}{8x^2} - \frac{2j+k+1}{4x} + \frac{1}{8}\right)\Phi_j^{(k)} = 0\,. \tag{20.62}$$

This is similar to the radial equation for R, Eq. (20.48), but not quite the same. The most important difference is that in Eq. (20.48) first derivative is not divided by 2. Let us eliminate the first derivative, with the transformation $\Phi_j^{(k)}(x) = x^{-1/2}\Upsilon_j^{(k)}(x)$, which leads to

$$-\frac{1}{2}\frac{d^2\Upsilon_j^{(k)}}{dx^2} + \left(\frac{k^2-1}{8x^2} - \frac{2j+k+1}{4x} + \frac{1}{8}\right)\Upsilon_j^{(k)} = 0\,. \tag{20.63}$$

This is very similar to the equation for $\chi = rR$, Eq. (20.53). The only differences are the constants in the terms in the parentheses. Making the change of variable $x = 2\alpha r$ then multiplying the whole equation by $(2\alpha)^2$, we obtain

$$-\frac{1}{2}\frac{d^2\Upsilon_j^{(k)}}{dr^2} + \left[\frac{k^2-1}{8r^2} - \frac{(2j+k+1)\alpha}{2r} - E\right]\Upsilon_j^{(k)} = 0 \tag{20.64}$$

(using the fact that $\alpha^2 = -2E$). This really is equivalent to Eq. (20.53).

Comparing the constants in Eqs. (20.53) and (20.64), we see that

$$\ell(\ell+1) = \tfrac{1}{4}(k^2-1) = \left(\frac{k}{2} - \frac{1}{2}\right)\left(\frac{k}{2} + \frac{1}{2}\right) = \frac{k-1}{2}\left(\frac{k-1}{2} + 1\right),$$

which implies that $k = 2\ell + 1$, and that

$$Z = \tfrac{1}{2}(2j + 2\ell + 2)\alpha = (j + l + 1)\sqrt{2|E|}\,. \tag{20.65}$$

Squaring this last result gives $2E = -Z^2/(j+\ell+1)^2$. Because j only appears in combination with $\ell + 1$, it is conventional to replace it with a new quantum number

$$n = j + \ell + 1\,. \tag{20.66}$$

Because the minimum degree of the associated Laguerre polynomial is $j = 0$, the minimum value of n is $\ell + 1$.

Thus we have derived Eq. (20.50) for E and we have proved that the eigenfunction $\chi(r)$ is equivalent to $\Upsilon_{n-\ell-1}^{(2\ell+1)}(2\alpha r)$ with $\alpha = Z/(j + \ell + 1) = Z/n$. The eigenfunction we seek is $R = r^{-1}\chi$. Including a normalization factor, we arrive at Eq. (20.49).

20.4 Hybrid Orbitals

For atoms or molecules with more than one electron, Schrödinger's equation cannot be solved exactly in terms of standard functions. Not long after Schrödinger published his solution for the one-electron atom, Pauling[10] used the one-electron orbitals $\psi_{n,\ell,m}$ as the building blocks of an approximate qualitative theory of chemical bonding, which made it possible to rationalize chemical bonding in terms of basic physics. Pauling's theory provides the conceptual foundation for modern organic chemistry.

[10] American theoretical chemist Linus Pauling (1901-1994) proposed hybrid orbitals in 1928. This concept, along with his ideas of bond resonance and electronegativity, were later presented as a comprehensive theory of chemical bonding in his extremely influential book *The Nature of the Chemical Bond*, originally published in 1938.

A covalent bond can be thought of as a pair of ions, consisting of nuclei surrounded by their core electrons, mutually attracted to a region between them of high valence-electron probability. This electron probability distribution is expressed as a linear combination of two valence-electron orbitals, one centered on one of the nuclei and the other on the other nucleus. Consider the methane molecule CH_4, which is known from experiment to have a tetrahedral bonding structure. The carbon valence orbitals are the $2s$ orbital, $\psi_{2,0,0}$, and the three $2p$ orbitals, $\psi_{2,1,m}$ with $m = -1$, 0, or 1. If the energy eigenvalues of these eigenfunctions depend only on n, then any linear combination of them will also be an eigenfunction with this same eigenvalue. Pauling argued that the appropriate linear combinations for carbon in CH_4 are those that produce a linearly independent set of eigenfunctions with lobes oriented in tetrahedral symmetry. The following, called sp^3 orbitals, are suitable:

$$\begin{aligned} h_1 &= \tfrac{1}{2}(s + p_x + p_y + p_z)\,, & h_2 &= \tfrac{1}{2}(s + p_x - p_y - p_z)\,, \\ h_3 &= \tfrac{1}{2}(s - p_x + p_y - p_z)\,, & h_4 &= \tfrac{1}{2}(s - p_x - p_y + p_z)\,, \end{aligned} \tag{20.67}$$

with $s = \psi_{2,0,0}$, $p_z = \psi_{2,1,0}$, and

$$p_x = \sqrt{\tfrac{1}{2}}\,(\psi_{2,1,1} + \psi_{2,1,-1})\,, \qquad p_y = \frac{1}{i}\sqrt{\tfrac{1}{2}}\,(\psi_{2,1,1} - \psi_{2,1,-1})\,. \tag{20.68}$$

(The p_x and p_y orbitals were discussed in detail in Example 14.4.) The axes of the four lobes are at the tetrahedral angle of arccos(−1/3), which is 109.47°. The $1s$ orbitals of the hydrogen atoms each form bonding orbitals through linear combination with one of these sp^3 orbitals.

For molecules in which the central atom comes from group VA or VIA the situation is more complicated. Nitrogen, from group VA, commonly forms molecules in which it bonds to just three other atoms, for example, NH_3. Spectroscopic data tell us that the H–N–H bond angles in NH_3 are 107.3°, which is slightly distorted from the tetrahedral angle. H_2O is more distorted, with a bond angle of 104.5°. Evidently, the appropriate hybrid orbitals are linear combinations $c_s s + c_x p_x + c_y p_y + c_z p_z$ in which the coefficients differ slightly from the coefficients in Eqs. (20.67). Molecules such as H_3Sb and H_2Te, with valence orbitals at the $n = 5$ level, have bond angles close to 90°, which suggests that their bonding can be reasonably described as resulting from overlap of a $1s$ orbital with an unhybridized $5p_x$, $5p_y$, or $5p_z$ orbital.

For the trigonal planar molecule BCl_3, the boron s, p_x, and p_y orbitals can be used to construct an orthonormal set with trigonal planar geometry, with bond angles of 120°:

$$h_1 = \sqrt{\tfrac{1}{3}}\,s + \sqrt{\tfrac{2}{3}}\,p_x\,,$$

$$h_2 = \sqrt{\tfrac{1}{3}}\,s - \sqrt{\tfrac{1}{6}}\,p_x + \sqrt{\tfrac{1}{2}}p_y\,, \quad h_3 = \sqrt{\tfrac{1}{3}}\,s - \sqrt{\tfrac{1}{6}}\,p_x - \sqrt{\tfrac{1}{2}}p_y\,. \tag{20.69}$$

These are called sp^2 orbitals. They can also be used to rationalize the approximately 120° bond angles in planar molecules such as CH_2O and H_2CCH_2.

Linear molecules such as HCCH are described with *sp* hybrid orbitals,

$$h_1 = \sqrt{\tfrac{1}{2}}\,(s + p_z), \qquad h_2 = \sqrt{\tfrac{1}{2}}\,(s - p_z). \tag{20.70}$$

For atoms bonded to more than four atoms, appropriate hybrid orbitals can be constructed by including *d* orbitals. The octahedral geometry of a molecule such as SF_6 can be described with the following set of sp^3d^2 orbitals,

$$\begin{aligned}
h_1 = \sqrt{\tfrac{1}{6}}\,s + \sqrt{\tfrac{1}{2}}\,p_z + \sqrt{\tfrac{1}{3}}\,d_{z^2}\,, \quad & h_2 = \sqrt{\tfrac{1}{6}}\,s - \sqrt{\tfrac{1}{2}}\,p_z + \sqrt{\tfrac{1}{3}}\,d_{z^2}\,, \\
h_3 &= \sqrt{\tfrac{1}{6}}\,s + \sqrt{\tfrac{1}{12}}\,d_{z^2} + \tfrac{1}{2}d_{x^2-y^2} + \sqrt{\tfrac{1}{2}}\,p_x\,, \\
h_4 &= \sqrt{\tfrac{1}{6}}\,s + \sqrt{\tfrac{1}{12}}\,d_{z^2} + \tfrac{1}{2}d_{x^2-y^2} - \sqrt{\tfrac{1}{2}}\,p_x\,, \\
h_5 &= \sqrt{\tfrac{1}{6}}\,s + \sqrt{\tfrac{1}{12}}\,d_{z^2} - \tfrac{1}{2}d_{x^2-y^2} + \sqrt{\tfrac{1}{2}}\,p_y\,, \\
h_6 &= \sqrt{\tfrac{1}{6}}\,s + \sqrt{\tfrac{1}{12}}\,d_{z^2} - \tfrac{1}{2}d_{x^2-y^2} - \sqrt{\tfrac{1}{2}}\,p_y\,,
\end{aligned} \tag{20.71}$$

with

$$d_{z^2} = \psi_{3,2,0}\,, \qquad d_{x^2-y^2} = \sqrt{\tfrac{1}{2}}\,(\psi_{3,2,+2} + \psi_{3,2,-2})\,. \tag{20.72}$$

The *s*, *p*, and *d* orbitals for sulfur have radial parts corresponding to $n = 3$. For trigonal bipyramidal molecule, such as PCl_5, the sp^2 hybrid orbitals can be used to describe the bonding in the trigonal plane while the linear combinations

$$h_4 = \sqrt{\tfrac{1}{2}}\,(d_{z^2} + p_z), \qquad h_5 = \sqrt{\tfrac{1}{2}}\,(d_{z^2} - p_z) \tag{20.73}$$

can describe the axial bonds. The resulting set is called sp^3d.

20.5 Antisymmetry*

For a full description of a many-electron atom or molecule, we must take into account electron spin. Spin angular momentum is an intrinsic property of electrons. The functions $\psi_{n,l,m}$ alone are not in general an adequate basis. We need to multiply them by ***spin eigenvectors***, $\boldsymbol{\alpha}$ and $\boldsymbol{\beta}$, which are eigenvectors of a matrix that, upon diagonalization, becomes

$$\widetilde{\mathbf{S}}_z = \begin{pmatrix} \hbar/2 & 0 \\ 0 & -\hbar/2 \end{pmatrix}. \tag{20.74}$$

The normalized eigenvectors of $\widetilde{\mathbf{S}}_z$ are

$$\boldsymbol{\alpha} = \begin{pmatrix} 1 \\ 0 \end{pmatrix}, \quad \boldsymbol{\beta} = \begin{pmatrix} 0 \\ 1 \end{pmatrix}, \qquad \widetilde{\mathbf{S}}_z\boldsymbol{\alpha} = \frac{\hbar}{2}\,\boldsymbol{\alpha}\,, \quad \widetilde{\mathbf{S}}_z\boldsymbol{\beta} = -\frac{\hbar}{2}\,\boldsymbol{\beta}\,. \tag{20.75}$$

The eigenvalues can be written $m_s\hbar$ with quantum number $m_s = 1/2$ for $\boldsymbol{\alpha}$ and $-1/2$ for $\boldsymbol{\beta}$. In atomic units, the eigenvalues are simply m_s. To avoid confusion, we will put a subscript ℓ on the previous quantum number m to distinguish it from m_s. The basis functions $\psi_{n,\ell,m_\ell}\boldsymbol{\alpha}$ and $\psi_{n,\ell,m_\ell}\boldsymbol{\beta}$, and linear combinations contructed from them, are called ***spin orbitals***. Note that they are column vectors.

A convenient notational convention is to write the orbital for electron k as $\psi_{n,\ell,m_\ell}(x_k, y_k, z_k)\,\boldsymbol{\alpha}(k)$ if the spin component of the electron is $+1/2$, and $\psi_{n\ell,m_\ell}(x_k, y_k, z_k)\,\boldsymbol{\beta}(k)$ if it is $-1/2$. Usually, the notation is further streamlined by simply writing $\psi_{n,\ell,m_\ell}(k)\,\boldsymbol{\alpha}(k)$ and $\psi_{n,\ell,m_\ell}(k)\,\boldsymbol{\beta}(k)$. The ground state of He, for example, might then be described with

$$\Psi_{1s^2} = \psi_{1,0,0}(1)\,\psi_{1,0,0}(2)\,\boldsymbol{\alpha}(1)\,\boldsymbol{\beta}(2). \tag{20.76}$$

This notation deserves some comment. $\boldsymbol{\alpha}(1)\boldsymbol{\beta}(2)$ is a double column vector $\binom{1}{0}\binom{0}{1}$. This is a way of saying "electron 1 has $m_s = +1/2$ while electron 2 has $m_s = -1/2$." When calculating an inner product, it is understood that column vector multiplication only occurs between columns corresponding to the same electron. For example,

$$\langle \psi_{1,0,0}(1)\psi_{1,0,0}(2)\boldsymbol{\alpha}(1)\boldsymbol{\beta}(2), \psi_{1,0,0}(1)\psi_{1,0,0}(2)\boldsymbol{\alpha}(1)\boldsymbol{\beta}(2)\rangle$$

$$= \langle \psi_{1,0,0}(1), \psi_{1,0,0}(1)\rangle \langle \psi_{1,0,0}(2), \psi_{1,0,0}(2)\rangle \begin{pmatrix}1 & 0\end{pmatrix}\begin{pmatrix}1\\0\end{pmatrix} \begin{pmatrix}0 & 1\end{pmatrix}\begin{pmatrix}0\\1\end{pmatrix} = 1$$

while

$$\langle \psi_{1,0,0}(1)\psi_{1,0,0}(2)\boldsymbol{\beta}(1)\boldsymbol{\alpha}(2), \psi_{1,0,0}(1)\psi_{1,0,0}(2)\boldsymbol{\alpha}(1)\boldsymbol{\beta}(2)\rangle$$

$$= \langle \psi_{1,0,0}(1), \psi_{1,0,0}(1)\rangle \langle \psi_{1,0,0}(2), \psi_{1,0,0}(2)\rangle \begin{pmatrix}0 & 1\end{pmatrix}\begin{pmatrix}1\\0\end{pmatrix} \begin{pmatrix}1 & 0\end{pmatrix}\begin{pmatrix}0\\1\end{pmatrix} = 0.$$

It is an observed fact that no two electrons in an atom or molecule can be in the same quantum state. This is the ***Pauli exclusion principle.*** It is an additional postulate, discovered by Pauli in 1928, that must be added to quantum mechanics in order to explain the structure of the periodic table. The exclusion principle turns out to be a consequence of a more general principle, proved by Pauli[11] in 1940:

> ***Pauli antisymmetry principle.*** Consider the wavefunction of a many-electron atom or molecule, $\Psi(\vec{x}_1, \vec{x}_2, \ldots, \vec{x}_N)$, where $\vec{x}_k$ is the coordinate vector (x_k, y_k, z_k) of electron k. If the coordinates of one electron are exchanged with the coordinates of another, then Ψ is transformed to $-\Psi$.

Ψ_{1s^2} of Eq. (20.76) is not physically reasonable, because exchange of the two electrons gives $\psi_{1,0,0}(2)\psi_{1,0,0}(1)\,\boldsymbol{\alpha}(2)\boldsymbol{\beta}(1)$, which is *not* equal to $-\Psi_{1s^2}$.

Example 20.3. *Antisymmetry of electron exchange.* The ground state of a He atom in the $1s^2$ configuration can be described (approximately) with a linear combination

$$\Psi_1(\vec{x}_1, \vec{x}_2) = \psi_{1,0,0}(1)\psi_{1,0,0}(2)[\boldsymbol{\alpha}(1)\boldsymbol{\beta}(2) - \boldsymbol{\alpha}(2)\boldsymbol{\beta}(1)]/\sqrt{2}\,.$$

---

[11] Austrian-American physicist Wolfgang Pauli (1900-1958). The exclusion principle was based on the empirical observation, noted earlier by American physical chemist G. N. Lewis (1875-1946), that electrons in atoms usually seem to be found in groups of two or eight. The antisymmetry principle, however, is a theoretically derived proof that results from combining Einstein's theory of special relativity with quantum mechanics.

If we interchange the electrons, we obtain

$$\mathcal{P}_{1,2}\Psi_1(\vec{x}_1,\vec{x}_2) = \psi_{1,0,0}(2)\psi_{1,0,0}(1)[\boldsymbol{\alpha}(2)\boldsymbol{\beta}(1) - \boldsymbol{\alpha}(1)\boldsymbol{\beta}(2)]/\sqrt{2} = -\Psi_1(\vec{x}_1,\vec{x}_2),$$

where $\mathcal{P}_{m,n}$ is the permutation operator—it represents the interchange of the coordinates of electrons m and n.

The first four excited states of He, all of which have the electron configuration $1s2s$, can be described with

$$\Psi_2 = [\psi_{1,0,0}(1)\psi_{2,0,0}(2) - \psi_{1,0,0}(2)\psi_{2,0,0}(1)][(\boldsymbol{\alpha}(1)\boldsymbol{\beta}(2) + \boldsymbol{\alpha}(2)\boldsymbol{\beta}(1)]/2,$$

$$\Psi_3 = [\psi_{1,0,0}(1)\psi_{2,0,0}(2) - \psi_{1,0,0}(2)\psi_{2,0,0}(1)]\boldsymbol{\alpha}(1)\boldsymbol{\alpha}(2)/\sqrt{2},$$

$$\Psi_4 = [\psi_{1,0,0}(1)\psi_{2,0,0}(2) - \psi_{1,0,0}(2)\psi_{2,0,0}(1)]\boldsymbol{\beta}(1)\boldsymbol{\beta}(2)/\sqrt{2},$$

$$\Psi_5 = [\psi_{1,0,0}(1)\psi_{2,0,0}(2) + \psi_{1,0,0}(2)\psi_{2,0,0}(1)][(\boldsymbol{\alpha}(1)\boldsymbol{\beta}(2) - \boldsymbol{\alpha}(2)\boldsymbol{\beta}(1)]/2.$$

Given a set $\{\chi_j\}$ of spin orbitals, a convenient way to construct an antisymmetric linear combination for an N-electron atom or molecule is to use a ***Slater determinant***,[12]

$$\frac{1}{\sqrt{N!}}\begin{vmatrix} \chi_1(1) & \chi_2(1) & \chi_3(1) & \cdots & \chi_N(1) \\ \chi_1(2) & \chi_2(2) & \chi_3(2) & \cdots & \chi_N(2) \\ \vdots & \vdots & \vdots & & \vdots \\ \chi_1(3) & \chi_2(3) & \chi_3(3) & \cdots & \chi_N(3) \end{vmatrix}.$$

Recall from Section 18.5 that interchanging any two rows of a determinant changes the value by a factor of -1. Each row in the Slater determinant corresponds to a particular electron coordinate. Interchanging two rows is equivalent to exchanging the coordinates of the two electrons. The Slater determinant is guaranteed to be antisymmetric.

Example 20.4. *Slater determinant for helium.* For the helium atom ground state,

$$\frac{1}{\sqrt{2}}\begin{vmatrix} \psi_{1,0,0}(1)\boldsymbol{\alpha}(1) & \psi_{1,0,0}(1)\boldsymbol{\beta}(1) \\ \psi_{1,0,0}(2)\boldsymbol{\alpha}(2) & \psi_{1,0,0}(2)\boldsymbol{\beta}(2) \end{vmatrix} = \psi_{1,0,0}(1)\boldsymbol{\alpha}(1)\psi_{1,0,0}(2)\boldsymbol{\beta}(2)/\sqrt{2} - \psi_{1,0,0}(1)\boldsymbol{\beta}(1)\psi_{1,0,0}(2)\boldsymbol{\alpha}(2)/\sqrt{2},$$

which agrees with Example 20.3.

20.6 Molecular Orbitals*

Strictly speaking, a molecule does not consist of atoms; it consists of a collection of nuclei and electrons governed by the molecular Schrödinger equation, Eq. (20.43). The electrons in a molecule cannot really be described as "belonging" to any one nucleus. With this in mind, perhaps we should be seeking *molecular* orbitals rather than atomic orbitals. A ***molecular orbital*** is a function $\psi_j(x, y, z)$ such that $|\psi_j|^2$ describes the probability distribution of a single electron in a molecule, in a coordinate system with origin fixed relative to the nuclear framework. Just as there were many different possible solutions for the atomic orbitals, there will also be many possible molecular orbitals. The subscript j labels the various different solutions.

[12]Introduced by American physicist John C. Slater (1900-1976).

This is only an approximation, because ψ_j is being written as a function only of the coordinates of a single electron. This greatly reduces the number of coordinates in the differential equation, from $3N_e$ down to just 3, which makes the mathematical analysis much more tractable. However, the behavior of any given electron in fact depends on the locations of all the other electrons, due to the repulsive Coulomb potential between them. This effect, which we are omitting from the analysis, is called ***electron correlation***. The error in calculated electronic energy levels from ignoring electron correlation tends to be on the order of a few percent. This is significant, because enthalpy changes for chemical reactions, which are energy *differences* between products and reactants, are often of this same order of magnitude. However, molecular geometries are usually reasonably accurate at this level of approximation.

In contrast to the one-electron atomic orbitals, which we were able to derive as the exact analytic solutions of a differential equation, orbitals for systems with two or more electrons (whether they be molecules or many-electron atoms) must be obtained using numerical approximation methods. The usual approach is to expand the molecular orbitals in terms of some basis set $\{\phi_\mu\}$,

$$\psi_j = \sum_{\mu=1}^{N} c_{\mu,j}\phi_\mu , \tag{20.77}$$

where the ϕ_μ comprise a linearly independent set of functions in the Hilbert space. The true Hilbert space is of infinite dimension. In practice we truncate the summation at some finite dimension N. Then the $c_{\mu,j}$ can be determined from a variational computation, diagonalizing the Hamiltonian matrix, as described in Section 19.5. Using a variational calculation with antisymmetrized one-electron molecular spin orbitals is called the ***Hartree-Fock method***.[13]

In principle, any basis set will do, but in practice the choice of an optimal basis is very important. One strategy is to choose the basis functions so as to try to minimize the dimension N of the basis, so that it can give a desired level of accuracy with a relatively small dimension. For this purpose, the ϕ_μ can be chosen to have functional forms resembling the one-electron atomic orbitals ψ_{n,ℓ,m_ℓ}. An alternative strategy is to choose the ϕ_μ so that the matrix elements can be computed efficiently. However, if the ϕ_μ are chosen for computational convenience rather than for resemblance to the true physical solution, then a relatively larger basis dimension will be needed to obtain a given level of accuracy. The computation of the matrix elements is in practice the most time consuming step in the algorithm. Therefore, this second strategy is the one usually used.

The most popular kinds of basis functions have the form of one-electron atomic orbitals but have the $e^{-(Z/n)r}$ factor replaced with $e^{-\zeta r^2}$, where ζ is an arbitrary parameter. These are called ***gaussians*** due to their resemblance

[13] It was Hartree who developed this approach but he did not know to use antisymmetric wavefunctions. The antisymmetric version of the theory was developed independently, and almost simultaneously, by Soviet physicist Vladimir Fock (1898-1974) and by Slater.

to the Gaussian probability distribution function. For example, an s-type gaussian is

$$g^{(s)}(r,\zeta) = (2\,\zeta/\pi)^{3/4} e^{-\zeta r^2}. \tag{20.78}$$

A p_z-type gaussian is

$$g^{(z)}(r,\theta,\zeta) = (128\,\zeta^5/\pi^3)^{1/4}\, r\cos\theta\, e^{-\zeta r^2}. \tag{20.79}$$

(The prefactors are for normalization.) Using gaussians, the integrals needed to calculate the elements of the Hamiltonian matrix can be expressed in terms of error functions and standard elementary functions,[14] which can be numerically approximated with high speed and accuracy by a computer. This is a great advantage. The difficult terms in the matrix elements are those with integrals that contain products of orbitals centered on different nuclei of a molecule. With other basis sets these must be computed with much slower numerical methods. Thus, for the same computational cost, one can use a larger basis dimension and in practice this can more than compensate for the fact that for given dimension the gaussians give a less accurate approximation than do functions that more closely resemble the one-electron atom solutions.

Gaussian bases can be optimized using a technique called ***contraction.*** Instead of using a single gaussian atomic orbital as a basis element, one uses a linear combination. The single gaussians are called "primitive" while the linear combinations are said to be "contracted." A contracted s-type basis function has the form

$$\phi^{(s)}(r) = \sum_i b_i^{(s)} g^{(s)}(r,\zeta_i)\,, \tag{20.80}$$

in which the coefficients $b_i^{(s)}$ and exponent parameters ζ_i have predefined fixed values taken from studies in which the Hartree-Fock results are compared with results from more accurate benchmark computations. Various contracted gaussian basis sets are available in the literature. Constructing the contractions is something of an art. Quantum chemistry software packages typically supply a variety of contracted basis sets from which to choose.

One can use Hartree-Fock wavefunctions as a basis set for a second diagonalization of the Hamiltonian matrix, in order to obtain a more accurate approximation. The new basis functions are then antisymmetric Hartree-Fock spin orbitals corresponding to different quantum states. For helium, for example, Hartree-Fock solutions for orbital configurations $1s^2$, $1s2s$, $2s^2$, $2p_z^2$, $1s3s$, and so on can be used as the basis. This procedure is called the ***configuration-interaction*** method. This method is dependable and can give extremely accurate solutions to Schrödinger's equation, but the computations require a great deal of computer memory and processing time. Various more efficient, but less accurate, methods are also available for improving the accuracy from an initial Hartree-Fock calculation.

---

[14]This important observation was made in 1950 by the British theoretical chemist S. F. Boys (1911-1972). See A. Szabo and N. S. Ostlund, *Modern Quantum Chemistry: Introduction to Advanced Electronic Structure Theory* (McGraw-Hill, New York, 1989), Appendix A, for details.

Exercises

20.1 $\Psi = e^{ij\phi}$ for a particle on a circle is a bound state while $\Psi = e^{ikx}$ for a particle on a line is not a bound state. What is the difference?

20.2 Derive Eqs. (20.12) and (20.13) for the angular momentum operators.

20.3* Prove that $\sum_k \psi_k^*(x)\psi_k(x_0)$, where $\{\psi_k\}$ is a basis for $L^2[-\infty, \infty]$, is a representation of the Dirac delta function $\delta(x - x_0)$.

20.4 Carry out the transformation of the one-electron atomic Schrödinger equation to atomic units. In other words, derive Eq. (20.40) from Eq. (20.39).

20.5 Why is the quantum number m not needed in Eq. (20.52) to label the electron's radial probability $P_{n,\ell}^{(r<r_0)}(r_0)$?

20.6* Derive Eq. (20.58) by substituting $\chi = u(r)e^{-\alpha r}$ into Eq. (20.53).

20.7 (a) For a $1s$ state of the hydrogen atom confirm that the orbital from Eqs. (20.49) and (20.51) is normalized, and then calculate the probability of the electron being within one Bohr radius of the nucleus. Express your answer for the probability as an incomplete gamma function and then use a computer to determine the numerical value. (b) Repeat for a $2p$ state. (c) Repeat for a $2s$ state.

20.8 Solve (using a computer) for the radial distance within which there is a 90% probability of finding the electron in a $2p$ state of the hydrogen atom and in a $2p$ state of the He^+ ion.

20.9 Plot the angular distribution (as in Example 14.4) for each of the sp^3 orbitals. Superimpose these plots in a single graph to show that the lobes are oriented as a tetrahedron.

20.10* Prove that a Slater determinant for an orbital configuration in which two electrons have completely identical quantum numbers will be identically zero.

20.11* Can the four $1s2s$ wavefunctions in Example 20.3 be expressed in terms of Slater determinants.

20.12* Write a Slater determinant for the ground state of the beryllium atom.

20.13 Pauling argued that his hybrid orbitals are just as fundamental as Schrödinger's orbitals $\psi_{n,\ell,m}$, and suggested that if chemists, not physicists, had discovered quantum mechanics, then the original theory might have been formulated in terms of hybrid orbitals. Demonstrate that the sp^3 orbitals comprise an orthonormal set and then express $\psi_{2,0,0}$, $\psi_{2,1,0}$, $\psi_{2,1,1}$, and $\psi_{2,1,-1}$ as linear combinations of the sp^3 orbitals.

20.14 Are all of the sets of hybrid orbitals given in Section 20.4 orthonormal?

20.15 Hybrid orbitals are just a qualitative model. What approximations are they based on?

Chapter 21

Fourier Analysis

21.1 The Fourier Transform

Suppose that a function $f(t)$, where t is time, is the sum of several waves, each with a different angular frequency ω_j, such as

$$f(t) = F_1 \frac{1}{\sqrt{2\pi}} e^{-i\omega_1 t} + F_2 \frac{1}{\sqrt{2\pi}} e^{-i\omega_2 t} + F_3 \frac{1}{\sqrt{2\pi}} e^{-i\omega_3 t} + F_4 \frac{1}{\sqrt{2\pi}} e^{-i\omega_4 t}. \quad (21.1)$$

This is illustrated by the left-hand panel of Fig. 21.1. f might, for example, represent the electric field of a wave of electromagnetic radiation. Plotting the real and imaginary parts of f vs. t is one way to represent the wave. Another way, which is more informative for most purposes, is to plot the amplitudes F_j vs. the frequencies ω_j, as in the panel on the right-hand side.

f is a linear combination of linearly independent functions $e^{-i\omega_j t}$. It resembles a Fourier series, except that the ω_j are arbitrary, not evenly spaced. Fourier series can be written

$$f(t) = \sum_{n=-\infty}^{\infty} F_n \frac{e^{-int}}{\sqrt{2\pi}}, \quad (21.2)$$

with evenly spaced frequencies $\omega_n = n$. [See Section 17.3. We are now explicitly including the factors of $(2\pi)^{-1/2}$ so that the basis functions will be normalized.] We can generalize this to allow for arbitrary frequencies by making ω_n a continuous variable ω, converting the summation into an integral,

$$f(t) = \frac{1}{\sqrt{2\pi}} \int_{-\infty}^{\infty} F(\omega) e^{-i\omega t} d\omega \,. \quad (21.3)$$

This gives us freedom to construct the wave however we wish, by choosing an

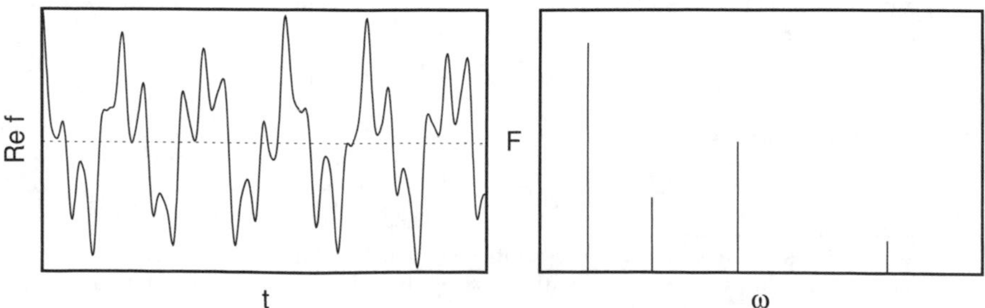

Figure 21.1: A sum of four waves with different amplitudes and frequencies, in the time domain and in the frequency domain.

appropriate amplitude function $F(\omega)$. For the sum of four waves in Eq. (21.1), we would choose[1]

$$F(\omega) = F_1\delta(\omega - \omega_1) + F_2\delta(\omega - \omega_2) + F_3\delta(\omega - \omega_3) + F_4\delta(\omega - \omega_4)\,, \tag{21.4}$$

where $\delta(x)$ is the Dirac delta function (defined in Section 4.4).

For this wave with known components the integral formalism does not tell us anything new. The utility of this approach is seen with complicated functions $f(t)$, representing, for example, an electromagnetic wave that has just passed through a sample in a spectrometer. It allows us to systematically determine the wave components that make up the observed radiation signal.

The following theorem provides the foundation for this kind of analysis:

Theorem 21.1.1. ***(Fourier's integral theorem.)*** *Let f be any piecewise continuous, differentiable function such that $\int_{-\infty}^{\infty} |f(x)|\,dx$ is finite. Then*

$$f(x) = \frac{1}{2\pi}\int_{-\infty}^{\infty}\int_{-\infty}^{\infty} e^{-i\omega x} f(t) e^{i\omega t} d\omega dt\,. \tag{21.5}$$

There are two immediate implications of Eq. (21.5). The first is a new representation of the Dirac delta function. Recall that the delta function has the property that

$$\int_{-\infty}^{\infty} f(t)\delta(t-x)\,dt = f(x)\,. \tag{21.6}$$

Comparing with Eq. (21.5), we find

$$\delta(t-x) = \frac{1}{2\pi}\int_{-\infty}^{\infty} e^{i\omega(t-x)} d\omega\,. \tag{21.7}$$

This delta function representation has widespread applications in physics.

The second implication is an expression for the amplitude function $F(\omega)$. If we change the symbol for the free variable in Eq. (21.3) from t to x and then compare the equation with Eq. (21.5), we see that

$$F(\omega) = \frac{1}{\sqrt{2\pi}}\int_{-\infty}^{\infty} f(t) e^{i\omega t} dt\,. \tag{21.8}$$

This is called the ***Fourier transform*** of f. The expression for $f(t)$ in terms of $F(\omega)$, Eq. (21.3), is called the ***inverse Fourier transform***. They look similar, except that the inverse transform has a minus sign in the exponential.

Example 21.1. *Fourier analysis of a wave packet.* Consider a *pulse* of an electromagnetic wave, for example,

$$f(t) = \begin{cases} e^{-i2\pi t}, & 0 \le t \le 3, \\ 0 & t < 0,\ t > 3. \end{cases} \tag{21.9}$$

Its Fourier transform is

$$F(\omega) = \frac{1}{\sqrt{2\pi}}\int_{-\infty}^{\infty} f(t)e^{i\omega t}dt = \frac{1}{\sqrt{2\pi}}\int_0^3 e^{i(\omega-2\pi)t}dt = \frac{1}{\sqrt{2\pi}}\frac{e^{i3(\omega-2\pi)}-1}{i(\omega-2\pi)}\,.$$

[1] Instead of lines of *heights* F_j in the right-hand panel of Fig. 21.1, we should actually have infinitely high and narrow peaks with *areas* F_j, but that is hard to show in a picture.

The real part of f is shown below in the left-hand panel, while the right-hand panel shows $|F(\omega)|^2$, which is proportional to the power density of the wave. (The power, energy per unit time, from wave components in the frequency range ω_1 to ω_2 is an integral over power density $\int_{\omega_1}^{\omega_2} P(\omega)d\omega$, with P proportional to $|F|^2$.)

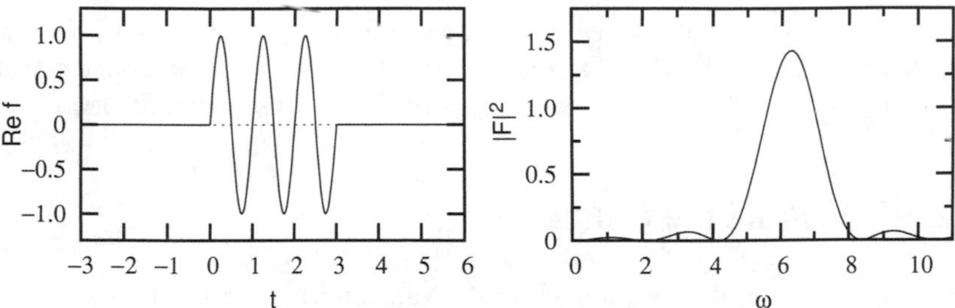

If we had a *continuous wave*, oscillating over the range $-\infty < t < \infty$ (in other words, lasting forever), only then would we have a *line spectrum*, with an infinitely narrow peak at the frequency $\omega = 2\pi$. We see instead that the wave can be resolved into a rather broad continuum of frequencies. (The word *resolve* is used here in the sense introduced in Section 16.7.) The maximum power density occurs at $\omega = 2\pi$ but there is significant power density over the range $2\pi \pm 1.3$.

The use of a single pulsed wave in order to simultaneously scan a range of frequencies is called ***Fourier-transform spectroscopy***. It can be much more efficient than using a sequence of many continuous waves each with a different frequency.

$f(t)$ and $F(\omega)$ are two different ways to express the same function. $f(t)$ is a representation in the time domain while $F(\omega)$ is a representation in the frequency domain. The Fourier transform and the inverse Fourier transform allow us to switch between the two representations at will. We can use whichever representation is more convenient for the problem at hand.

There are different conventions for where to put the factors of 2π. We are using the symmetric convention, with both transforms divided by $\sqrt{2\pi}$. Two other conventions are also widely used, as summarized in Table 21.1. One convention uses no factor in front of $F(\omega)$ but divides the inverse transform by 2π. The other uses

$$\nu = \omega/2\pi \tag{21.10}$$

as the frequency variable, with no prefactors in either transform. Strictly speaking, ν is the "frequency" while ω is the "*angular* frequency." In common usage "angular" is often omitted, and ω also is just called the "frequency."

Table 21.1: Different conventions for the Fourier transform and its inverse.

Fourier transform	Inverse Fourier transform
$F(\omega) = \frac{1}{\sqrt{2\pi}} \int_{-\infty}^{\infty} f(t)e^{i\omega t}dt$	$f(t) = \frac{1}{\sqrt{2\pi}} \int_{-\infty}^{\infty} F(\omega)e^{-i\omega t}d\omega$
$F(\omega) = \int_{-\infty}^{\infty} f(t)e^{i\omega t}dt$	$f(t) = \frac{1}{2\pi} \int_{-\infty}^{\infty} F(\omega)e^{-i\omega t}d\omega$
$F(\nu) = \int_{-\infty}^{\infty} f(t)e^{i2\pi\nu t}dt$	$f(t) = \int_{-\infty}^{\infty} F(\nu)e^{-i2\pi\nu t}d\omega$

In general, any mapping of a function f to a function F via an integral of the form

$$F(u) = \int_a^b f(x)K(u,x)dx, \tag{21.11}$$

where $K(u,x)$ is some arbitrary specified function, is called an ***integral transform***. K is called the ***kernel*** of the transform. The Fourier transform, with kernel e^{iux}, is the most commonly used integral transform.

21.2 Spectral Line Shapes*

According to the Bohr frequency condition, the angular frequency ω_0 of a photon emitted or absorbed during a quantum transition between a state of higher energy E_{high} and a state of lower energy E_{low} is determined by[2]

$$\hbar\omega_0 = E_{\text{high}} - E_{\text{low}}\,. \tag{21.12}$$

This would seem to suggest that the spectrum should consist of a series of vertical lines corresponding to the various possible energy differences, with zero intensity at all other frequencies. In fact, it is observed in practice that the spectrum does not consist of vertical lines, but rather of peaks that are centered at or near the predicted ω_0 values and include a continuum of nearby frequencies. The shapes of these peaks are called ***line shapes***.[3]

Consider an emission spectrum. A sample of molecules in an excited state at time $t = 0$ is allowed to relax to the ground state. The molecules will not all relax instantaneously; some will relax immediately while others will linger in the excited state. It is possible to show that the probability of relaxation is highest immediately after absorption and decreases exponentially with time, so that the intensity of emitted radiation will decrease in proportion to $e^{-t/\tau}$, where τ is a constant called the ***lifetime*** of the excited state.[4] The general expression for the emission wave with frequency ω_0 is a linear combination

$$f(t) = \begin{cases} \left(ae^{i\omega_0 t} + be^{-i\omega_0 t}\right)e^{-t/\tau}, & t \geq 0\,, \\ 0, & t < 0\,. \end{cases} \tag{21.13}$$

a and b are arbitrary and ω_0 is determined from Eq. (21.12). The spectrum

[2]Or $h\nu_0 = E_{\text{high}} - E_{\text{low}}$. h is Planck's constant while $\hbar = h/2\pi$.

[3]Interpreted literally, this would of course be a contradiction in terms, as the "shape" of a "line" is a line. However, "line shape" nicely expresses the idea that while in principle we should have a line, due to other factors we have a peak with a broadened shape.

[4]"Lifetime" is clearly something of a misnomer. To be precise, τ is the time at which the fraction of molecules still in the excited state has dropped to $e^{-1} \approx 0.37$. The exponential decrease follows from the assumption that relaxation is a unimolecular process, obeying first-order kinetics. (See Section 15.2.) Some authors define the lifetime as the time needed for the fraction to drop to $e^{-1/2}$. In that case 2τ sould be substituted for τ in our equations.

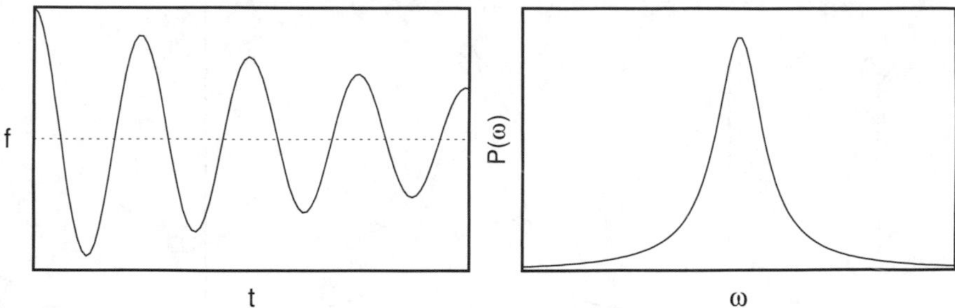

Figure 21.2: Lorentzian line shape, in the time domain (left-hand panel), Eq. (21.13), and in the frequency domain (right-hand panel), Eq. (21.17).

is given by the Fourier transform of $f(t)$,

$$F(\omega) = \frac{1}{\sqrt{2\pi}}\left[a\int_0^\infty e^{[i(\omega+\omega_0)-1/\tau]t}dt + b\int_0^\infty e^{[i(\omega-\omega_0)-1/\tau]t}dt\right]$$
$$= -\frac{1}{\sqrt{2\pi}}\left[\frac{a}{-1/\tau + i(\omega+\omega_0)} + \frac{b}{-1/\tau + i(\omega-\omega_0)}\right]. \tag{21.14}$$

The power spectrum is proportional to $|F(\omega)|^2$.

It is often the case that the oscillation period, $2\pi/\omega_0$, is much smaller than the lifetime of the excited state. Then $1/\tau \ll \omega_0$ in the denominators of Eq. (21.14). This implies that $|F(\omega)|^2$ will be relatively small except for values of ω in the vicinity of ω_0, where the absolute value of the second term in Eq. (21.14) becomes very large, as we divide by a number that is relatively close to zero. In practice, it is usually reasonable to completely ignore the first term in Eq. (21.14). That leads to the simple expression

$$P(\omega) \approx \frac{b^2}{2\pi}\frac{1}{1/\tau^2 + (\omega-\omega_0)^2}. \tag{21.15}$$

We would like to interpret this as the probability density for observing an emission wave (a photon) of frequency ω. In that case $P(\omega)$ needs to be normalized,

$$\frac{b^2}{2\pi}\int_{-\infty}^{\infty}\frac{1}{1/\tau^2 + (\omega-\omega_0)^2}\,d\omega = 1, \tag{21.16}$$

which implies that $b^2 = 2/\tau$. The normalized power spectrum is

$$P(\omega) \approx \frac{2/(\pi\tau)}{1/\tau^2 + (\omega-\omega_0)^2}. \tag{21.17}$$

This power spectrum is called the ***Lorentzian line shape***.[5] $f(t)$ and the corresponding $P(\omega)$ are shown in Fig. 21.2. The frequency of the emitted wave is exactly ω_0; the broadening of the observed spectral signal comes from the fact that the wave amplitude is not constant.

[5]Named for the Dutch physicist Hendrik Lorentz (1853-1928), who derived it for spectral line shapes. However, this probability distribution had been studied earlier by Cauchy. Mathematicians and statisticians refer to it as the *Cauchy distribution*.

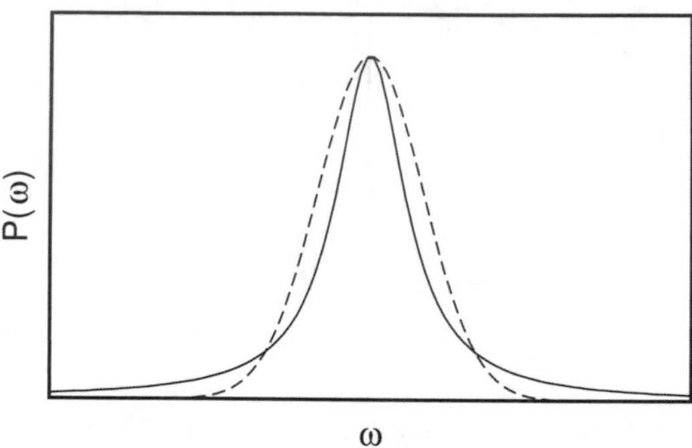

Figure 21.3: Gaussian (dashed curve) and Lorentzian (solid curve) line shapes, with the same height and area.

Lorentzian line shapes are a common phenomenon. They can be seen in absorption spectra as well as in emission spectra. In the gas phase, it is often the case that the excited state's relaxation is induced by a collision with another molecule. This is called ***pressure broadening***. It gives a pressure-dependent lifetime τ that increases the peak width with increasing pressure.

However, the Lorentzian is not the only line shape seen in practice. Often a Gaussian line shape,

$$P(\omega) = \frac{1}{\sqrt{2\pi s^2}}\, e^{-(\omega-\omega_0)^2/2s^2}, \tag{21.18}$$

is observed. The parameter s determines the width of this line shape. Fig. 21.3 compares a Gaussian and a Lorentzian with the same height. They are qualitatively different. The Lorentzian is narrower but has higher tails.

Example 21.2. *Doppler broadening.* Gaussian line shapes are expected from random processes. Consider the effect on the spectrum of the random velocity distribution of the molecules. The probability distribution for a single component v_x of the velocity vector of a molecule of mass m in a sample of gas at temperature T is[6]

$$P(v_x) = \left(\frac{m}{2\pi k_B T}\right)^{1/2} e^{-mv_x^2/2k_B T}. \tag{21.19}$$

Consider a wave with angular frequency ω_0. Suppose that the wavefront is approaching a molecule from the x direction. The wavefront has speed c and the molecule has speed v_x. The speed of a wave is the product of wavelength and frequency, $c = \lambda\nu = \lambda\omega_0/2\pi$. However, from the molecule's point of view, the speed of the wave is just $c - v_x$:

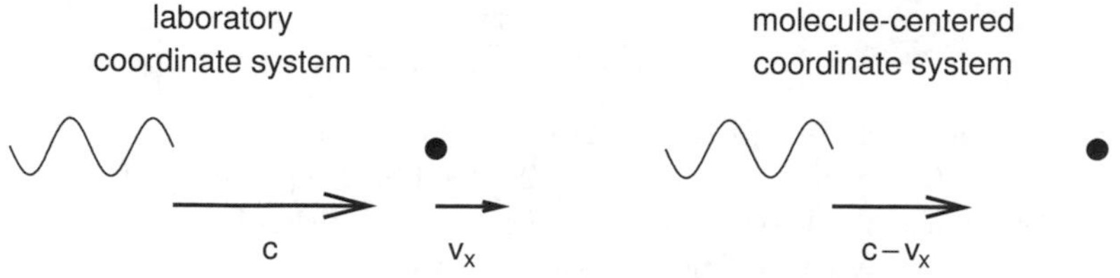

[6]This was derived by the Scottish physicist James Clerk Maxwell (1831-1879). We can derive it from Boltzmann's distribution, Eq. (8.31), by replacing E_j with the kinetic energy $\frac{1}{2}m(v_x^2 + v_y^2 + v_z^2)$ and replacing the normalization summation with an integral.

When the wavefront reaches the molecule the electromagnetic field will seem, to the molecule, to be oscillating not with the true angular frequency ω_0 but with the apparent angular frequency $\omega = 2\pi(c - v_x)/\lambda = (c - v_x)\omega_0/c$. Solving for v_x gives $v_x = (\omega_0 - \omega)c/\omega_0$. Substituting this into Eq. (21.19), and normalizing, leads to a probability distribution for ω in the form of Eq. (21.18). This effect is called ***Doppler broadening.***[7]

21.3 Discrete Fourier Transform*

The signals from spectroscopic instruments usually come in a digitized form. Instead of a continuous function $f(t)$, $-\infty < t < \infty$, we have a set of values $f(t_k)$ measured at discrete evenly spaced times t_k over a finite time period. Suppose we have N measurements starting at $t_0 = 0$. We can write

$$t_k = (t_N/N)k\,, \qquad k = 0, 1, 2, \ldots, N-1\,. \tag{21.20}$$

t_N is the total time period and t_N/N is the time interval between measurements. The Fourier transform is an integral over $f(t)e^{i\omega t}$. We will construct the discrete version as a *sum* of $f(t_k)e^{i\omega t_k}$. Similarly, we will construct a discrete inverse Fourier transform as a sum over $F(\omega_j)e^{-i\omega_j t}$ for a set of discrete frequency values ω_j.

Because we have only N points in t-space, we will be able to solve for only N points in ω-space. The range of the ω_j values is arbitrary but it turns out to be convenient to choose the range $0 \leq \omega < 2\pi N/t_N$, with evenly spaced frequencies

$$\omega_j = (2\pi/t_N)j\,, \qquad j = 0, 1, 2, \ldots, N-1\,. \tag{21.21}$$

We then define the ***discrete Fourier transform*** as

$$F(\omega_j) = \frac{1}{\sqrt{N}} \sum_{k=0}^{N-1} f(t_k)e^{i\omega_j t_k} \tag{21.22}$$

and the ***discrete inverse Fourier transform*** as

$$f(t_k) = \frac{1}{\sqrt{N}} \sum_{j=0}^{N-1} F(\omega_j)e^{-i\omega_j t_k}. \tag{21.23}$$

Fig. 21.4 shows a discretized Lorentzian wave in the time domain and its corresponding power spectrum as given by the discrete Fourier transform. The spacing between points in $P(\omega)$ is

$$\Delta\omega = \omega_{j+1} - \omega_j = 2\pi/t_N\,. \tag{21.24}$$

Thus, the resolution of the frequency spectrum is improved by increasing the

[7] It is analogous to the perceived shift in pitch of a train whistle as the train approaches and then passes the listener. This was first explained by Christian Doppler (1803-1853), an Austrian mathematician.

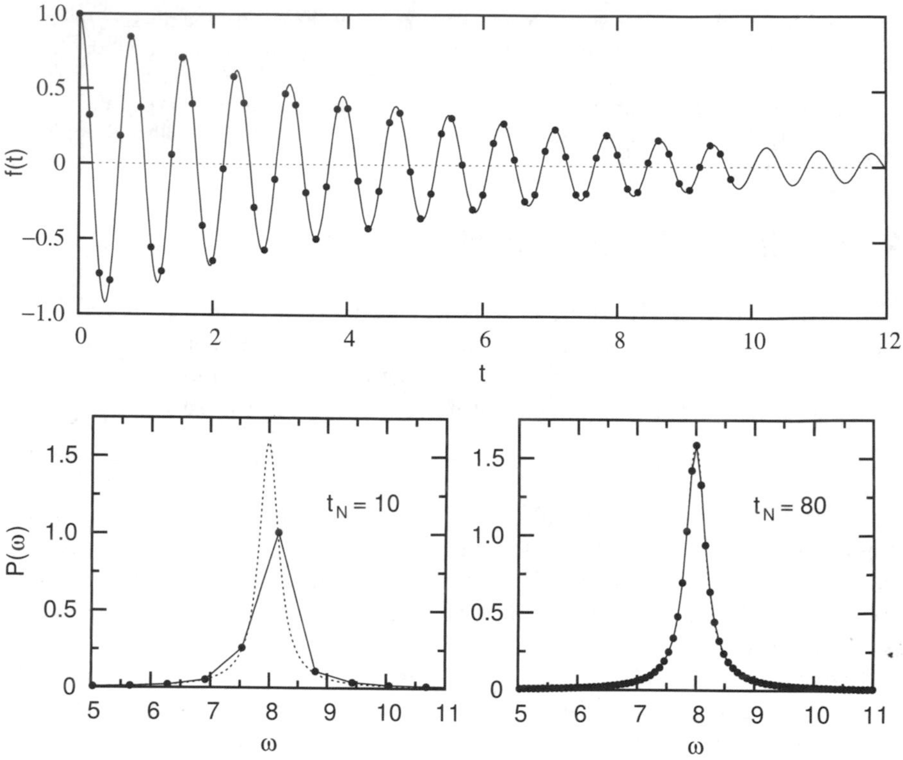

Figure 21.4: Discrete Fourier transform of a Lorentzian signal. The upper panel shows the original signal, sampled at the indicated points. The lower left-hand panel is the corresponding Fourier transform, with the continuous transform shown by the dotted curve. The right-hand panel shows the improvement in resolution from increasing the sampling period from 10 to 80.

sampling time t_N. A very efficient computer algorithm, called *fast Fourier transform*, exists for computing the $F(\omega_j)$ given a set of measured values $f(t_k)$.

Example 21.3. *Discrete Fourier transform of a Lorentzian signal.* Let us use *Mathematica* to compute the power spectrum of a discretized Lorentzian signal $f(t) = e^{-t/2\tau}\cos(\omega_0 t)$. We begin by assigning values for N, τ, ω_0, and t_N, and then simulating a set of measured $f(t_k)$ values:

```
f[t_] := Exp[-t/(2*tau)]*Cos[w0*t];

n = 500; tau = 2.5; w0 = 8.0; per = 80.0;

fdata = Table[f[(r-1)*per/n],{r,1,n}];
```

The fast Fourier transform is computed with the *Mathematica* `Fourier` command:

```
ft = Fourier[fdata];
```

This yields a list of the F_j values. Let us construct ordered pairs $(\omega_j, |F_j|^2)$:

```
rawpowerspec = Table[{(r-1)*(2*Pi)/per, Abs[ ft[[r]] ]^2}, {r,1,n}];
```

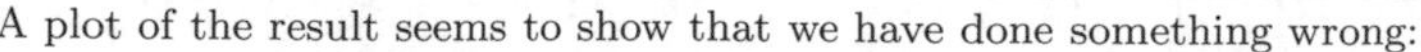

A plot of the result seems to show that we have done something wrong:

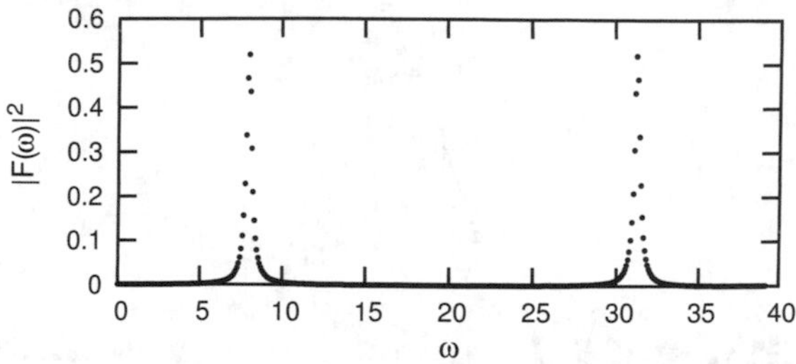

The surprise is that we have *two* peaks. One is at $\omega_0 \approx 8$, where we expect it. The other peak is spurious. It results from the fact that our indices j and k range from 0 to $N-1$. We are discretizing integrals that cover the range $-\infty$ to ∞. Should the index values, then, be $-N/2, -N/2+1, \ldots, 0, \ldots N/2-1$? No—"negative" frequency has no physical meaning and, in any case, $|F(-\omega)|$ is equal to $|F(\omega)|$. (See Exercise 21.4.) It is conventional to ignore the negative-frequency region and start the indices at 0. One can prove (Exercise 21.6) that $F(\omega_{-j}) = F(\omega_{N-j})$. The left-hand peak in this example is at $\omega_{102} = 8.01$ and the right-hand peak is at $\omega_{500-102} = \omega_{398} = 31.26$. There is a critical frequency in the middle,

$$\omega_c = \omega_{N/2} = N\pi/t_N \,, \tag{21.25}$$

beyond which we have just the exact reflection of the $0 \le \omega \le \omega_c$ region. ω_c is called the **Nyquist frequency**.[8] In this example, $\omega_c = \omega_{250} = 19.63$. It is senseless to plot points past this value, as they contain no additional information. The only way to obtain information about higher frequencies would be to decrease the spacing between the sampling times by increasing the ratio N/t_N.

Another problem with our result is that the peak is too small. The area under the spectrum ought to be unity but it is evident from the plot that the actual area is smaller than this. We have neglected to normalize the spectrum. We can compute the normalization factor using quadrature over the range $0 \le \omega \le \omega_c$:

```
quadrature[data_, xmax_] := Module[{dx, int, kmax},
    (* Quadrature using trapezoids, integration range 0 to xmax.*)
    (* Data set assumed evenly spaced and ordered according to x value.*)
    dx = data[[2, 1]] - data[[1, 1]]; n = Length[data]; kmax = 1;
    While[data[[kmax, 1]] <= xmax, kmax++]; (* kmax is last index.*)
    int = Sum[dx*(data[[k, 2]] + data[[k + 1, 2]])/2.0, {k, kmax}]
];

powernorm = quadrature[rawpowerspectrum, n*Pi/per];

powerspectrum = Table[
    {rawpowerspectrum[[r, 1]], rawpowerspectrum[[r, 2]]/powernorm},
    {r, 1, Length[rawpowerspectrum]/2} ];
```

This generates the points shown in the lower right-hand panel of Fig. 21.4. The value of t_N determines the spacing of points in ω-space, according to Eq. (21.21), while N/t_N determines the upper range of the frequencies, according to Eq. (21.25).

[8]Named after American electrical engineer Harry Nyquist (1889-1976). If the signal contains frequency components higher than ω_c, then the reflected frequency spectrum will overlap the $\omega < \omega_c$ part of the spectrum and the result will be distorted. It is important to choose N/t_N large enough to avoid this problem.

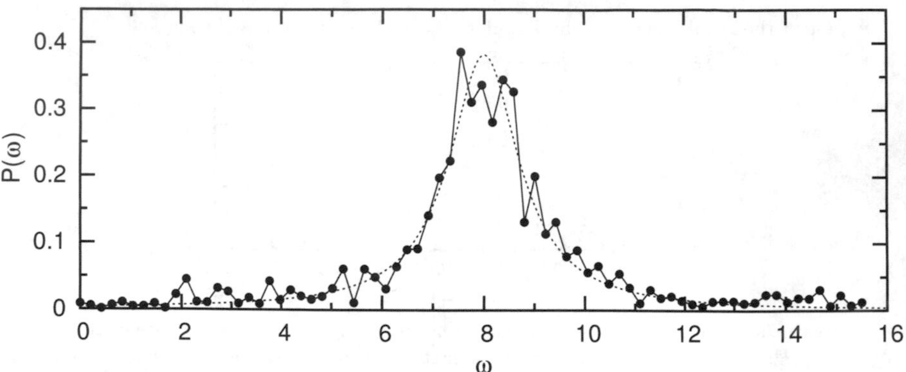

Figure 21.5: A discrete noisy power spectrum compared with the true power spectrum (the dotted curve). Random noise was added in the time domain to a Lorentzian signal.

The reason that Eq. (21.21) is chosen for the discrete frequencies is that this choice leads to an orthonormal set of basis functions $e^{i\omega_j t_k}$ with inner product

$$\langle e^{i\omega t_{k'}}, e^{i\omega t_k} \rangle = \frac{1}{N} \sum_{j=0}^{N} \left(e^{i\omega_j t_{k'}} \right)^* e^{i\omega_j t_k} = \frac{1}{N} \sum_{j=1}^{N} e^{i\omega_j (t_k - t_{k'})} = \delta_{k,k'} \quad (21.26)$$

for different time steps k and k'. This is just a discrete version of the integrals for the orthonormality of the Fourier series basis set, Eqs. (17.31) and (17.32).

21.4 Signal Processing

Signal processing is the use of mathematical methods to enhance and to analyze the numerical output (the *signal*) from a measuring instrument. To *process* the signal is to apply some kind of mathematical transformation that makes the correspondence between the signal and the underlying physical phenomenon more apparent. The principal signal processing problems of analytical chemistry are random noise and overlapping peaks.

21.4.1 Noise Filtering*

Fig. 21.5 compares an idealized Lorentzian spectral signal with no noise and the same signal with random noise added to it. The noisy signal was generated by adding random fluctuations to a discretized wave in the time domain and then computing the Fourier transform. We would like to transform the noisy signal into something more closely resembling the true signal. Any algorithm that accomplishes this is called a ***noise filter***.

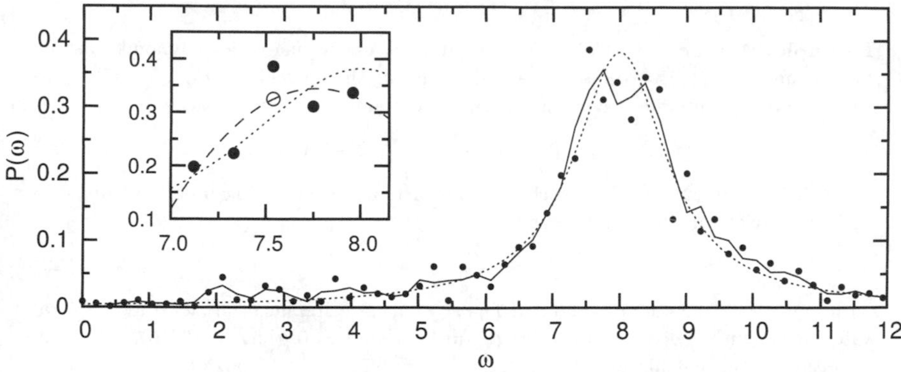

Figure 21.6: Savitzky-Golay noise filtering, with a five-point window and quadratic fitting function. The inset shows the fitting function (dashed curve) for a single window. The central measured point is replaced with the point shown by the open circle. A separate fit is used for each window.

The simplest approach to reducing the effects of random variation is to take an average over neighboring points. The more points included in the average, the stronger the noise suppression. However, if too many points are included, the shape of the peak will be significantly distorted. The number of points being included in the average is called the ***window*** of the filter. Consider a window of five points, a typical choice. The averaged signal at the frequency ω_j is then

$$\langle P_j \rangle = \sum_{n=-2}^{2} P_{j+n} \,, \tag{21.27}$$

with $\langle P_j \rangle$ here representing the smoothed discrete power spectrum.

However, it is unwise to use a simple average such as Eq. (21.27). Consider what it does to the true spectrum in the absence of noise. At the top of a peak the surrounding points, in the noiseless spectrum, are all lower than the central point. Replacing this central point with the average has the effect of decreasing the peak height. Thus, Eq. (21.27) distorts the shape of the spectral peak, flattening it.

An alternative strategy, which was developed by Savitzky and Golay,[9] is to perform a least-squares fit on the points in each window and then replace the measured point with the value given by the fitting function at the central frequency value of the window. This is illustrated in Fig. 21.6, which shows the effect of fitting a quadratic polynomial in five-point windows. The filtered spectrum is still somewhat choppy but it does a better job of modeling the true spectrum than did the original unfiltered spectrum. The Savitzky-Golay filter introduces no particular bias into the peak shape.

[9]American chemist Abraham Savitzky and Swiss-born American physicist Marcel Golay, employees of the analytical instruments company Perkin-Elmer, proposed this method in an influential paper published in 1964.

Example 21.4. *Savitzky-Golay filtering.* Let us use a quadratic polynomial as the fitting function, $P(\omega) = \alpha_0 + \alpha_1\omega + \alpha_2\omega^2$, and fit the coefficients α_0, α_1, and α_2 to the five points centered at some given ω_j. The points are evenly spaced,

$$\omega = \omega_j + (2\pi/t_N)\,n, \quad n = -2, -1, 0, 1, 2, \tag{21.28}$$

according to Eq. (21.21). Let us change the variable of the fitting function from ω to n, and write the fitting function as

$$\widetilde{P}^{(j)}(n) = a_0 + a_1 n + a_2 n^2, \tag{21.29}$$

with new coefficients a_0, a_1, and a_2. The strategy is to replace the measured spectral value at ω_j with the value of the fitting function at $n = 0$, that is, $\widetilde{P}^{(j)}(0) = a_0$. All we need to do is calculate a_0 for each ω_j.

Let us carry out a least-squares fit to the data set

$$\{(-2,\ P_{j-2}),\ (-1,\ P_{j-1}),\ (0,\ P_j),\ (1,\ P_{j+1}),\ (2,\ P_{j+2})\}$$

using the basis set $\{1, n, n^2\}$. In matrix notation,

$$\widetilde{P}^{(j)}(n) = \mathbf{a}^{\mathbf{T}}\mathbf{u}, \quad \mathbf{a}^{\mathbf{T}} = (a_0 \quad a_1 \quad a_2), \quad \mathbf{u} = \begin{pmatrix} 1 \\ n \\ n^2 \end{pmatrix}. \tag{21.30}$$

The design matrix (see Example 18.7) has the elements $A_{k,i} = x_k^{i-1}$, where $x_1 = -2$, $x_2 = -1$, $x_3 = 0$, $x_4 = 1$, and $x_5 = 2$, with $A_{k,1} = 1$ and $A_{3,i>1} = 0$; that is,

$$\mathbf{A} = \begin{pmatrix} 1 & -2 & 4 \\ 1 & -1 & 1 \\ 1 & 0 & 0 \\ 1 & 1 & 1 \\ 1 & 2 & 4 \end{pmatrix}. \tag{21.31}$$

The solution for the coefficients is

$$\mathbf{a} = \mathbf{A}^{+}\,\mathbf{b}^{(j)}, \tag{21.32}$$

where $\mathbf{A}^{+} = (\mathbf{A}^{\mathbf{T}}\mathbf{A})^{-1}\mathbf{A}^{\mathbf{T}}$ is the pseudoinverse of the design matrix and $b^{(j)}$ is the column matrix with elements $P_{j-2}, \ldots, P_{j+2}$. The only coefficient we need to know is a_0, the first element of $\mathbf{a}$, given by the dot product of the first row of $\mathbf{A}^{+}$ and the vector $\mathbf{b}^{(j)}$,

$$\widetilde{P}^{(j)}(0) = a_0 = \sum_{k=1}^{5} A^{+}_{1,k} b^{(j)}_k = \sum_{n=-2}^{2} A^{+}_{1,n+3} P_{j+n}\,. \tag{21.33}$$

This is our smoothed spectrum. We can write it in a form analogous to the expression for the average over the window, Eq. (21.27):

$$\langle P_j \rangle = \sum_{n=-2}^{2} c_n P_{j+n}\,, \qquad c_n = A^{+}_{1,n+3}\,. \tag{21.34}$$

This is a weighted average. Note that the weight factors depend only on the value of n; they are the same for all values of j. They only need to be computed once. For a quadratic fit with a five-point window, the weight coefficients are

$$c_{\pm 2} = -3/35, \quad c_{\pm 1} = 12/35, \quad c_0 = 17/35\,. \tag{21.35}$$

If this does not provide sufficient smoothing, as in the case shown in Fig. 21.6, then a larger window might be used. To beter detect very narrow peaks, a fitting function of higher degree than quadratic could be used.

A more general expression for a noise filter in the frequency domain is

$$\langle P_j \rangle = \sum_{n=-m_1}^{m_2} c_n P_{j+n} \,, \tag{21.36}$$

with arbitrary choices for m_1, m_2, and for the weights c_n. The Savitzky-Golay filter is a popular choice, but other choices are sometimes used. It can be proved that the Savitzky-Golay filter conserves the height, area, and width of spectral peaks in the absence of noise. Because of this, and because of its computational efficiency, this method is widely used to smooth spectra and chromatographs.

21.4.2 Convolution*

The ***convolution*** operator, represented by the symbol "$*$," maps a pair of functions $f(x)$ and $g(x)$ to a new function of x, represented by the symbol $f * g$, according to

$$f * g = \frac{1}{\sqrt{2\pi}} \int_{-\infty}^{\infty} f(x-\xi) g(\xi) d\xi \,. \tag{21.37}$$

$f * g$ is called the "convolution of f and g." It has a variety of applications in instrumental analysis and signal processing.

Convolution can be used to model the response of an instrument to a physical phenomenon. $f(x)$ is the "true" signal, an accurate representation of the physical phenomenon, while $f * g$ is the output of the instrument. g describes the conversion from input to output. Consider a chromatograph. Then f is the true chromatograph signal, a function of retention time t, with symmetric peaks centered at the retention times of each component of the mixture. In principle, these peaks should have a Gaussian shape, due to diffusion of the solute in both directions about the advective trajectory. $f * g$ is the actual chromatograph produced by the instrument, in which peaks may be unsymmetrical due to the way the sample was injected into the chromatographic column. g describes the injection process. For a "perfect" instrument, injection would be instantaneous. We would have $g(t) = \delta(t)$, a Dirac delta function. In practice it might be more like

$$g(t) = \begin{cases} 0, & t < 0 \,, \\ b^{-1} e^{-t/b}, & t \geq 0 \,, \end{cases} \tag{21.38}$$

where b is some small but nonzero positive constant. This results in a smeared asymmetric peak, as shown in Fig. 21.7.

The Fourier transform of $f * g$ is *not* equal to the convolution of the Fourier transforms, $F * G$. In fact it is much simpler:

Theorem 21.4.1. ***(The convolution theorem.)*** *The Fourier transform of $f * g$ is FG, and the inverse Fourier transform of $F * G$ is fg, where F and G are the Fourier transforms of f and g, respectively.*

Thus, Fourier transform converts convolution into simple multiplication. This property is responsible for the utility of convolution in signal processing. For

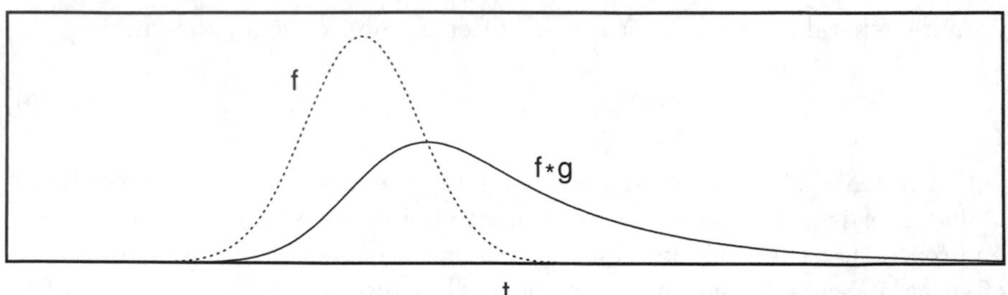

Figure 21.7: A simulated chromatograph peak. The dotted curve is what would result from instantaneous injection of the sample. The solid curve, the convolution with $g(t)$ of Eq. (21.38), is a more realistic representation of the instrument signal.

example, if we know the function $g(t)$ that describes the sample injection, then we can construct the desired signal $f(t)$ as follows:

$$F(\omega) = \left[\frac{1}{\sqrt{2\pi}}\int_{-\infty}^{\infty} f(t) * g(t) e^{i\omega t} dt\right] \Big/ G(\omega), \tag{21.39}$$

$$\begin{aligned} f(t) &= \frac{1}{\sqrt{2\pi}}\int_{-\infty}^{\infty} F(\omega) e^{-i\omega t} d\omega \\ &= \frac{1}{\sqrt{2\pi}}\int_{-\infty}^{\infty}\left\{\left[\int_{-\infty}^{\infty} f(t) * g(t) e^{i\omega t} dt\right] \Big/ \left[\int_{-\infty}^{\infty} g(t) e^{i\omega t} dt\right]\right\} e^{-i\omega t} d\omega . \end{aligned} \tag{21.40}$$

This procedure is called ***deconvolution***.

The proof of the convolution theorem is straightforward:

Proof. (Of convolution theorem.)
For arbitrary functions $f(t)$ and $g(t)$,

$$f * g = \frac{1}{\sqrt{2\pi}}\int_{-\infty}^{\infty} f(t-\xi) g(\xi)\, d\xi .$$

$f(t-\xi)$ can be expressed as an inverse Fourier transform,

$$f(t-\xi) = \frac{1}{\sqrt{2\pi}}\int_{-\infty}^{\infty} F(\omega)\, e^{-i\omega(t-\xi)} d\omega .$$

It follows from Fubini's theorem, Eq. (1.33), that

$$\begin{aligned} f * g &= \frac{1}{\sqrt{2\pi}}\int_{-\infty}^{\infty}\left[\frac{1}{\sqrt{2\pi}}\int_{-\infty}^{\infty} F(\omega) e^{-i\omega t + i\omega\xi} d\omega\right] g(\xi)\, d\xi \\ &= \frac{1}{\sqrt{2\pi}}\int_{-\infty}^{\infty} F(\omega)\left[\frac{1}{\sqrt{2\pi}}\int_{-\infty}^{\infty} g(\xi) e^{i\omega\xi} d\xi\right] e^{-i\omega t} d\omega \\ &= \frac{1}{\sqrt{2\pi}}\int_{-\infty}^{\infty} F(\omega) G(\omega) e^{-i\omega t} d\omega . \end{aligned} \tag{21.41}$$

This proves that $f * g$ is the inverse Fourier transform of FG, which implies that FG is the Fourier transform of $f * g$.

The convolution of the Fourier transforms $F(\omega)$ and $G(\omega)$ is

$$F * G = \frac{1}{\sqrt{2\pi}} \int_{-\infty}^{\infty} F(\omega - u) G(u)\, du \,. \tag{21.42}$$

It follows that

$$F * G = \frac{1}{\sqrt{2\pi}} \int_{-\infty}^{\infty} f(t) g(t) e^{i\omega t} dt \,, \tag{21.43}$$

which implies that fg is the inverse Fourier transform of $F * G$. □

The presentation thus far has been for a continuous signal. Consider now a discrete signal. Let us replace t with $t_j = (t_N/N)j$ and ξ with $\xi_k = (t_N/N)k$, and ω with $\omega_j = (2\pi/t_N)j$ and u with $u_k = (2\pi/t_N)k$, and introduce the streamlined notation

$$\begin{aligned} g(\xi_k) &= g_k, \qquad f(t_j - \xi_k) = f\big((j-k)t_N/N\big) = f_{j-k} \,, \\ G(u_k) &= G_k, \qquad F(\omega_j - u_k) = F\big((j-k)t_N/N\big) = f_{j-k} \,. \end{aligned}$$

The discrete convolutions are given by

$$(f * g)_j \;=\; \sum_{k=-N/2+1}^{N/2} f_{j-k}\, g_k \,, \tag{21.44}$$

$$(F * G)_j \;=\; \sum_{k=-N/2+1}^{N/2} F_{j-k}\, G_k \,. \tag{21.45}$$

An important application of discrete convolution is as a noise filter. Compare Eq. (21.45) with the general expression for a noise filter as a weighted average, given at the end of the previous section, which we can write as

$$\langle P_j \rangle = \sum_{k=-m_1}^{m_2} G_k\, P_{j-k} \,. \tag{21.46}$$

The substitutions $n = -k$ and $c_{-k} = G_k$ have been made in Eq. (21.36), to make it apparent that the two expressions are equivalent. Thus, we see that discrete convolution is equivalent to weighted averaging.

Now suppose that instead of applying the noise filter to the power spectrum $P(\omega) \propto |F(\omega)|^2$, we apply it directly to the Fourier transform of the electromagnetic wave, $F(\omega)$,

$$\langle F_j \rangle = \sum_{k=-N/2+1}^{N/2} G_k\, F_{j-k} = (F * G)_j \,. \tag{21.47}$$

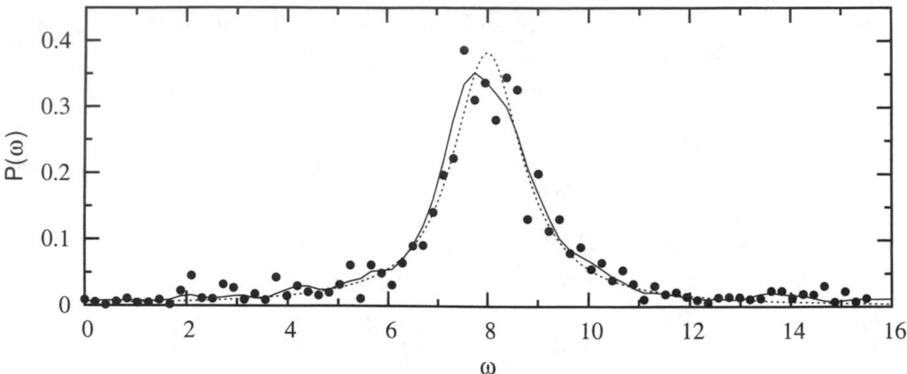

Figure 21.8: A noisy spectrum filtered in the time domain by multiplying with a Gaussian, Eq. (21.49), with $a = 0.02$. The dotted curve shows the true spectrum.

Thanks to the convolution theorem (which also holds for discrete convolution), we can obtain $\langle F_j \rangle$, the filtered Fourier transform, by multiplying the original signal $f(t)$ by an arbitrary function $g(t)$ and then computing the discrete Fourier transform of $f(t)g(t)$ using a discretized version of Eq. (21.43),

$$\langle F_j \rangle = (F * G)_j = \frac{1}{\sqrt{N}} \sum_{k=0}^{N-1} f(t_k)\, g(t_k)\, e^{i\omega_j t_k}. \qquad (21.48)$$

This is called ***time-domain filtering*** of a spectrum.

Example 21.5. *Time-domain filtering.* A demonstration is shown in Fig. 21.8. The noisy signal $f(t)$ whose power spectrum was shown in Fig. 21.5 has here been multiplied by a Gaussian time-domain filter

$$g(t) = \sqrt{\frac{a}{2\pi}}\, e^{-at^2}. \qquad (21.49)$$

Why this function? The amplitude of the signal $f(t)$ decreases with time, as in Fig. 21.2. Eventually the signal amplitude drops below the average noise amplitude. From then on we obtain no useful information, just random fluctuations due to the noise. Multiplying $f(t)$ by this Gaussian factor serves to damp out the large-t part of the spectrum while leaving the small-t portion relatively unaffected.

An advantage of time-domain filtering is that it can simultaneously accomplish not just noise filtering but also resolution of overlapping peaks. In Section 5.2 we used numerical differentiation to resolve a pair of overlapping peaks. The problem with that appoach is the fact that differentiation magnifies noise. Derivatives measure change. It is in the nature of random noise to change rapidly, and this can lead to large fluctuations in the derivative even if the magnitude of the noise is small. The resolution of peaks by numerical differentiation came about because the peaks in the second derivative of the spectrum were narrower than the peaks in the original spectrum. This suggests that we choose a time-domain filter that not only smoothes but also narrows the peaks.

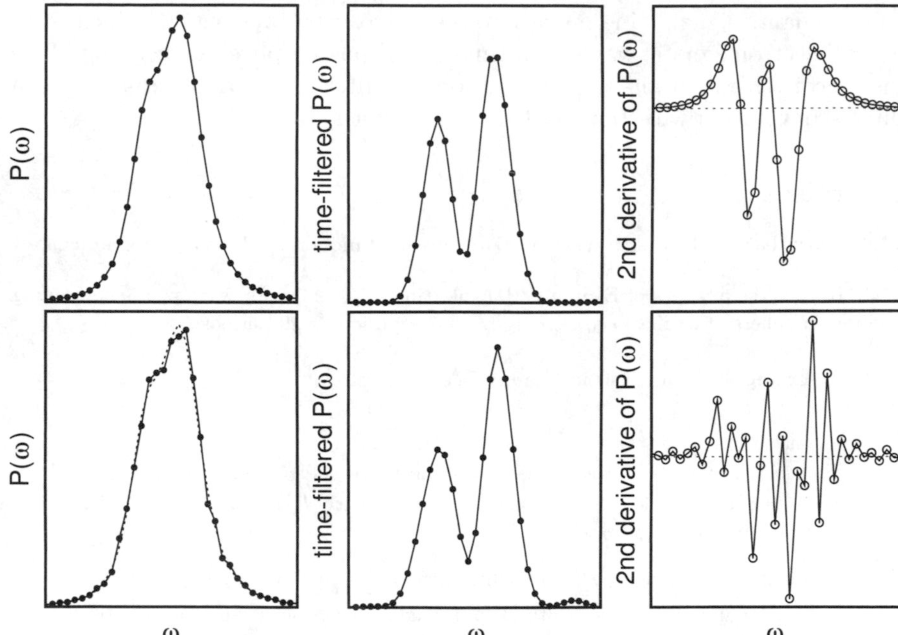

Figure 21.9: Resolution of overlapping peaks using a time-domain filter, in the form of Eq. (21.50), compared with using the numerical second derivative. For the lower panels random noise was added to the time-domain signal.

Example 21.6. *Simultaneous noise filtering and resolution of overlapping peaks.* We saw in Section 21.2 that Lorentzian broadening resulted from exponential decay of the signal amplitude; the faster the decay, the broader the peaks. Let us counteract this effect by multiplying the signal by an *increasing* exponential. This ought to narrow the peaks. We can simultaneously filter noise and resolve overlap using a time-domain filter in the form

$$g(t) = e^{-at^2+bt}. \tag{21.50}$$

Appropriate values for a and b can be found by trial and error, so that the peaks are clearly resolved and the noise is sufficiently damped down. This is a more dependable method than the numerical second derivative, as demonstrated in Fig. 21.9. Without noise both methods nicely resolve the two overlapping Lorentzian peaks, but in the presence of even just a small amount of random noise, barely enough to see in the plot, the second derivative is useless while the time-domain filter continues to perform very well.

Time-domain filtering works best with Lorentzian peaks. It is commonly used with NMR spectroscopy, because in that case the line shapes do tend to be Lorentzian. However, it can also work well with other line shapes. It is even sometimes used to resolve overlapping peaks in chromatographs. This might seem surprising, as the chromatograph has peaks in a *time* domain. However, we can formally treat the solute retention time as if it were a frequency variable ω and then compute an inverse Fourier transform to obtain a corresponding time-domain "signal." In contrast to a spectroscopy time signal, which is an actual electromagnetic wave, this time signal has no physical meaning. It is just a convenient mathematical method for filtering, as an alternative to a direct convolution with the Savitzky-Golay method.

Exercises

21.1 Show that the Fourier transform of a Gaussian function e^{-at^2} is also a Gaussian.

21.2 Repeat the analysis of Example 21.1, plotting $|F|^2$ vs. ω, for wave pulses containing different numbers of oscillations, ranging from one period to 20 periods.

21.3 According to the analysis in Example 21.1, we get

$$F(\omega) = \frac{1}{\sqrt{2\pi}} \frac{e^{in(\omega-2\pi)} - 1}{i\,(\omega - 2\pi)}$$

as the Fourier transform of a wave pulse containing n periods of oscillation.
(a) Derive the expression for the normalized power density, $P(\omega)$.
[Hint: $\int_{-\infty}^{\infty} \frac{1-\cos x}{x^2}\, dx = \pi$.*]*

(b) Calculate the peak height, $P(2\pi)$. (Obtain the result analytically.)

(c) Determine (numerically) the width at half height for various n and plot the results.

21.4 Derive the following symmetry properties of the Fourier transform:
(a) If $f(t)$ is real, then $F(-\omega) = F^*(\omega)$.
(b) If $f(t)$ is even [that is, $f(-t) = f(t)$], then $F(\omega)$ is even.

21.5 Demonstrate that the two alternative conventions for where to put the 2π in the Fourier transform are consistent with the Fourier integral theorem.

21.6* Prove that the discrete Fourier transform has the symmetry property $F(\omega_{-j}) = F(\omega_{N-j})$.

21.7* In Example 21.3 if the number of measurements is reduced to 220 with the time period still at 80.0, so that the time interval between measurements is increased, where will the two peaks be found?

21.8* Why is there a factor of b^{-1} in front of the exponential in Eq. (21.38)?

21.9* (a) Simulate a spectrum containing a Gaussian peak with random noise. (b) Smooth the spectrum using the Savitzky-Golay method. (c) Smooth the spectrum using time-domain filtering.

21.10* (a) Simulate a spectrum containing a pair of overlapping Lorentzian peaks (with no noise). (b) Resolve the peaks using a time-domain filter. (c) Resolve the peaks using numerical second derivatives.

21.11* Repeat Exercise 21.10 using Gaussian peaks instead of Lorentizians.

21.12* Repeat Exercise 21.10 with noise added to the spectrum. (Try also to use Savitzky-Golay filtering to reduce the noise in the second-derivative spectrum.)

Appendix A

Computer Programs

A.1 Robust Estimators

The *Mathematica* function `src`, as defined below, computes the Rousseeuw-Croux scale estimator, according to Eq. (8.36).

```
src[data_] :=
   Median[
     Table[
       Median[
         Table[Abs[ data[[i]]-data[[j]] ],{j,1,Length[data]}]
       ], {i,1,Length[data]}
     ]
   ]*1.1926
```

For example:

```
sample1={3.45302, 6.2983, 4.54182, 2.97837, 4.57128, 5.48347}
```

```
src[sample1]
  1.26341
```

The following computes Huber location and scale estimators:

```
huber[data_,mu0_:0,s0_:0,n_:40,iprint_:0] :=
   Module[{winsormin,winsormax,size,newdata,mu,s},
     size = Length[data];
     If[s0==0, mu0 = Median[data]; s0 = MedianDeviation[data] ];
     mu = mu0; s = s0;
     For[m=1, m<= n, m=m+1,
       newdata = data;
       winsormin = mu - 1.5*s; winsormax = mu + 1.5*s;
       For[j=1,j<=size,j=j+1,
         If[newdata[[j]] < winsormin, newdata[[j]] = winsormin ];
         If[newdata[[j]] > winsormax, newdata[[j]] = winsormax ];
       ];
       mu = Mean[newdata];
       s = 1.134*StandardDeviation[newdata];
       If[iprint != 0, Print["m = ",m," mu = ",mu," s = ",s] ];
     ];
     {mu,s}
   ]
```

The syntax is `huber[sample,mean0,sd0,iter,1]`, where `sample` is the data list, `mean0` and `sd0` are initial guesses for the mean and standard deviation, respectively, and `iter` is the number of iterations. If the final argument is any nonzero number, then the result from each iteration will be displayed, to assess convergence. If the final argument is zero (or omitted), this display will be suppressed. For example:

```
huber[sample1, 6.0, 0.5, 10]
    {4.55438, 1.39825}
```

A.2 FREML

On the following page is a *Mathematica* function for obtaining the FREML solution for a straight-line fit with arbitrary standard deviations for x_k and y_k. The input is a list of data points $\{\{x_1, y_1, \lambda_1, \sigma_1\}, \{x_2, y_2, \lambda_2, \sigma_2\}, \ldots\}$. The output is a list of the two coefficients, $\{a, b\}$, of the straight-line fit with $y(x) = a + bx$; for example:

```
datasample = {{1.92, 4.90, 0.76, 0.41}, {0.92, 2.58, 0.58, 1.21},
   {5.58, 9.12, 4.81, 8.73}, {9.68, 12.21, 8.13, 6.55}};
```

```
freml[datasample]
{0.930113, 1.81487}
```

A.3 Nelder-Mead Simplex Optimization

Here is a demonstration of Nelder-Mead optimization in three dimensions. The needed *Mathematica* functions are given in Fig. A.2

Begin with a list containing a single arbitrary simplex. The syntax for each point is `{x,y,z,f}`, where f is the measured response. Here is the initial simplex, sorted in order of the response:

```
s0 = {{7.0, 4.0, 3.0, 20.8}, {8.0, 4.0, 3.0, 34.7},
      {7.0, 5.0, 3.0, 42.1}, {7.0,4.0,4.0,47.0}};
```

Reflection:

```
f1 = nmreflect[s0]
 {7.66667, 4.66667, 3.66667}
```

Perform experiment with these factor values. Result is 59.3, the best so far.

```
p1 = Append[f1, 59.3];
```

Expand in direction of reflection:

```
f2 = nmexpand[s0,p1]
 {8., 5., 4.}
```

A new experiment with these factor values gives worse result, so ignore it. The new simplex is (sorted in order of responses):

```
s1 = {s0[[2]], s0[[3]], s0[[4]], p1}
 {{8, 4, 3, 34.7}, {7, 5, 3, 42.1},
     {7, 4, 4, 47.}, {7.66667, 4.66667, 3.66667, 59.3}}
```

Figure A.1: *Mathematica* function for FREML straight-line fitting.

```
freml[data_]:=
  Module[{wk,w,afn,bfn,xydata,bguess,a,b,bb,afreml,bfreml},
    (* Weight factors.  *)
    wk[b_, sigx_, sigy_] := 1/(sigy^2 + b^2 * sigx^2);
    (* Sum of weight factors, W(b).  *)
    w[b_, list_]:=
      Module[{wklist, wktotal},
        wklist = Table[wk[b, list[[j,3]], list[[j,4]]],
          {j, 1, Length[list]}];
        wktotal = Total[wklist]
      ];
    (* Function for evaluating the FREML y-intercept.  *)
    afn[b_, list_]:=
      Module[{alist, atotal},
        alist = Table[
          wk[b, list[[j,3]], list[[j,4]]]*
            (list[[j,2]]-b*list[[j,1]]),
            {j, 1, Length[list]}];
        alist = alist/w[b, list]; atotal = Total[alist]
      ];
    (* Function for evaluating chi-square.  *)
    chisquare[b_, list_]:=
      Module[{chisqlist, chisqtotal},
        chisqlist = Table[
          wk[b, list[[j,3]], list[[j,4]]]*
            (list[[j,2]]-afn[b,list]-b*list[[j,1]])^2,
          {j, 1, Length[list]}];
        chisqtotal = Total[chisqlist]
      ];
    (* Use the conventional LS fit as initial guess.  *)
    xydata = Table[{data[[j,1]],data[[j,2]]}, {j,1,Length[data]}];
    bguess = b /. FindFit[xydata, a + b*x, {a,b}, x][[2]];
    (* Minimize chi^2 for FREML solution.  *)
    bfreml=bb /. FindMinimum[chisquare[bb, data],{bb,bguess}][[2]];
    afreml = afn[bfreml, data];
    {afreml, bfreml}
  ]
```

Continue with a reflection:

```
f3 = nmreflect[s1]
 {6.44444, 5.11111, 4.11111}
```

New experiment gives 41.2, which is second-worst. Try an inside contraction:

```
f4 = nmcontractin[s1]
 {7.61111, 4.27778, 3.27778}
```

And so on.

Figure A.2: *Mathematica* functions for the Nelder-Mead optimization algorithm.

```
(* Nelder-Mead uphill simplex optimization.  *)
(* A single-step-at-a-time implementation.  *)

(* Centroid of the N best points.  *)

nmcentroidbest[p_] := Module[{n,psort,pbest},
   n = Length[p]-1; psort = Sort[p, #1[[-1]]<#2[[-1]] &];
   pbest = Drop[psort,1];
   Drop[ Sum[ pbest[[i]], {i,n} ]/n, -1]
]

(* Operations *)
(* Input is an ordered set of N+1 points and, for expansion or
   for outside contraction, one additional test point.
   Output is a list of factor values for new expt.
*)

(* Reflection *)

nmreflect[p_] := Module[{alpha,qw,c,qr}, alpha= 1;
   c = nmcentroidbest[p]; qw = Drop[ p[[1]], -1 ];
   qr = c + alpha*(c-qw)
]

(* Expansion *)

nmexpand[p_,pr_] := Module[{gamma,c,qr,qe}, gamma = 2;
   c = nmcentroidbest[p]; qr = Drop[pr,-1];
   qe = c + gamma*(qr-c)
]

(* Contract inside *)

nmcontractin[p_] := Module[{beta,c,qw,qci}, beta = 0.5;
   c = nmcentroidbest[p]; qw = Drop[ p[[1]], -1 ];
   qci = c - beta*(c-qw)
]

(* Contract outside *)

nmcontractout[p_,pr_] := Module[{beta,c,qr,qco}, beta = 0.5;
   c = nmcentroidbest[p]; qr = Drop[pr,-1];
   qco = c + beta*(qr-c)
]
```

Appendix B

Answers to Selected Exercises

Chapter 1

1.1 No. (Not if x is a real number, but for complex numbers it is multiple valued—see Section 6.3.)

1.2 $f'(7) \approx \frac{f(7.0001)-f(7)}{0.0001} = \frac{6.94651951-6.94641838}{0.0001} = 1.0113.$

1.3 $df = \frac{\partial f}{\partial p}dp + \frac{\partial f}{\partial q}dq + \frac{\partial f}{\partial r}dr = qrdp + prdq + pqdr.$

1.4 $df = \frac{df}{dx}dx$ because f is a function of just a single variable, x. Using the chain rule repeatedly, $\frac{df}{dx} = \frac{dp}{dq}\frac{dq}{dr}\frac{dr}{dx}$.

1.5 $df = \frac{dy}{dx}\frac{dp}{dy}dx = 2x\frac{dp}{dy}dx.$

1.6 (a) $\frac{\partial f}{\partial x} = 1 + 3y$, $\frac{\partial f}{\partial y} = 3x + 2y$, $\frac{\partial^2 f}{\partial x \partial y} = \frac{\partial}{\partial x}(3x + 2y) = 3$, $\frac{\partial^2 f}{\partial x^2} = 0$, $\frac{\partial^2 f}{\partial y^2} = 2$.

(b) $\frac{\partial^2 f}{\partial y \partial x} = \frac{\partial}{\partial y}(1 + 3y) = 3.$

(c) Here f is a function of only one variable;

$$\frac{d}{dx}(1 + x + 3xy + y^2) = 0 + 1 + 3x\frac{dy}{dx} + y\frac{d(3x)}{dx} + 2y\frac{dy}{dx} = 1 + 3y + (3x + 2y)y'.$$

1.7 $\frac{\partial g}{\partial c_n} = \sum_{j=0}^{23} y_j^j \frac{\partial c_j}{\partial c_n}$. The partial derivative is zero if $j \neq n$. $\frac{\partial g}{\partial c_n} = y_n^n$ and

$$\frac{\partial f}{\partial c_n} = \sum_{k=0}^{23} 2(y_k - g)\frac{\partial}{\partial c_n}(y_k - g) = 2\sum_{k=0}^{23}(y_k - g)\left(0 - \frac{\partial g}{\partial c_n}\right) = 2y_n^n\left(24g - \sum_{k=0}^{23} y_k\right).$$

1.8

$$\int_a^b x^3 e^x dx = x^3 e^x \big|_a^b - 3\int_a^b x^2 e^x dx = x^3 e^x \big|_a^b - 3x^2 e^x \big|_a^b + 6\int_a^b x e^x dx$$
$$= (x^3 - 3x^2 + 6x - 6)e^x \big|_a^b .$$

1.10 Consider each term individually. $\lim_{x\to 0} 1 = 1$, $\lim_{x\to 0}(-2x) = 0$, and for the last term,

$$\lim_{x\to 0} \frac{\frac{d}{dx}x}{\frac{d}{dx}\sin(3x)} = \lim_{x\to 0} \frac{1}{3\cos(3x)} = \frac{1}{3};$$

it follows that $\lim_{x\to 0} f = 4/3$.

1.11 $\lim_{x\to 0} f = \lim_{x\to 0} 5\cos(5x)2\sin(5x)/2x$. Note that $\cos(5x)$ is nonzero in the limit. $\lim_{x\to 0} f = 5\cos(0)\lim_{x\to 0} \frac{\sin(5x)}{x} = 5\lim_{x\to 0} \frac{5\cos(5x)}{1} = 25.$

1.12 Consider the limit x^n/e^x for any positive n. If this limit is zero, then we know that e^x goes to infinity faster than does x^n.

$$\lim_{x\to\infty} \frac{x^n}{e^x} = \lim_{x\to\infty} \frac{nx^{n-1}}{e^x} = \lim_{x\to\infty} \frac{n(n-1)x^{n-2}}{e^x} = \lim_{x\to\infty} \frac{n(n-1)(n-2)x^{n-3}}{e^x} = \cdots$$
$$= \lim_{x\to\infty} \frac{[n(n-1)(n-2)\cdots 1]x^{n-n}}{e^x} = [n(n-1)(n-2)\cdots 1] \lim_{x\to\infty} \frac{1}{e^x} = 0.$$

1.13 Applying L'Hospital's rule to $e^{-1/u}/u^2$ doesn't work—it continues to give 0/0 no matter how many times the rule is applied. Instead, we analyze $u^{-2}/e^{1/u}$,

$$\lim_{u\to 0} \frac{u^{-2}}{e^{1/u}} = \lim_{u\to 0} \frac{-2u^{-3}}{-u^{-2}e^{1/u}} = \lim_{u\to 0} \frac{-2u^{-1}}{e^{1/u}} = \lim_{u\to 0} \frac{2u^{-2}}{-u^{-2}e^{1/u}} = \lim_{u\to 0} \frac{-2}{e^{1/u}} = -2/\infty = 0.$$

Chapter 2

2.1 1/6

2.2 (a) $(x + dx)^n = \binom{n}{0}x^n + \binom{n}{1}x^{n-1}dx + \cdots$, where additional terms are multiply infinitesimal. $\binom{n}{0} = 1$ and $\binom{n}{1} = n$. The derivative is the factor multiplying dx, which is nx^{n-1}.
(b) Let $m = -n$. We want $\dfrac{1}{(x+dx)^m} = \dfrac{1}{x^m + mx^{m-1}dx} = x^{-m} + gdx$, where g is the derivative. Multiply through by $(x + dx)^m$ and collect terms proportional to dx. The result is $g = (-m)x^{-m-1} = nx^{n-1}$.

2.3 Let $x = q^2$; then $x^2 - 13x + 36 = 0$, $x = 4$ or 9, $q = \pm 2$ or ± 3.

2.4 $x = e^{\ln x} = 16^{\log_{16} x}$, $\ln(e^{\ln x}) = \ln x = \ln(16^{\log_{16} x}) = \log_{16} x \ln 16$, $\log_{16} x = \ln x/\ln 16$.

2.6 $\gamma(2, 10) = \int_0^{10} e^{-t}tdt = -e^{-t}t\big|_0^{10} + \int_0^{10} e^{-t}dt = -11e^{-10} + 1$.

2.7 (a) $-12/x^4$ (b) $\frac{4}{3}x^{-2/3}$ (c) $-12(4x-5)^{-4}$ (d) $4/(4x+3)$ (e) $-\frac{2}{3}xe^{-x^2/3}$

2.8 (a) $\frac{2}{3}\cos(\theta/3)\sin(\theta/3)$ (b) $-\sin(\arctan x)/(1+x^2)$ (c) $-1/(2x\sqrt{x^{-1}-1})$
(d) $3\sin\theta\cos^{-4}\theta$. Use $(1-\sin^2\theta)^{-3/2} = (\cos^2\theta)^{-3/2} = \cos^{-3}\theta$.

2.9 (a) $(x\ln 10)^{-1}$ (b) $(x\ln 47)^{-1}$ (c) $(2x\sqrt{\ln x}\,)^{-1}$ (d) $(\ln x + 1)x^x$
(e) $(\ln x + 1 - \frac{1}{x})x^{x-1}$

2.10 (a) $(\frac{5}{2})(\frac{3}{2})(\frac{1}{2})\Gamma(\frac{1}{2}) = \frac{15}{8}\sqrt{\pi}$. (b) $\Gamma(\frac{1}{2})/[(-\frac{7}{2})(-\frac{5}{2})(-\frac{3}{2})(-\frac{1}{2})] = \frac{16}{105}\sqrt{\pi}$

2.11 (a) $\Gamma(x+1) = x\Gamma(x)$. $\Gamma(x)$ behaves as $\frac{1}{x}$ for $x \to 0$. $\lim_{x\to 0}\Gamma(x+1) = \Gamma(1) = 1$.
(b) $\Gamma(x+n) = \frac{1}{x+n}\Gamma(x+n+1) \to 1/(x+n)$ for $x \to -n$.

2.12 (a) $\int_b^\infty e^{-t}(\ln t)t^{a-1}dt$ (b) $-e^{-b}b^{a-1}$ (c) $-e^{-b}(\ln b)b^{a-1}$

2.13 $\binom{a}{b} = 1/B(b+1, a-b+1)$

2.14 $x! = x\Gamma(x) \approx \sqrt{2\pi}x^{1/2}x^xe^{-x}$, $\ln(x!) \approx x\ln x - x + \frac{1}{2}\ln x + \ln\sqrt{2\pi}$. For large x the last two terms are insignificant compared with the first two.

Chapter 3

3.1 (a) $\rho = \sqrt{10}$, $\phi = \arctan 3 = 71.565°$, $z = 2$.
(b) $r = \sqrt{14}$, $\theta = \arctan(\sqrt{10}/2) = 57.688°$, $\phi = \arctan 3$.

3.2 (a) $x = -12.021$, $y = -12.021$, $z = 0$. (b) $\rho = 17$, $\phi = 5\pi/4 = 225°$, $z = 0$.

3.3 $z_2 = r\cos\theta_2$, $z_1 = r\sin\theta_2\cos\theta_1$, $x = r\sin\theta_2\sin\theta_1\cos\phi$, $y = r\sin\theta_2\sin\theta_1\sin\phi$.

3.5 $\dot\phi = 7.27 \times 10^{-5}\ \mathrm{s}^{-1}$ and $T = \frac{1}{2}mr^2(\sin^2\theta)\dot\phi$, where θ is your latitude and m is your mass, in kg.

3.6 $T = \frac{1}{2}m(\dot x^2 + \dot y^2 + \dot z^2) = \frac{1}{2}m(\dot\rho^2 + \rho^2\dot\phi^2 + \dot z^2)$.

3.7 $(-1.79\,\text{Å}, -0.35\,\text{Å}, 0)$ for H and $(0.72\,\text{Å}, -0.70\,\text{Å}, 0)$ for Cl.

3.8 0.066 Å for the oxygen and 0.919 Å for the hydrogens.

3.9 $4\sin\theta\sin\phi\, e^{-2r}\left(1 - \frac{1}{r} - \frac{1}{2r^2}\right)$ **3.10** $\nabla^2 = \frac{1}{\rho}\frac{\partial}{\partial\rho}\left(\rho\frac{\partial}{\partial\rho}\right) + \frac{1}{\rho^2}\frac{\partial^2}{\partial\phi^2} + \frac{\partial^2}{\partial z^2}$

3.11 $\frac{\partial}{\partial r} = \frac{1}{\sqrt{x^2+y^2+z^2}}\left(x\frac{\partial}{\partial x} + y\frac{\partial}{\partial y} + z\frac{\partial}{\partial z}\right)$, $\frac{\partial}{\partial\phi} = -y\frac{\partial}{\partial x} + x\frac{\partial}{\partial y}$,
$\frac{\partial}{\partial\theta} = \frac{z}{\sqrt{x^2+y^2}}\left(x\frac{\partial}{\partial x} + y\frac{\partial}{\partial y}\right) - \sqrt{x^2+y^2}\,\frac{\partial}{\partial z}$.

Chapter 4

4.1 (a) $\frac{1}{6}(e^{2b^3} - e^{2a^3})$ (b) $\frac{8}{9}\ln 4 - 2/3$
(c) Let $u = \cos\theta$. Then $I = -\int_{\cos a}^{\cos b} u^2 du = \frac{1}{3}(\cos^3 a - \cos^3 b)$.

4.2 (a) $(\frac{1}{3}a + \frac{1}{9})e^{-3a} - (\frac{1}{3}b + \frac{1}{9})e^{-3b}$ (b) $(\frac{1}{3}a^2 + \frac{2}{9}a + \frac{2}{27})e^{-3a} - (\frac{1}{3}b^2 + \frac{2}{9}b + \frac{2}{27})e^{-3b}$
(c) Let $u = x^2$. $I = \frac{1}{2}\int_{a^2}^{b^2} ue^{-3u}du = (\frac{1}{6}a^2 + \frac{1}{18})e^{-3a^2} - (\frac{1}{6}b^2 + \frac{1}{18})e^{-3b^2}$.

4.3 $\gamma = \alpha$, $\beta = (a - 2b)\alpha$, $\alpha = (a - b)^{-2}$,
$I = \frac{\beta + b\gamma}{b - x} + \gamma\ln(b - x) - \alpha\ln(a - x) + c = \frac{1}{(a-b)(b-x)} + \frac{1}{(a-b)^2}\ln\frac{b-x}{a-x} + c$,
where c is an arbitrary constant of integration.

4.4 (a) e (b) divergent (c) 20/81 (d) divergent (e) divergent

4.6 $\frac{\sqrt{\pi}}{2}e^{b^2/4}[1 + \mathrm{erf}(b/2)]$ **4.7** e^{47} **4.8** $dxdy = \frac{\tau^2 - \sigma^2}{(\sigma^2+\tau^2)^{1/2}}\,d\sigma d\tau$

4.13 (a) $32\,B_{1/4}(\frac{5}{2}, \frac{7}{4})$ (b) $2^{13/4}B_{1/2}(\frac{5}{2}, \frac{7}{4})$ **4.15** 2

Chapter 5

5.1 $-0.5833x^2 + 3.7500x - 2.167$

5.2 $-0.292x^3 + 0.875x^2 + 1.417x - 1.00$, $0.146x^3 - 1.750x^2 + 6.667x - 4.50$

5.3 (c) $\frac{4}{3}2^{-2/3}e^{2^{1/3}}$

Chapter 6

6.1 $2e^{i\pi/4}$ **6.2** 47.0, −0.927 **6.3** 2

6.4 $61.0^{1/6}e^{i0.695/3} = 1.93 + 0.46i$, $61.0^{1/6}e^{i(0.695/3+2\pi/3)} = -1.36 + 1.44i$, $61.0^{1/6}e^{i(0.695/3+4\pi/3)} = -0.57 - 1.90i$.

6.5 $0.9065 + 0.2255i$ **6.6** $0.7528 + 0.1872i$ **6.7** (b) $\cosh b \sin a + i \sinh b \cos a$

6.9 $c + id = e^{i3\phi} = \cos 3\phi + i \sin 3\phi$, $\phi = \frac{1}{3}\arcsin d$.

6.10 $\frac{e^{-ax}}{a^2+b^2}(b \sin bx - a \cos bx) + c$ **6.11** $-\frac{e^{-ax}}{a^2+b^2}(a \sin bx + b \cos bx) + c$

6.15 $(2e^{i\phi})^{2/3} = 2^{2/3}e^{i2\phi/3}$; for $\phi = 2\pi$, $2^{2/3}e^{i4\pi/3}$. **6.16** (a) $\pm\sqrt{2}\,i$ **6.17** $\tau/2$

6.18 The denominator can be factored into $(x^2+1)(x^2+4) = (x+i)(x-i)(x^2+4) = (x^2+1)(x+2i)(x-2i)$. The pole at $+i$ has residue $\rho_1 = 1/6i$ while the pole at $+2i$ has residue $\rho_2 = -1/12i$. Using the integration contour of Example 6.10, we find that the value of the integral is $2\pi i(\frac{1}{6i} - \frac{1}{12i}) = \pi/6$.

Chapter 7

7.1 (a) $e^2 + e^2x + \frac{1}{2}e^2x^2 + \mathcal{O}(x^3)$ (b) $1 - x^2 + \mathcal{O}(x^4)$

(c) $\pi - \frac{1}{6}\pi^3x^2 + \mathcal{O}(x^4)$ (d) $-\frac{1}{3}\pi x - \frac{1}{9}\pi x^2 + \mathcal{O}(x^3)$

7.2 (a) $e^5 + e^5(x-3) + \frac{1}{2}e^5(x-3)^2 + \mathcal{O}\big((x-3)^3\big)$

(b) $e^{-9} - 6e^{-9}(x-3) + 17e^{-9}(x-3)^2 + \mathcal{O}\big((x-3)^3\big)$

(c) $-\frac{1}{3}\pi(x-3) + \frac{1}{9}\pi(x-3)^2 + \mathcal{O}\big((x-3)^3\big)$ (d) $-\pi + \frac{1}{6}\pi^3(x-3)^2 + \mathcal{O}\big((x-3)^4\big)$

7.4 $1 - \frac{1}{2}x + \frac{3}{8}x^2 - \frac{5}{16}x^3 + \frac{35}{128}x^4 + \mathcal{O}(x^5)$

7.6 (a) $x - \frac{1}{3}x^3 + \frac{1}{5}x^5 - \frac{1}{7}x^7 + \cdots$. (b) 1

(c) $\pm i$, there $\frac{d}{dx}\arctan x = 1/(1+x^2)$ becomes infinite.

7.7 (a) $1+x+y+x^2+2xy+y^2+x^3+3x^2y+3xy^2+y^3+x^4+4x^3y+6x^2y^2+4xy^3+y^4$

(b) $\binom{m+n}{m}$, a binomial coefficient.

7.8 $\theta^{-2} - \frac{2}{3} + \frac{1}{15}\theta^2 + \mathcal{O}(\theta^4)$ **7.9** (a) $Da^2(R-R_e)^2 + \mathcal{O}\big((R-R_e)^3\big)$ (d) $a = \sqrt{v_2/2D}$

7.10 $-\frac{x^2(1+x^{-2})}{x^2(1-x^{-1}-x^{-2})} \sim -1 - \frac{1}{x} - \frac{3}{x^2} - \frac{4}{x^3} + \mathcal{O}(x^4)$.

7.11 $e^{-ax}\big(1 + \frac{d}{dx} + \frac{1}{2}\frac{d^2}{dx^2} + \frac{1}{6}\frac{d^3}{dx^3} + \cdots + \frac{1}{n!}\frac{d^n}{dx^n} + \cdots\big)e^{ax} = e^{-ax}\big(\sum_{n=0}^{\infty}\frac{1}{n!}a^n\big)e^{ax}$
$= e^{-ax}e^ae^{ax} = e^a$.

7.13 (a) $\ln 2 + \frac{1}{2}\delta - \frac{1}{8}\delta^2 + \frac{1}{24}\delta^3 - \frac{1}{64}\delta^4 + \mathcal{O}(\delta^5)$, where $\delta = x - 2$.

(b) $\ln x$ is singular at $x = 0$. Therefore, the radius of convergence of the series in powers of $\delta = x - 2$ is $r = |2 - 0| = 2$.

Chapter 8

8.1 7.61 **8.3** $3\sigma^4 + 6\mu^2\sigma^2 + \mu^4$ **8.4** $\langle x^2\rangle = \sigma^2 + \mu^2$, $\langle x\rangle^2 = \mu^2$; $\langle x^2\rangle - \langle x\rangle^2 = \sigma^2$.

8.5 $\langle x^2 - 2\langle x\rangle x + \langle x\rangle^2\rangle = \langle x^2 - 2\langle x\rangle\langle x\rangle - \langle x\rangle^2 = \langle x^2\rangle - \langle x\rangle^2 = \sigma^2$.

8.6 $P_{5\le x\le 12} = \frac{1}{2}\text{erf}(5/\sqrt{18}) + \frac{1}{2}\text{erf}(2/\sqrt{18}) = 0.700$. **8.9** (a) 32 (b) 0.048

8.11 The breakdown point is 0%.

8.12 (a) The suspected outlier is 0.354 ml/dL, which gives $Q = 0.450$. The critical Q value is 0.412 at 90% confidence and 0.466 at 95%. We conclude that it is an outlier at 90% but not at 95%.

(b) The suspected outlier is $\ln(0.032) = -3.44$, with $Q = 0.29$, well below the critical value 0.412. We conclude it is *not* an outlier. It seems to be consistent with a log-normal distribution.

(c) $\bar{x} = 0.134$, med $x = 0.1045$; $\hat{\sigma} = 0.092$, MAD $= 0.047$, $S_{\text{RC}} = 0.051$; $\mu_{\text{Hub}} = 0.122$, $S_{\text{Hub}} = 0.072$.

(d) Analyzing the set of logarithms and then taking the exponential of the result gives $\overline{\ln x} = -2.201$, med $\ln x = -2.259$, and $\bar{x} = 0.111$, med $x = 0.1045$. The scale estimators for the logarithms are $\hat{\sigma} = 0.66$, MAD $= 0.55$, and $S_{\text{RC}} = 0.55$. Huber estimation gives $\mu_{\text{Hub}} = -2.19$ and $S_{\text{Hub}} = 0.64$ for $\ln x$ and 0.112 for x.

(e) If we think the data follow a log-normal distribution with no outliers (in fact, that's how the set was generated), then the best 95% confidence intervals are from the mean and standard deviation of the logarithm, $-2.672 < \ln x < -1.729$, and from Huber estimation of the logarithm, $-2.648 < \ln x < -1.735$. The corresponding results for x itself are $0.069 < x < 0.177$ with best estimate 0.111, and $0.071 < x < 0.176$ with best estimate 0.112. Note, however, that the robust estimate from Huber estimation of x itself, $0.070 < x < 0.173$ and $x = 0.122$, gives a remarkably good result without the need to assume a log-normal distribution.

Chapter 9

9.1 0.500 ± 0.028 mol/L

9.2 Median and MAD: 0.509 ± 0.038 mol/L. Median and S_{RC}: 0.509 ± 0.029 mol/L. Huber: 0.513 ± 0.033 mol/L. Note that the extreme values at either edge of the data set are in rather close pairs. It is unlikely that just one member of one of the pairs is an outlier—more likely, neither or both are outliers. If two elements of the set are outliers, then this is 2/7=29%, which is beyond the breakdown point of Huber estimation. Huber estimation would not be robust for this data set. The best approach in this case is the median and S_{RC}.

9.4 $t_{0.09,5} = 1.558$, $t_{0.18,5} = 1.007$

9.5 $P_\nu^{(\chi^2<\chi_1^2)} = \gamma(\nu/2, \chi_1^2/2)/\Gamma(\nu/2) = P(\nu/2, \chi_1^2/2)$, where the last P is one of the regularized incomplete gamma functions.

9.6 The confidence interval is $4.95 < \sigma < 13.14$.

9.8 (b) $F = 2.66$. It passes the F-test, but just barely.

9.9 $\delta_f^2 \sim \left(\frac{\partial f}{\partial x}\right)^2 \delta_x^2 + \left(\frac{\partial f}{\partial y}\right)^2 \delta_y^2 + \left[\frac{1}{4}\left(\frac{\partial^2 f}{\partial x^2}\right)^2 + \frac{1}{3}\frac{\partial f}{\partial x}\frac{\partial^3 f}{\partial x^3}\right]\delta_x^4$

$$+\left[\left(\frac{\partial^2 f}{\partial x \partial y}\right)^2 + \frac{\partial f}{\partial x}\frac{\partial^3 f}{\partial x \partial y^2} + \frac{\partial f}{\partial y}\frac{\partial^3 f}{\partial x^2 \partial y} + \frac{1}{2}\frac{\partial^2 f}{\partial x^2}\frac{\partial^2 f}{\partial y^2}\right]\delta_x^2\delta_y^2 + \left[\frac{1}{4}\left(\frac{\partial^2 f}{\partial y^2}\right)^2 + \frac{1}{3}\frac{\partial f}{\partial y}\frac{\partial^3 f}{\partial y^3}\right]\delta_y^4.$$

9.10 (a) $\langle (x - \langle x \rangle)^2 \rangle = \langle x^2 - 2\langle x \rangle x + \langle x \rangle^2 \rangle$, but $\langle x \rangle$ is a constant. Therefore, this is equal to $\langle x^2 \rangle - 2\langle x \rangle\langle x \rangle + \langle x \rangle^2 = \langle x^2 \rangle - \langle x \rangle^2$.

(b) $\sigma_{\bar{x}}^2 = \langle (\bar{x} - \langle \bar{x} \rangle)^2 \rangle = \langle \bar{x}^2 \rangle - \langle \bar{x} \rangle^2$. $\langle \bar{x} \rangle = \frac{1}{N}\sum_{n=1}^N \langle x_n \rangle = \frac{1}{N}\sum_{n=1}^N \langle x \rangle = \langle x \rangle$ because all the $\langle x \rangle_n$ are equal—they are measurements of the same sample population. Expanding $\langle \bar{x}^2 \rangle = \frac{1}{N}\langle (\sum_{n=1}^N x_n)^2 \rangle$ gives N terms $\langle x_n^2 \rangle = \langle x^2 \rangle$ and $N(N-1)$ terms $\langle x_m \rangle\langle x_n \rangle = \langle x \rangle^2$, $m \neq n$. It follows that $\langle \bar{x}^2 \rangle = \frac{N}{N^2}\langle x^2 \rangle + \frac{N(N-1)}{N^2}\langle x \rangle^2$ and $\langle \bar{x}^2 \rangle - \langle \bar{x} \rangle^2 = \frac{1}{N}(\langle x^2 \rangle - \langle x \rangle^2) = \sigma^2/N$.

9.13 The Q-test *assumes* that the underlying distribution is normal and removes the point that is least consistent with normality. A non-outlier that comes from a non-normal distribution might be rejected as if it were an outlier and this could cause the Shapiro-Wilk test to erroneously conclude the distribution is normal.

9.14 For the normally distributed set, $W = 0.963$. This is greater than the critical value 0.881, as expected. For the log-normal set, $W = 0.844$.

9.15 Use Bernoulli's formula, Eq. (2.18).

Chapter 10

10.1 $\hat{b} = (\overline{xf} - \bar{x}\bar{f})/(\overline{x^2} - \bar{x}^2)$, $\bar{a} = \bar{f} - \hat{b}\bar{x}$, where $\bar{f} = \sum_{k=1}^N e^{2f_k} f_k$, $\overline{xf} = \sum_{k=1}^N e^{2f_k} x_k f_k$, etc.

10.2 $\overline{xf} - \bar{x}\bar{f} = (\overline{x^2} - \bar{x}^2)\hat{b} + \hat{c}(\overline{xy} - \bar{x}\bar{y})$, $\overline{yf} - \bar{y}\bar{f} = \hat{b}(\overline{xy} - \bar{x}\bar{y}) + (\overline{y^2} - \bar{y}^2)\hat{c}$, $\hat{a} = \bar{f} - \hat{b}\bar{x} - \hat{c}\bar{y}$. It follows that $\hat{b} = (s_{yy}s_{xf} - s_{xy}s_{yf})/(s_{xx}s_{yy} - s_{xy}^2)$ and $\hat{c} = (s_{xx}s_{yf} - s_{xy}s_{xf})/(s_{xx}s_{yy} - s_{xy}^2)$, using the notation of Eqs. (10.36).

10.3 (a) $\sum_i u_k v_{i,k} = u_k \sum_i v_{i,k}$ because the factor u_k is a constant (independent of the summation index i).

(b) Using Eqs. (10.3) and the fact that $\sum_{i=0}^M \sum_{k=1}^N \hat{c}_i x_k^{j+i} = \sum_{i=0}^M \hat{c}_i \sum_{k=1}^N x_k^{j+i}$, and then dividing through by N, we obtain $\overline{x^j f} = \sum_{i=0}^M \hat{c}_i \overline{x^{j+i}}$.

10.8 (a) There are various ways to choose the dependent variable. One possibility is to choose it as y, according to $y = c_0 + \sum_{i=1}^4 c_i u_i$, where $u_1 = x$, $u_2 = x^2$, $u_3 = xy$, and $u_4 = x^2 y$.

(b) $y = c_0 + c_1 x$ with $c_0 = e^a$ and $c_1 = -b$.

Chapter 11

11.1 Confidence intervals for y-intercept: 1.03 ± 0.49 (95%), 1.03 ± 0.33 (90%). Confidence intervals for slope: 4.91 ± 0.90 (95%), 4.91 ± 0.61 (90%).

11.2 $x(y) = A + By = -0.2075 + 0.2029y$. The confidence intervals for these parameters at 90% are $\delta_A = \pm 0.029$, $\delta_B = \pm 0.025$; at 95%, $\delta_A = \pm 0.043$, $\delta_B = \pm 0.037$. The fit for $y(x)$ is then $a + bx$ with $a = -A/B$ and $b = 1/B$. Using propagation of error,

$$\delta_a^2 = \left(\frac{\partial a}{\partial A}\right)^2 \delta_A^2 + \left(\frac{\partial a}{\partial B}\right)^2 \delta_B^2 = \left(\frac{1}{B}\right)^2 \delta_A^2 + \left(\frac{A}{B^2}\right)^2 \delta_B^2, \quad \delta_b = \left|\frac{db}{dB}\right| \delta_B = \frac{1}{B^2}\,\delta_B\,.$$

The confidence intervals are $a = 1.02 \pm 0.19$, $b = 4.93 \pm 0.61$ at 90% and $a = 1.02 \pm 0.28$, $b = 4.93 \pm 0.90$ at 95%.

11.4 The F-ratio is 138, which corresponds to a P value of 0.054. This means that we can state with 94.6% confidence that the fit with a quadratic function is better than the fit with a linear function.

11.7 One divides $\hat{\sigma}$ by $\sqrt{N}$ to obtain the standard deviation of the mean. The parameters $\hat{a}$ and $\hat{b}$ are individual values, not means. However, if we were to make M independent determinations of $\hat{a}$ and $\hat{b}$, from M independent runs of the experiment, then we would divide by $\sqrt{M}$.

11.8 (a) 5 (b) 4 (c) 7 (d) 6 (e) 11 (f) 11

Chapter 12

12.1 For $N = 10$ we obtain $\delta_{0.025} = 52.7$ mg/L and $\beta = 0.090$. β first drops below 0.05 at $N = 12$.

12.2 One possible protocol is as follows: Choose a small number N_0 of pills for measurement and estimate σ (assuming a normal distribution) from $\hat{\sigma}$. Using a two-sided t-test with $\alpha = 0.001$, calculate $\delta_{\alpha/2}$ and the corresponding β for various N values until you find the minimum N such that $\beta \leq 0.001$. Sacrifice some more pills so that the total of sacrificed pills is somewhere between N_0 and your calculated N in order to obtain a more reliable estimate of $\hat{\sigma}$ to refine your prediction of the optimal N.

12.3 $W = 10$, and the actual labeling gives the eighth-highest Δ value. [Only $(+,+,-,-,-)$ and $(-,+,-,+,-)$ would give higher values.] With such a small W value, the sensitivity is low. The highest Δ would correspond to P between 0 and 10%, the second-highest between 10 and 20%, and so on. In this case, for the third-highest, we estimate P of approximately 25%.

12.4 $W = 32$. The only way to improve Δ would be to exchange #1 and #2. Therefore, the actual Δ ranks 31 out of 32, and $P = 1 - 30.5/32 = 4.7\%$.

12.6 (a) $\hat{\sigma}_T^2 = 387.08$, $\hat{\sigma}_R^2 = 65.82$, $F = 5.88$, $F_{\text{crit}} = 4.737$. $F > F_{\text{crit}}$ implies that there is a significant effect at $\alpha = 0.05$. (b) According to Dunnett's test, we cannot conclude that any effect that gives $\bar{y}_m < 523.8$ is significant. The effect of treatment 3, with $\bar{y}_3 = 531$ is significant while the effect of treatment 2, with $\bar{y}_2 = 519$, is not. (c) According to Tukey's test, any difference between $\bar{y}_m$'s is insignificant unless it is greater than 19.49. In this case, $|\bar{y}_2 - \bar{y}_3| = 12$. We cannot draw any conclusion regarding treatment 2. Its effect is *not* significantly different from that of treatment 3 or from no treatment at all (even though the effect of treatment 3 *is* significantly different from no treatment).

12.7 0.967%

12.8 The simplex after the 10th experiment consists of the 6th, 8th, and 10th factor values, $(5.5, 15.0)$, $(6.0, 7.0)$, and $(6.125, 13.0)$, respectively. The worst result is #8. The centroid of the two better points is $(5.813, 14.0)$. The vector pointing to it from the worst point is $(5.813, 14.0) - (6.0, 7.0) = (-0.188, 7.0)$. Adding this vector to the centroid gives the reflection, $(5.625, 21.0)$. We can tell from the contours that this would give an even worse result. Therefore, the 12th experiment will correspond to the contraction, $(6.0, 7.0) + \frac{1}{2}(-0.188, 7.0) = (5.906, 10.5)$. Judging from the contour plot, this would give a result quite close to the maximum.

Chapter 13

13.1 $c_j = \nu_j \int d\xi = \nu_j \xi + \gamma$, where γ is a constant of integration. By definition, the extent of reaction is zero at $t = 0$. Therefore, $\gamma = \lim_{t\to 0} c_j(t) = c_j(0)$.

13.2 It depends on the value of $c_3 = [\mathrm{HBr}]$. If $[\mathrm{Br_2}] \gg k_\mathrm{b}[\mathrm{HBr}]$ then $\dot{\xi}$ is proportional to $[\mathrm{Br_2}]^{1/2}$ and $[\mathrm{H_2}]^1$, which is more sensitive to $[\mathrm{H_2}]$. If $[\mathrm{Br_2}] \ll k_\mathrm{b}[\mathrm{HBr}]$ then $\dot{\xi}$ is proportional to $[\mathrm{Br_2}]^{3/2}$ and $[\mathrm{H_2}]^1$, which is more sensitive to $[\mathrm{Br_2}]$.

13.3 $c_1(t) = c_1(0) - \xi(t)$, where $c_1(0) = n_{1,\mathrm{initial}}/V = p_{1,\mathrm{initial}}/RT = 0.18$ mol/L.

$c_2(t) = c_2(0) + f(t) - \xi(t)$, where $c_2(0) = 0$ and $f(t) = (0.020 \text{ mol L}^{-1}\,\text{h}^{-1})t$.

$c_3(t) = c_3(0) + 2\xi(t)$, with $c_3(0) = 0$. Substitute these into Eq. (13.5).

13.4 $\ddot{\rho} - \rho\dot{\phi}^2 = 0, \quad 2\dot{\rho}\dot{\phi} + \rho\ddot{\phi} = 0, \quad \ddot{z} = 0.$

13.5 Lagrange's equations: $\dot{\theta}^2 + \dot{\phi}^2 \sin^2\theta = 0, \quad \ddot{\theta} - \dot{\phi}^2 \sin\theta\cos\theta = 0, \quad \ddot{\phi}\sin^2\theta + 2\dot{\phi}\dot{\theta}\sin\theta\cos\theta = 0.$

Hamilton's equations: $p_r = 0, \quad 0 = p_\theta^2 + \frac{1}{\sin^2\theta}p_\phi^2$, and the equations p_θ, p_ϕ, $\dot{p}_\theta$, and $\dot{p}_\phi$ are unchanged from Eqs. (13.42) except r is replaced by R.

13.6 $\mathcal{L} = \frac{1}{2}m(\dot{r}^2 + r^2\dot{\phi}^2) - \frac{1}{2}(k_r r^2 + k_\phi \phi^2)$; Lagrange's equations are:

$$m\ddot{r} - mr\dot{\phi}^2 + k_r r = -\gamma\dot{r} \quad \text{and} \quad mr^2\ddot{\phi} + 2mr\dot{r}\dot{\phi} + k_\phi\phi = 0.$$

Chapter 14

14.2 It is not linear in y. However, note that $yy' = \frac{1}{2}\frac{d}{dx}y^2$. If we make the substitution $u = y^2/2$ then we obtain the linear equation $u' = g(x)$ for $u(x)$.

14.4 (b), (c), and (e). In each of these cases the eigenvalue is 7.

14.5 $x(t) = (67 \text{ cm}) \cos(1.4 \text{ s}^{-1}\ t)$

14.6 $[\text{A}]_{60\,\text{s}} = [\text{A}]_0 - \frac{1}{2}\Delta[\text{B}] = 0.0854 \text{ mol/L}, \quad k = 0.0026 \text{ s}^{-1}$.

14.7 (a) $\xi(t) = \frac{ab(e^{2akt} - e^{bkt})}{2ae^{2akt} - be^{bkt}}$, where $a = c_{\text{A},0}$, $b = c_{\text{B},0}$.

14.8 (a) $dy/dx + (1+y)x = 0$, $dy/(1+y) = -x dx$, $y(x) = 3e^{(1-x^2)/2} - 1$
(b) $y(x) = 8 - 6x$ (c) $y(x) = (2\ln x - 4x + x^2 + 13/4)^{-1/2}$ (d) $y(x) = 3 - 1/(2-x)$

14.9 Let $x = cu$. Then the choice $c = \alpha^{1/3}$ will give the Airy differential equation with independent variable u. It follows that $y(x) = a\,\text{Ai}(\alpha^{-1/3}x) + b\,\text{Bi}(\alpha^{-1/3}x)$ where a and b are constants of integration.

14.10 (a) 2 (b) 3 constants of integration. Any solution can be written in the form $ap_x + bp_y + cp_z$ with arbitrary constants a, b, and c. (c) There is no inconsistency. This is a *partial* differential equation. The theorem does not apply.

14.11 The substitution $z(x,y) = u(x)v(y)$ leads to the ODE's $u''(x) = \kappa u(x)$ and $v''(y) = (\kappa - 4)v(y)$, where κ is an arbitrary separation constant. We find that $u(x) \propto e^{\pm\kappa^{1/2}x}$ and $v(y) \propto e^{\pm(\kappa-4)^{1/2}y}$. A general solution is $z(x,y) = Ae^{\kappa^{1/2}x + (\kappa-4)^{1/2}y} + Be^{-\kappa^{1/2}x + (\kappa-4)^{1/2}y} + Ce^{\kappa^{1/2}x - (\kappa-4)^{1/2}y} + De^{-\kappa^{1/2}x - (\kappa-4)^{1/2}y}$.

14.12 First make the substitution $c(x,y,t) = W(x,y)T(t)$, which gives

$$\frac{1}{T(t)}\frac{dT}{dt} = -\lambda = \frac{1}{W(x,y)} D\left(\frac{\partial^2 W}{\partial x^2} + \frac{\partial^2 W}{\partial y^2}\right)$$

with separation constant $-\lambda$. Then make the substitution $W(x,y) = X(x)Y(y)$. This gives

$$\frac{1}{X(x)}\frac{d^2X}{dx^2} + \frac{\lambda}{D} = \mu = \frac{1}{Y(y)}\frac{d^2Y}{dy^2}$$

with a separation constant μ. The three ODE's are

$$\frac{dT}{dt} = -\lambda T\,, \qquad \frac{d^2X}{dx^2} = (\mu - \lambda/D)X\,, \qquad \frac{d^2Y}{dy^2} = \mu Y\,.$$

Chapter 15

15.1

	$y(0)$	$y(1)$	$y(2)$	$y(3)$
Euler, $s = 1$	0	−13.218	−36.115	−22.896
RK2, $s = 1$	0	−22.035	−32.001	0.191
RK4, $s = 3$	0			−10.599

Comparing with Fig. 15.1, we see that in this case second-order Runge-Kutta gives the best result for $y(3)$.

15.2 (a) $d\xi/dt = k(c_{\text{A},0} - 3\xi)^2/(c_{\text{E},0} + 2\xi)$
(b) $t(\xi) = \frac{2}{9k}\ln\frac{c_{A,t}}{c_{A,0}} - \frac{1}{9k}(3c_{E,0} + 2c_{A,0})(c_{A,t}^{-1} - c_{A,0}^{-1}), \quad c_{A,t} = c_{A,0} - 3\xi$.

15.3 (a) Let x_1 be the particle on the left side. The equations are $\dot{x}_1 = u_1$, $\dot{u}_1 = m_1^{-1}[2(5 - x_1) + (x_2 - x_1)^{-2}]$, $\dot{x}_2 = u_2$, $\dot{u}_2 = m_2^{-1}[2(5 - x_2) - (x_2 - x_1)^{-2}]$.

Chapter 16

16.1 (a) Not closed; e.g., $\frac{1}{1+x} + \frac{1}{2+x} = \frac{3+2x}{2+3x+x^2}$ is not in the set. (b) Closed. (c) Not closed. If a function in the set is subtracted from a function with the same residue, then it will be nonsingular, and therefore not an element of the set.

16.3 $\langle\vec{x},\vec{y}\rangle = \langle\vec{y},\vec{x}\rangle \implies qx_2y_1 - rx_1y_2 = qy_2x_1 - ry_1x_2 \implies q = r$. $\langle\mathbf{0},\vec{x}\rangle = 0$, $\langle\vec{x},\vec{y}+\vec{z}\rangle = \langle\vec{x},\vec{y}\rangle + \langle\vec{x},\vec{z}\rangle$, and $\langle\vec{x},a\vec{y}\rangle = a\langle\vec{x},\vec{y}\rangle$, are true if and only if $t = 0$.

$0 \le \langle\vec{x},\vec{x}\rangle = px_1^2 - q(2x_1x_2) + sx_2^2 = (\sqrt{p}\,x_1 - \sqrt{s}\,x_2)^2 + 2(\sqrt{ps} - q)x_1x_2$. We know that $(\sqrt{p}\,x_1 - \sqrt{s}\,x_2)^2 \ge 0$. Therefore, we require $q < \sqrt{ps}$.

16.4 $\langle\hat{u}_1, (1,0,2)\rangle = \frac{1}{\sqrt{2}}$, $\langle\hat{u}_2, (1,0,2)\rangle = \frac{1}{\sqrt{6}} + 2\sqrt{\frac{2}{3}} = \frac{5}{\sqrt{6}}$, $\langle\hat{u}_3, (1,0,2)\rangle = \frac{1}{\sqrt{3}}$.

16.5 For any basis vector $\hat{u}_j$, we have $\langle\hat{u}_j, (0,0,0)\rangle = (\hat{u}_{j,1}, \hat{u}_{j,2}, \hat{u}_{j,3})\cdot(0,0,0) = 0$. Therefore, all three coordinates will always be zero.

16.7 $\arccos(-1/3) = 109.47°$

16.8 $E^{-1} = E$, $C_3^{-1} = C_3^2$, $(C_3^2)^{-1} = C_3$, $(\sigma^{(1)})^{-1} = \sigma^{(1)}$, $(\sigma^{(2)})^{-1} = \sigma^{(2)}$, $(\sigma^{(3)})^{-1} = \sigma^{(3)}$.

16.9 No. It is not closed to scalar multiplication. For example, $\sqrt{2}\,(1,1,1) \notin \mathbb{Q}^3$.

Chapter 17

17.1 (a) $\langle u, (-1)v\rangle = (-1)\langle u,v\rangle = 0$ if $\langle u,v\rangle = 0$. $\langle(-1)u, (-1)u\rangle = (-1)^2\langle u,u\rangle = 1$ if $\langle u,u\rangle = 1$.

(b) $\langle u, iv\rangle = i\langle u,v\rangle = 0$ if $\langle u,v\rangle = 0$. $\langle iu, iu\rangle = i^*i\langle u,u\rangle = 1$ if $\langle u,u\rangle = 1$.

17.2 (a) $\phi_0 = 3^{-1/2}(1,1,1)$, $\phi_1 = 2^{-1/2}(1,0,-1)$, $\phi_2 = 6^{-1/2}(1,-2,1)$.

(b) $\phi_0\cdot(-5,1,-7) = -11/\sqrt{3}$, $\phi_1\cdot(-5,1,-7) = 2/\sqrt{2}$, $\phi_2\cdot(-5,1,-7) = -14/\sqrt{6}$. $(-11/\sqrt{3})3^{-1/2}(1,1,1) + (2/\sqrt{2})2^{-1/2}(1,0,-1) - (14/\sqrt{6})6^{-1/2}(1,-2,1)$
$= (-5,1,-7)$.

(c) $(1,0,0) = (1/\sqrt{3})\phi_0 + (1/\sqrt{2})\phi_1 + (1/\sqrt{6})\phi_2$,
$(0,1,0) = (1/\sqrt{3})\phi_0 + 0\phi_1 - (2/\sqrt{6})\phi_2$, $(0,0,1) = (1/\sqrt{3})\phi_0 - (1/\sqrt{2})\phi_1 + (1/\sqrt{6})\phi_2$.

17.3 Using $v_0 = x^3$, $v_1 = 1$, $v_2 = x$, $v_3 = x^2$, one obtains the orthnormal vectors

$$\phi_0 = \frac{1}{\sqrt{720}}x^3, \quad \phi_1 = \sqrt{\frac{20}{19}}\left(1 - \frac{1}{120}x^3\right), \quad \phi_2 = \sqrt{\frac{19}{10}}\left(-\frac{16}{19} + x - \frac{1}{38}x^3\right),$$

$$\phi_3 = \sqrt{\frac{5}{2}}\left(\frac{4}{5} - \frac{11}{5}x + x^2 - \frac{1}{10}x^3\right).$$

17.4 90°

17.5 The straightforward approach is to use Eq. (16.29b) to calculate the coordinates individually. A quicker approach is to start with the recursion relation, $4P_4 = 7xP_3 - 3P_2 = 7x(5x^3 - 3x)/2 - 3P_2$. It follows immediately that $35x^4 = 8P_4 + 6P_2 + 21x^2$. One can see from Table 17.1 that $3x^2 = 2P_2 + 1$, and hence, $x^4 = \frac{8}{35}P_4 + \frac{4}{7}P_2 + \frac{1}{5}P_0$.

Chapter 18

18.2 $\mathbf{u}\mathbf{v}^{\mathrm{T}} = 28, \quad \mathbf{u}^{\mathrm{T}}\mathbf{v} = \begin{pmatrix} 14 & 14 & 21 & 28 \\ 2 & 2 & 3 & 4 \\ 0 & 0 & 0 & 0 \\ 6 & 6 & 9 & 12 \end{pmatrix}.$

18.3 (a) $\begin{pmatrix} 15 \\ 12 \\ 11 \\ 8 \end{pmatrix}$ (b) 15 (c) (7 2 27 10) (d) $\begin{pmatrix} 6 & 2 & 11 & 8 \\ 2 & 2 & 4 & 4 \\ 11 & 4 & 51 & 16 \\ 8 & 4 & 16 & 14 \end{pmatrix}$ (e) 7 (f) 22 (g) 22

18.6 $(\mathbf{B}^{-1}\mathbf{A}^{-1})(\mathbf{AB}) = \mathbf{B}^{-1}(\mathbf{A}^{-1}\mathbf{A})\mathbf{B} = \mathbf{B}^{-1}\mathbf{IB} = \mathbf{B}^{-1}\mathbf{B} = \mathbf{I}$. Therefore, $\mathbf{B}^{-1}\mathbf{A}^{-1}$ is the inverse of $\mathbf{AB}$.

18.7 (a) $\mathbf{R}_z(\phi) = \begin{pmatrix} \cos\phi & -\sin\phi & 0 \\ \sin\phi & \cos\phi & 0 \\ 0 & 0 & 1 \end{pmatrix}$ (b) $\mathbf{R}_x(\phi) = \begin{pmatrix} 1 & 0 & 0 \\ 0 & \cos\phi & -\sin\phi \\ 0 & \sin\phi & \cos\phi \end{pmatrix}$

(c) $\mathbf{R}_x(\phi_x)\mathbf{R}_z(\phi_z) = \begin{pmatrix} \cos\phi_z & -\sin\phi_z & 0 \\ \cos\phi_x \sin\phi_z & \cos\phi_x \cos\phi_z & -\sin\phi_x \\ \sin\phi_x \sin\phi_z & \sin\phi_x \cos\phi_z & \cos\phi_x \end{pmatrix}$

Chapter 19

19.2 The eigenvalue is $\frac{49}{2}\frac{\hbar^2}{mr^2}$.

19.4 (*i*) $q_0 = 1$, $q_1 = 0$; Hermitian. (*ii*) $q_0 = 1$, $q_1 = e^{-x}$; not Hermitian. (*iii*) $q_0 = x$, $q_1 = 1$; Hermitian.

19.6 No. Because $\{f_j\}$ is a basis, any other possible eigenfunction would have to be a linear combination of a least two basis functions with different eigenvalues, which would not satisfy the eigenvalue equation.

19.7 Consider the first term of the operator and use integration by parts:

$$\int_0^\pi f^* \frac{\partial}{\partial\theta}\left(\sin\theta \frac{\partial g}{\partial\theta}\right) d\theta = f^* \sin\theta \frac{\partial g}{\partial\theta}\bigg|_0^\pi - \int_0^\pi \frac{\partial f^*}{\partial\theta} \sin\theta \frac{\partial g}{\partial\theta}\, d\theta$$

$$= 0 - \frac{\partial f^*}{\partial\theta} \sin\theta\, g\bigg|_0^\pi + \int_0^\pi \frac{\partial}{\partial\theta}\left(\sin\theta \frac{\partial f^*}{\partial\theta}\right) g\, d\theta = \int_0^\pi \frac{1}{\sin\theta}\frac{\partial}{\partial\theta}\left(\sin\theta \frac{\partial f^*}{\partial\theta}\right) \sin\theta\, d\theta\,.$$

Consider the second term, again integrating by parts twice:

$$\int_0^{2\pi} f^* \frac{\partial^2 g}{\partial\phi^2}\, d\phi = \left(f^* \frac{\partial g}{\partial\phi} - g\frac{\partial f^*}{\partial\phi}\right)\bigg|_0^{2\pi} + \int_0^{2\pi} \frac{\partial^2 f^*}{\partial\phi^2} g\, d\phi\,.$$

If $f^*(\phi)g'(\phi) - g(\phi)f^{*\prime}(\phi) = f^*(\phi+2\phi)g'(\phi+2\phi) - g(\phi+2\phi)f^{*\prime}(\phi+2\phi)$ then the operator will be Hermitian.

Chapter 20

20.1 $\langle e^{ikx}|e^{ikx}\rangle = \int_{-\infty}^{\infty} dx = \infty$. The probability density, $P(x) = \Psi^*\Psi/\sqrt{\langle\Psi|\Psi\rangle}$, is therefore zero at any given x. This is reasonable. Because the particle is free, it can be anywhere with equal probability. $P(x)$ at any point is $1/\mathcal{X}$ where $\mathcal{X}$ is the length of the universe, and $\mathcal{X}$ is quite large! $\langle e^{ij\phi}|e^{ij\phi}\rangle = r\int_0^{2\pi} d\phi = 2\pi r$ is finite.

20.5 The ϕ dependence in $Y_\ell^{(m)}(\theta,\phi)$ is exclusively in the multiplicative factor $e^{im\phi}$. When the absolute value is taken, in $|R_{n,\ell}Y_\ell^{(m)}|^2$, one gets $e^{-im\phi}e^{im\phi} = 1$.

20.7 (a) $\alpha = 1$, $R_{1,0} = 2e^{-r}$, $P_{1,0}^{(r<r_0)}(r_0) = \frac{1}{2}\gamma(3, 2r_0)$. In atomic units, one Bohr radius corresponds to $r_0 = 1$. $P_{1,0}^{(r<r_0)}(1) = \gamma(3,2)/2 = 0.323 = 32.3\%$.
(b) $\alpha = 1/2$, $R_{2,1} = (24)^{-1/2}re^{-r/2}$, $P_{2,1}^{(r<r_0)}(1) = \gamma(5,1)/24 = 0.004 = 0.4\%$.
(c) $\alpha = 1/2$, $R_{2,0} = (8)^{-1/2}(2-r)e^{-r/2}$, $P_{2,0}^{(r<r_0)}(1) = \frac{1}{8}[4\gamma(3,1) - 4\gamma(4,1) + \gamma(5,1)] = 0.034 = 3.4\%$.

20.8 $P(r_0) = \gamma(5, Zr_0)/24$ with $Z = 1$ or 2. $P = 0.90$ for $Zr_0 = 8.0\ a_0$. For H, $r_0 = 8.0\ a_0$ and for He^+, $r_0 = 4.0\ a_0$.

20.10 Two columns of the determinant would then be identical. It follows from Theorem 18.5.7 that the determinant is zero.

20.11 Ψ_3 and Ψ_4 can be expressed as Slater determinants. Ψ_2 and Ψ_5 cannot be expressed in terms of a single determinant, but they can be expressed as linear combinations of two Slater determinants.

20.12
$$\frac{1}{\sqrt{24}}\begin{vmatrix} \psi_{1,0,0}(1)\boldsymbol{\alpha}(1) & \psi_{1,0,0}(1)\boldsymbol{\beta}(1) & \psi_{2,0,0}(1)\boldsymbol{\alpha}(1) & \psi_{2,0,0}(1)\boldsymbol{\beta}(1) \\ \psi_{1,0,0}(2)\boldsymbol{\alpha}(2) & \psi_{1,0,0}(2)\boldsymbol{\beta}(2) & \psi_{2,0,0}(2)\boldsymbol{\alpha}(2) & \psi_{2,0,0}(2)\boldsymbol{\beta}(2) \\ \psi_{1,0,0}(3)\boldsymbol{\alpha}(3) & \psi_{1,0,0}(3)\boldsymbol{\beta}(3) & \psi_{2,0,0}(3)\boldsymbol{\alpha}(3) & \psi_{2,0,0}(3)\boldsymbol{\beta}(3) \\ \psi_{1,0,0}(4)\boldsymbol{\alpha}(4) & \psi_{1,0,0}(4)\boldsymbol{\beta}(4) & \psi_{2,0,0}(4)\boldsymbol{\alpha}(4) & \psi_{2,0,0}(4)\boldsymbol{\beta}(4) \end{vmatrix}$$

20.13 Use Eqs. (16.29); for example,
$$p_z = \langle t_1|p_z\rangle t_1 + \langle t_2|p_z\rangle t_2 + \langle t_3|p_z\rangle t_3 + \langle t_4|p_z\rangle t_4 = \tfrac{1}{2}(t_1 - t_2 - t_3 + t_4),$$
$$s = \tfrac{1}{2}(t_1 + t_2 + t_3 + t_4),\ p_x = \tfrac{1}{2}(t_1 + t_2 - t_3 - t_4),\ p_y = \tfrac{1}{2}(t_1 - t_2 + t_3 - t_4),$$
$$\psi_{2,1,1} = \langle p_x|\psi_{2,1,1}\rangle p_x + \langle p_y|\psi_{2,1,1}\rangle p_y = 2^{-1/2}p_x + \tfrac{1}{i}2^{-1/2}p_y = 2^{-1/2}(p_x - ip_y)$$
$$= 2^{-3/2}[(1-i)t_1 + (1+i)t_2 + (-1-i)t_3 + (-1+i)t_4]\,.$$
$\psi_{2,1,-1}$ can be treated similarly, or one can simply note that
$$\psi_{2,1,-1} = \psi_{2,1,1}^* = 2^{-3/2}[(1+i)t_1 + (1-i)t_2 + (-1+i)t_3 + (-1-i)t_4]\,.$$

20.14 No. The octahedral orbital pairs are not all orthogonal; e.g., $\langle h_1|h_3\rangle = \frac{1}{3} \neq 0$.

20.15 The basic assumption is that the s, p, and d orbitals for given n have the same energy. This allows us to use arbitrary linear combinations. However, due to interelectron repulsion, the energy also depends on ℓ. Furthermore, the functional forms of the atomic orbitals are correct only for a one-electron atom.

Chapter 21

21.3 (a) $\frac{1-\cos[n(\omega-2\pi)]}{n\pi(\omega-2\pi)^2}$ (b) $P(2\pi) = n/2\pi$ (Use L'Hospital's rule.)

21.7 The left-hand peak will be at approximately the same frequency as before, $\omega \approx \omega_0 = 8.0$. The Nyquist frequency is now $\omega_c = 200\pi/80.0 = 8.6$ (in appropriate inverse-time units). The separation between ω_0 and ω_c is 0.6. The spurious reflection peak is the same distance from ω_c but on the opposite side, at $\omega_c + 0.6 = 9.2$. Actually, both of the peaks will appear to be slightly displaced toward ω_c because the shoulder of each peak will overlap the center of the other peak.

21.8 $g(t)$ is a probability distribution for the injection process. $\int_0^{t_1} g(t)dt$ is the fraction of the sample injected by time t_1. At infinite t, the entire sample has been injected. This means $\int_0^\infty g(t)dt = 1$. The normalization factor b^{-1} makes this so.

Appendix C

Bibliography

General References

Much of the material in this book is covered in more detail in mathematical methods texts for physicists and engineers such as the following:

G. B. Arfken and H. J. Weber, *Mathematical Methods for Physicists*, 6th ed. (Elsevier, Boston, 2005).

M. L. Boas, *Mathematical Methods in the Physical Sciences* (Wiley, New York, 2005).

J. T. Cushing, *Applied Analytical Mathematics for Physical Scientists* (Wiley, New York, 1975).

D. A. McQuarrie, *Mathematical Methods for Scientists and Engineers* (University Science Books, Sausalito, CA, 2003).

W. H. Press, S. A. Teukolsky, W. T. Vetterling, and B. P. Flannery, *Numerical Recipes: The Art of Scientific Computing*, 3rd ed. (Cambridge University Press, Cambridge, 2007).

The first four emphasize analytical methods while Press *et al.* focuses on numerical methods.

The following textbooks provide good introductions to analytical chemistry:

G. D. Christian, *Analytical Chemistry*, 6th ed. (Wiley, New York, 2003).

D. C. Harris, *Quantitative Chemical Analysis*, 7th ed. (Freeman, New York, 2006).

Of the many physical chemistry textbooks available, the following are among the more comprehensive:

P. W. Atkins, *Physical Chemistry*, 8th ed. (Freeman, New York, 2006).

R. S. Berry, S. A. Rice, and J. Ross, *Physical Chemistry*, 2nd ed. (Oxford University Press, New York, 2000).

The following are more approachable, although more modest in scope:

D. W. Ball, *Physical Chemistry* (Brooks/Cole, Pacific Grove CA, 2003).

C. E. Dykstra, *Physical Chemistry: A Modern Introduction* (Prentice-Hall, Upper Saddle River, NJ, 1997).

Bibliography for Part I

Most of the material in Chapters 1 through 7 is covered in university calculus textbooks. See, for example:

G. B. Thomas, *Calculus and Analytic Geometry*, 4th ed. (Addison-Wesley, Reading, MA, 1968).

M. Spivak, *Calculus*, 2nd ed. (Publish or Perish, Berkeley CA, 1980).

For a rigorous but relatively unintimidating development of the underlying theory of calculus, see

M. Bridger, *Real Analysis: A Constructive Approach* (Wiley, Hoboken, NJ, 2007).

For a refresher course in elementary calculus, with applications to physical chemistry, see either of the following:

D. A. McQuarrie, *Mathematics for Physical Chemistry: Opening Doors* (University Science Books, Sausalito, CA, 2008).

R. G. Mortimer, *Mathematics for Physical Chemistry* (Academic Press, San Diego, 2005).

The following contain convenient summaries of calculus techniques:

CRC Handbook of Chemistry and Physics, 89th ed., ed. D. R. Lide (CRC Press, Boca Raton, FL, 2008).

I. S. Gradshteyn and I. M. Ryzhik, *Table of Integrals, Series, and Products*, ed. A. Jeffrey and D. Zwillinger (Academic Press, San Diego, 2000).

For more on numerical methods, the best place to start is *Numerical Recipes* by Press *et al.* Although ostensibly a collection of computer programs, this book is especially valuable for the descriptions of the theoretical ideas that precede each of the programs. They are clearly written, without too much mathematical jargon, have a wealth of practical advice on which algorithms are most appropriate for different kinds of problems, and include references for more detailed study.

There are various mathematical methods books for physicists or engineers that provide more detailed introductions to the more advanced topics such as special functions, coordinate transformations, and singularities; for example, the books by Arfken and Weber, by Boas, and by McQuarrie listed above. For more on functions of a complex variable, those three texts have brief introductions, and a number of texts devoted solely to that subject are available, such as:

J. W. Brown and R. V. Churchill, *Functions of Complex Variables* (McGraw-Hill, Boston, 2004).

P. Dienes, *The Taylor Series: An Introduction to the Theory of Functions of a Complex Variable* (Dover, New York, 1957).

For more on Taylor series, see:

C. M. Bender and S. A. Orszag, *Advanced Mathematical Methods for Scientists and Engineers* (McGraw-Hill, New York, 1978).

G. A. Baker, *Essentials of Padé Approximants* (Academic Press, New York, 1975).

G. A. Baker and P. R. Graves-Morris, *Padé Approximants* (Cambridge University Press, Cambridge, 1996).

For more on special functions:

N. N. Lebedev, *Special Functions and their Applications*, trans. R. A. Silverman (Dover, New York, 1972).

Handbook of Mathematical Functions, ed. M. Abramowitz and I. A. Stegun (Dover, New York, 1965).

Bibliography for Part II

A wide variety of books are available on the subject of statistical methods of data analysis. However, many of them tend toward one of two extremes: Either they give a superficial treatment of the mathematics, presenting statistical techniques essentially as black-box recipes, or they present the mathematics in a rigorous and abstract manner using cryptic notation that is difficult for non-mathematicians to decipher. Two notable exceptions that focus on applications to chemistry are:

J. N. Miller and J. C. Miller, *Statistics and Chemometrics for Analytical Chemistry*, 5th ed. (Pearson, Harlow, UK, 2005).

E. A. McBean and F. A. Rovers, *Statistical Procedures for Analysis of Environmental Monitoring Data and Risk Assessment* (Prentice-Hall, Upper Saddle River, NJ, 1998).

They strike a middle ground, describing practical applications along with at least some mathematical motivation. Miller and Miller provides an extensive bibliograhy.

The following texts are somewhat more theoretical but are aimed toward scientists. Each takes a distinct approach.

P. R. Bevington, *Data Reduction and Error Analysis for the Physical Sciences* (McGraw-Hill, New York, 1969).

G. E. P. Box, W. G. Hunter, and J. S. Hunter, *Statistics for Experimenters: An Introduction to Design, Data Analysis, and Model Building* (Wiley, New York, 1978).

J. Mandel, *The Statistical Analysis of Experimental Data* (Dover, New York, 1964).

J. Stevens, *Intermediate Statistics: A Modern Approach* (Laurence Erlbaum Associates, Mahwah, NJ, 1999).

J. R. Taylor, *An Introduction to Error Analysis*, 2nd ed. (University Science Books, Sausalito, CA, 1997).

The book by Stevens is intended for social scientists and Taylor's book is written for physicists, but they contain much of value to chemists as well. More thorough treatments of experiment design, optimization, and related topics can be found in Box *et al.*, and Stevens, and the following:

R. Brereton, *Chemometrics for Pattern Recognition* (Wiley, Chichester, UK, 2009).

R. E. Bruns, I. S. Scarminio, and B. de Barros Neto, *Statistical Design—Chemometrics* (Elsevier, Amsterdam, 2006).

W. J. Diamond, *Practical Experiment Designs for Engineers and Scientists* (Wiley, New York, 2001).

R. Mead, *The Design of Experiments: Statistical Principles for Practical Application* (Cambridge University Press, Cambridge, 1990).

For more detailed examples of specific applications, see:

D. M. Allen and F. B. Cady, *Analyzing Experimental Data by Regression* (Lifetime Learning, Belmont, CA, 1982).

A. Cornish-Bowden, *Fundamentals of Enzyme Kinetics* (Portland Press, London, 1995).

D. Livingstone, *A Practical Guide to Scientific Data Analysis* (Wiley, Chichester, UK, 2009). (This book focuses on pharmaceutical chemistry.)

P. C. Meier and R. E. Zünd, *Statistical Methods in Analytical Chemistry* (Wiley, New York, 2000).

For a convenient summary of statistical tests, see:

G. K. Kanji, *100 Statistical Tests* (SAGE, London, 1993).

A good introduction to robust fitting is:

P. J. Rousseeuw and A. M. Leroy, *Robust Regression and Outlier Detection* (Wiley, New York, 1987).

For an introduction to surface fitting, see:

P. Lancaster and K. Šalkauskas, *Curve and Surface Fitting: An Introduction* (Academic Press, San Diego, 1986).

For a detailed description of the bootstrap and related techniques, see:

B. Efron and R. J. Tibshirani, *An Introduction to the Bootstrap* (Chapman & Hall, New York, 1993).

Bibliography for Part III

Many introductory textbooks on differential equations are available. They tend to focus more on analytic methods than on numerical methods. See, for example:

M. Braun, *Differential Equations and Their Applications*, 4th ed. (Springer-Verlag, New York, 1993).

E. A. Coddington, *An Introduction to Ordinary Differential Equations* (Dover, New York, 1989).

S. J. Farlow, *Partial Differential Equations for Scientists and Engineers* (Dover, New York, 1993).

A well-organized and reasonably comprehensive reference, with extensive tables of ordinary and partial differential equations, concise descriptions of analytic and numerical solution methods, and extensive citations, is:

D. Zwillinger, *Handbook of Differential Equations*, 2nd ed. (Academic Press, San Diego, 1992).

The CRC *Handbook of Chemistry and Physics* also has a brief but convenient article on differential equation with tables of ordinary differential equations, including some simpler equations that are not included in Zwillinger's tables. A much more comprehensive reference is:

A. D. Polyanin and V. F. Zaitsev, *Handbook of Exact Solutions for Ordinary Differential Equations*, 2nd ed. (Chapman and Hall/CRC, Boca Raton, FL, 2003).

For more on numerical methods for solving differential equations, a good place to start is *Numerical Recipes* by Press *et al.* An excellent textbook on analytic approximation methods for solving differential and finite difference equations is:

C. M. Bender and S. A. Orszag, *Advanced Mathematical Methods for Scientists and Engineers* (McGraw-Hill, New York, 1978).

For an introduction to molecular dynamics simulations, see:

M. P. Allen and D. J. Tildesley, *Computer Simulation of Liquids* (Oxford University Press, Oxford, 1989).

For more detailed treatments of classical mechanics, see:

G. R. Fowles and G. L. Cassiday, *Analytical Mechanics*, 6th ed. (Saunders, Fort Worth, TX, 1999).

H. Goldstein, *Classical Mechanics*, 2nd ed. (Addison-Wesley, Reading, MA, 1980).

For more systematic treatments of chemical kinetics and transport, see:

P. L. Houston, *Chemical Kinetics and Reaction Dynamics* (Dover, Mineola, NY, 2006).

R. F. Probstein, *Physicochemical Hydrodynamics: An Introduction*, 2nd ed. (Wiley, New York, 1994).

Bibliography for Part IV

A variety of introductory linear algebra texts are available. One with a relatively more comprehensive coverage of numerical calculation methods is:

C. D. Meyer, *Matrix Analysis and Applied Linear Algebra* (Society for Industrial and Applied Mathematics, Philadelphia, 2000).

An accessible introduction to numerical linear algebra, with references for further study, can be found in *Numerical Recipes.* The following are more detailed:

B. N. Parlett, *The Symmetric Eigenvalue Problem* (Society for Industrial and Applied Mathematics, Philadelphia, 1998).

L. N. Trefethen and David Bau III, *Numerical Linear Algebra* (Society for Industrial and Applied Mathematics, Philadelphia, 1997).

J. H. Wilkinson, *The Algebraic Eigenvalue Problem* (Oxford University Press, Oxford, 1988).

Cushing's textbook (in the General References) provides a nice treatment of functional analysis. For an authoritative and rigorous presentation of the linear algebra of spaces of functions, see:

M. Reed and B. Simon, *Methods of Modern Mathematical Physics, I: Functional Analysis* (Academic Press, New York, 1980).

Very many textbooks on quantum mechanics are available, although most are designed more for physicists than for chemists. The following are a representative sample, each with a rather different approach:

G. Baym, *Lectures on Quantum Mechanics* (Benjamin/Cummings, Menlo Park, CA, 1969).

D. Bohm, *Quantum Theory* (Dover, Mineola, NY, 1989).

K. Gottfried and T.-M. Yan, *Quantum Mechanics: Fundamentals*, 2nd ed. (Springer-Verlag, New York, 2003).

The following present the fundamentals of quantum mechanics from more of a chemistry-oriented perspective:

J. Avery, *The Quantum Theory of Atoms, Molecules, and Photons* (McGraw-Hill, London, 1972).

V. Magnasco, *Methods of Molecular Quantum Mechanics* (Wiley, New York, 2010).

F. L. Pilar, *Elementary Quantum Chemistry* (Dover, Mineola, NY, 2001).

For further reading on applied quantum chemistry, see:

F. Jensen, *Introduction to Computational Chemistry* (Wiley, Chichester, England, 1999).

W. Koch and M. C. Holthausen, *A Chemist's Guide to Density Functional Theory* (Wiley-VCH, Weinheim, Germany, 2001).

I. N. Levine, *Quantum Chemistry*, 6th ed. (Prentice-Hall, Upper Saddle River, NJ, 2008).

E. Lewars, *Computational Chemistry: Introduction to the Theory and Applications of Molecular and Quantum Mechanics* (Kluwer Academic, Boston, 2003).

A. Szabo and N. S. Ostlund, *Modern Quantum Chemistry: Introduction to Advanced Electronic Structure Theory* (McGraw-Hill, New York, 1989).

For more detailed treatments of the mathematics of spectroscopy, see:

P. F. Bernath, *Spectra of Atoms and Molecules* (Oxford University Press, New York, 1995).

H. Haken and H. C. Wolf, *Molecular Physics and Elements of Quantum Chemistry* (Springer-Verlag, Berlin, 1995).

L. A. Woodward, *Introduction to the Theory of Molecular Vibrations and Vibrational Spectroscopy* (Oxford University Press, London, 1972).

For chemical applications of group theory, see:

F. A. Cotton, *Chemical Applications of Group Theory*, 2nd ed. (Wiley, New York, 1971).

D. J. Willock, *Molecular Symmetry* (Wiley, Chichester, UK, 2009).

See also the books by Bernath and by Woodward.

For the mathematical description of chromatography, see:

A. Felinger, *Data Analysis and Signal Processing in Chromatography* (Elsevier, Amsterdam, 1998).

For more on Fourier analysis, see:

E. A. González-Velasco, *Fourier Analysis and Boundary Value Problems* (Academic Press, San Diego, 1995).

A. G. Marshall and F. R. Verdun, *Fourier Transforms in NMR, Optical, and Mass Spectrometry: A User's Handbook* (Elsevier, New York, 1990).

History of Mathematics and Science

I have included historical footnotes throughout the text to provide context and to entertain readers interested in such details. My principal source has been the online *MacTutor History of Mathematics Archive*:

http://www-history.mcs.st-andrews.ac.uk/history/

This site, created by John J. O'Connor and Edmund F. Robertson, is sponsored by the School of Mathematics and Statistics, University of St. Andrews, Scotland.

To see some of the original papers (in translation) of Cauchy, Fourier, Abel, etc., with commentary to put them in their historical context, see:

G. Birkhoff, *A Source Book in Classical Analysis* (Harvard University Press, Cambridge, 1973).

Computer Algebra Software

The computer algebra software used in examples in this book is:

Mathematica, version 7.0.0 (Wolfram Research Inc.,1988-2008, http://www.wolfram.com).

Other popular computer algebra software systems are:

Maple (Waterloo Maple, Inc., http://www.maplesoft.com),

Matlab (The Mathworks, Inc., http://www.mathworks.com).

A noncommercial open-source alternative, available free of charge, is:

Maxima (http://maxima.sourceforge.net).

Index